VERTEBRATE ENDOCRINOLOGY:
FUNDAMENTALS AND BIOMEDICAL
IMPLICATIONS

Volume 1
Morphological Considerations

VERTEBRATE ENDOCRINOLOGY: FUNDAMENTALS AND BIOMEDICAL IMPLICATIONS

Volume 1
Morphological Considerations

Edited by

Peter K. T. Pang
Department of Physiology
School of Medicine
University of Alberta
Edmonton, Alberta, Canada

Martin P. Schreibman
Department of Biology
Brooklyn College
City University of New York
Brooklyn, New York

Consulting Editor

Aubrey Gorbman
Department of Zoology
University of Washington
Seattle, Washington

1986

ACADEMIC PRESS, INC.
Harcourt Brace Jovanovich, Publishers
Orlando   San Diego   New York   Austin
Boston   London   Sydney   Tokyo   Toronto

ACADEMIC PRESS, INC.
Orlando, Florida 32887

United Kingdom Edition published by
ACADEMIC PRESS INC. (LONDON) LTD.
24–28 Oval Road, London NW1 7DX

Library of Congress Cataloging in Publication Data

Vertebrate endocrinology.

Includes index.
Contents: v. 1. Morphological considerations /
consulting editor, Aubrey Gorbman.
1. Endocrinology, Comparative—Collected works.
2. Vertebrates—Physiology—Collected works. I. Pang,
P. K. T. (Peter Kai To), Date . II. Schreibman,
Martin P. [DNLM: 1. Hormones—physiology.
2. Vertebrates—physiology. WK 102 V567]
QP187.V564 1986   596'.0142   86-10731
ISBN 0–12–544901–1 (v. 1 : alk. paper)

The continuing support, endurance, and understanding of the other Pangs (Rosemary, Naomi, Aidan, and Marianna) and Schreibmans (Alice, Jill, and Abbe) have made this project feasible.

Contents

4   The Pineal Organ

Horst-W. Korf and Andreas Oksche

5   The Caudal Neurosecretory System in Fishes

Hideshi Kobayashi, Kyoko Owada, Chifumi Yamada,
and Yuji Okawara

6   The Thyroid Gland

James Dent

7 The Parathyroid Glands

Nancy B. Clark, Karen Kaul, and Sanford I. Roth

8 The Ultimobranchial Body

Douglas R. Robertson

9 Gastrointestinal Tract

Steven R. Vigna

10 Pancreatic Islets

A. Epple and J. E. Brinn

11 The Adrenal and Interrenal Glands

I. Chester Jones and J. G. Phillips

12 The Ovary

J. M. Dodd

13 Testis

Yoshitaka Nagahama

14 Stannius Corpuscles

Sjoerd Wendelaar Bonga and Peter K. T. Pang

15 Evolutionary Morphology of Endocrine Glands

Aubrey Gorbman

Preface

Studies of vertebrate endocrine systems have made significant contributions to basic biological and biomedical research. This importance is generally not recognized, as the ideas, concepts, and revelations have not been adequately presented and developed.

The intent in this work is to provide an overall view of vertebrate endocrinology and to underscore the impetus that comparative endocrinological studies have provided to clinical and biomedical research. In doing so, the wealth of basic information and the practical implications of these findings will become apparent. In many cases, a sound understanding of an endocrine process in lower vertebrates has provided fundamental information about the comprehension of mammalian and human systems. Many lower vertebrate endocrine tissues, through their simplicity, serve as useful animal models for biomedical research. Some hormones used for therapeutic purposes originate from lower vertebrates. Sometimes, actions of hormones are first described in lower vertebrates and only later demonstrated in humans and mammals.

This volume on the morphology of vertebrate endocrine systems is the first of a multivolume treatise. Each volume will deal with a specific topic in endocrinology (e.g., regulation of water and electrolytes, regulation of calcium and phosphate, reproduction, and growth and development). The basic aims and goals of the treatise are to be accomplished by requiring contributors in each volume to present a comprehensive review of the state of the science in a specific area, discuss in some detail biomedical implications, suggest possible new areas of research, and introduce novel animal models for study. The lower vertebrate animal models so identified may also serve as important alternates to mammalian models, an important consideration in view of recent public concern on the use of some mammals in biomedical research. The first chapter in each volume, which introduces the contributions of comparative endocrinology to the specific topic discussed, is written by a recognized medical endocrinologist. The final chapter deals with evolutionary considerations and is prepared by a prominent comparative endocrinologist.

It is our intention that this work will serve to strengthen the bridge between comparative and medical endocrinologists and to increase the interaction between basic biological and biomedical researchers by calling attention to the mutual benefits that can be derived by this association.

Peter K. T. Pang
Martin P. Schreibman

VERTEBRATE ENDOCRINOLOGY:
FUNDAMENTALS AND BIOMEDICAL
IMPLICATIONS

Volume 1
Morphological Considerations

1

Introduction to Vertebrate Endocrinology

IVOR M. D. JACKSON
Division of Endocrinology
Brown University, Rhode Island Hospital
Providence, Rhode Island 02902

Full understanding of endocrine function in humans and other mammals cannot be accomplished without knowledge of the homologous systems in lower vertebrates. The editors of this book have brought together a series of experts to discuss the comparative morphologic aspects of endocrine organs in vertebrate species, with particular consideration of potential model systems that may aid the understanding of basic biological and biomedical issues in humans. In this introduction, I shall highlight some of the concepts which are dealt with at greater length in subsequent chapters, as well as provide a personal overview of those principles that seem of importance in this context.

The modern era of neuroendocrinology was ushered in, some 15 years ago, with the isolation and characterization of a tripeptide amide, thyrotropin-releasing hormone (TRH), from the hypothalami of the sheep and pig. Somewhere in the neighborhood of a quarter of a million ovine and porcine brains were utilized in the laboratories of Guillemin and Schally (Wade, 1981), respectively, in order to isolate this hypothalamic releasing factor. The same amount of peptide could well have been obtained from the skin of 20 frogs. More about this later!

Following the isolation of TRH, an unexpected finding was the widespread extrahypothalamic distribution of this material in other regions of the central nervous system (Jackson and Reichlin, 1974). Function of TRH in such locations has not been determined for sure. However, TRH is widely distributed throughout the brain of submammalian vertebrates in extraordinarily high concentrations, despite the fact that TRH has no clear role in the regulation of the pituitary–thyroid axis in species lower than the Aves (Jackson and Mueller, 1982). These comparative studies attest to the ubiquitous distribution of TRH in all vertebrate species so studied and strongly support

1

1
AGCACAGAGCAGCACAAGGACACACTCTGCATATTGTGCTGCCGGACAAGGAGGTGACAGCCAGTCAGGCTGAGACAAAGGA
                      109
ACTTCCAGACCTCTGACAGCAGGAAAG ATG GTG TCT GTC TGG TGG TTG CTG CTT CTC GGT ACA ACC GTA TCT
                            Met-Val-Ser-Val-Trp-Trp-Leu-Leu-Leu-Leu-Gly-Thr-Thr-Val-Ser-
                            1   15
CAC ATG GTG CAC ACA CAA GAG CAG CCT TTA CTG GAG GAG GAC ACA GCA CCA TTA GAT GAC TCG GAT
-His-Met-Val-His-Thr-Gln-Glu-Gln-Pro-Leu-Leu-Glu-Glu-Asp-Thr-Ala-Pro-Leu-Asp-Asp-Ser-Asp-

GTT CTT GAG AAA GCC AAA GGT ATC CTG ATC CGC AGT ATC CTG GAG GGA TTT CAA GAA GGG CAA CAA
-Val-Leu-Glu-Lys-Ala-Lys-Gly-Ile-Leu-Ile-Arg-Ser-Ile-Leu-Glu-Gly-Phe-Gln-Glu-Gly-Gln-Gln-

AAC AAT AGA GAT CTA CCA GAT GCA ATG GAA ATT ATA TCT AAG CGC CAG CAC CCA GGG AAA CGA TTC
-Asn-Asn-Arg-Asp-Leu-Pro-Asp-Ala-Met-Glu-Ile-Ile-Ser-Lys-Arg-Gln-His-Pro-Gly-Lys-Arg-Phe-

CAG GAG GAG ATA GAA AAG AGA CAA CAC CCT GGA AAG AGG GAT CTG GAA GAT CTG AAT CTA GAG CTT
-Gln-Glu-Glu-Ile-Glu-Lys-Arg-Gln-His-Pro-Gly-Lys-Arg-Asp-Leu-Glu-Asp-Leu-Asn-Leu-Glu-Leu-
                                                                                    478
TCC AAA AGG CAA CAC CCC GGA AGA AGA TTT GTG GAT GAT GTA GAG AAG AGG CAA CAT CCA CCCCC...
-Ser-Lys-Arg-Gln-His-Pro-Gly-Arg-Arg-Phe-Val-Asp-Asp-Val-Glu-Lys-Arg-Gln-His-Pro .....
                                                                                    123

Fig. 1. Nucleotide sequence of a cDNA clone derived from the skin of *Xenopus laevis* showing the coding strand and the predicted amino acid sequence of the amino terminal regions of pre-proTRH. The four copies of the TRH sequence, along with the flanking basic and C-terminal glycine amino acids, are underlined (from Richter *et al.*, 1984).

the view that the primary role of TRH is as neurotransmitter or neuromodulator and that only later in the process of evolution was TRH "co-opted" as a thyroid-stimulating hormone (TSH) releasing factor (Jackson, 1981).

During the course of studies designed to examine the origin of TRH circulating in the blood of the amphibian *Rana pipiens,* it was found that the source of this material was located in the skin; indeed one frog skin contained as much as 50 μg of this material (Jackson and Reichlin, 1977). There is evidence that the skin of the African clawed toad, *Xenopus laevis,* contains TRH in its cutaneous tissue in even higher concentrations. Based on studies reporting large quantities of TRH in amphibian cutaneous tissue, Richter *et al.* (1984) investigated biosynthesis of TRH in the skin of *Xenopus laevis* by use of molecular biological techniques. These workers were able to identify a cDNA clone with an insert of 478 nucleotides coding for pre-proTRH (the large-molecular-weight precursor hormone from which TRH is subsequently cleaved). The deduced TRH precursor of 123 amino acids contains three copies of the sequence Lys–Arg–Gln–His–Pro–Gly–Lys (Arg)–Arg and a fourth incomplete copy (Fig. 1). Posttranslational processing would yield the mature TRH (pGlu–His–ProNH₂). These findings demonstrate that TRH arises by the processing of a large precursor protein and not by a nonribosomal enzymatic mechanism as had been postulated. Thus the mode of biosynthesis of TRH, the first mammalian hypothalamic releasing factor to be characterized, was elucidated from the cutaneous tissue of an amphibian.

In addition to TRH, amphibian cutaneous tissue is a veritable storehouse of peptides identical or closely related to neural peptides in the central nervous system (CNS) and/or gastrointestal tract of rat and humans (Jackson and Mueller, 1982). Within the skin these peptides are located in the poison or serous glands, which are embryologically derived from the neural crest, and thus form part of the amine precursor uptake decarboxylation (APUD) or diffuse neuroendocrine system (Pearse and Takor, 1979). Peptides isolated from amphibian cutaneous tissue include caerulein (*Hyla caerulea*), related to gastrin and cholecystokinin (CCK); sauvagine (*Phyllomedusa sauvagei*), a 40-amino-acid peptide which has significant homology with corticotropin-releasing factor (CRF); and bombesin (*Bombina bombina*), a tetradecapeptide (Erspamer and Melchiorri, 1973).

Bombesin first isolated from amphibian skin has been recognized in neuroectodermally derived tissue in the human lung. Of particular interest is the fact that a bombesin-like peptide is actively secreted by human small-cell carcinoma (SCC) of the lung, a devastating malignancy that is rapidly progressive. These tumor cells secrete a bombesin-like peptide which may function as an autocrine mitogenic factor in this cancer (Weber *et al.*, 1985). A number of studies are currently under way to elucidate the role of bombesin-like peptide in this human malignancy and may lead to important therapeutic advances. Such developments illustrate the biomedical relevance of studies on the anatomical distribution and function of a neural peptide in an "ectopic" tissue such as skin of an amphibian.

The distribution of the mammalian hypothalamic releasing hormones in the CNS of lower vertebrates is reviewed *in extenso* by Peter (Chapter 3). Such a distribution suggests a functional role for these substances in neural transmission. However, the most convincing evidence comes from studies on the luteinizing hormone-releasing hormone (LHRH) decapeptide in the sympathetic nervous system of the bullfrog, *Rana catesbeiana*. In the sympathetic ganglia of this amphibian, a distinct group of preganglionic nerve fibers releases an LHRH-like material which functions as a neurotransmitter of the late, slow excitatory postsynaptic potential (EPSP) variety (Jan and Jan, 1981). This ganglionic LHRH is chemically distinct from mammalian LHRH, and some studies have suggested its identity with the LHRH decapeptide of the salmon (Sherwood *et al.*, 1983).

The gross and microscopic anatomy of the pituitary gland in vertebrates is extensively detailed by Schreibman (Chapter 2), who draws attention to the similarity, if not always the identicalness, of endocrine function that frequently occurs throughout vertebrates, despite a variability in the structure of endocrine tissues in different species. This variability may present a simpler model for studying basic biological processes than may be found in the usual mammalian experimental animals. Thus, the anatomic location of the pituitary gland in the teleost makes it readily accessible for removal through the buccal cavity. Additionally, the regulation of the teleost lactotrope by osmolality provides a mechanism for the study of prolactin biosynthesis and regulation that does not appear to occur in mammalian species.

Recent studies in my laboratory on the neuronal pathways of human growth hormone-releasing factor (hGRF) in the teleost codfish (*Gadus morhua*) have shed light on the pituitary-regulating role of this material in the fish, with implications regarding the extent of its action in humans. We performed immunohistochemical studies on cod brain and pituitary using two distinct antisera directed against hGRF(1–40) OH and hGRF(1-44)NH$_2$, respectively (Pan *et al.*, 1985). Two topograpically distinct peptide neuronal systems were present in the brain–pituitary (Fig. 2). With antiGRF(1–40)OH, intense immunostaining of axons arising in the preoptic region (NPO), coursing in the preoptic hypophysial pathway was observed (Fig. 2A). These fibers entered the pituitary, ramifying extensively in the pars nervosa (PN), with some fibers entering the proximal (caudal) pars distalis (PPD) (Fig. 2C). In contrast, when anti-GRF(1–44)NH$_2$ was used, no immunostaining was observed anywhere in the hypothalamus (Fig. 2B). However, a reaction product was seen in characteristic cell bodies in the rostral pars distalis (RPD) (Fig. 2D) but not in the PPD or PN.

The apparent presence of two separate neuronal systems of hGRF-related peptides in the codfish brain suggests different roles for these substances in teleost pituitary regulation. One system (IR-hGRF(1–40)OH) (IR = immuno-reactive) innervating the PN may have a role in neurohypophysial function, such as water regulation. Additionally, these fibers entering the PPD where somatotrophs reside may function to regulate the release of growth hormone (GH). The other system (IR-hGRF(1–44)NH$_2$) with cell bodies in RPD, the location of the lactotrophs, may function in prolactin rather than GH regulation in the teleost. Interestingly, somatostatin has been shown to inhibit prolactin secretion in *Tilapia* (Grau *et al.*, 1982), raising the speculation that the primitive role of GRF and somatostatin in pituitary function was directed at prolactin as well as GH regulation. Such comparative studies provide a rational basis for reports that hGRF is capable of stimulating prolactin release in the rat as well as in humans.

The comparative anatomy of the vertebrate pineal is reviewed by Korf and Oksche (Chapter 4). As discussed by these workers, despite anatomic variations in this neuroendocrine structure, its major secretory product, the indoleamine melatonin, is under a similar regulatory mechanism in both amphibians and mammals. The rate-limiting enzyme in melatonin synthesis, serotonin *N*-acetyltransferase (NAT), is under inhibitory control by photoillumination in vertebrates. The pinealocytes are photoneuroendocrine cells which convert photic stimuli (directly in the case of poikilotherms, indirectly via sympathetic innervation in mammals) into a neuroendocrine response.

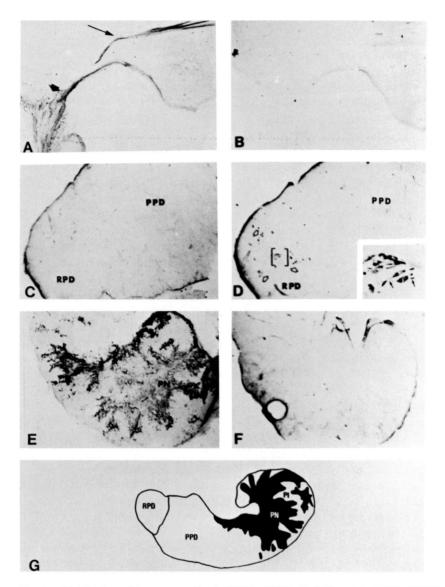

Fig. 2. Distribution of immunoreactive h$_p$GRF(1–40)OH (A,C,E) and h$_p$GRF(1–44)NH$_2$ (B,D,F) in sagittal, 50-μm sections of teleost brain (A,B) and pituitary (C–F). Within the brain, h$_p$GRF(1–40) OH (A), but not h$_p$GRF(1–44)NH$_2$ (B), is seen in axons within the preopticohypophysial pathway (long arrow in A) and infundibular region (broad arrow in A). In the pituitary gland (see diagram in G for orientation) h$_p$GRF(1–40)OH is seen in fine varicosities in the proximal pars distalis (PPD) (C) and heavily innervates pars nervosa (PN) tissue (E). In contrast, h$_p$GRF(1–44)NH$_2$ stains only clusters of cell bodies (open arrows in D) in the rostral pars distalis (RPD). Note their characteristic long, cystoplastic processes (small arrows in inset D). These cells were ovoid with prominent unstained nuclei and had several processes with terminal swellings which could be found in the region of capillary beds. No immunostaining is seen in the pars nervosa (F). PI, pars intermedia. [From Pan *et al.*, 1985.]

These variations in structure and mode of response, combined with a commonality in secretory product, suggest the potential usefulness of a comparative approach in the study of melatonin regulation both *in vivo* and *in vitro*.

The APUD system of Pearse (Pearse and Takor, 1979) has provided a rational basis for the seemingly random distribution of neural peptides in multiple tissues. However, it has been uncertain that peptide cells of the enteropancreatic axis arise from the neural crest. In their chapter on the pancreas, Epple and Brinn (Chapter 10) describe the role of transplantation experiments with early embryonic material in resolving the embryologic origin of the islets of Langerhans. While this tissue is part of the APUD system of Pearse and may be viewed as "neuroendocrine programed," quail–chicken chimeras demonstrate that islet cell tissue is of endodermal and not neural crest origin (Andrew, 1984).

Certain species of teleosts, especially of the genera *Lophius* (anglerfish) and *Ictalurus* (catfish), have large discrete areas of islet tissue (Brockmann bodies) with little associated exocrine tissue (see Epple and Brinn, Chapter 10). The Brockmann bodies constitute a highly important tissue for the study of the biosynthesis and molecular biology of islet cell hormones. The nucleotide sequences of the various forms of somatostatin have been derived from studies of anglerfish and catfish islet tissue (Oyama *et al.*, 1980; Warren and Shields, 1984). Most recently a new peptide homologous to porcine peptide YY and neuropeptide YY has been isolated from the anglerfish Brockmann body (Andrews *et al.*, 1985). Additionally, the Brockmann bodies are innervated by VIP (vasoactive intestinal polypeptide) fibers, illustrating the role of the CNS in the regulation of pancreatic endocrine function. These findings emphasize the richness of this tissue as a model system for the investigation of pancreatic endocrine function (Ronner and Scarpa, 1982)

The anatomic distribution of cells within the pancreatic islet varies in different species. In the rat, the insulin-containing B cells are surrounded by a rim of A (glucagon) and D (somatostatin) cells, whereas in the horse, a core of A cells is surrounded by B and D cells. How does this anatomic relationship impinge upon the regulation of carbohydrate metabolism in these different mammals? Such anatomic variations provide novel opportunities to study the regulation of intermediary metabolism in a physiological setting.

The role of the thyroid gland in amphibian metamorphosis provides a classical model system for studying the role of thyroid hormone in morphogenesis. As discussed by Dent (Chapter 6), the potential for studying the effect of iodothyronines on growth and development in different tissues is open to exploitation. The separation of the calcitonin-secreting (C) cells as ultimobranchial bodies from the rest of the thyroid gland in submammalian vertebrates makes such species particularly useful for the study of calcitonin regulation *in vitro*, as compared with mammals in which the "C" cells are interspersed with the follicular cells in a parafollicular location. Robertson (Chapter 8) has reviewed the embryology of the ultimobranchial tissue and points out that in fish and amphibians the ultimobranchial bodies may contain tissue of endodermal as well as neural crest origin. However, it seems likely that the calcitonin-secreting cells in mammals are of neural ectoderm derivation. Studies of ultimobranchial bodies in lower vertebrates have the potential of providing important information on the physiology of calcitonin secretion. These studies would help to elucidate the nature of medullary carcinoma of the thyroid—a severe malignancy of the "C" cells—in humans (Graze *et al.*, 1978). The C-cell hyperplasia in the Long-Evans rat and the progression to medullary carcinoma in this strain of rodent (DeLellis *et al.*, 1979) have already been used as a model system for the study of this condition in humans.

The ultimobranchial body of birds consists of different cell types (Treilhou-Lahille *et al.*, 1984) which show variability in their staining for calcitonin. It would be of much interest to determine the nature of the secretory products in these cell types, particularly to explore whether the calcitonin gene transcribes both calcitonin and calcitonin-gene-related peptide in this tissue (Jonas *et al.*, 1985). Such studies on the calcitonin gene products may provide information analagous to that provided by the Brockmann bodies of the pancreas in the elucidation of the molecular biology of somatostatin.

The nature of the *C*-thyroglobulin, a component of thyroglobulin found in dog parafollicular cells (Kameda and Ikeda, 1980) coexisting with calcitonin, is unclear. This relationship is intriguing and may be elucidated by studies of these cells in other mammalian and submammalian species.

Clark and associates (Chapter 7) provided a detailed review of the parathyroid glands in vertebrates. The glands consist of one or two pairs derived from the third and/or fourth branchial pouches. The chief cell secretes the calcium-elevating hormone parathormone, and its regulation by calcium is remarkably similar in all terrestial vertebrates. In some species, especially in the Aves, evidence is available that shows a seasonal regulation in the activity of the parathyroid gland. Whether this regulation is neural or endocrine has not been defined, but its elucidation may lead to an increased understanding of the function of this organ.

In addition to the ultimobronchial bodies and parathyroid glands, holostean and teleostean fishes possess an additional calcium-regulating organ, termed Stannius corpuscles, present in the kidney. This tissue produces a calcium-lowering substance called hypocalcin which is around 4,000 daltons (Pang *et al.*, 1980). The significance of this material in vertebrate calcium metabolism has not been established.

The adrenal gland in all mammalian species consists of an outer core (cortex) of steroid-producing cells and an inner core (medulla) of chromaffin-staining cells. The function of the adrenal gland, reviewed by Chester Jones and Phillips (Chapter 11), has been best characterized in mammals in which the steroids cortisol (or corticosterone) and aldosterone are secreted, respectively, in response to stress and salt depletion. Immunohistochemical and radioimmunoassay studies have demonstrated that the adrenal medulla, a neural-crest-derived organ, contains a number of neural peptides (Lundberg *et al.*, 1979) in addition to the monoamines norepinephrine and epinephrine. These peptides include the enkephalins (members of the opioid group of peptides), somatostatin, VIP, and most recently, corticotropin-releasing factor (CRF), which has also been identified. The function of these peptides in the adrenal gland is unknown, and it would be desirable as a means of exploring their functional role to determine whether they are also to be found in the interrenal gland, which is the adrenal homologue present in early vertebrates. Such a distribution may shed light on their significance in mammalian species.

While the basic process of spermatogenesis is similar in all vertebrates, there is much difficulty in studying this in the mammal *in vitro,* primarily because of the close functional relationship between the male germ cells and Sertoli cells *in vivo.* While coculture of rat spermatogenic cells and Sertoli cells has enabled spermatogenesis to occur, problems still exist. However, the germ cells of certain amphibians such as *Xenopus laevis* do not have a similar requirement for Sertoli cells *in vitro*—though these cells are present within the testis of these species—so that cell culture of the amphibian gonadal germ cells may provide important information on the physiology of spermatogenesis in vertebrates. This information would be of relevance to higher species (Risley, 1983).

All major classes higher than Agnatha possess testicular Leydig cells with the characteristics of steroid-producing cells, indicating that these cells are the primary source of testicular androgen production in vertebrates. However, subtle differences do exist. For example, in some species there is a distinct segregation of Leydig cells within the testis, whereas there is a variable relationship between germ cells and Sertoli cells in other species. Thus, the testes of anamniotes differ from amniotes in their being cystic; germ cell development within each cyst is synchronous. As emphasized by Nagahama (Chapter 13), these species provide potentially valuable models for teasing out the molecular biological processes underlying the anatomic and functional relationship of Sertoli cells to spermatogenesis.

CONCLUSIONS

These examples illustrate how mammalian hormonal substances found in particular locations in lower vertebrates allow specific biological studies to be undertaken or provide information unobtainable from mammalian species. Thus the high concentrations of thyrotropin-releasing hormone (TRH) in amphibian skin allows this tissue to be used as a model system for the study of TRH regulation and has permitted the mRNA to be cloned and the cDNA and the predicted amino acid sequence of pre-proTRH to be determined.

The isolation of bombesin from amphibian skin and the recognition of a related material as an autocrine secretory product from a form of human lung cancer is a finding of much biomedical significance. The significance of other amphibian skin peptides in human cancer requires further study. The separation of the islets of Langerhans or Brockmann bodies as structures anatomically distinct from pancreatic exocrine tissue has much potential in molecular biology, peptide biosynthesis, and hormonal secretion. Certain organs found in lower vertebrates but not in adult mammals [organs such as ultimobranchial bodies, Stannius corpuscles, and the urophysis—a neurosecretory system in the caudal portion of the teleost spinal cord (Pearson *et al.*, 1980)] have the potential for yielding basic biological information that may be of significance for better understanding of endocrine function in humans.

ACKNOWLEDGMENTS

Work reported from the author's laboratory was supported in part by NSF grant PCM 8343244 and NIH grant AM 34540. Secretarial assistance was provided by Carol Desjardins.

REFERENCES

Andrew, A. (1984). The development of the gastro-entero-pancreatic neuroendocrine system in birds. *In* "Evolution and Tumour Pathology of the Neuroendocrine System" (S. Falkmer, R. Hakanson, and F. Sundler, eds.), pp. 91–109. Elsevier, Amsterdam.

Andrews, P. C., Hawke, D., Shively, J. E., and Dixon, J. E. (1985). A nonamidated peptide homologous to porcine peptide YY and neuropeptide YY. *Endocrinology (Baltimore)* **116**, 2677–2681.

DeLellis, R. A., Nunnemacher, G., Bitman, W. R., Gagel, R. F., Tashjian, A. H., Jr., Blount, M., and Wolfe, H. J. (1979). C cell hyperplasia and medullary thyroid carcinoma in the rat. *Lab. Invest.* **40**, 140–154.

Erspamer, V., and Melchiorri, P. (1973). Active polypeptides of the amphibian skin and their synthetic analogues. *Pure Appl. Chem.* **35**, 463–494.

Grau, E. G., Nishioka, R. S., and Bern H. A. (1982). Effects of somatostatin and urotensin II on Tilapia pituitary prolactin release and interaction between somatostatin, osmotic pressure Ca++, and Adenosine 3′, 5′-monophosphate in prolactin release in vitro. *Endocrinology (Baltimore)* **110**, 910–915.

Graze, K., Spiler, I. J., Tashjian, A. H., Melvin, K. E. W., Cervi-Skinner, S., Gagel, R. F., Miller, H. H., Wolfe, H. J., DeLellis, R. A., Leape, L., Feldman, Z. T., and Reichlin, S. (1978). Natural history of familial medullary thyroid carcinoma. *N. Engl. J. Med.* **299**, 980–985.

Jackson, I. M. D. (1981). Evolutionary significance of the phylogenetic distribution of the mammalian hypothalamic releasing hormones. *Fed. Proc., (Fed. Am. Soc. Exp. Biol.)* **40**, 2545–2552.

Jackson, I. M. D., and Mueller, G. P. (1982). Neuroendocrine interrelationships. *In* "Biologic Regulation and Development" (R. F. Goldberger and K. R. Yamamoto, eds.), Vol. 3A, pp. 127–200. Plenum, New York.

Jackson, I. M. D., and Reichlin, S. (1974). Thyrotropin-releasing hormone (TRH): Distribution in hypothalamic and extrahypothalamic brain tissues of mammalian and submammalian chordates. *Endocrinology (Baltimore)* **95**, 854–862.

Jackson, I. M. D., and Reichlin, S. (1977). Thyrotropin-releasing hormone: Abundance in the skin of the frog, *Rana pipiens. Science* **198**, 414–415.

Jan, L. Y., and Jan, Y. N. (1981). Role of the LHRH-like peptide as a neurotransmitter in sympathetic ganglia of the frog. *Fed. Proc., (Fed. Am. Soc. Exp. Biol.)* **40**, 2560–2564.

Jonas, V., Lin, C. R., Kawashima, E., Semon, D., Swanson, L. W., Mermod, J., Evans, R. M., and Rosenfeld, M. G. (1985). Alternative RNA processing events in human calcitonin/calcitonin gene-related peptide gene expression. *Proc. Natl. Acad. Sci. U.S. A.* **82**, 1994–1998.

Kameda, Y., and Ikeda, A. (1980). Immunohistochemical study of the C-cell complex of dog thyroid glands with reference to the reactions of calcitonin, C-thyroglobulin and 19S thyroglobulin. *Cell Tissue Res.* **208**, 405–415.

Lundberg, J. M., Hamberger, B., Schultzberg, M., Hokfelt, T., Granberg, P., Efendic, S., Terenius, L., Goldstein, M., and Luft, R. (1979). Enkephalin- and somatostatin-like immuno reactivities in human adrenal medulla and pheochromocytoma. *Proc. Natl. Acad. Sci. U.S.A.* **76**, 4079–4083.

Oyama, H., Hirsch, H. J., Gabbay, K. H., and Permutt, A. (1980). Isolation and characterization of immunoreactive somatostatin from fish pancreatic islets. *J. Clin. Invest.* **65**, 993–1002.

Pan, J. X., Lechan, R. M., Lin, H. D., Sohn, J., Reichlin, S., and Jackson, I. M. D. (1985). Multiple forms of human pancreatic growth hormone-releasing factor like immunoreactivity in teleost brain and pituitary. *Endocrinology (Baltimore)* **116**, 1663–1665.

Pang, P. K. T., Kenny, A. D., and Oguro, C. (1980). Evolution of the endocrine control of calcium metabolism. *In* "Evolution of Endocrine Vertebrate Systems" (P. K. T. Pang and A. Epple, eds.), pp. 323–356. Texas Tech University Press, Lubbock.

Pearse, A. G. E., and Takor, T. (1979). Embryology of the diffuse neuroendocrine system and its relationship to the common peptides. *Fed. Proc., (Fed. Am. Soc. Exp. Biol.)* **38**, 2288–2294.

Pearson, D., Shively, J. E., Clark, B. R., Geschwind, I. I., Barkley, M., Nishioka, R. S., and Bern, H. A. (1980). Urotensin II: A somatostatin-like peptide in the caudal neurosecretory system of fishes. *Proc. Natl. Acad. Sci. U.S.A.* **77**, 5021–5025.

Richter, K., Kawashima, E., Egger, R., and Kreil, G. (1984). Biosynthesis of thyrotropin releasing hormone in the skin of *Xenopus laevis:* Partial sequence of the precursor deduced from cloned cDNA. *EMBO J.* **3**, 617–621.

Risley, M. S. (1983). Spermatogenic cell differentiation in vitro. *Gamete Res.* **4**, 331–346.

Ronner, P., and Scarpa, A. (1982). Isolated perfused Brockman body as a model for studying pancreatic endocrine secretion. *Am. J. Physiol.* **243**, E352–E359.

Sherwood, N., Eiden, L., Brownstein, M., Spiess, J., Rivier, J., and Vale, W. (1983). Characterization of a teleost gonadotropin-releasing hormone. *Proc. Natl. Acad. Sci. U.S.A.* **80**, 2794–2798.

Treilhou-Lahille, F., Jullienne, A., Aziz, M., Beaumont, A., and Moukhtar, M. S. (1984). Ultrastructural localization of immunoreactive calcitonin in the two cell types of the ultimobranchial gland of the common toad (*Bufo bufo L.*). *Gen. Comp. Endocrinol.* **53**, 241–251.

Wade, N. (1981). "The Nobel Duel." Anchor Press/Doubleday, New York.

Warren, T. G., and Shields, D. (1984). Cell-free biosynthesis of multiple preprosomatostatins: Characterization of hybrid selection and amino-terminal sequencing. *Biochemistry* **5**, 2684–2690.

Weber, S., Zuckerman, J. E., Bostwick, D. G., Bensch, K. G., Sikic, B. I., and Raffin, T. A. (1985). Gastrin releasing peptide is a selective mitogen for small cell lung carcinoma in vitro. *J. Clin. Invest.* **75**, 306–309.

2

Pituitary Gland

MARTIN P. SCHREIBMAN
Department of Biology
Brooklyn College, City University of New York
Brooklyn, New York 11210

I. INTRODUCTION—GENERAL COMMENTS

The pituitary gland (or hypophysis) is generally considered to be structurally and functionally the most complex organ of the endocrine system. All vertebrates, beginning with the two extant classes of cyclostomes, have a pituitary gland. There are those who would suggest that Hatschek's pit of amphioxi and the subneural gland of tunicates are progenitors of the vertebrate gland, because of similarities in embryogenesis and in the purported origin of some of their cells from a mucoid cell line (see review by Schreibman, 1980).

It was Galen (131–201 A.D.) who suggested that phlegm (*pituita* in Latin, and one of the four humors) was produced in the brain, transported down the third ventricle, stored briefly in the pituitary gland, and then released along the olfactory nerve into the nose. This is not such an astounding proposal even for present-day endocrinologists, considering the new concepts being developed that deal with interactions between olfactory and central nervous systems. Despite this long history of study, it is entirely reasonable to assume that there is much more to learn about the pituitary gland if one merely considers the inherent diversity of structure that resides in the more than 60,000 species of vertebrates (more than 20,000 teleost species alone). The use of animals never studied before, the application of new tools of investigation, and the new questions being posed have kept morphology, the most basic and oldest of the medical disciplines, from becoming anything but a static science. In recent years, exciting new information has been presented, and new questions have been posed which challenge or redefine well-accepted tenets of hypophysial structure and function. These include such basic phenomena as the direction of blood flow between brain and pituitary

11

VERTEBRATE ENDOCRINOLOGY:
FUNDAMENTALS AND BIOMEDICAL IMPLICATIONS
Volume 1

gland, the direction of axonal transport in neurosecretory cells, the number of hormones that can be produced by a single endocrine cell, the presence of unexpected peptides in pituitary and brain cells, the question of hypophysial cell lineage, the embryological origin of specific adenohypophysial regions, the existence of multiple forms of particular hormones, the concept of endocrine activity for cells generally considered to be nonendocrine types, and suggestions of new roles for pituitary cells and regions.

Although fish, frogs, fowl, and ferrets may differ in their appearance, may live in different places, and must deal with different environmental demands, there is a common basic structural pattern shared by the endocrine systems and the pituitary glands of all vertebrates. There are, however, marked class, and even species, variations in pituitary morphology that reflect these differences. In addition, a basic phenomenon too frequently overlooked by students of morphology is that considerable variation in structure and function of the pituitary gland of a particular species can be associated with sex, age, genotype and/or physiological state of the organism examined.

It is the intent of this chapter to summarize the essential morphological features of the pituitary gland in the various classes of vertebrates and, in passing, to briefly consider the evolutionary and functional aspects that are significant to our understanding of gland structure. It is hoped that this survey, derived largely from the studies by an ever-growing band of comparative endocrinologists, will reveal the wealth of untapped basic information available and will serve to stimulate biomedical researchers and clinicians (as well as students of comparative endocrinology) to view these nonmammalian animals as new and important experimental models to complement, supplement, and/or replace the traditional ones in future research and clinical application.

II. DEVELOPMENT (EMBRYOLOGY)

There is an underlying pattern of development for the pituitary gland that is common to all vertebrates, despite the diversity one notes in the structure of the adult gland. In all vertebrates the hypophysis has a dual origin. The neuropophysis, which provides a neural component to the pituitary gland and a means of suspending and connecting the adenohypophysis in close proximity to the base of the brain, develops from a downgrowth of the diencephalon. The adenohypophysis has its origin in the primitive buccal epithelium (stomodeum), known as ''Rathke's pouch,'' which is generally hollow but may be solid, as it is in fishes and amphibians. An epithelial stalk that connects the adenohypophysial anlage with the buccal epithelium commonly disappears during development; however, it may persist to form an open ciliated duct, as in some teleosts (e.g., *Polypterus* and *Calamoichthys*) or a solid cord, as in some birds [e.g., sparrows and swifts (Wingstrand, 1951)].

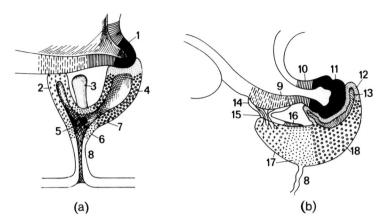

(a)   (b)

Fig. 1. Idealized sagittal sections through a developing (a) and fully differentiated (b) amniote pituitary gland showing the presumed embryonic origins of the neurohypophysis and adenohypophysis. Equivalent structures at the two stages are suggested by similar patterns of shading (dots, circles, x's, etc.). 1, Infundibulum; 2, anterior process of Rathke's pouch; 3, lateral lobe of Rathke's pouch (becomes 14 and 16); 4, aboral lobe of Rathke's pouch; 5, opening of lateral lobe into Rathke's pouch; 6, oral lobe of Rathke's pouch; 7, constriction of Rathke's pouch; 8, epithelial stalk of Rathke's pouch; 9, median eminence; 10, infundibulum; 11, pars nervosa; 12, pars intermedia; 13, hypophysial cleft; 14, juxtaneural part of the pars tuberalis; 15, portal vessels; 16, internal part of pars tuberalis; 17 and 18, differentiated zones of the pars distalis. [From Wingstrand, 1966a, as modified by Gorbman *et al.*, 1983.]

The formation of specific lobes from Rathke's pouch during development (Fig. 1) accounts for class-related differences in the size and position of pituitary regions in adult glands. Tracing the comparative development of these lobes is useful in determining homologies. According to Wingstrand (1966a), the lateral lobes always form the pars tuberalis, and the pars intermedia develops from the portion of the aboral lobe that first makes contact with the infundibulum. The failure of the bird pars intermedia to develop is explained by the failure of the adenohypophysis and neurohypophysis to meet. A similar explanation can probably be applied to explain the absence of the pars intermedia in whales, dolphins, elephants, and armadillos (cf. Hanström, 1966; Holmes and Ball, 1974). There is, however, no such apparent embryological explanation for the regression of the human pars intermedia after it has formed. The distinct zonation characteristic of the pars distalis of an adult hypophysis is explained by the fact that the pars distalis forms from both the oral and aboral lobes.

It is especially interesting to note that although the general pattern of pituitary gland morphology of the adult lamprey resembles that of other vertebrates, its developmental pattern is quite different (cf. Gorbman *et al.*, 1983; Wingstrand, 1966a). In lampreys, the posterior portion of the nasohypophysial pit (which forms from ectodermal folds on the head) differentiates as the adenohypophysis. The adult lamprey adenohypophysis remains ana-

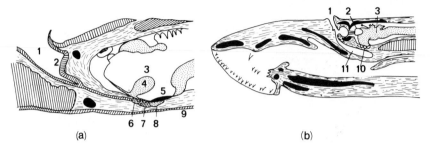

Fig. 2. Median sagittal sections of (a) the pituitary region of a larval lamprey (*Petromyzon*) and (b) the entire head region of an adult *Petromyzon*. 1, Nasohypophysial opening; 2, olfactory organ; 3, brain; 4, optic chiasma; 5, pars nervosa (colored black); 6 and 7, differentiated regions of the pars distalis; 8, pars intermedia; 9, epithelium of roof of mouth; 10, pituitary; 11, nasohypophysial sac. [From Wingstrand, 1966a, as modified by Gorbman *et al.*, 1983.]

tomically close to the olfactory organ, which opens to the top of the head by way of the nasohypophysial duct from which it was originally derived (Fig. 2).

Our knowledge of pituitary gland development is based on the study of a very limited number of highly selected vertebrates. However, this information has become firmly ensconced in embryology textbooks at every level of education. The examination of additional representatives of vertebrate classes and the application of newer methods of analysis is certain to modify our concepts. For example, recent studies of the two extant cyclostomes have demonstrated that developmental patterns among closely related animals may differ significantly. Gorbman (1983a, 1984) has pointed out that although the later development of the adenohypophysis is similar in lamprey and hagfish, their early embryology is not. The nasopharyngeal duct forms by a horizontal splitting of a single layer of yolk endoderm in hagfish, whereas it forms from head ectoderm in the lamprey. In both hagfish and lamprey the subsequent differentiation of the adenohypophysis occurs through acinar outgrowths of the dorsal nasopharyngeal epithelium which make contact with the infundibulum. There is then a horizontal delamination which results in the formation of the adenohypophysis. The difference in the early development of the adenohypophysis of lamprey and hagfish may have important implications in determining the evolutionary paths followed from these primitive organisms (Gorbman, 1983a, 1984).

The recent review of pituitary gland development in reptiles by Pearson (1985) serves to illustrate that the gaps in our knowledge of the embryology of the pituitary gland result from the paucity of samples studied within a particular class of vertebrates compared to the number of species available. Her studies also emphasize that the development of the shape of the embryonic gland is not, as implied in most texts, a passive process in which only the adenohypophysial anlage assumes an active role in reaching the infundibular process. Pituitary gland development is a dynamic phenomenon that involves Rathke's pouch formation and migration (see preceding paragraphs) and the intimate contact of neural and buccal components. The development of the neurohypophysis by active growth of neural tissue occurs concomitantly with the proliferation of secretory epithelium to establish the morphology of the adult gland.

III. STRUCTURAL ASSOCIATION OF ADENOHYPOPHYSIS AND NEUROHYPOPHYSIS

The structural intimacy of neurohypophysis and adenohypophysis that is established early during the embryology of the pituitary gland reflects the direct functional interaction between the central nervous and endocrine systems. Vascular and/or neuronal pathways provide the means of exchanging chemical signals, thus enabling hypophysiotropic centers to exert control over the synthesis and release of adenohypophysial hormones. The neurohypophysis serves, too, as a holding station for neurohypophysial hormones synthesized in the brain before they are released into the general circulation to act at endocrine and nonendocrine sites.

The extent of the anatomical intimacy between neuropituitary and adenopituitary components ranges considerably among the vertebrate classes, from no apparent contact in cyclostomes to intimate interdigitation in teleosts (Fig. 3). It should be pointed out that these variations in the structural association between the neurohypophysis and the adenohypophysis can also be seen within each class. The differences observed in these patterns and the extent to which neurohypophysial tissue differentiates (i.e., the development of a median eminence, neurointermediate lobe, or neural lobe) presumably reflect the different physiological demands imposed by the natural habitat and the diversity of means by which the hypothalamus controls adenohypophysial function.

Neurohormones which are synthesized in specific regions of the brain are conveyed to the neurohypophysis by way of axonal tracts where they may be stored in distended axonal endings as Herring bodies. Axons may also contact blood vessels and discharge neurosecretory products into the systemic circulation (e.g., octapeptides) or into a portal system leading to the adenohypophysis, or they may directly innervate pituitary gland cells (Fig. 4.).

Needless to say, the histology of the neurohypophysis reflects the preponderance of neuronal tracts conveying neurosecretory products and networks of vascular elements. Pituicytes in the neurohypophysis are a class of neuroglial elements whose cytoplasmic processes envelop secretory axon terminals and, along with the nerve terminals, may also make contact with the perivascular space. Whether pituicytes serve in some metabolic capacity in the secretory process or are merely supportive requires further study.

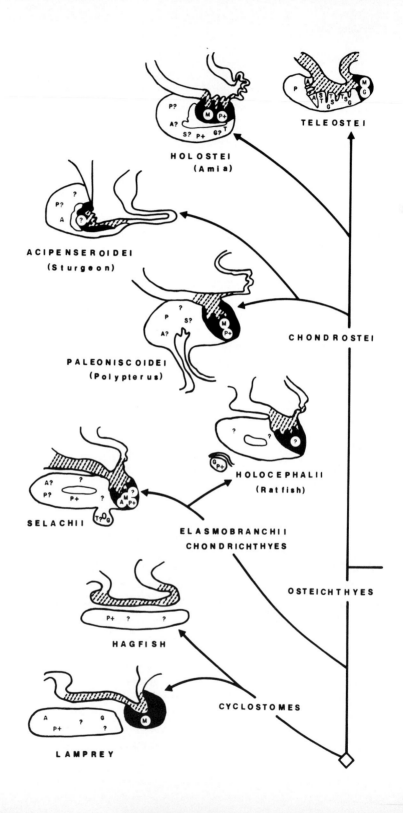

TELEOSTEI

HOLOSTEI
(Amia)

ACIPENSEROIDEI
(Sturgeon)

CHONDROSTEI

PALEONISCOIDEI
(Polypterus)

HOLOCEPHALII
(Ratfish)

SELACHII

ELASMOBRANCHII
CHONDRICHTHYES

OSTEICHTHYES

HAGFISH

CYCLOSTOMES

LAMPREY

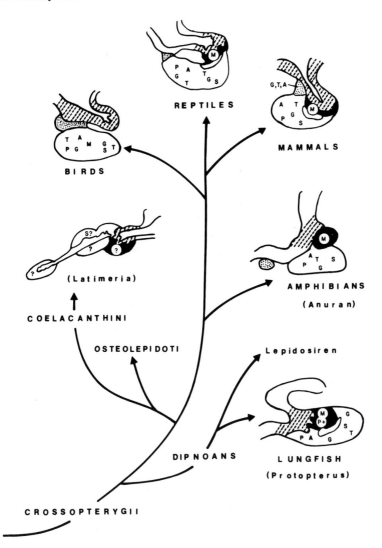

Fig. 3. A phylogenetic tree of generalized diagrams of vertebrate pituitary glands as seen in midsagittal section; anterior is to the left. Cell distribution, where indicated, is represented by tinctorial affinity for PAS (P$^+$) or by function suggested by immunocytochemical or physiological studies. P, prolactin; S, somatotropin; G, gonadotropin; T, thyrotropin; A, adrenocorticotropin; M, melanocyte-stimulating hormone; ?, unidentified function. Neurohypophysis is indicated by oblique dashed lines; pars tuberalis is represented by dots; pars intermedia is colored black. Drawings are not to scale.

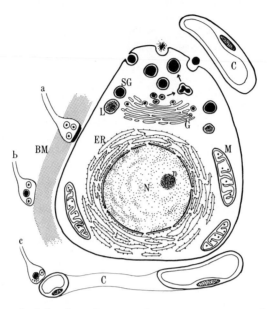

Fig. 4. A diagram depicting the various ways that neurosecretory material may be delivered to the adenohypophysis. (a) By direct synaptic contact with a pituitary cell membrane. (b) By diffusion through a basement membrane, BM, or extra vascular space. (c) By release into a blood vessel, C, or portal system. The cell depicted in the diagram represents a generalized pituitary cell: N, nucleus; n, nucleolus; ER, granular endoplasmic reticulum; L, lysosome; M, mitochondria; G, golgi; SG, secretory granule; the arrows suggest pathways of hormone product processing and release. [From Schreibman and Holtzman, 1975.]

Ependymal cells, which form a boundary of the neurohypophysis by lining the cavity of the third ventricle, are in a strategic position for affecting the transfer of materials between the central nervous system and pituitary gland cells. For example, tanycytes are modified ependymal cells that were first described in elasmobranchs (Horstmann, 1954). Their potential importance is strongly suggested by the functional cytological bridge they form between the ventricular systems and portal and pituitary elements, as well as from changes in their cytology that occur in association with specific endocrine-regulated processes (Akmayev and Fedelina, 1976). We need to clarify the role of these cells in pituitary gland functions and in information transfer between central nervous and endocrine systems.

IV. MICROSCOPIC ANATOMY/CYTOLOGY

The adenohypophysis of all vertebrates is essentially a conglomeration of cells that range in order from masses highly segregated by cell type (teleosts) to more randomized mixtures of endocrinotropic cells (birds and mammals).

Essential to any analysis of pituitary gland structure and function is the need to identify the various cells according to the function they serve. This was attempted in classical studies by the application of stains, either singly or in combination, and the recognition of cells on the basis of their tinctorial affinity (Table I). Staining methodologies have progressed from the fundamental hematoxylin and eosin to the trichrome and tetrachrome procedures of more recent years. Ultimately, stains were used to localize specific chemical characteristics of the hormones elaborated through the application of such procedures as periodic acid Schiff (PAS), alcian blue (AB), aldehyde fuchsin (AF) and lead hematoxylin (PbH). Electron microscopy confirmed the heterogeneity of cells seen with stains and light microscopy, and on the basis of granule size and morphology it was possible to recognize five or six different types which correlated well with the number of hormones suspected to originate in the pars distalis (Table I). However, only when these fundamental methods of analysis were applied to experimental animals that were subjected to classical endocrinological challenges (surgical or chemical ablation of endocrine glands, hormone replacements, drug administration, etc.) could cells be identified on the basis of physiological function. The use of radiolabeled precursor substances and autoradiography has enabled investigators to localize and to assess synthetic activity in glands in response to the experimental manipulations. Results from the studies just discussed have permitted investigators to replace the cumbersome system of nomenclature by Greek letters with one that identifies a cell's functional role (i.e., hormone elaborated).

In recent years the application of immunologic procedures at the cell level (immunocytochemistry) has been used with increasing frequency to yield valuable information related to the identification of pituitary cells and the evaluation of their activity. In this method, antibodies are generated against specific antigens (hormones) which have been isolated and purified or artificially synthesized. The specific antiserum generated, when applied to tissue sections, forms a typical antigen–antibody complex which can then be localized at the light or electron microscopic level if made visible by attaching to the antibody either a fluorescent dye or an enzyme that can form a colored product when reacted with its specific substrate. Aside from inherent problems related to specificity of antisera and the use of suitable controls (Petrusz, 1983), this approach has been most valuable in confirming the identification of cell types and in studying pituitary cell function related to the synthesis and release of hormones.

The use of immunocytochemical methods has also served to raise new questions regarding pituitary structure and function. For example, the presence of immunoreactive material in a cell does not necessarily imply that it was synthesized there, for it is conceivable that the substance was elaborated elsewhere and then transported to the localization site. Thus, one should interpret with caution reports of the localization of more than one

TABLE I

Comparative Cytology of the Pars Distalis from Representatives of Different Vertebrate Groups[a]

Hormone	Vertebrate group	Cellular names or types[b]	Cytoplasmic granule size (nm)[c]
Thyroid-stimu- lating hor- mone (TSH)	Chondrichthyes: Selachii[d]	Type 1, basophil	90–120
	Osteichthyes: Holostei[e]	Type 1, basophil	—
	Osteichthyes: Teleostei[f]	Type 1, basophil, delta	400
	Amphibia: Anura[g]	Type 1, basophil	150–400
	Reptilia[h]	Type 1, basophil	300–400 × 200–250
	Aves[i]	Type 1, basophil, delta	50, 100, 200
	Mammalia	Type 1, basophil, beta	150
Gonadotropin (GTH)	Chondrichthyes: Selachii	Type 2, basophil	100–700
	Osteichthyes: Holostei	Type 2, basophil	
	Osteichthyes: Teleostei	Type 2, basophil, beta, gamma	60–160; 80–240
	Amphibia: Anura	Type 2, basophil	Polymorphous to 900
	Reptilia	Type 2, basophil	150–270; 600–800
	Aves	Type 2, basophil, beta, gamma	120–400
	Mammalia	Type 2, basophil, delta	200
Adrenocortico- tropin (ACTH)	Chondrichthyes: Selachii	Type 3, basophil	140
	Osteichthyes: Holostei	acidophil(?)	—
	Osteichthyes: Teleostei	Type 3, basophil, epsilon	110–250
	Amphibia: Anura	Type 3, basophil	100–200
	Reptilia	Type 3, basophil	—
	Aves	Type 3, basophil, epsilon	150–300
	Mammalia	Type 3, basophil	200
Prolactin (PRL)	Chondrichthyes: Selachii	Type 1, acidophil	263
	Osteichthyes: Holostei	Type 1, acidophil	—
	Osteichthyes: Teleostei	Type 1, acidophil, epsilon	Polymorphic; 170–350
	Amphibia: Anura	Type 1, acidophil	180–500
	Reptilia	Type 1, acidophil	—
	Aves	Type 1, acidophil, eta	Polymorphic; 250–300
	Mammalia	Type 1, acidophil, epsilon	Polymorphic; 600–900
Somatotropin (GH)	Chondrichthyes: Selachii	Type 2, acidophil	200
	Osteichthyes: Holostei	Type 2, acidophil	—
	Osteichthyes: Teleostei	Type 2, acidophil, alpha	—

TABLE I　(*Continued*)

Hormone	Vertebrate group	Cellular names or types[b]	Cytoplasmic granule size (nm)[c]
	Amphibia: Anura	Type 2, acidophil	180–250
	Reptilia	Type 2, acidophil	310
	Aves	Type 2, acidophil, alpha	250–300
	Mammalia	Type 2, acidophil, alpha	350

[a] Modified from Norris (1980).

[b] Acidophils: Type 1 (stain with azocarmine, acid fuchsin, and erythrosin); Type 2 (stain with orange G). Basophils: Type 1 (stain with PAS, AF, alcian blue, and aniline blue); Type 2 (stain with PAS, alcian blue, aniline blue; some classes AF^+, sometimes orange G granules); Type 3 (almost chromophobic and often PbH^+).

[c] Determined by electron microscopy.

[d] *Scyliorhinus caniculus.*

[e] *Amia calva.*

[f] *Zoarces* sp.

[g] *Rana temporaria.*

[h] General summary from Holmes and Ball (1974).

[i] General summary based upon domestic duck and Japanese quail.

immunoreactive product within a cell. Such findings have fueled discussions of whether one cell may produce more than one hormone. This topic will be discussed later in this chapter. The variations in the immunoreactive staining responses that have been observed when utilizing the same antiserum or different batches of antisera to the same antigen should make us aware that the chemical structure of a particular hormone may vary during different stages of its "life history" (synthesis, processing, storage, and translocation), as well as with the species, age, and physiological state of the animal.

V. COMPARATIVE MORPHOLOGY

A. Class Agnatha

Agnathans are the most primitive living vertebrates and are represented by two orders: Petromyzoniformes (lamprey eels) and Myxiniformes (hagfish). The great disparity between the way hagfishes and lampreys regulate reproductive functions (see Gorbman, 1983b) may be reflected, in part, by variations in the structure and function of their pituitary glands (Fig. 3).

The hagfish pituitary gland appears to be the most primitive of the two. Its neurohypophysis is highly developed, flattened, saclike and contains many aldehyde-fuchsin-positive (AF$^+$) neurosecretory endings (Gorbman *et al.*, 1983). The hypophysial–portal system of hagfish is usually markedly reduced but may show species variation (see Gorbman, 1983b). There is no developed portal system in the Eastern Pacific hagfish, *Eptatretus (Polistotrema) stouti* (Gorbman, 1965). In *E. burgeri*, the Western Pacific hagfish, some 20 portal vessels have been recorded and the suggestion is presented that the ventral floor of the neurohypophysis is a primitive median eminence structurally similar to that of tetrapods (Kobayashi and Uemura, 1972). Among the many questions still to be resolved is the determination of the direction of blood flow and whether the neurohypophysis exercises control over the adenohypophysis. It is interesting to note that *E. stouti* and the Atlantic hagfish, *Myxine glutinosa*, do not migrate from a consistently cold and dark environment and do not display seasonal cycles (Gorbman *et al.*, 1963). These organisms have few (if any) portal blood vessels. *Eptatretus burgeri*, on the other hand, has a seasonal gonadal cycle, moves seasonally into shallow water (Kobayashi *et al.*, 1972), and may have a better developed "median eminence," with greater potential for hypothalamic control of pituitary functions, although this has not yet been demonstrated. This may very well be a case of a structure appearing before its function has developed. The suggestion that neurosecretions may reach the adenohypophysis by diffusion through connective tissue rather than by a vascular route (Nozaki *et al.*, 1975) is supported by the recent observations of Tsukahara *et al.* (1986). They found that trypan blue, horseradish peroxidase, and ferric ion, when injected into the third ventricle of the hagfish *E. burgeri*, could diffuse across the wall of the neurohypophysis and the brain–pituitary connective tissue septum within several minutes. In addition, ferric ion and horseradish peroxidase continued to diffuse between the follicles to the ventral border of the adenohypophysis. Thus, there exists in the hagfish an effective method of distributing neuropeptides to adenohypophysis cells [a "diffusional median eminence," according to Tsukahara *et al.* (1986), see Chapter 15 for additional discussion].

The adenohypophysis is separated from the neurohypophysis and is not differentiated into zones except, perhaps, in the posterior region of older and larger animals (*Myxine*) (Holmes and Ball, 1974). Generally, islets of cells are embedded in a poorly vascularized, loose connective tissue which is continuous with the layer separating the neurohypophysis from the adenohypophysis (Fernholm, 1972; Olsson, 1969; Tsukahara *et al.*, 1986). Some cells form follicles and contain periodic acid Schiff positive (PAS$^+$) colloid.

Cell types of the hagfish pituitary are difficult to characterize, but there may be several types if they are classified by their affinities for standard pituitary stains. Rare acidophils and PAS$^+$ cells have been noted, and these may correspond to cells containing granules 100–200 nm in diameter (see Norris, 1980). The function of the various cells is obscure, and without a clear-cut pattern of cell distribution, the pars distalis lacks a zonation typical of other vertebrates. Application of bioassay (Nicoll, 1974) and immunocytochemistry (Aler *et al.,* 1971) failed to demonstrate the presence of a prolactin-like substance in the adenohypophysis of the hagfish. A thyrotropic substance may be present in *E. stouti* (Dickhoff and Gorbman, 1977); however, attempts to demonstrate a gonadotropic factor in the hagfish have not been successful (Matty *et al.,* 1976). It may be that a single class of cells secretes more than one hormone with similar chemical structures. The functional significance of immunocytochemical localization of FMRFamide (phenylalanyl-methionyl-arginyl-phenylalanine amide; a transmitter substance in pituitary cells as well as in the brain) is unknown (Jirikowski *et al.,* 1984).

The lamprey neurohypophysis consists of a thin anterior portion and a thickened posterior part which, because it posesses neurosecretory neurons ending in a neurohemal structure, has been termed a "pars nervosa" (Fig. 3). The adenohypophysis is compact and is differentiated into two regions, a pars distalis and a pars intermedia. A well-developed pars intermedia forms a neurointermediate lobe with the pars nervosa; however, there are no direct nervous or vascular connections between the two components (Tsuneki and Gorbman, 1975a).

The lamprey pars distalis is separated from the infundibulum by connective tissue, and its cells display a regional distribution, at least according to their stain affinity, ultrastructure, and purported function (see Ball and Baker, 1969; Larsen and Rothwell, 1972; Holmes and Ball, 1974; Nozaki, 1986). The rostral pars distalis (RPD) contains either chromophobes or PAS$^+$ and AF$^+$ basophils. Cells containing immunoreactive adrenocorticotropin (ACTH) have been identified (Dores *et al.,* 1984), and ACTH activity in the pars distalis has been reported (Baker and Buckingham, 1983). The caudal pars distalis (CPD) contains both acidophils and fewer basophils. The functions of the pars distalis cells are not clear, despite attempts to relate them to metamorphosis, spawning, hypophysectomy, the action of inhibiting agents, and immunochemical analysis. The presence of cells in the CPD that cross-react with anti-rat luteinizing hormone (LH) (Wright, 1983) and luteinizing hormone releasing hormone (LHRH) immunoreactivity in brains (Crim *et al.,* 1979a,b) suggests that typical control mechanisms may operate in lampreys. Reports for the pars intermedia vary, but generally one cell type is present that is carminophilic, either PAS$^+$ or PAS$^-$ and lead-hematoxylin-positive (PbH$^+$). These cells undergo cytological change with varying illumination, migration, and spawning, and presumably they secrete melanocyte-stimulating hormone (MSH) (van Oordt, 1968; Ball and Baker, 1969; Larsen and Rothwell, 1972); Baker and Buckingham (1983) have demonstrated the presence of MSH activity in the neurointermediate of *Lampetra fluviatilis* in an *Anolis* skin bioassay. Antiserum to α-MSH cross-reacts with all cells in the pars intermedia (Dores *et al.,* 1984).

There is little evidence to suggest that the hypothalamus exerts any influence on the pars distalis. Larsen (1969) showed that even with an ectopic adenohypophysis, complete gonadal maturation may take place in male lampreys. This is supported by ultrastructural studies, which fail to demonstrate the presence of nervous or vascular connections between the anterior neurohypophysis and the pars distalis, even though this region in lampreys is similar to the median eminence of higher vertebrates (Tsuneki and Gorbman, 1975b). The suggestion that neurosecretions diffuse across the thin connective tissue separation into the adenohypophysis (Tsuneki and Gorbman, 1975b) is supported by the experiments of Tsukahara *et al.* (1986). This could represent the most primitive of hypothalamus-pituitary associations.

Unlike hagfish, adult marine lampreys generally lead an active predatory life. Changes in the amount of neurosecretory material in their neural lobe may reflect accomodations to varying environments brought about by their migrations between rivers and the sea (Ball and Baker, 1969; Holmes and Ball, 1974), variations in daylight (Öztan and Gorbman, 1960), and a variety of other seasonally related phenomena (Sterba, 1969). What remains a mystery is how, in the absence of anatomical links between sensory receptors and reproductive structures, seasonal changes and reproductive functions are related in lampreys (Gorbman, 1983b). It is interesting to speculate that changes in reproductive processes could be related to patterns of feeding. Variations in quantity and type of food consumed may reflect alterations in environmental conditions. The pituitary may be affected directly when food passes through the buccal cavity or, perhaps, indirectly by gut hormones that are released upon feeding to affect brain, pituitary, and/or gonad activity.

B. Class Chondrichthyes

The cartilaginous fish diverged early and pursued an evolution independent of all other fish groups. There are two extant subclasses—the sharks, skates, and rays, comprising the Elasmobranchii, and the ratfishes or rabbit fishes and chimaeroids of the Holocephali.

In elasmobranch fishes, the neurohypophysis consists of a thin-walled anterior portion that is a true median eminence, which is connected to the pars distalis by a hypothalamo-hypophysial portal system. The median eminence is divided into anterior and posterior portions in at least some species (Knowles *et al.,* 1975). The posterior region receives both hypothalamic aminergic and peptidergic axons and is linked by the portal system to the CPD and perhaps indirectly to the ventral lobe (Follénius, 1965; Dodd and Dodd, 1985); the anterior region contains less neurosecretory material and is linked to the RPD. It has often been suggested that this arrangement could reflect a more rapid and efficient method for controlling different pituitary cell types. This situation is similar to that of the median eminence in birds, which also is divided into anterior and posterior portions and serves in a "point-to-point" delivery of vessels to the adenohypophysis. Since, in contrast to teleosts, there is no direct innervation of the pars distalis, the portal system is the only route for hypothalamic control. There is also a portal supply to the neurointermediate lobe. A saccus vasculosus appears for the first time in elasmobranchs. It is not, however, quite as developed as its homologue in teleosts. It develops from the hypothalamus as a balloon-like structure just above the neurointermediate lobe, but since it is considered to be sensory and nonendocrine, it will not be discussed further.

The adenohypophysis is divided into four areas. There is an elongated pars distalis (the "dorsal lobe"), with its characteristic RPD and CPD; a ventral lobe attached by a stalk to the CPD; and a large pars intermedia that is heavily penetrated by neurohypophysial tissue to form a typical neurointermediate lobe. A pars tuberalis is not seen (Fig. 3).

Follicles and spaces are characteristic of the chondrichthyian pituitary and range from a highly developed system of vesicles and tubules that communicates with a hollow pars distalis in sharks and dogfish to a situation that is found in skates and rays, in which the hypophysial cavity is small or lost entirely. The ventral lobe, which forms from the fusion of the lateral lobes of Rathke's pouch, is also hollow and often vesicular (Ball and Baker, 1969). All of the cavities and spaces, including a persistent hypophysial cleft, are apparently remnants of Rathke's pouch. (Elasmobranchii are the only fishes to have a hollow Rathke's pouch during development.) The cavity frequently contains a PAS^+, AF^+, and alcian blue-positive colloid. The concept that this material represents stored hormones is no longer tenable (Holmes and Ball, 1974). It is likely that these secretions emanate from nonendocrine, chromophobic cells that line the walls of the cavity. The structure of these cells and the chemistry of their secretion suggests that they are similar to the mucus-secreting epithelial cells of the buccal cavity (Mellinger, 1969; Alluchon-Gerard, 1971). They have been homologized with the stellate cells that are found in the adenohypophysis of most vertebrates (Vila-Porcile, 1972).

Although certain tropic hormone functions have been defined clearly and have been associated with specific areas of the gland, others remain elusive. Attempts to link hormones to specific cells have spawned contradictions. ACTH and prolactin presumably are produced in the RPD (Fontaine and Olivereau, 1975). There is one cell type present that stains with PAS and orange G but not with AF or alcian blue. This cell type, therefore, is unlike the corticotropin- and prolactin-producing cells of teleosts. Earlier reports that ACTH may be found in the pars intermedia as well as in the RPD (deRoos and deRoos, 1967) have been confirmed by the use of immunofluorescence. There is, however, no cross-reactivity with anti-α- or anti-β-MSH antibodies or with porcine anti-ACTH (Mellinger and Dubois, 1973).

The CPD cells are acidophils that have a varied response to PAS and an unknown function. The cells of the ventral lobe are large and stain with AF, PAS, and alcian blue and are presumed to represent thyrotropes and gonadotropes (Ball and Baker, 1969; Dodd et al., 1960; Gorbman et al., 1983; Dodd and Dodd, 1985). Scanes et al. (1972) reported that gonadotropic activity, as judged by ^{32}P uptake in the chick gonad, occurs only in the ventral lobe. Immunofluorescence confirms LH activity in this region (Mellinger, 1972). In addition, removal of the ventral lobe from the dogfish results in the degeneration and phagocytosis of a specific stage of spermatogenesis (Dobson and Dodd, 1977a,b,c; Dodd et al., 1960; Mellinger, 1965).

The holocephalian neurohypophysis (as, for example, in the ratfish) has a pars nervosa and a prominent median eminence that sends many short blood vessels into the RPD and CPD. A typical neurointermediate lobe is seen. In holocephalians, the entire pars distalis is hollow and is not as clearly divisible into rostral and caudal regions as in the elasmobranchs (Fig. 3). A ventral lobe, characteristic of the elasmobranchs, is not present. In adult holocephalians, however, there is a large compact follicular structure called the Rachendachhypophyse (the pharyngeal lobe). It lies outside the cranium in the roof of the buccal cavity and has an independent blood supply (Jasinski and Gorbman, 1966). It is derived from the embryonic pars distalis rudiment and is ultimately separated from it when a connecting stalk disappears. In young Hydrolagus, the pharyngeal lobe is vesicular and composed of a single layer of follicular epithelium (Sathyanesan and Das, 1975). According to Ball and Baker (1969), the Rachendachhypophyse contains two kinds of basophils; both are PAS$^+$ but only one is AF$^+$. Recently, gonadotropic activity has been identified in the Rachendachhypophyse of the rabbit fish (Dodd et al., 1982), which would make this buccal lobe comparable to the elasmobranch ventral lobe. However, it is not likely that the Rachendachhypophyse and the ventral lobe are homologues, because their embryogenesis differs (Honma, 1969). Rather, a "homology" has been suggested with the buccal part of the Latimeria pituitary gland (Holmes and Ball, 1974). The cell types of the Hydrolagus pituitary have been described (Sathyanesan and Das, 1975), but their function(s) remain(s) unknown. An immunocytochemical study would be useful here.

C. Class Osteichthyes

1. Subclass Actinopterygii

The subclass Actinopterygii (fish with a bony skeleton and paired ray fins) includes the infraclasses Teleostei, Chondrostei, and Holostei (the latter two taxa are often referred to as the ganoid fishes).

a. Infraclass Chondrostei. The Chondrostei contains two orders, Polypteriformes (e.g., *Polypterus* and *Calamoichthys*) and Acipenseriformes (paddlefish and sturgeons). The pituitary glands of *Polypterus* and *Calamoichthys* are apparently similar in structure. Type A fibers (peptidergic and AF$^+$) originate in the preoptic nucleus and end in the pars nervosa. The origin of the type B fibers (aminergic and AF$^-$) is unknown (some suspect it is also the preoptic nucleus), but they terminate in the median eminence. A typical neurointermediate lobe is present. *Calamoichthys* and *Polypterus* both possess a typical median eminence and portal system and, in this respect, are different from the phylogenetically more recent teleosts (Lagios, 1968). There is no direct innervation of the pars distalis cells, a feature common to teleosts. The most remarkable feature of the Chondrostei is the presence of a duct or canal that connects a cavity in the ventral region of the RPD to the roof of the mouth (the buccohypophysial or orohypophysial duct). In section, they sometimes appear as follicles within the RPD. The canal presumably arises from a solid Rathke's pouch.

The gland subdivisions are easily discernible and resemble those seen in teleosts. The only cell that has been identified with any certainty by structure, stain, and the use of immunofluorescence has been the one which produces prolactin. Fluorescent-labeled antibodies to ovine prolactin are bound by cells scattered throughout the RPD of *Polypterus* and *Calamoichthys* (Aler, 1971; Aler *et al.*, 1971). No specific somatotrope has been identified, although growth hormone activity has been demonstrated (Hayashida, 1971). Basophils, chromophobes, and other acidophils have been identified, but no specific hormones have been ascribed to them. The pars intermedia contains a single cell type that is PAS$^+$ but PbH$^-$ (Norris, 1980). If this cell produces MSH, as is suspected, then its staining characteristics would be at variance with other vertebrates. The duct cells do not stain with the labeled antibodies and are, therefore, not homologous with the prolactin cells that surround the follicles of the Salmonidae and the Clupeidae. The generally chromophobic duct cells, which contain PAS$^+$ and alcian blue-positive granules, are mucus-producing (Lagios, 1968) and nonendocrine.

In the order Acipenseriformes, the most notable anatomic feature is a large, central hypophysial cavity with many tubular extensions. Unlike the palaeoniscoids, however, there is no hypophysial duct. There are also many "follicles" scattered throughout the RPD and CPD that contain PAS$^+$ colloid and are surrounded by acidophils, basophils, and chromophobes. These acidophils may be the source of the prolactin detected by bioassay (Sage and Bern, 1972). All these observations have led to the speculation that these vesicles are homologous to the RPD follicles in holosteans and some "primitive" teleosts (Holmes and Ball, 1974).

A thin-walled neural lobe penetrates into the deepest regions of a large pars intermedia (Fig. 3). There is no direct innervation of the pars distalis (Sathyanesan and Chavin, 1967), but there is a well-developed median eminence and portal system (Ball and Baker, 1969; Hayashida and Lagios, 1969).

b. Infraclass Holostei. The Holostei are the closest living primitive relatives of the teleosts. Neither genus of the two living holosteans, *Amia* (bowfin) and *Lepisosteus* (*Lepidosteus*) (garpike), has a hypophysial cavity or duct as adults. The gland is attached along most of its length to the infundibular floor, and a vascular connective tissue is found in between. There is a large saccus vasculosus (Sathyanesan and Chavin, 1967).

In contrast to teleosts, the pars distalis is not penetrated by large strands of neurohypophysial tissue. Lagios (1970) has shown, however, that there are small numbers of fibers with neurosecretory material (alcian blue-positive) that extend into and abut against the cell cords of the CPD. Innervation of the pars intermedia of *Amia* is similar to that of other ganoids and primitive teleosts, in that type A fibers contact the basement membrane of the intervascular space that separates the neural processes from endocrine cells (Holmes and Ball, 1974). Occasionally, type A fibers appear to synapse with pars intermedia cells (Lagios, 1970). This type of novel contact is rare in other ganoids but common in teleosts. Holosteans also maintain a median eminence that is somewhat better developed than their phylogenetic predecessors, the acipenseroids.

Thus, it appears from an examination of the pituitary in *Amia* that innervation of the adenohypophysial cells, a characteristic that typifies teleosts, may have preceded the change in the median eminence/portal system that we see in the bony fishes.

The adenohypophysis consists of an RPD, a CPD, and a pars intermedia (Fig. 3). The RPD contains follicles that enclose PAS$^+$, alcian blue-positive, and AF$^+$ material and are lined with the acidophils that may be the source of prolactin (Sage and Bern, 1972). These cells are in close association with an erythrosin-, aniline blue-, and PbH-positive cell that may be the corticotrope. At the boundary of RPD and CPD are the so-called basophils (PAS$^+$, AF$^+$, PbH$^+$, and alcian blue-positive), which also stain with erythrosin and could be (although no experimental evidence is available) the gonadotropes (Ball and Baker, 1969; Holmes and Ball, 1974). In the CPD, there is an acidophil that differs from acidophils of the RPD; it stains with orange G and is, most likely, a somatotrope. Two types of basophils, one of which is a thyrotrope and the other unknown, are present also in the CPD. In the pars intermedia, a prominent PbH$^+$ cell thought to produce MSH and a less common PAS$^+$ cell of unknown function are present.

c. Infraclass Teleostei. Reviews of the structure and function of the teleostean pituitary gland have appeared with a regular periodicity (Ball and Baker, 1969; Sage and Bern, 1971; Schreibman *et al.*, 1973; Follénius *et al.*, 1978; van Oordt and Peute, 1983). Teleosts represent the most diverse group of vertebrates, with more than 20,000 extant species recorded. Inasmuch as they occupy every conceivable habitat, it would not be surprising to find this variability reflected in a number of different patterns of pituitary organization and function. However, this is generally not the case, for within all this diversity there are common anatomical, histological, and cytological denominators. The most notable of these is that the adenohypophysis is clearly divided into three regions, the RPD, CPD, and pars intermedia (Fig. 3). (There is no pars tuberalis.) The clarity of this separation is essentially due to a restriction of specific cell types to distinct regions of the gland (see page 30). In teleosts, there is a structural intimacy of the neurohypophysis with both RPD and CPD as well as with the pars intermedia, and a unique feature is that the adenohypophysial cells are innervated directly. A few exceptions to the general rule have been noted (Fridberg and Ekengren, 1977; Knowles and Vollrath, 1966), and in these forms an elaborate system of extravascular circulation serves to convey neurosecretions to pituitary cells (Abraham *et al.*, 1982). There is generally no true median eminence or portal system present in teleosts.

Variations in the structure of the pituitary gland appear in the axis formed between the pituitary gland and the brain and, thus, in the orientation of the various adenohypophysial regions. Specific cell types may not be found in the same region of the gland in all fishes. This is especially true for the thyrotropes. The arrangement of cells may also vary. For example, follicles that are common in the Salmonidae (salmon and trout), Clupeidae (herrings), and Anguillidae (eels) generally are not found in other fish. A persistent orohypophysial duct may be seen in adult *Chanos chanos* (Olsson, 1974) and in some Clupeidae (Misra and Sathyanesan, 1959). Different hypothalamic control mechanisms also have been noted among the teleosts (see Holmes and Ball, 1974, for discussion).

The neurohypophysis receives neuronal endings from hypothalamic and extrahypothalamic nuclei. Axons may originate from perikarya as close as the nucleus lateralis tuberis or as distant as the nucleus olfactoretinalis (nervus terminalis: Schreibman *et al.*, 1982b; Halpern-Sebold and Schreibman, 1983). The neurohypophysis typically can be divided into two parts, an anterior region with AF^-, type B fibers (and a modicum of pituicytes) in contact with the pars distalis, and a posterior region composed mainly of AF^+, type A fibers (but with many pituicytes) in direct association with the pars intermedia. It has been suggested from functional, anatomical, and embryological evidence that these two regions correspond to the median eminence and the neural lobe, respectively (Ball and Baker, 1969; Perks, 1969; Vollrath, 1972). However, there has been considerable discussion as to whether a true median eminence exists in teleosts that is similar in structure and function to that seen in other groups, and opposing conclusions have been reached (see discussion by Munro and Dodd, 1983). Some maintain that the median eminence is represented by the vascular system resulting from the close anatomical association of the anterior neurohypophysis and the pars distalis and by evidence of synaptoid contacts on perivascular spaces (Fridberg and Ekengren, 1977). Other evidence to support the presence of a median eminence comparable to tetrapods is frequently based on india ink injections that are difficult to evaluate and on reconstructions from serial histological sections. This reviewer is inclined to agree with Gorbman (Gorbman et al., 1983) that although there is the possibility that neurosecretions could be released into blood vessels in the neurohypophysis that course through the pars distalis and thus reach pituitary secretory cells, there is no vascular pattern seen that is comparable to that of other vertebrates. A true median eminence system is composed of a primary capillary plexus, which then drains to a secondary plexus by way of a portal vein. From a functional point of view, it would be a redundant, unnecessary system, since pituitary cells are directly (or almost directly) innervated by these nerve fibers.

In general, hormone-producing cells of a particular type are restricted to specific regions of the adenohypophysis (Fig. 5). Nomenclature for pituitary cells has been derived from studies involving the application of standard staining methods to fish whose physiological homeostasis has been challenged (castration, interference with thyroid gland function, alteration of ionic content, administration of exogenous humoral agents, etc.). In recent years the functional accuracy of these terms has been examined using immunocytochemical methods with antisera from homologous and heterologous organisms, including antisera generated against human pituitary hormones (Fig. 6) (Margolis-Kazan and Schreibman, 1981). The RPD has a preponderance (essentially a single mass) of prolactin cells (erythrosinophilic). A narrow band of corticotropes forms the posterior boundary of the RPD. In the CPD, thyrotropes and somatotropes are intermingled in islets formed by pervading neurohypophysial tissue. The external boundary of the CPD is made up of several cell layers of gonadotropes. In midsagittal section these gonadotropes form a distinct ventral border in the CPD, which does not appear in poeciliids until the process of sexual maturation commences (Schreibman et al., 1982a) (Fig. 7). In platyfish the age at which this gonadotropic zone develops is under genetic control (Kallman and Schreibman, 1973; Kallman et al., 1973; Schreibman and Kallman, 1977) and is dependent on the prior maturation of specific LHRH systems in the brain (Schreibman et al., 1982b; Halpern-Sebold and Schreibman, 1983; Halpern-Sebold et al., 1984).

The number of gonadotropins produced by teleosts appears to be an irreconcilable problem (see van Oordt and Peute, 1983); however, based on immunocytochemistry, one gonadotrope cell type is present in the pars distalis (Follénius et al., 1978). The pars intermedia generally contains two cell types. One is PbH$^+$, cross-reacts with antiserum to ACTH, and is presumed to be the source of MSH. The other cell is strongly PAS$^+$ and its function remains obscure. It would appear that the differences reported for the func-

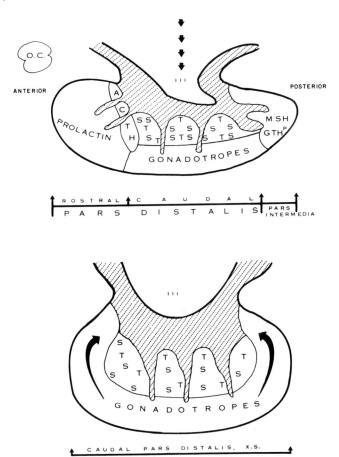

Fig. 5. A diagrammatic representation of the pituitary gland from a sexually mature platy-fish as seen in a midsagittal section (upper panel) and in a cross section through the caudal pars distalis (lower panel). The major regions of the gland and the hormone-producing cells that populate the various zones are indicated. The arrows in the upper diagram indicate the plane of section for the cross section depicted below. O.C., optic chiasma; ACTH, adrenocorticotropes; T, thyrotropes; S, somatotropes; III, third ventricle; MSH, melanocyte-stimulating hormone producing cells; GTH^P, gonadotropes (PAS$^+$ cells) of the pars intermedia; oblique dashed lines, neurohypophysis. [From Schreibman and Margolis-Kazan, 1979.]

tion of this cell in the fish examined are probably species-related. We suggest that in platyfish these PAS$^+$ pars intermedia cells are a second type of gonadotrope that we propose calling the GTH^P cell (Fig. 5). Because these cells are present at birth in great number and are immunoreactive with antiserum to the beta, but not the alpha, subunit of carp GTH, we have suggested that they serve to regulate early developmental events (from birth to sexual maturity) in the brain–pituitary–gonad axis and may produce a second type of gonadotropin (Schreibman and Margolis-Kazan, 1979; Schreibman *et al.*, 1982a; Halpern-Sebold and Schreibman, 1983; Halpern-

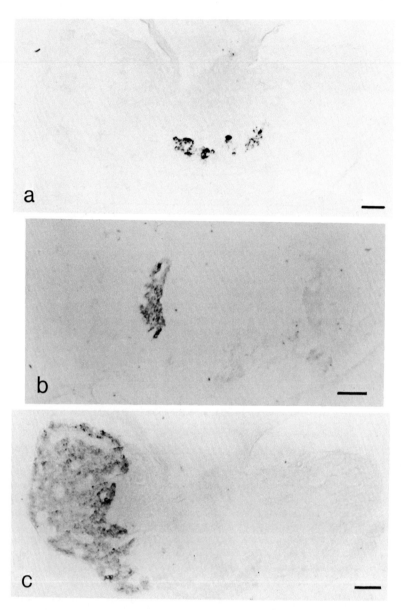

Fig. 6. Sagittal sections through the pituitary gland of sexually mature male platyfish to illustrate the distribution of immunoreactivity to antisera to several human (h) pituitary hormones. Use Fig. 5 as a guide to identify the distribution of immunoreactive cell types. In all figures anterior is to the left and the bar represents 50 μm. (a) Immunoresponse to the beta subunit of anti-hTSH (1 : 5000). (b) Immunoresponse to anti-ACTH (1 : 2000). (c) Immunoresponse to anti-hPRL (1 : 4000).

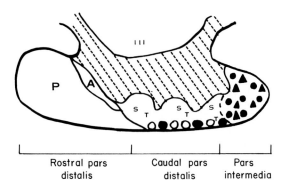

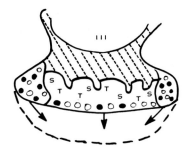

Fig. 7. Drawings of the pituitary gland of a neonatal fish in midsagittal section (upper panel) and in cross section through the caudal pars distalis (lower panel). P, prolactin cells; A, ACTH cells; S, somatotropes; T, thyrotropes; ○, chromophobes without immunoreactive material; ●, PAS⁺ and immunoreactive LHRH- and GTH-containing cells; ▲, PbH⁺ cells in the pars intermedia; oblique lines, neurohypophysis; III, third ventricle. The arrows suggest the possible origin of gonadotropes that populate the ventral caudal pars distalis (dashed lines, lower panel) of sexually mature platyfish. Compare with Fig. 5, and see text.

Sebold *et al.*, 1984). These cells are similar to the gonadotropes of the CPD in that they contain immunoreactive GTH and LHRH and bind sex steroid hormones (Schreibman *et al.*, 1979; Margolis-Kazan and Schreibman, 1981; Schreibman *et al.*, 1982c). GTH[P] cells differ from CPD gonadotropes in their morphology (Schreibman, 1964; Margolis-Kazan *et al.*, 1981), temporal pattern of proliferation (Schreibman *et al.*, 1982a; Halpern-Sebold *et al.*, 1984), and immunoreactive response to serotonin (Margolis-Kazan *et al.*, 1985) and tyrosine hydroxylase (Halpern-Sebold *et al.*, 1985), as well as to GTH subunit (Schreibman and Margolis-Kazan, 1979), antisera.

2. Subclass Sarcopterygii

The subclass Sarcopterygii consists of the orders Dipnoi and Crossopterygii. The single Coelacanth genus, *Latimeria,* and the half dozen species of lungfish are interesting to consider in the study of the evolution of the tetrapod pituitary gland.

a. Order Dipnoi. The neurohypophysis of lungfish is divisible into a median eminence, infundibular stalk, and neural lobe (Perks, 1969). The preoptic nucleus is the major hypothalamic nucleus projecting to the neurohypophysis, and it stains with the usual neurosecretory material methods. The projecting neurons are peptidergic and terminate around the primary portal vessels, the plexus intermedius, the neural lobe, and in the pars intermedia. In *Lepidosiren,* Zambrano and Iturriza (1973) have described many small aminergic neurons whose perikarya lie just below the ependyma of the median eminence and whose axons end in the pars distalis, pars intermedia, and the neural lobe close to the pars intermedia. These neurons resemble the nucleus lateralis tuberis of teleosts and the infundibular and arcuate nuclei of tetrapods. Thus, we see here a system of type A fibers terminating on capillaries anteriorly, but on an avascular connective tissue posteriorly.

The adenohypophysis of lungfish is more similar in structure to that of amphibians than it is to other fishes (Holmes and Ball, 1974). Among the amphibian characteristics of the lungfish pituitary are a well-formed neural lobe, a prominent median eminence, the absence of a saccus vasculosus, and a less obvious regional distribution of cell types (Kerr and van Oordt, 1966). Nevertheless, certain fishlike characteristics persist. Most notable is the direct aminergic innervation of the pars distalis cells (Zambrano and Iturriza, 1973), the interdigitation of the neurohypophysis and pars intermedia, the prominent hypophysial cleft between the pars distalis and the pars intermedia, the absence of a pars tuberalis, and the presence of follicles, especially in *Neoceratodus.*

The five cell types that have been defined in the pars distalis using standard staining procedures are intermingled and lack the distinct zonation seen in teleosts; however, differences in relative proportions of cells permit delineating pars distalis regions (Fig. 3). The absence of immunocytochemical evidence and the lack of physiological experiments preclude assigning specific hormonal roles to pituitary cells. Nevertheless, investigators have extrapolated from the data gathered in amphibians to assign function to pituitary cells in the dipnoans on the grounds that cells in these two groups show similarities of staining responses, ontogenesis, and distribution. According to van Oordt (1968) and Kerr and van Oordt (1966), one type of acidophil that stains with alizarine BT is distributed throughout the gland and secretes prolactin. A second acidophil type, orangeophilic and faintly PAS$^+$, is restricted to the CPD and is purported to be a somatotrope. Three types of basophils, distinguishable by size and distribution, are the putative sources for thyrotropin, gonadotropin, and adrenocorticotropin. The size of the pars intermedia, the arrangement of its cells, and its structural association with the neural lobe processes varies among the lungfish; however, in all cases it is separated from the pars distalis by the hypophysial cleft. Two cell types are found in the pars intermedia; one is weakly PAS$^+$, and alcian blue- and aniline blue-positive. The second, which is more abundant in young fish, is strongly PAS$^+$, alcian blue-positive, and orangeophilic (Kerr and van Oordt, 1966). Both types are directly innervated by type A and a few type B fibers (Zambrano and Iturriza, 1973). The results reviewed here relating tinctorial cell types to function are not to be construed as being confirmatory, but rather they are merely meant to stimulate thought and to serve as a basis for further, more sophisticated cytological analyses.

There is still much to be learned about the lungfish pituitary. We need to know what variations there are in hypophysial structure, aside from the relative number of follicles, size of the pars intermedia, and innervation of cells that would account for differences in the physiological activities of the three living genera (*Protopterus*, African; *Neoceratodus*, Australian; and *Lepidosiren*, South American) and the six species of the Dipnoi. This would be especially interesting in regard to the differences in the ability of lungfish to estivate (Thomson, 1969).

b. Order Crossopterygii. *Latimeria chaulumnae* is the only living representative of the suborder Coelacanthini. Lagios (1972, 1975) and van Kemenade and Kremers (1975) have described single specimens collected in 1970 and 1972, respectively. The orthodox features of the coelacanth pituitary include a neurointermediate lobe (typically fishlike) ventral to the brain, direct contact between pars distalis and pars intermedia, follicles and tubules containing glycoprotein colloid seen in many fish, a well-developed hypothalamo-hypophysial portal system, and a probable median eminence (Lagios, 1972). Other features that are more "fishlike" include a small saccus vasculosus, penetration of neurohypophysial tissue into the pars distalis, and regional differentiation of the gland based on cell type.

Among the exceptional features is a greatly extended pars distalis that is tripartite in nature (Fig. 3). An orthodox proximal portion lies close to (even interdigitating with) a neurointermediate lobe and can be separated on the basis of vascular supply and cell type into dorsal (rostral) and posterior (caudal) regions. The rostral portion contains orangeophils and erythrosinophils, and orangeophils and basophils populate the caudal region. The presence of growth hormone has been reported (Hayashida, 1977); however, its cellular origin is unknown. An elongate extension of the pars distalis comprises its most rostral lobe (buccal portion). Relatively few basophils were found in the buccal rostral lobe in the one immature female studied (Lagios, 1975). What is unique is a long (up to 12 cm), cylindrical, tubular hypophysial cavity that extends from the rostral lobe of the pars distalis and contains vascularized masses of adenohypophysial cells, the so-called rostral islets or pars buccalis. These structures are comparable with the Rachendachhypophyse of holocephalians rather than with the ventral lobe of elasmobranchs insofar as their anatomical connection is concerned; however, cell types of the islets are similar to those in the ventral lobe. The need to study additional specimens is obvious.

Thus, the pituitary gland of *Latimeria* shows many specializations and peculiarities. It possesses many features of the teleost and elasmobranchiomorph gland rather than those of dipnoans and amphibians.

D. Class Amphibia

The amphibian gland is very much like the hypophysis of the other tetrapods. The adenohypophysis is generally composed of a compact pars distalis that is differentiated into rostral and caudal regions. In anurans, it lies ventral and posterior to the neural lobe. The pars distalis is flattened and often wider than long (Fig. 3). Dorsally it continues with the pars intermedia without an intervening cleft, and anteriorly it is connected to the median eminence by a strand of connective tissue through which the portal vessels pass. The pars tuberalis is seen in amphibians for the first time in vertebrate phylogeny (Fitzgerald, 1979). It is represented by a pair of epithelial plaques attached on either side of the tuber cinereum anterior to the median eminence. In urodeles, the pars distalis occupies the same position but is more elongated.

The neurohypophysis is differentiated into a neural lobe and a median eminence. The neural lobe is generally the largest in the more terrestrial amphibians (Dodd *et al.*, 1971). In anurans, the median eminence varies in its complexity and can be as elaborate in structure as one sees in mammals (Holmes and Ball, 1974). The median eminence forms in the ventral portion of the neurohypophysis. The thin-walled structure of this region facilitates surgical manipulations and invites special observation and study of median eminence function (Fig. 8). AF^+ neurons, with cell bodies in the prominent preoptic area, terminate in the pars nervosa (thickened dorsal component of the neurohypophysis). Neurons terminating in the primary plexus of the median eminence have a wider distribution between the preoptic area and the midbrain. The pars intermedia receives blood from the neural lobe, the median eminence, and the general circulation; the pars distalis receives blood only from the median eminence (Rodriguez and Piezzi, 1967).

The scattering of specific cell types precludes separation of the pars distalis into distinct zones, although regionalization is somewhat easier to recognize in urodeles than in anurans. Unfortunately, disagreement among researchers prevents assigning definitive cellular sources for specific hormones. According to van Oordt (1974) five stainable cell types are identifiable in *Rana temporaria*; acidophils that are large, stain with erythrosin, orange G, and anti-sheep prolactin (Doerr-Schott, 1976a). These presumptive prolactin cells are scattered throughout the pars distalis, although they are in greatest number in the anterior region. Purported somatotropes are orangeophilic and are restricted to the caudal portion of the pars distalis (Schreibman and Holtzman, 1975). Type 1 basophils that occur in small clusters and stain positively with PAS, AF, and aniline blue are thought to be

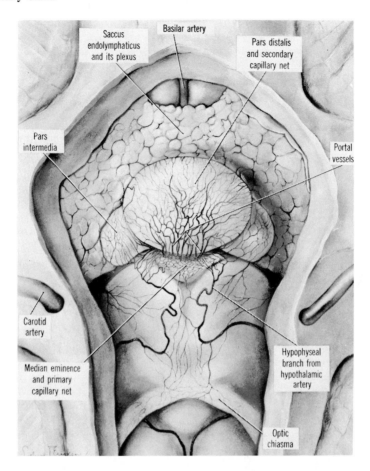

Fig. 8. Ventral view of the brain, pituitary gland, and hypophysial portal circulation of the adult bullfrog *Rana catesbeiana*. [From Gorbman *et al.*, 1983; photograph of a drawing supplied by J. D. Green, University of California.]

thyrotropes. Cells having a similar, but less intense, staining profile and which show an affinity for orange G are believed to be gonadotropes. ACTH-producing cells are generally found in the RPD and their staining reaction is similar to type 1 basophils, except with alcian blue.

There is little information on the cytology of the pars tuberalis except that on the basis of electron microscopic analysis of granule size and cytological characteristics, there may be two cell types present and that they are presumably secretory in function (cf. Fitzgerald, 1979).

The pars intermedia forms a distinct and separable band of tissue between the pars distalis and pars nervosa. A PbH$^+$ cell in the pars intermedia predominates and is probably the source of MSH. Release of MSH is under neural control from catecholaminergic fibers that directly innervate pars intermedia cells.

E. Class Reptila

The morphology of the reptilian hypophysis is well known from the works of Wingstrand (1966a,b) and Saint-Girons (1963, 1967, 1970) and, in general, its structure is felt to strongly resemble both the amphibian and the avian types.

The reptilian gland differs from that of the amphibian in that the pars distalis and pars intermedia are structurally more intimate, the pars tuberalis may not be present in some groups, the pars intermedia may be larger than in any other vertebrate, and a globular pars nervosa is suspended from a narrow infundibular stalk (Fig. 3). A median eminence is found on the infundibular stalk just anterior to the pars distalis. Portal vessels connect primary and secondary capillary plexuses. Peptidergic nerve endings from the supraoptic and paraventicular nuclei terminate in the pars nervosa. A separate infundibular nucleus, which is presumably homologous to the nucleus of the same name in amphibians and birds, sends aminergic and peptidergic fibers which terminate in the median eminence (Dodd et al., 1971). The reptilian pars distalis and pars nervosa enjoy separate blood supplies (Enemar, 1960; cf. Holmes and Ball, 1974).

As is typical in tetrapods, the reptilian pars distalis is made up of branching cords of cells that are highly vascularized by intervening sinusoids. Occasionally, follicular structures may form. Immunocytochemical studies by Doerr-Schott (1976b) have demonstrated a similarity in the distribution of hypophysial cell types in reptiles and amphibians. That is, ACTH- and prolactin-producing cells are predominant in the rostral zone, STH-producing cells are found in the caudal zone, and gonadotropes and thyrotropes are found in both lobes (see Doerr-Schott, 1976a,b; Schreibman and Holtzman, 1975). Bioassay of the RPD and CPD regions confirms the distribution of prolactin and growth hormone (Sage and Bern, 1972). Similarly, bioassay for thyroid-stimulating hormone (TSH) correlates well with the preponderance of type 1 basophils in the caudal lobe (Licht and Rosenberg, 1969). These are the same basophils that are cytologically responsive to thyroidectomy (cf. Licht and Pearson, 1978).

Type 2 basophils (PAS+ and alcian-blue- and aniline-blue-positive) have a scattered distribution. In some snakes and lizards they are ventrally located in the midline; in other lizards they are restricted to the rostral lobe. These cells respond to fluctuating levels of sex steroids, thus suggesting a gonadotropic function for them (cf. Holmes and Ball, 1974). Type 3 basophils are found in the rostral lobe and respond to experimental alterations of corticosteroid synthesis (Del Conte, 1969). The one stainable cell type in the pars intermedia is the presumed source of MSH (Saint-Girons, 1970).

In *Sphenodon,* turtles, and crocodiles there is a prominent pars intermedia and pars tuberalis and a pars distalis that has clearly defined rostral and caudal lobes. In some reptiles (e.g., lizards) an epithelial stalk remnant of Rathke's pouch may persist (Wingstrand, 1966a). The pars tuberalis is the most variable component of the gland. In turtles it is relatively large, with tuberalis tissue in association with the hypothalamus and the pars distalis; in lizards it is reduced and limited to the hypothalamus or is absent; and in snakes it never develops (cf. Pearson and Licht, 1982). Generally, portal vessels pass from the median eminence to the pars distalis within connective tissue but without involving pars tuberalis tissue; in turtles, however, portal vessels do pass through the pars tuberalis. In some adult lizards, the pars tuberalis, which forms plaques on either side of the median eminence, invades the hypothalamus and median eminence when the basement membrane and pia separating the tuberalis disappear (cf. Holmes and Ball, 1974). Recent immunocytochemical studies of several species of turtles have demonstrated the presence of gonadotropes and thyrotropes in the pars tuberalis tissue associated with both the hypothalamus and the pars distalis (Pearson and Licht, 1982).

In *Typhlops* and *Leptotyphlops* (blind, burrowing snakes) and in burrowing lizards, there is no pars intermedia, and no intermedia cells are found in the pars distalis (Saint-Girons, 1963, 1967). This is unlike the situation in birds and some mammals, which, although they lack a discrete pars intermedia, have intermedia cells in their pars distalis. In most snakes, however, the pars intermedia is massive.

F. Class Aves

The hypophysis of birds has been described by Wingstrand (1951), and several features distinguish the avian gland from that of other vertebrates. In many birds, the oral part of Rathke's pouch gives rise to an epithelial stalk that persists in some birds as a connection between the pars distalis and the buccal epithelium. The separation of the adenohypophysis from the neurohypophysis in birds makes the gland a less compact unit than, for example, that in mammals.

A well-developed pars tuberalis is located between the median eminence and the pars distalis, although in some species (e.g., *Diomeda melanophris*) it may lie within the median eminence (Wingstrand, 1951). In those birds in which the pars tuberalis is not closely applied to the neural downgrowth, an isolated portotuberal tract may form that serves to bridge neural and glandular regions (Benoit and Assenmacher, 1953). There is no organized pars intermedia, although PbH$^+$ cells [so-called Kappa cells (Tixier-Vidal and Follett, 1973)] can be found in the pars distalis, and these apparently secrete MSH. Positive bioassays for MSH activity in the avian, as well as in the mammalian, pars distalis are difficult to evaluate, since there is a second

PbH$^+$ cell (epsilon) which is presumably the source of ACTH, a hormone that with the use of appropriate antibodies also gives a positive α-MSH response. The distribution of prolactin cells (in the RPD), somatotropes (in the CPD), gonadotropes (in the RPD and CPD), adrenocorticotropes (in the RPD), and thyrotropes (in either the RPD or CPD, depending on species) is expected, based on cell distribution in other groups (Fig. 3). The suggestion, based on ultrastructural observations, that there may be two types of gonadotropes is not supported by immunocytochemical and physiological studies (Mikami, 1983).

The median eminence differs from that of mammals in that in birds the primary capillaries of the portal system are more superficial and form a somewhat less complex arrangement (Wingstrand, 1951; cf. Holmes and Ball, 1974). The avian median eminence can be regionally differentiated into the so-called point-to-point portal system, which comprises a rostral portion (AF$^+$), which supplies blood to the RPD, and a caudal area, which delivers blood to the CPD (Vitums *et al.*, 1964; Sharp and Follett, 1969). This arrangement of the vessels may have considerable importance in the light of the distribution of cell types into the different lobes, for it may provide a mechanism for the separation and delivery of specific stimuli from the central nervous system to the adenohypophysis (see Section V). The differentiation of the median eminence into two regions is comparable to the situation in reptiles and elasmobranchs. A detailed study of the hypothalamo-hypophysial system of the white-crowned sparrow (Oksche and Farner, 1974) has demonstrated that the two zones of the median eminence which vascularize different areas of the pars distalis are also innervated by different populations of neurosecretory neurons (Fig. 9). The pars nervosa receives separate innervation. The degree to which the phenomenon of two median eminences is expressed among the birds remains to be determined.

G. Class Mammalia

The mammalian pituitary gland probably has been studied more often and used in more hypophysial-related research than that of any other vertebrate group. Despite the fact that there is a basic similarity in the structure of the pituitary gland among the mammals, departures from the general pattern can be found in various genera (see reviews by Hanström, 1966; Holmes and Ball, 1974). Usually, these differences reside in the degree of development of the parts of the neurohypophysis, the axis of orientation of the neurohypophysis in relation to the brain, the presence of an intraglandular cleft, and the extent of development of the three parts of the adenohypophysis and their relationships to the neurohypophysis (Fig. 3). In some mammals, including humans, a pharyngeal hypophysis (craniopharyngeal) develops from a remnant of Rathke's pouch. It is found in the roof of the mouth and is vascularized, has staining and immunoreactive properties similar to the sel-

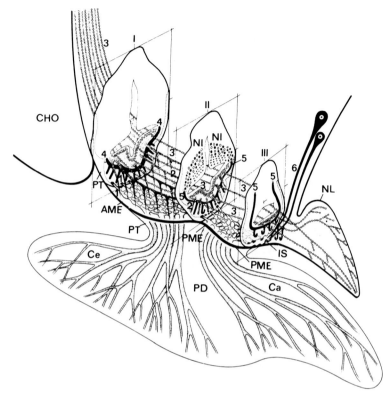

Fig. 9. Three-dimensional organization of neurosecretory pathways and vascular channels in the hypothalamo-hypophysial system of the white-crowned sparrow. Illustration depicts the two anatomically distinct median eminences (AME and PME) and their "point-to-point" portal systems. Only blood vessels in the pars distalis (PD) are shown. I, II, and III are cross sections at three levels of the hypothalamus, showing the kinds of neurosecretory fibers found at those levels. The beaded-appearing fibers are AF+ and proceed to the pars nervosa. The fine-stippled fiber groups are AF+ and AF− and go to the median eminence, where some of them form loops. Fiber groups shown in black are AF− and most go to the median eminence. Anatomic names of the neuronal fiber tracts are as follows: 1 and 2, tractus hypophyseus anterior; 3, tractus supraoptico-(paraventriculo-)hypophyseus; 4 and 5, tractus tuberohypophyseus; 6, tractus hypophyseus posterior. Note that in the basal infundibular nucleus (NI) there are clusters of neurons of different types. AME and PME, anterior and posterior median eminence; Ca and Ce, differentiated zones of the pars distalis; CHO, optic chiasma; IS, infundibular stalk; NL, pars nervosa; PT, pars tuberalis. The primary capillary network and hypophysial portal vessels are shown but are not labeled. [From Oksche and Farner, 1974.]

lar hypophysis (McGrath, 1971; Ciocca *et al.*, 1985), and exhibits cytological changes related to physiological state (Boyd, 1956; McGrath, 1968, 1971).

Another structural variation is the failure of the pars intermedia to form in cetaceans (whales and dolphins), elephants, and armadillos, presumably as a result of a connective tissue barrier that separates the neurohypophysis and the adenohypophysis (cf. Hanström, 1966). Although a pars tuberalis is characteristic of the mammalian gland, in some animals such as the sloth (Wislocki, 1938) and the pangolin (Herlant, 1958) it may be absent. In more primitive forms (for example, members of the order Monotremata, the duckbilled platypus and spiny anteaters) a typical portotuberal tract is present which is similar to that of many birds and reptiles. Recently, immunocytochemical methods have been used to demonstrate two cell types in the pars tuberalis of mammals—one containing gonadotropin and the other, thyrotropin (Baker et al., 1977; Baker and Yu, 1975; Gross, 1984). ACTH-containing cells have also been noted (Mikami, 1980). It is no longer acceptable to consider that the pars tuberalis in mammals, as well as in all other vertebrates where it is found, as a nonfunctional structure (Fitzgerald, 1979). Because of the anatomical position of the pars tuberalis and of the hormones it elaborates, the hypophysectomized mammal cannot be considered an animal devoid of adenohypophysial hormones (Gross, 1983). These findings emphasize the need to investigate further the suggestion that the pars tuberalis may be an extension of the pars distalis as well as the question of the ontogenetic relationships between these two regions.

Most mammalian adenohypophysial cells occur in cords or small groups and in close proximity to blood vessels. The regional distribution of cells seen so frequently in nonmammalian vertebrates may sometimes also occur in mammals. For example, a differential distribution of basophils has been noted in the pars distalis of rats (Purves and Griesbach, 1951). In other mammals, like the tree shrew and the potto (Holmes and Ball, 1974), typical follicular structures containing PAS+ colloid are clearly evident. It appears, therefore, that the arrangement of pituitary cells into follicles maybe considered a ubiquitous feature in vertebrates. Perhaps it is an inherent property of adenohypophysial cells to be able to reverse polarity and to secrete their product into a central storage area that is bounded by pituitary cells.

The mammalian pars distalis contains a variety of cells which can be distinguished by selective staining procedures, including immunocytochemistry, as well as by ultrastructural analysis. These cells are responsible for elaborating approximately seven hormones. It is still not certain whether LH and follicle-stimulating hormone (FSH) originate from different cells or from a common one (Girod, 1984)—a question not unique for mammals. Recent immunocytochemical studies in rats suggests that LH and FSH are produced in a single cell type; however, cells vary in the ratios of the several hormones to each other (Inoue and Kurosumi, 1984). The source of lipotropin (LPH) remains unclear.

It is generally accepted that direct innervation of the adenohypophysis does not play an important role, if, indeed, it plays any role at all in mammals. However, the innervation of blood vessels that irrigate the adenohypophysis may be an important control mechanism. The dominant regulatory mechanism of the mammalian adenohypophysis is via its vascular supply.

This feature shows phylogenetic constancy, for a portal system is found not only in mammals, but also in birds, reptiles, amphibians, and some fishes. The adult mammalian neurohypophysis is a neurohemal organ which receives aminergic and peptidergic neurons. The median eminence is also prominent, and it forms a primary capillary plexus around much of the infundibular stalk in most species. Most often, portal vessels pass through a well-developed pars tuberalis on their way to the pars distalis. The pioneering work of Wislocki and King (1936) and Green and Harris (1947) has demonstrated that blood that carries pituitary-regulating substances flows from the median eminence to the adenohypophysis. Later information suggests that there is an anatomic basis for retrograde flow from adenohypophysis to the hypothalamus (Fig. 10) (Bergland and Page, 1979). This pattern of flow would explain how pituitary hormones could directly affect the central nervous system. This concept is particularly intriguing when considering the analgesic effect on the brain of β-endorphin, whose source is in the pars distalis and/or pars intermedia (Bloom *et al.,* 1977; Rossier, 1982; Lowry *et al.,* 1985). Retrograde axonal transport in neurosecretory cells may provide yet another two-way system of communication between pituitary and brain and makes the story of information transfer between the endocrine and nervous systems even more intriguing.

VI. COMPARATIVE CONTRIBUTIONS AND BIOMEDICAL IMPLICATIONS

One of the basic goals of this volume, and of all the volumes in this series on vertebrate endocrinology, is to call attention to the wealth of fundamental information that has been derived from endocrinological studies of nonmammalian forms and to the potential that this knowledge holds for future investigations. This chapter summarizes the morphological features of the pituitary gland in the various classes of vertebrates. It is hoped that from the information presented, biomedical investigators and clinicians whose studies have been restricted to traditional laboratory animals (rats, cats, rabbits, mice, etc.) will be made aware that many other vertebrates, through their simplicity of structure (or function) or because of unique morphological features, may serve as useful, novel experimental models to answer fundamental and applied questions related to endocrine-regulated physiological processes. The examples which follow have been selected to demonstrate the usefulness that this information has served, as well as to suggest its potential for future investigations.

Our current knowledge about the role of prolactin in salt and water regulation and related processes stems from early studies in teleosts on the role of this hormone in osmoregulation. The anatomical position of the teleost pituitary gland just above a thin parasphenoid bone and just posterior to the optic chiasma makes it readily accessible for removal by a buccal cavity approach

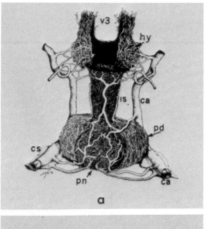

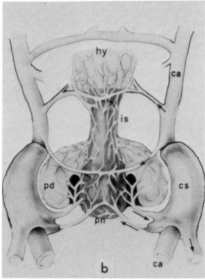

Fig. 10. Drawings of a vascular cast (a) and an interpretation (b) of the blood vessel connections between the pituitary and hypothalamus of a monkey (viewed from the posterior side). The arterial supply to the pituitary region is mostly from two branches of the carotid arteries (ca). The venous drainage is into the cavernous sinuses (cs). Directions of flow are indicated in b. The vascular spikes of the median eminence would be within the tissue of the infundibular stalk (is) and pars nervosa (pn). In b, a network of vessels is shown, extending from the neurohypophysis (is and pn) to the pars distalis (pd). (The pars distalis is wrapped around the anterior aspect of the pars nervosa but extends beyond it at the sides.) However, both figures show that this network is continuous with vessels that extend all the way to the hypothalamus (hy). [From Gorbman *et al.,* 1983; drawings by R. M. Bergland; see Bergland and Page, 1979.]

(Schreibman and Kallman, 1966). Observations that euryhaline teleosts which had been hypophysectomized could survive only by replacement therapy with mammalian prolactin or fish prolactin cell grafts or by placing the fish into dilute seawater set the stage for additional further studies (cf. Ball, 1969; Schreibman and Holtzman, 1975). In teleosts, the restriction of prolactin cells to a specific, easily identifiable region of the gland (Figs. 5 and 6) facilitated the functional cytological evaluation of prolactin cell activity in response to various treatments as, for example, the modification of the salinity and specific ionic content of the environment (Sage and Bern, 1971; Schreibman *et al.*, 1973), the administration of prolactin and other pituitary and nonhypophysial hormones, and the administration of drugs that were being tested for osmoregulatory functions. These studies in teleosts, which exploited the knowledge of the location of prolactin cells and the salt-conserving properties of prolactin, helped to lay the basis for our understanding of the roles served by ergot alkaloids (e.g., CB-154) and neurotransmitters in controlling the cellular release of prolactin in mammals. This information has had tremendous value in the treatment of basic human clinical conditions such as breast cancer and aberrant reproductive physiological processes and has extended our understanding of other prolactin-regulated processes such as, for example, lactation (cf. Bern *et al.*, 1980).

The ability to separate specific cell types by surgical means has facilitated studies of the effects of hormone preparations from closely related animals on specific physiological processes. For example, analysis of the data derived from experiments in which the prolactin portion of the carp pituitary was administered to killifish (Pang *et al.*, 1978) has led to a better understanding of the role of prolactin in calcium regulation in fish. This information is now being evaluated by clinicians for possible application to human disorders.

The exchange of surgically removed regions of the pituitary, and even of entire glands, within members of genetically related stocks has furthered our comprehension of the hypothalamic control of hypophysial function (Ball, 1965; Ball *et al.*, 1965; Schreibman and Kallman, 1964). Similarly, *in vivo* analyses of the effects of surgical or chemical ablation of other endocrine organs, with or without hormone replacement, on particular teleost pituitary cell types which are segregated according to function (truly a "cytological cell separation") has clarified the nature of adenohypophysial regulation of endocrine gland development and function and feedback control mechanisms.

For now and for the immediate future, the cellular "floor plan" of the teleost pituitary gland could serve as an excellent bioassay model with which to assess the effectiveness of neurotransmitters, releasing factors, their analogues, and drugs that could affect a variety of endocrine processes (Fig. 5). What better vertebrate model than the teleost pituitary gland, in which hypophysiotropic neurons directly innervate adenohypophysial cells, could provide us with such a unique biological tool for studying the direct interaction between the nervous and endocrine systems and the specific, immediate actions of releasing substances on secretory cells? Similarly, the usefulness of the teleost pituitary model in the study of malignancies (e.g., melanomas and thyroid tumors) has been suggested previously (Schreibman, 1964).

The rationale presented for analysis of the teleost pituitary gland could be extended to cartilaginous fishes, reptiles, or birds (homeotherms in phylogenetic proximity to mammals) in order to study the hypothalamic control of adenohypophysial function. In each of these vertebrates, and especially in birds where it is most clearly defined, the median eminence is divided into two anatomically distinct regions. These two areas receive innervation from different hypophysiotropic regions and are drained by separate portal systems which supply different zones of the pars distalis (Fig. 9). The pars nervosa is separately innvervated and vascularized. This morphological pattern presents a model system with which to study the interaction between distinct regions of neurohypophysis and adenohypophysis and the specific role of the vascular system which connects them physiologically. The elegantly designed and readily accessible median eminence in amphibia is well suited for surgical intervention and for the administration of humoral and chemical agents into the portal system and could, therefore, be useful to further investigate central nervous system and endocrine system interactions.

There is currently much emphasis placed on the study of age-related physiological phenomena in mammals. The anatomical characteristics of the vertebrates discussed previously make them novel and simple systems in which to examine the popular suggestions that a basic mechanism for aging may reside in the faltering of communications between the nervous and endocrine systems (cf. Finch and Hayflick, 1977; Timiras, 1978, 1983).

The failure of the pars tuberalis to develop in some vertebrates, the incorporation of tuberalis tissue into the brain of birds and lizards, and the absence of a pars intermedia in some reptiles, birds, and mammals presents enticing material for the endocrinologist with interests in embryology to study basic mechanisms of induction and to trace the ontogeny of pituitary cell lines. To the basic endocrinologist, these developmental anomalies present elegant systems with which to study the effects atypical anatomical positions have on the structure and function of specific cells. It is most interesting to note that of all the adenohypophysial regions, it is the pars tuberalis and pars intermedia which display the most class and species variations in morphology and physiology.

To the developmental endocrinologist one may also offer the use of the foundation of information available on the embryology of reptiles. New studies on the development of the brain–pituitary axis using immunocytochemical methods could shed valuable information on the differentiation of function during embryogenesis of specific neuroendocrine components. Because of the critical evolutionary position the Reptilia enjoy as early ("primitive") amniotes, the information derived could have considerable application toward our understanding of development in birds and mammals.

The Rachendachhypophyse in holocephalians and the ventral lobe in elasmobranchs are masses of adenohypophysial tissue that are anatomically removed from the pars distalis proper (Fig. 3). As such, they present material for executing surgical partial hypophysectomy, a procedure that can only be done with some difficulty (in teleosts) or not at all (in other vertebrates). These observations, coupled with the evidence that now exists for gonadotropic activity in these structures, suggest some basic experiments that could be performed to study interactions among components of the brain–pituitary–gonad axis. The buccal part of the pars distalis of *Latimeria* and the pharyngeal hypophysis in humans have similar anatomical positions; however, lack of availability of these animals would make them an impractical choice for this type of study.

The examples of hypophysial systems from diverse vertebrates just discussed are obviously selective and incomplete. They are presented to introduce the endocrinologist, at any level of expertise and interest, to the variety of structures that exist among the vertebrates. It is my hope that the reader will be struck by some unique feature of a vertebrate class of species that will permit a novel or more suitable approach to an old problem or, perhaps, that will suggest some new areas for investigation. The potential uses of the information presented is restricted only by the availability of animals and one's imagination!

ACKNOWLEDGMENTS

I am grateful to Drs. Henrietta Margolis-Nunno and Leslie Halpern-Sebold for their comments and constructive criticism of the manuscript. Original research of the author has been supported by the National Science Foundation, the National Institute on Aging (AGO 1938), and the City University of New York (PSC-CUNY Award).

REFERENCES

Abraham, M., Kieselstein, M., Hilge, V., and Lison, S. (1982). Extravascular circulation in the pituitary of *Mugil cephalus* (Teleostei). *Cell Tissue Res.* **225**, 567–579.

Akmayev, I. G., and Fidelina, O. V. (1976). Morphological aspects of the hypothalamic-hypophyseal system. VI. The tanycytes: Their relation to the sexual differentiation of the hypothalamus. An enzyme-histochemical study. *Cell Tissue Res.* **173**, 407–416.

Aler, G. M. (1971). Prolactin-producing cells in *Clupea harengus membras, Polypterus palmas* and *Calamoichthys calabaricus* studied by immuno-histochemical methods. *Acta Zool. (Stockholm)* **52**, 275–286.

Aler, G. M., Bage, G., and Fernholm, B. (1971). On the existence of prolactin in cyclostomes. *Gen. Comp. Endocrinol.* **16**, 498–503.

Alluchon-Gérard, M. J. (1971). Types cellulaires et étapes de la differenciation de l'adenohypophyse chez l'embryon de roussette (*Scyllium canicula, Chondrichthyens*). Étude au microscope électronique. *Z. Zellforsch. Mikrosk. Anat.* **120,** 525–545.

Baker, B. I., and Buckingham, J. C. (1983). A study of corticotrophic and melanotrophic activities in the pituitary and brain of the lamprey, *Lampetra fluviatilis. Gen. Comp. Endocrinol.* **52,** 283–290.

Baker, B. L., and Yu, Y.-Y. (1975). Immunocytochemical analysis of cells in the pars tuberalis of the rat hypophysis with antisera to hormones of the pars distalis. *Cell Tissue Res.* **156,** 443–449.

Baker, B. L., Karsch, F. J., Hoffman, D. L., and Beckman, W. C., Jr. (1977). The presence of gonadotropic and thyrotropic cells in the pituitary pars tuberalis of the monkey (*Macaca mulatta*). *Biol. Reprod.* **17,** 232–240.

Ball, J. N. (1965). Partial hypophysectomy in the teleost *Poecilia:* Distinction between growth hormone and prolactin. *Gen. Comp. Endocrinol.* **5,** 654–661.

Ball, J. N. (1969). Prolactin and osmoregulation in teleost fishes: A review. *Gen. Comp. Endocrinol., Suppl.* **2,** 10–25.

Ball, J. N., and Baker, B. I. (1969). The pituitary gland: anatomy and histophysiology. *In* "Fish Physiology" (W. S. Hoar and D. J. Randall, eds.), Vol. 2, pp. 1–110. Academic Press, New York.

Ball, J. N., Olivereau, M., Slicher, A., and Kallman, K. D. (1965). Functional capacity of ectopic pituitary transplants in the teleost, *Poecilia formosa,* with a comparative discussion on the transplanted pituitary. *Philos. Trans. R. Soc. London, Ser. B* **249,** 66–99.

Benoit, J., and Assenmacher, I. (1953). Rapport entre la stimulation sexuelle prehypophysaire et la neurosécrétion chez l'oiseau. *Arch. Anat. Microsc. Morphol. Exp.* **42,** 334–386.

Bergland, R. M., and Page, R. B. (1979). Pituitary-brain vascular relations: A new paradigm. *Science* **204,** 18–24.

Bern, H. A., Loretz, C. A., and Biskee, C. A. (1980). Prolactin and transport in fishes and mammals. *Prog. Reprod. Biol.* **6,** 166–171.

Bloom, F., Battenberg, E. L. F., Rossier, J., Ling, N., Leppaluoto, J., Vargo, T. M., and Guillemin, R. (1977). Endorphins are located in the intermediate and anterior lobes of the pituitary gland, not in the neurohypophysis. *Life Sci.* **20,** 43–47.

Boyd, J. D. (1956). Observations on the human pharyngeal hypophysis. *J. Endocrinol.* **14,** 66–77.

Ciocca, D. R., Puy, L. A., and Stati, A. O. (1985). Identification of seven hormone-producing cell types in the human pharyngeal hypophysis. *J. Clin. Endocrinol. Metab.* **60,** 212–216.

Crim, J. W., Urano, A., and Gorbman, A. (1979a). Immunocytochemical studies of luteinizing hormone-releasing hormone in brains of agnathan fishes. I. Comparisons of adult Pacific lamprey (*Entosphenus tridentia*) and the Pacific hagfish (*Eptatretus stouti*). *Gen. Comp. Endocrinol.* **37,** 294–305.

Crim, J. W., Urano, A., and Gorbman, A. (1979b). Immunocytochemical studies of luteinizing hormone-releasing hormone in brains of agnathan fishes. II. Patterns of immunoreactivity in larval and maturing Western brook lamprey (*Lampetra richardson*). *Gen. Comp. Endocrinol.* **38,** 290–299.

Del Conte, E. (1969). The corticotroph cells of the anterior pituitary gland of a reptile: *Cnemidophorus lemniscatus* (Sauria, Teiidae). *Experientia* **25,** 1330–1332.

deRoos, R., and deRoos, C. C. (1967). Presence of corticotropin activity in the pituitary gland of chrondrichthyean fish. *Gen. Comp. Endocrinol.* **9,** 267–275.

Dickhoff, W. W., and Gorbman, A. (1977). *In vitro* thyrotropic effect of the pituitary of the Pacific hagfish, *Eptatretus stouti. Gen. Comp. Endocrinol.* **31,** 75–79.

Dobson, S., and Dodd, J. M. (1977a). Endocrine control of the testes in the dogfish *Scyliorhinus canicula* L. I. Effects of partial hypophysectomy on gravimetric, hormonal biochemical aspects of testes function. *Gen. Comp. Endocrinol.* **32,** 41–52.

Dobson, S., and Dodd, J. M. (1977b). Endocrine control of the testes in the dogfish *Scyliorhinus canicula* L. II. Histological and ultrastructural changes in the testes after partial hypophysectomy (ventral lobectomy). *Gen. Comp. Endocrinol.* **32**, 53–71.

Dobson, S., and Dodd, J. M. (1977c). The roles of temperature and photoperiod in the response of the testes of the dogfish, *Scyliorhinus canicula* L. to partial hypophysectomy (ventral lobectomy). *Gen. Comp. Endocrinol.* **32**, 114–115.

Dodd, J. M., and Dodd, M. H. I. (1985). The ventral lobe and rachendachhypophyse (buccal lobe) of the pituitary gland of cartilagenous fishes. *In* "Current Trends in Comparative Endocrinology" (B. Lofts and W. N. Holmes, eds.), Vol. 1, pp. 127–130. Hong Kong Univ. Press, Hong Kong.

Dodd, J. M., Evenett, P. J., and Goddard, C. K. (1960). Reproductive endocrinology in cyclostomes and elasmobranchs. *Symp. Zool. Soc. London* **1**, 77–103.

Dodd, J. M., Follett, B. K., and Sharp, P. J. (1971). Hypothalamic control of pituitary function in submammalian vertebrates. *Comp. Physiol. Biochem.* **4**, 113–223.

Dodd, J. M., Dodd, M. H. I., Sumpter, J. P., and Jenkins, N. (1982). Gonadotropic activity in the buccal lobe (Rachendachhypophyse) of the pituitary gland of the rabbit fish. *Gen. Comp. Endocrinol.* **48**, 174–180.

Doerr-Schott, J. (1976a). Immunohistochemical detection, by light and electron microscopy, of pituitary hormones in cold-blooded vertebrates. I. Fishes and amphibians. *Gen. Comp. Endocrinol.* **28**, 487–512.

Doerr-Schott, J. (1976b). Immunohistochemical detection, by light and electron microscopy, of pituitary hormones in cold-blooded vertebrates. II. Reptiles. *Gen. Comp. Endocrinol.* **28**, 513–529.

Dores, R. M., Finger, T. E., and Gold, M. R. (1984). Immunohistochemical localization of enkephalin- and ACTH-related substances in the pituitary of the lamprey. *Cell Tissue Res.* **235**, 107–115.

Enemar, A. (1960). The development of the hypophysial vascular system in the lizards *Lacerta a. agilis* (Linnaeus) and *Anguis fragilis* (Linnaeus) and in the snake *Natrix n. natrix*, with comparative remarks on the amniota. *Acta Zool. (Stockholm)* **41**, 141–237.

Fernholm, G. (1972). The ultrastructure of the adenohypophysis of *Myxine glutinosa*. *Z. Zelloforsch. Mikrosk. Anat.* **132**, 451–472.

Finch, C. E., and Hayflick, L. (1977). *In* "Handbook of the Biology of Aging." p. 771. Van Nostrand-Reinhold, Princeton, New Jersey.

Fitzgerald, K. T. (1979). The structure and function of the pars tuberalis of the vertebrate adenohypophysis. *Gen. Comp. Endocrinol.* **37**, 383–399.

Follénius, E. (1965). Bases structurales et ultrastructurales des correlations diencéphalo-hypophysaires chez les sélaciens et les téléostéens. *Arch. Anat. Microsc. Morphol. Exp.* **54**, 195–216.

Follénius, E., Doerr-Schott, J., and Dubois, M. P. (1978). Immunocytology of pituitary cells from teleost fishes. *Int. Rev. Cytol.* **54**, 193–223.

Fontaine, M., and Olivereau, M. (1975). Some aspects of the organization and evolution of the vertebrate pituitary. *Am. Zool.* **15**, Suppl., 61–79.

Fridberg, G., and Ekengren, B. (1977). The vascularization and the neuroendocrine pathways of the pituitary gland in the Atlantic salmon, *Salmo salar*. *Can. J. Zool.* **55**, 1284–1296.

Girod, C. (1984). Fine structure of the pituitary pars diatalis. *In* "Ultrastructure of Endocrine Cells and Tissues" (P. M. Motta, ed.), pp. 12–28. Martinus Nijhoff, The Hague.

Gorbman, A. (1965). Vascular relations between the neurohypophysis and adenohypophysis of clyclostomes and the problem of evolution of hypothalmic endocrine control. *Arch. Anat. Microsc. Morphol. Exp.* **54**, 163–194.

Gorbman, A. (1983a). Early development of the hagfish pituitary gland. Evidence for the endodermal orgin of the adenohypophysis. *Am. Zool.* **23**, 639–654.

Gorbman, A. (1983b). Reproduction in cyclostome fishes and its regulation. *In* "Fish Physiology" (W. S. Hoar, D. J. Randall, and E. M. Donaldson, eds.), Vol. 10, pp. 1–29. Academic Press, New York.

Gorbman, A. (1984). Pituitary development in cyclostomes compared to higher vertebrates. *Int. Symp. Pituitary Gland, 1984,* Abstr. No. 8-1, p. 51.

Gorbman, A., Kobayashi, H., and Uemura, H. (1963). The vascularization of the hypophysial structures of the hagfish. *Gen. Comp. Endocrinol.* **3,** 505–514.

Gorbman, A., Dickhoff, W. W., Vigna, S. R., Clark, N. B., and Ralph, C. L. (1983). "Comparative Endocrinology." Wiley, New York.

Green, J. D., and Harris, G. W. (1947). The neurovascular link between the neurohypophysis and adenohypophysis. *J. Endocrinol.* **5,** 136–146.

Gross, D. S. (1983). Hormone production in the hypophysial pars tuberalis of intact and hypophysectomized rats. *Endocrinology (Baltimore)* **112,** 733–744.

Gross, D. S. (1984). The mammalian hypophysial pars tuberalis: A comparative immunocytochemical study. *Gen. Comp. Endocrinol.* **56,** 283–298.

Halpern-Sebold, L., and Schreibman, M. P. (1983). Ontogeny of luteinizing hormone-releasing hormone containing centers in the brain of platyfish (*Xiphophorus maculatus*) as determined by immunocytochemistry. *Cell Tissue Res.* **229,** 75–84.

Halpern-Sebold, L., Schreibman, M. P., and Margolis-Kazan, H. (1984). Ontogeny of immunoreactive luteinizing hormone releasing hormone (ir-LHRH) containing centers in the brain and their relationship to sexual maturation. *Neuroscience* **10,** 456 (abstr.).

Halpern-Sebold, L., Margolis-Kazan, H., Schreibman, M. P., and Joh, T. H. (1985). Immunoreactive tyrosine hydroxylase in the brain and pituitary gland of the platyfish. *Proc. Soc. Exp. Biol. Med.* **178,** 486–489.

Hanström, B. (1966). Gross anatomy of the hypophysis of mammals. *In* "The Pituitary Gland: Anterior Pituitary" (G. W. Harris and B. T. Donovan, eds.), pp. 1–57. Univ. of California Press, Berkeley.

Hayashida, T. (1971). Biological and immunochemical studies with growth hormone in pituitary extracts of holostean and chondrostean fishes. *Gen. Comp. Endocrinol.* **17,** 275–280.

Hayashida, T. (1977). Immunoassay and biological studies with growth hormone in a pituitary extract of the coelacanth, *Latimeria chalumnae* Smith. *Gen. Comp. Endocrinol.* **32,** 221–229.

Hayashida, T., and Lagios, M. D. (1969). Fish growth hormone: A biological immunochemical and ultrastructural study of sturgeon and paddlefish pituitaries. *Gen. Comp. Endocrinol.* **13,** 403–411.

Herlant, M. (1958). L'hypophyse et le système hypothalamo-hypophysaire du pangolin (*Manis tricuspis* et *Manis tetradactyla*). *Arch. Anat. Microsc. Morphol. Exp.* **47,** 1–23.

Holmes, R. L., and Ball, J. N. (1974). "The Pituitary Gland: A Comparative Account." Cambridge Univ. Press, London and New York.

Honma, Y. (1969). Some evolutionary aspects of the morphology and role of the adenohypophysis in fishes. *Gunma Symp. Endocrinol.* **6,** 19–36.

Horstmann, E. (1954). Die Faserglia des Selachiergehirns. *Z. Zellforsch. Mikrosk. Anat.* **39,** 588–617.

Inoue, K., and Kurosumi, K. (1984). Ultrastructural immunocytochemical localization of LH and FSH in the pituitary of the untreated male rat. *Cell Tissue Res.* **235,** 77–83.

Jasinski, A., and Gorbman, A. (1966). Hypothalamo-hypophysial vascular and neurosecretory links in the ratfish. *Hydrolagus colliei* (Lay and Bennett). *Gen Comp. Endocrinol.* **6,** 476–490.

Jirikowski, G., Erhart, G., Grimmelikhuijzen, C. J. P., Triepel, J., and Patzner, R. A. (1984). FMRF-amide-like immunoreactivity in brain and pituitary of the hagfish, *Eptatretus burgeri* (Cyclostomata). *Cell Tissue Res.* **237,** 363–366.

Kallman, K. D., and Schreibman, M. P. (1973). A sex-linked gene controlling gonadotrop differentiation and its significance in determining the age of sexual maturation and size of the platyfish, *Xiphophorus maculatus*. *Gen. Comp. Endocrinol.* **21**, 287–304.

Kallman, K. D., Schreibman, M. P., and Borkoski, V. (1973). Genetic control of gonadotrop differentiation in the platyfish, *Xiphophorus maculatus* (Poeciliidae). *Science* **181**, 678–680.

Kerr, T., and van Oordt, P. G. W. J. (1966). The pituitary of the African lungfish *Protopterus sp. Gen. Comp. Endocrinol.* **7**, 549–558.

Knowles, F., and Vollrath, L. (1966). Neurosecretory innervation of the pituitary of the eels *Anguilla* and *Conger*. II. The structure and innervation of the pars distalis at different stages of the life-cycle. *Philos. Trans. R. Soc. London, Ser. B* **250**, 329–342.

Knowles, F., Vollrath, L., and Meurling, P. (1975). Cytology and neuroendocrine relations of the pituitary of the dogfish, *Scyliorhinus canicula*. *Proc. R. Soc. London* **191**, 507–525.

Kobayashi, H., and Uemura, H. (1972). The neurohypophysis of the hagfish, *Eptatretus burgeri* (Girard). *Gen. Comp. Endocrinol., Suppl.* **3**, 114–124.

Kobayashi, H., Ichikawa, T., Suzuki, H., and Sekimoto, M. (1972). Seasonal migration of the hagfish, *Eptatretus burgeri*. *Jpn. J. Ichthyol.* **19**, 191–194.

Lagios, M. D. (1968). Tetrapod-like organization of the pituitary gland of the polypteriformid fishes, *Calamoichthys calabaricus* and *Polypterus palmas*. *Gen. Comp. Endocrinol.* **11**, 300–315.

Lagios, M. D. (1970). The median eminence of the bowfin, *Amia calva*, L. *Gen. Comp. Endocrinol.* **15**, 453–463.

Lagios, M. D. (1972). Evidence for a hypothalamo-hypophysial portal vascular system in the coelacanth *Latimeria chalumnae* Smith. *Gen. Comp. Endocrinol.* **18**, 73–82.

Lagios, M. D. (1975). The pituitary gland of the coelacanth *Latimeria chalumnae*. Smith. *Gen. Comp. Endocrinol.* **25**, 126–146.

Larsen, L. O. (1969). Hypophyseal functions in river lampreys. *Gen. Comp. Endocrinol. Suppl.* **2**, 522–527.

Larsen, L. O. and Rothwell, B. (1972). Adenohypophysis. *In* "The Biology of Lampreys" (M. W. Hardisty and I. E. Potter, eds.), Vol. 2, pp. 1–67. Academic Press, New York.

Licht, P., and Pearson, A. K. (1978). Cytophysiology of the reptilian pituitary gland. *Int. Rev. Cytol. Suppl.* **7**, 239–286.

Licht, P., and Rosenberg, L. L. (1969). Presence and distribution of gonadotropin and thyrotropin in the pars distalis of the lizard, *Anolis carolinensis*. *Gen. Comp. Endocrinol.* **13**, 439–454.

Lowry, P. J., Jackson, S., Estivariz, F. E., and Al-Dujaili, E. A. S. (1985). Biosynthesis and release of peptides derived from pro-opiocortin. *In* "Current Trends in Comparative Endocrinology" (B. Lofts and W. N. Holmes, eds.) Vol. 1, pp. 159–161. Hong Kong Univ. Press, Hong Kong.

Margolis-Kazan, H., and Schreibman, M. P. (1981). Cross-reactivity between human and fish pituitary hormones as demonstrated by immunocytochemistry. *Cell Tissue Res.* **221**, 257–267.

Margolis-Kazan, H., Peute, J., Schreibman, M. P., and Halpern, L. R. (1981). Ultrastructural localization of gonadotropin and luteinizing hormone-releasing hormone in the pituitary gland of a teleost fish (the platyfish). *J. Exp. Zool.* **215**, 99–102.

Margolis-Kazan, H., Halpern-Sebold, L. R., and Schreibman, M. P. (1985). Immuno-cytochemical localization of serotonin in the brain and pituitary gland of the platyfish, *Xiphophorus maculatus*. *Cell Tissue Res.* **240**, 311–314.

Matty, A. J., Tsuneki, K., Dickhoff, W., and Gorbman, A. (1976). Thyroid and gonadal function in hypophysectomized hagfish, *Eptatretus stouti*. *Gen. Comp. Endocrinol.* **30**, 500–516.

McGrath, P. (1968). Prolactin activity and human growth hormone in pharyngeal hypophysis from embalmed cadavers. *J. Endocrinol.* **42**, 205–212.

McGrath, P. (1971). The volume of the human pharyngeal hypophysis in relation to age and sex. *J. Anat.* **110**, 275–282.

Mellinger, J. C. A. (1965). Stades de la spermatogénese chez *Scyliorhinus caniculus* (L): Description données histochimiques, variation normales et expérimentales. *Z. Zellforsch. Mikrosk. Anat.* **67**, 653–673.

Mellinger, J. C. A. (1969). Développement post-embryonnaire de l'adénohypophyse de la torpille (*Torpedo marmorata,* Chondichthyens); évolution du système des cavitiés et manifestations du dimorphisme sexuel. *Ann. Univ. ARERS* **7**, 33–48.

Mellinger, J. C. A. (1972). Types cellulaires et function de l'adénohypophyse de la Torpile (*Torpedo marmorata*). *Gen. Comp. Endocrinol.* **18**, 608.

Mellinger, J. C. A., and Dubois, M. P. (1973). Confirmation par l'immunofluorescence, de la fonction cortiocotrope du lobe rostral et de la fonction gonadotrope du lobe ventral de l'hypophyse d'un poisson cartilagineux, la Torpile marbrée (*Torpedo marmorata*). *C. R. Hebd. Seances Acad. Sci., Ser. D* **267**, 1879–1881.

Mikami, S. (1980). Comparative anatomy and evolution of the hypothalamo-hypophysial systems in higher vertebrates. *In* "Hormones Adaptation and Evolution" (S. Ishii, T. Hirano, and M. Wada, eds.), pp. 57–70. Jpn. Sci. Soc. Press, Toyko.

Mikami, S. (1983). Avian adenohypophysis: Recent progress in immunocytochemical studies. *In* "Avian Endocrinology" (S. Mikami, K. Honma, and M. Wada, eds.), pp. 39–56. Jpn. Sci. Soc. Press, Toyko.

Misra, A. B., and Sathyanesan, A. G. (1959). On the persistence of the orohypophysial duct in some clupeoid fishes. *Proc. Int. Zool. Congr. 15th, 1958,* pp. 5–27.

Munro, A. D., and Dodd, J. M. (1983). Forebrain of fishes: Neuroendocrine control mechanisms. *In* "Progress in Nonmammalian Brain Research" (G. Nistico and L. Bolis, eds.), Vol. 3, pp. 1–78. CRC Press, Boca Raton, Florida.

Nicoll, C. S. (1974). Physiological actions of prolactin. *In* "Handbook of Physiology" (D. W. Hamilton and R. O. Greep, eds.), Sect. 7, Vol. IV, Part 2, pp. 253–292. Williams & Wilkins, Baltimore, Maryland.

Norris, D. O. (1980). "Vertebrate Endocrinology." Lea & Febiger, Philadelphia, Pennsylvania.

Nozaki, M. (1986). Tissue distribution of hormonal peptides in primitive fishes. *In* "Evolutionary Biology of Primitive Fishes" (A. Gorbman, R. Olsson, and R. E. Foreman, eds.). Plenum, New York. (in press).

Nozaki, M., Fernholm, B., and Kobayashi, H. (1975). Ependymal absorption of peroxidase into the third ventricle of the hagfish *Eptatretus burgeri* (Gerard). *Acta Zool. (Stockholm)* **56**, 265–269.

Oksche, A., and Farner, D. S. (1974). Neurohistological studies of the hypophyseal system of *Zonotrichia leucophrys gambelli* (Aves, Passeriformes). *Adv. Anat. Embryol. Cell Biol.* **48**, 1–136.

Olsson, R. (1969). General review of the endocrinology of the Protochordata and Myxinoidea. *Gen. Comp. Endocrinol., Suppl.* **2**, 485–499.

Olsson, R. (1974). Fine structure of the pituitary eta cells of larval *Chanos chanos* (Teleostei). *Gen. Comp. Endocrinol.* **22**, 364.

Öztan, N., and Gorbman, A. (1960). The hypophysis and hypothalamo-hypophyseal neurosecretory system of larval lampreys and their response to light. *J. Morphol.* **103**, 243–261.

Pang, P. K. T., Schreibman, M. P., Balbontin, F., and Pang, R. K. (1978). Prolactin and pituitary control of calcium regulation in the killfish, *Fundulus heteroclitus. Gen. Comp. Endocrinol.* **36**, 306–316.

Pearson, A. K. (1985). Development of the pituitary in reptiles. *In* "Biology of the Reptilia" (C. Gans, F. Billet, and P. F. A. Maderson, eds.), Vol. 14A, pp. 679–719. Wiley, New York.

Pearson, A. K., and Licht, P. (1982). Morphology and immunocytochemistry of the turtle pituitary gland with special reference to the pars tuberalis. *Cell Tissue Res.* **222,** 81–100.

Perks, A. M. (1969). The neurohypophysis. *In* "Fish Physiology" (W. S. Hoar and D. J. Randall, eds.), Vol. 2, pp. 111–205. Academic Press, New York.

Petrusz, P. (1983). Essential requirements for the validity of immunocytochemical staining procedures. *J. Histochem. Cytochem.* **31,** 177–180.

Purves, H. D., and Griesbach, W. E. (1951). The site of the thyrotrophin and gonadotrophin production in the rat pituitary studied by the McManus-Hotchkiss staining for glycoprotein. *Endocrinology (Baltimore)* **49,** 244–264.

Rodriguez, E. M., and Piezzi, R. S. (1967). Vascularization of the hypophysial region of the normal and adenohypophysectomized toad. *Z. Zellforsch. Mikrosk. Anat.* **83,** 207–218.

Rossier, J. (1982). Functions of beta-endorphin and enkephalins in the pituitary. *In* "Frontiers in Neuroendocrinology" (W. F. Ganong and L. Martini, eds.), Vol. 7, pp. 191–210. Raven Press, New York.

Sage, M., and Bern, H. A. (1971). Cytophysiology of the teleost pituitary. *Int. Rev. Cytol.* **31,** 339–376.

Sage, M., and Bern, H. A. (1972). Assay of prolactin in vertebrate pituitaries by its dispersion of xanthophore pigment in the teleost *Gillichthys mirabilis. J. Exp. Zool.* **180,** 169–174.

Saint-Girons, H. (1963). Histologie comparée de l'adénohypophyse chez les reptiles. *Colloq. Int. C. N. R. S.* **128,** 275–285.

Saint-Girons, H. (1967). Morphologie comparée de l'hypophyse chez les squamata. *Ann. Sci. Nat. Zool. Biol. Anim.* [12] **9,** 229–308.

Saint-Girons, H. (1970). The pituitary gland. *In* "Biology of the Reptilia" (C. Gans and T. S. Parsons, eds.), Vol. 3, pp. 135–199. Academic Press, New York.

Sathyanesan, A. G., and Chavin, W. (1967). Hypothalamo-hypophyseal neurosecretory system in the primitive actinopterygian fishes (Holostei and Chondrostei). *Acta Anat.* **68,** 284–299.

Sathyanesan, A. G., and Das, R. C. (1975). Pituitary cytology of the chimaeroid fish *Hydrolagus colliei* (Lay and Bennett). *Z. Mikrosk.-Anat. Forsch.* **89,** 715–726.

Scanes, C. G., Dobson, S., Follett, K., and Dodd, J. M. (1972). Gonadotrophic activity in the pituitary gland of the dogfish (*Scyliorhinus canicula*). *J. Endocrinol.* **54,** 343–344.

Schreibman, M. P. (1964). Studies on the pituitary gland of *Xiphophorus maculatus* (the platyfish). *Zoologica (N.Y.)* **49,** 217–244.

Schreibman, M. P. (1980). Adenohypophysis: Structure and function. *In* "Evolution of Vertebrate Endocrine Systems" (P. K. T. Pang and A. Epple, eds.), pp. 107–131. Texas Tech Univ. Press, Lubbock.

Schreibman, M. P., and Holtzman, S. (1975). The histophysiology of the prolactin cell in nonmammalian vertebrates. *Am. Zool.* **15,** 867–880.

Schreibman, M. P., and Kallman, K. D. (1964). Functional pituitary grafts in freshwater teleosts. *Am. Zool.* **4,** 417 (abstr.).

Schreibman, M. P., and Kallman, K. D. (1966). Endocrine control of freshwater tolerance in teleosts. *Gen. Comp. Endocrinol.* **6,** 144–155.

Schreibman, M. P., and Kallman, K. D. (1977). The genetic control of the pituitary-gonadal axis in the platyfish, *Xiphophorus maculatus J. Exp. Zool.* **200,** 277–294.

Schreibman, M. P., and Margolis-Kazan, H. (1979). Immunocytochemical localization of gonadotropin, its subunits and thyrotropin in the teleost, *Xiphophorus maculatus. Gen. Comp. Endocrinol.* **39,** 467–474.

Schreibman, M. P., Leatherland, J. F., and McKeown, B. A. (1973). Functional morphology of the teleost pituitary gland. *Am. Zool.* **13,** 719–742.

Schreibman, M. P., Halpern, L. R., Goos, H. J. Th., and Margolis-Kazan, H. (1979). Identification of luteinizing hormone-releasing hormone (LH-RH) in the brain and pituitary gland of a fish by immunocytochemistry. *J. Exp. Zool.* **200,** 153–160.

Schreibman, M. P., Margolis-Kazan, H., and Halpern-Sebold, L. R. (1982a). Immunoreactive gonadotropin and luteinizing hormone-releasing hormone in the pituitary gland of neonatal fish. *Gen. Comp. Endocrinol.* **47**, 385–391.

Schreibman, M. P., Margolis-Kazan, H., Halpern-Sebold, L., and Goos, H. J. Th. (1982b). The functional significance of the nucleus olfactoretinalis in the platyfish *Xiphophorus muculatus*. *Proc. Int. Symp. Reprod. Physiol. Fish 2nd, 1982*, p. 59.

Schreibman, M. P., Pertschuk, L. P., Rainford, E. A., Margolis-Kazan, H., and Gelber, S. J. (1982c). The histochemical localization of steroid binding sites in the pituitary gland of a teleost (the platyfish). *Cell Tissue Res.* **226**, 523–530.

Sharp, P. J., and Follett, B. K. (1969). The blood supply of the pituitary and the basal hypothalamus in the Japanese quail (*Coturnix coturnix japonica*). *J. Anat.* **104**, 227–232.

Sterba, G. (1969). Endocrinology of the lampreys. *Gen. Comp. Endocrinol., Suppl.* **2**, 500–509.

Thomson, K. S. (1969). The biology of the lobe-finned fishes. *Biol. Rev. Cambridge Philos. Soc.* **44**, 91–154.

Timiras, P. S. (1978). Biological perspectives on aging. *Am. Sci.* **66**, 605–613.

Timiras, P. S. (1983). Neuroendocrinology of aging: Retrospective, current and prospective views. *In* "Neuroendocrinology of Aging" (J. Meites, ed.), Chapter 2, pp. 5–30. Plenum, New York.

Tixier-Vidal, A., and Follett, B. K. (1973). The adenohypophysis. *In* "Avian Biology" (D. S. Farner and J. R. Kings, eds.). Vol. 3, pp. 109–182. Academic Press, New York.

Tsukahara, T., Gorbman, A., and Kobayaski, H. (1986). Median eminence equivalence of the neurohypophysis of the hagfish *Eptatretus burgeri*. *Gen. Comp. Endocrinol.* (in press).

Tsuneki, K., and Gorbman, A. (1975a). Ultrastructure of pars nervosa and pars intermedia of the lamprey, *Lampetra tridentata*. *Cell Tissue Res.* **157**, 165–184.

Tsuneki, K., and Gorbman, A. (1975b). Ultrastructure of the anterior neurohypophysis and the pars distalis of the lamprey, *Lampetra tridentata*. *Gen. Comp. Endocrinol.* **25**, 487–508.

van Kemenade, J. A. M., and Kremers, D. W. (1975). The pituitary gland of the coelacanth fish *Latimeria chalumnae* Smith: General structure and adenohypophysial cell types. *Cell Tissue Res.* **163**, 291–311.

van Oordt, P. G. W. J. (1968). The analysis and identification of the hormone producing cells of the adenohypophysis. *In* "Perspectives in Endocrinology" (E. J. W. Barrington and C. B. Jorgensen, eds.), pp. 405–467. Academic Press, New York.

van Oordt, P. G. W. J. (1974). Cytology of the adenohypophysis. *In* "Physiology of the Amphibia" (B. Lofts, ed.), Vol. 2, pp. 53–106. Academic Press, New York.

van Oordt, P. G. W. J., and Peute, J. (1983). The cellular origin of pituitary gonadotropins in teleosts. *In* "Fish Physiology" (W. S. Hoar, D. J. Randall, and E. M. Donaldson, eds.), Vol. 9, pp. 137–186. Academic Press, New York.

Vila-Porcile, E. (1972). Le reseau des cellules folliculo-stellaires et les follicules de l'adénohypophyse du rat (pars distalis). *Z. Zellforsch. Mikrosk. Anat.* **129**, 328–369.

Vitums, A., Mikami, S., Oksche, A., and Farner, D. S. (1964). Vascularization of the hypothalamo-hypophysial complex in the white-crowned sparrow, *Zonotrichia leucophrys gambellii*. *Z. Zellforsch. Mikrosk. Anat.* **64**, 541–569.

Vollrath, L. (1972). Morphological correlates of releasing factors in the pituitary of the eel, *Anguilla anguilla*. *In* "Brain-Endocrine Interaction. Median Eminence: Structure and Function" (K. M. Knigge, D. E. Scott, and A. Weindl, eds.), pp. 154–163. Karger, Basel.

Wingstrand, K. G. (1951). "The Structure and Development of the Avian Pituitary." C. W. K. Gleerup, Lund, Sweden.

Wingstrand, K. G. (1966a). Comparative anatomy and evolution of the hypophysis. *In* "The Pituitary Gland: Anterior Pituitary" (G. W. Harris and B. T. Donovan, eds.), Vol. 1, pp. 58–126. Univ. of California Press, Berkeley.

Wingstrand, K. G. (1966b). Microscopic anatomy, nerve supply and blood supply of the pars intermedia. *In* "The Pituitary Gland: Pars Intermedia and Neurophyophysis" (G. W. Harris and B. T. Donovan, eds.), Vol. 3, pp. 1–27. Butterworth, London.

Wislocki, G. B. (1938). The topography of the hypophysis in the Xenarthra. *Anat. Rec.* **70,** 451–471.

Wislocki, G. B., and King, L. S. (1936). The permeability of the hypophysis and the hypothalamus to vital dyes, with a study of the hypophysial vascular supply. *Am. J. Anat.* **58,** 421–472.

Wright, G. M. (1983). Immunocytochemical study of luteinizing hormone in the pituitary of the sea lamprey, *Petromyzon marinus* L., during its upstream migration. *Cell Tissue Res.* **230,** 225–228.

Zambrano, D., and Iturriza, F. C. (1973). Hypothalamo-hypophysial relationships in the South American lungfish, *Lepidosiren paradoxa. Gen. Comp. Endocrinol.* **20,** 256–273.

3

Vertebrate Neurohormonal Systems

R. E. PETER
Department of Zoology
The University of Alberta
Edmonton, Alberta, Canada T6G 2E9

I. INTRODUCTION

The number of brain peptides, listed near 40 in a recent review (Krieger, 1983), continues to expand. What is more impressive than the length of this list is the lack of understanding of the functions of this array of peptides. The purpose of this chapter is to assess, in the vertebrate brain, origins of peptides and classical neurotransmitters that may serve as neurohormones. However, covering all aspects of the brain distribution of a peptide or transmitter that may serve as a neurohormone is not practical within the limits of this chapter. The intent is to review the known and probable brain origins of the nerve fibers that terminate in the pituitary or median eminence, on the assumption that the neurosecretion from these fibers has an endocrine function. Accordingly, the literature coverage in this review is selective and is not intended to be an exhaustive review of the distribution throughout the nervous system of the various peptides and transmitters that have some neurohormonal function.

In general, more is known about the distribution of neurohormones in the brain of selected mammals than in species of any other class of vertebrate. This raises a problem of how to approach the topic of brain distribution of neurohormones in vertebrates; should the topic be approached ascending or descending the vertebrate phylogenetic tree? The approach I have taken is to start with the class of vertebrates for which the most information is available and to use that as the basis for comparison with other classes; in essence, my approach will be to descend the phylogenetic tree.

By comparing the brain origins of the peptides and neurotransmitters that impinge on the pituitary in vertebrates, it is hoped that new approaches to the study of function of neurohormones can be realized. Comparative ana-

57

tomical studies of vertebrate neuroendocrine systems may reveal special adaptations in some species that could be particularly useful as a model for investigations. On the other hand, the apparent conservatism, in the anatomical sense, of most of the brain neuroendocrine systems in vertebrates suggests that nonmammalian animals are, in general, useful and appropriate models for investigation.

II. NEUROHORMONES OF THE POSTERIOR PITUITARY

A. Mammals

The literature in this area has been reviewed recently (Buijs et al., 1983; Sofroniew, 1983; Silverman and Pickard, 1983; Zimmerman, 1983; Verbalis and Robinson, 1985). The paraventricular and supraoptic nuclei are well established as the primary sources of vasopressin and oxytocin in mammals; the principal projections of these nuclei are to the posterior pituitary. Vasopressin and oxytocin fibers are also found in the median eminence, the origin of these latter fibers being the paraventricular nucleus. More recently the suprachiasmatic nucleus, bed nucleus of the stria terminalis, amygdala, locus coeruleus, and dorsomedial nucleus of the hypothalamus have also been demonstrated to contain vasopressin-immunopositive perikarya. Buijs et al. (1983) suggest that vasopressin-immunopositive perikarya will be found to be more widespread than previously assumed when immunohistochemical techniques with greater sensitivity are applied.

The paraventricular nucleus, particularly its parvocellular part, has broad projections in addition to those to the posterior pituitary and median eminence; it alone may project oxytocin and vasopressin fibers to the organum vasculosum of the lamina terminalis (OVLT), limbic structures, stria medullaris and lateral habenular nuclei, dorsal hippocampus, dorsal raphe nucleus, nucleus of the tractus solitarius, the dorsal–vagal complex, nucleus ambiguus, and the spinal cord, as well as other possible locations (Buijs et al., 1983; Silverman and Pickard, 1983; Sofroniew, 1983; Zimmerman, 1983). The supraoptic nucleus, in which all or nearly all of the perikarya are immunopositive to oxytocin- or vasopressin-neurophysin, projects to the posterior pituitary and possibly to the median eminence. The suprachiasmatic nucleus contains vasopressin-immunopositive perikarya which are thought to project to the OVLT, dorsomedial hypothalamic nucleus, posterior periventricular hypothalamus, mediodorsal thalamus, and possibly other locations (Buijs et al., 1983; Sofroniew, 1983; Zimmerman, 1983); projections to some of the other brain regions containing vasopressin-immunopositive fibers have also been proposed.

Immunohistochemical evidence has been presented for colocalization of a number of peptides, including angiotensin II, glucagon, somatostatin, corticotropin-releasing factor (CFR), enkephalin, cholecystokinin, and dynorphin, with oxytocin and/or vasopressin in magnocellular neurons in both the supraoptic and paraventricular nuclei (for review, see Silverman and Pickard, 1983; Sofroniew *et al.*, 1984). Dynorphin is particularly associated with vasopressin in perikarya of the supraoptic nucleus and is also present in the posterior pituitary. Cholecystokinin is particularly associated with oxytocin in paraventricular nucleus perikarya. The original suggestion of colocalization of CRF with oxytocin in the paraventricular nucleus of the rat (Swanson *et al.*, 1983) was confirmed by Burlet *et al.* (1983), who provided direct evidence for colocalization of CRF in about 35% of the oxytocin- and oxytocin-neurophysin-immunopositive perikarya of the supraoptic and paraventricular nuclei in the rat; also, most CRF-immunopositive perikarya in these nuclei contained oxytocin and oxytocin-neurophysin. Dreyfuss *et al.* (1984) presented evidence for colocalization of oxytocin and CRF in the secretory granules of some neurosecretory fiber endings in the median eminence in rats, providing additional support for colocalization of these two peptides. Likewise, colocalization in secretory granules in the rat posterior pituitary of methionine-enkephalin and cholesystokinin with oxytocin and of leucine-enkephalin with vasopressin (Martin *et al.*, 1983) confirms that these peptides are colocalized at the level of the perikarya. The neuroendocrinological implications of such colocalizations remain to be determined.

B. Nonmammalian Vertebrates

Birds, reptiles, amphibians, and lungfishes have arginine vasotocin and mesotocin as the principal peptides of the posterior pituitary (Gorbman *et al.*, 1983). In birds and reptiles the nonapeptides primarily originate from the supraoptic and paraventricular nuclei; however, in amphibians and other vertebrates lower on the phylogenetic scale the preoptic nucleus is the homologue of the two nuclei found in higher vertebrates. The nonapeptides found in the various groups of fishes are arginine vasotocin and a second neutral peptide related to oxytocin, including oxytocin in some cases; cyclostomes are unique in their possession of the single peptide arginine vasotocin.

The distribution of arginine vasotocin and mesotocin in the brain has been investigated in several bird species, including starling, *Sturnus vulgaris;* collared turtledove, *Streptopelia decaocto;* zebra finch, *Taeniopygia castanotis* (Goossens *et al.*, 1977); Japanese quail, *Coturnix coturnix japonica;* mallard duck, *Anas platyrhynchos* (Goossens *et al.*, 1977; Bons, 1980); pigeon, *Columba livia* (Goossens *et al.*, 1977; Berk *et al.*, 1982); and domestic chicken (Blähser, 1983; Tennyson *et al.*, 1985). Both peptides are found in separate neurons in the supraoptic and paraventricular nuclei; Tennyson *et al.* (1985) report arginine vasotocin- and mesotocin-immunoreactive perikarya in some areas adjacent to these nuclei in the chicken. Immunopositive fibers for both peptides are found in the posterior pituitary and the OVLT; however, only arginine vasotocin has been localized in the median eminence (Blähser, 1984). Cerebrospinal-fluid-contacting fibers, immunoreactive with an antibody to arginine vasotocin and mesotocin, are present in the chicken (Tennyson *et al.*, 1985).

In the frog *Rana temporaria* immunocytochemical studies demonstrate arginine vasotocin, mesotocin, and somatostatin (SRIF) in separate peri-karya in the preoptic nucleus (Vandesande and Dierickx, 1980; Van Vossel-Daeninck *et al.*, 1981). Similarly, neurophysin-immunopositive perikarya are distinct from SRIF perikarya in the preoptic nucleus of the newt *Triturus carnifex* (Fasolo and Gaudino, 1981). Jokura and Urano (1985) described, in the Japanese toad, *Bufo japonicus,* arginine vasotocin-immunopositive peri-karya in the preoptic nucleus and immunopositive fibers in the posterior pituitary and, to a lesser extent, in the median eminence. In the frog *Rana ridibunda,* arginine vasotocin, mesotocin, and CRF were located separately in perikarya of the preoptic nucleus (Tonon *et al.*, 1985); arginine vasotocin-immunopositive fibers, prominent in the posterior pituitary, were also found in a number of areas in the forebrain. In the goldfish, *Carassius auratus,* separate immunohistochemical-positive arginine vasotocin, isotocin, and enkephalin perikarya were described in the preoptic nucleus (Reaves and Hayward, 1980; Cumming *et al.*, 1982). In the platyfish, *Xiphophorus macu-latus,* an immunocytochemical study by Schreibman and Halpern (1980) demonstrated arginine vasotocin and neurophysin in cell bodies of the pars parvocellularis and pars magnocellularis of the preoptic nucleus, in the pos-terior neurohypophysis (=posterior pituitary) and in tracts between the pre-optic nucleus and neurohypophysis. Additional studies on nonmammalian vertebrates are required to determine whether perikarya containing the clas-sical posterior pituitary nonapeptides are located outside the preoptic nu-cleus, and whether fibers from arginine-vasotocinergic and isotocinergic perikarya terminate in locations other than the neurohypophysis of the pos-terior pituitary.

III. NEUROHORMONES AND THE ANTERIOR PITUITARY

A. Routes to the Anterior Pituitary

There are two established routes for neurohormones to the anterior pitui-tary, by the blood and by direct innervation. See Chapter 2 on pituitary morphology for a discussion of variabilities in each vertebrate class.

B. Growth Hormone Releasing Factor (GRF)

1. Mammals

The primary structures of human, porcine, bovine, and rat hypothalamic growth hormone releasing factor (GRF) have recently been established

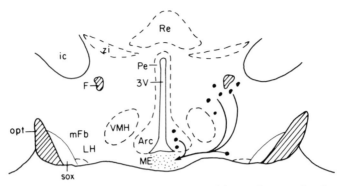

Fig. 1. Schematic illustration of the major GRF-containing perikarya and pathways in the rat hypothalamus. The majority of the GRF-immunoreactive perikarya are located in the arcuate nucleus (Arc) and the medial perifornical region. Fewer perikarya can be seen in the lateral basal hypothalamus. Fibers from the perifornical cell group form a fanlike projection to the median eminence (ME). Scattered perikarya seen in the paraventricular and dorsomedial nuclei are not illustrated. Other abbreviations as follows: ic, internal capsule; F, fornix; mFb, medial forebrain bundle; opt, optic tract; Pe, periventricular nucleus; Re, reuniens nucleus of the thalamus; sox, supraoptic decussation; VMH, ventromedial nucleus; 3V, third ventricle; zi, zona incerta. [From Merchentaler *et al.*, 1984c. Reproduced from *Endocrinology* **114**, 1082–1085, by permission of the Endocrine Society.]

(Guillemin *et al.*, 1984). The native peptides consist of either 44 (human, bovine, porcine) or 43 (rat) amino acids, and show a very high degree of homology.

Immunohistochemical studies on the brain of human (Bloch *et al.*, 1983b, 1984a; Bresson *et al.*, 1984; Fellmann *et al.*, 1985; Sawchenko *et al.*, 1985); macaque, squirrel monkey (Bloch *et al.*, 1983a,b); rat (Jacobowitz *et al.*, 1983; Tramu *et al.*, 1983a; Bloch *et al.*, 1984b; Merchenthaler *et al.*, 1984c); cat (Bugnon *et al.*, 1983b); dormouse (Fellmann *et al.*, 1983); guinea pig (Tramu *et al.*, 1983a); and rhesus monkey (Lechan *et al.*, 1984) all describe immunopositive perikarya in the arcuate (=infundibular) and ventromedial nucleus, the greatest concentration of perikarya being in the arcuate nucleus. GRF-positive perikarya have also been found outside this area in the ventrolateral and dorsolateral hypothalamus in the rat (Fig. 1) (Bloch *et al.*, 1984b; Merchenthaler *et al.*, 1984c; Fellmann *et al.*, 1985); the hippocampus, amygdala, and putamen in the cat (Bugnon *et al.*, 1983b); and the anteroventral and ventrolateral hypothalamus in the human (Bloch *et al.*, 1984a). The distribution of GRF-positive perikarya was shown to be exclusive to luteinizing hormone releasing hormone (LHRH), corticotropin-releasing factor (CRF), and somatostatin (SRIF) in the rat (Bloch *et al.*, 1984b) and humans (Bloch *et al.*, 1984a), and to SRIF and thyrotropin-releasing hormone (TRH) in the rhesus monkey (Lechan *et al.*, 1984). However, Sawchenko *et al.* (1985) found colocalization of GRF with neurotensin, and Okamura *et al.* (1985) reported colocalization with tyrosine hydroxylase-like immunoactivity in arcuate nucleus neurons in the rat.

In the rhesus monkey (Lechan *et al.*, 1984), GRF is specifically distributed rostrally in the lateral margins of the median eminence, merging more caudally in the midline. This is distinct from the distribution of TRH and SRIF. In the rat and guinea pig, GRF is located in the external layer of the median eminence; in the rostral median eminence the GRF-positive fibers are located laterally, whereas posteriorly the fibers are distributed uniformly throughout the external layer of the median eminence (Tramu *et al.*, 1983a; Bloch *et al.*, 1984b). Notably, all GRF-positive fibers in the median eminence disappear following destruction of arcuate nucleus neurons in the rat by treatment of neonates with monosodium glutamate (Bloch *et al.*, 1984b; Fellmann *et al.*, 1985); GRF-positive perikarya were still present in the dorsolateral and ventrolateral hypothalamus of glutamate-treated animals. These results demonstrate that the arcuate nucleus is the source of the GRF-positive fibers in the median eminence of the rat. Recently, Sawchenko *et al.* (1985) reported a second GRF pathway ascending from the ventrobasal hypothalamus to the preoptic region in the rat.

GRF-positive perikarya and fibers first appear in the hypothalamus at week 18 of gestation in the human (Bresson *et al.*, 1984). In contrast, other neuropeptide systems appear earlier in development; LHRH appears at week 11 (Bugnon *et al.*, 1978), SRIF at week 14 (Bugnon *et al.*, 1977), and CRF at week 16 (Bugnon *et al.*, 1982b). The functional implications of the delayed development of the GRF system remain to be investigated.

2. *Nonmammalian Vertebrates*

Pan *et al.* (1985a,b) reported that antisera to human GRF-44-NH$_2$ and GRF-40-OH cross-react with fractions from brain extracts from codfish. However, only the antiserum to GRF-40-OH showed any immunohistochemical cross-reactivity in codfish. Immunoreactive perikarya were located in both the parvocellular and magnocellular regions of the preoptic nucleus. Two GRF-immunopositive pathways to the pituitary originate from the preoptic nucleus; one pathway that appears to arise from the ventral parvocellular neurons courses ventrally to the horizontal commissure, whereas the second pathway arises from the dorsal parvocellular and magnocellular neurons and courses dorsally to the horizontal commissure. GRF-immunopositive perikarya are also located in the nucleus lateralis tuberis (homologous to arcuate nucleus), and in the midline of the midbrain tegmentum. GRF-immunopositive fibers are present in the proximal pars distalis, in the region of the growth hormone cells, and to a greater extent, in the pars nervosa of the neurointermediate lobe (=posterior pituitary). What is of particular interest is that in codfish the GRF-immunoreactive perikarya that innervate the pituitary are located in the preoptic region instead of in the ventrobasal hypothalamus as in mammals. Additional studies are needed to clarify this situation.

C. Somatostatin (SRIF)

1. The Nature of SRIF

Biochemical studies have demonstrated that in mammals, SRIF is synthesized as both a 14-amino-acid peptide (SRIF-14), or as a 28-amino-acid peptide (SRIF-28) that embodies SRIF-14 in its C-terminal end (for review, see Reichlin, 1983; Benoit et al., 1985). SRIF-14 and SRIF-28 are identical in all mammals investigated to date. SRIF-14 is present in birds and teleost fishes; however, in anglerfish and catfish, respectively, the structures of two variants of SRIF-28 and a unique SRIF-22 have also been found (for review, see Reichlin, 1983). These variants each have a high degree of homology with mammalian SRIF-28, suggesting on a biochemical basis that this peptide has been conserved during evolution.

2. Mammals

The literature on the distribution of SRIF in the brain of mammals has been reviewed on a number of occasions, and I shall refer to some such reviews as an aid in covering the large volume of information available (McQuillan, 1980; Rorstad et al., 1980; Krisch, 1981; Reichlin, 1983).

In the rat (Rorstad et al., 1980; Finley et al., 1981; Markara et al., 1983; Reichlin, 1983), dog (Hoffman and Hayes, 1979), rhesus monkey, and in humans (Fig. 2) (Filby and Gross, 1983) the most prominent and concentrated group of hypothalamic SRIF-immunopositive perikarya is located in the anterior periventricular nucleus and the adjacent periventricular (parvocellular) portion of the paraventricular nucleus. SRIF-immunopositive perikarya are also scattered in the arcuate and ventromedial nuclei and in the lateral hypothalamic and preoptic areas. Outside the rat hypothalamus, SRIF-immunopositive perikarya have also been found in the subthalamic entopeduncular nucleus and a number of sites in the telencephalon, mesencephalon, and myelencephalon.

In the rat (Bennett-Clarke et al., 1980; Rorstad et al., 1980; Finley et al., 1981; Markara et al., 1983; Reichlin, 1983), dog (Hoffman and Hayes, 1979), cat (Bugnon et al., 1983b), rhesus monkey, and in humans (Filby and Gross, 1983) SRIF-immunopositive fibers and endings are abundant in the external zone, and a few fibers and endings occur in the internal zone of the median eminence. The distribution of SRIF in the rat median eminence is different from that of LHRH (Baker and Yu, 1976). The posterior pituitary of mammals also contains some SRIF-immunopositive fibers and endings. SRIF-immunopositive fibers are found throughout the hypothalamus, with a relatively dense network coursing through the suprachiasmatic and lateral retrochiasmatic area, and a dense bed in the arcuate–ventromedial nuclei complex. The OVLT, subfornical organ, and pineal gland also have dense

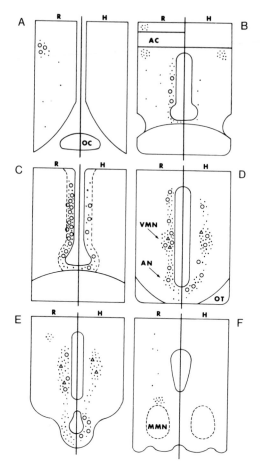

Fig. 2. Distribution of immunoreactive somatostatin in neuronal perikarya (circles), fibers (black dots), and presumptive synaptic terminals (triangles) in diagrammatic representation of coronal sections. On left, distribution in rhesus monkey (R), on right, in human (H). For clarity, minor structural differences in rhesus and human sections are omitted. (A) Ventral septal area. OC, optic chiasm. (B) Preoptic area at level of optic chiasm and anterior commissure (AC). (C) Anterior hypothalamic area, including anterior periventricular region (dashed line). (D) Anterior tuberal region, including ventromedial (VMN) and arcuate (AN) nuclei. OT, optic tract. (E) Infundibular region, including periventricular region, VMN, AN, and median eminence. (F) Mammillary region, including medial mammillary nucleus (MMN). [From Filby and Gross, 1983. Reproduced by permission.]

accumulations. SRIF-immunopositive fibers are widely distributed in the other diencephalic regions, the telencephalon, mesencephalon, and myelencephalon.

To determine the brain origin of the SRIF-immunopositive fibers and endings in the median eminence in rats, transections in various locations of the hypothalamus have been made, followed by SRIF immunohistochemistry. Epelbaum *et al.* (1981) and Markara *et al.* (1983) found evidence that fibers from the SRIF-immunopositive perikarya in the anterior periventricular nucleus and parvocellular portion of the paraventricular nucleus course laterally and ventrally, and enter the mediobasal hypothalamus near the ventral surface as they pass posteriorly to the median eminence. They found no evidence for any sizeable number of fibers taking a more direct periventricular route. However, studies by Jew *et al.* (1984) indicate a periventricular route as being of major importance for both median eminence and arcuate nucleus SRIF-immunopositive fibers. Although additional studies are needed to resolve this conflict, it is likely that both routes contribute to median eminence and arcuate nucleus innervation.

Since SRIF-14 and SRIF-28 have both been demonstrated to exist in biochemical studies (Reichlin, 1983; Benoit *et al.*, 1985), an important question is whether, on the basis of immunohistochemical studies, both peptides have the same distribution in the hypothalamus. Immunohistochemical evidence indicates that SRIF-14, SRIF-28, and the 12-amino-acid fragment remaining after processing SRIF-28 to SRIF-14, here termed SRIF-28(1–12), are widely distributed in the brains of rats and other mammals (Benoit *et al.*, 1985). In the rat hypothalamus, the relative distribution of immunopositive perikarya in the periventricular region staining for SRIF-14, SRIF-28, and SRIF-28(1–12) is similar to the apparent distribution of these peptides in the median eminence (Iwanaga *et al.*, 1983; Bakhit *et al.*, 1984; Guy *et al.*, 1985). Bakhit *et al.* (1984) further found that after treatment with cysteamine, which causes specific depletion of SRIF-14 in the brain, there was complete depletion of immunoreactive staining for SRIF-14 in perikarya and fibers and a decreased staining in the median eminence; another antiserum that cross-reacts with both SRIF-14 and SRIF-28 also demonstrated decreased immunoreactive material. However, there was no depletion of staining for SRIF-28(1–12) following cysteamine treatment. This suggests that there is selective release of SRIF-14 and SRIF-28(1–12) from the rat median eminence.

While it is apparent that SRIF released from the median eminence has a neurohormonal function, SRIF in the arcuate nucleus is most likely serving as a neurotransmitter. Ohtsuka *et al.* (1983) found SRIF-immunopositive terminals in the arcuate nucleus of the rat. Furthermore, treatment of neonatal mice with monosodium glutamate causes depletion of SRIF–immunopositive fibers in the arcuate nucleus but produces no apparent effects on the levels of SRIF-immunohistochemical staining in the median eminence or the hypothalamic periventricular area (Romagnano *et al.*, 1982). SRIF and neuropeptide Y were found colocalized in rat and human cortical neurons and occasionally in rat hypothalamic neurons (Chronwall *et al.*, 1984). However, colocalization at the level of the median eminence has not yet been reported.

SRIF-immunopositive perikarya first appear in the brain of rats at gestation days 16–17; immunopositive perikarya first appear in the periventricular region of the hypothalamus at days 18–19 (Ibata *et al.*, 1982; Shiosaka *et al.*, 1982). The number of SRIF-immunopositive structures in the diencephalon and telencephalon increases between gestation day 21 and postnatal day 28 to a point at which many more sites are present at that stage of development as compared to the adult. The functional significance of these changes is not known. SRIF-immunopositive perikarya first appear at week 14 of human gestation (Bugnon *et al.*, 1977).

3. Nonmammalian Vertebrates

The distribution of SRIF-immunopositive perikarya and fibers was investigated in the brain of the mallard duck (Blähser *et al.*, 1978; Blähser and Heinrichs, 1982); grass parakeet, *Melopsittacus undulatus* (Takatsuki *et al.*, 1981); Japanese quail (Mikami and Yamada, 1984); and embryos of mallard, chicken, and Japanese quail (Blähser and Heinrichs, 1982). Blähser *et al.* (1978) described SRIF-immunopositive perikarya within, and in areas between, the supraoptic and paraventricular nuclei and along the hypothalamo-hypophysial tract originating from these nuclei in adult mallard ducks; they also found that SRIF perikarya are distinct from those containing vasotocin and mesotocin. In the grass parakeet, SRIF perikarya were found in the anterolateral and ventromedial hypothalamus (Takatsuki *et al.*, 1981); SRIF perikarya were also found in a number of other brain regions (Takatsuki *et al.*, 1981; Shiosaka *et al.*, 1981). Mikami and Yamada (1984) reported three groups of SRIF perikarya [the periventricular preoptic region, lateral hypothalamus, and anterior infundibular (=arcuate) nucleus in Japanese quail]; SRIF-immunopositive fibers projected to the median eminence from all three groups of perikarya. The distribution of SRIF perikarya in the avian hypothalamus varies somewhat from that in mammals. The median eminence of the bird species investigated is heavily innervated by SRIF-immunopositive fibers, and SRIF fibers are widely distributed in other parts of the brain.

SRIF-immunopositive perikarya and fibers have been described in several reptilian species [the lizards *Lacerta muralis* (Doerr-Schott and Dubois, 1977), *L. sicula* (Fasolo and Gaudino, 1982), and *Ctenosauria pectinata* (Goossens *et al.*, 1980), and the turtles *Pseudemys scripta* (Bear and Ebner, 1983), and *Testudo hermanni* (Weindl *et al.*, 1984)]. Each of these studies demonstrated SRIF-immunopositive perikarya in a continuous laminar arrangement in the periventricular nucleus and the adjacent ventral periventricular (parvocellular) portion of the paraventricular nucleus, and in the infundibular nucleus (homologous to arcuate nucleus). SRIF perikarya were shown to be distinct from those containing neurohypophysial hormones (mesotocin and vasotocin) in the lizard *Ctenosauria pectinata* (Goossens *et al.*, 1980). In some species the SRIF-immunopositive cells in the infundibular nucleus were noted to have extensions between ependymal cells so as to contact the cerebrospinal fluid (Goossens *et al.*, 1980; Weindl *et al.*, 1984). All studies on reptiles reported that the median eminence was heavily laden with SRIF-immunopositive fibers, and in several cases, fibers in the posterior pituitary were also noted. The locations of SRIF-immunopositive perikarya and fibers in other parts of the brain appear to be homologous to that in mammals (Weindl *et al.*, 1984).

SRIF-immunopositive perikarya and fibers have been described in tadpoles of the frog *Alytes obstetricans* (Rémy and Dubois, 1978); the toad *Bufo bufo* (Doerr-Schott and Dubois, 1978); the frogs *Rana temporaria* (Vandesande and Dierickx, 1980; Dierickx *et al.*, 1981; Van Vossel-Daeninck *et al.*, 1981), *R. catesbiana* (Inagaki *et al.*, 1981), and *R. ridibunda* (Olivereau *et al.*, 1984c); the newts *Triturus cristatus carnifex* (Fasolo and Gaudino, 1981) and *Pleurodeles waltlii* (Olivereau *et al.*, 1984c); the axolotl *Ambystoma mexicanum* (Olivereau *et al.*, 1984c); and tadpoles of the toad *Xenopus laevis* (Blähser *et al.*, 1982). In the hypothalamus of *R. temporaria*, SRIF-immunopositive perikarya were demonstrated within the magnocellular preoptic nucleus, as well as laterally and dorsally to it, and in the pars ventralis of the tuber cinereum, the homologue to the infundibular or arcuate nucleus, as illustrated in Fig. 3 (Vandesande and Dierickx, 1980); a number of other brain loci containing SRIF-immunopositive perikarya were also described. The orientation of the SRIF-immunopositive fibers and fiber tracts suggests that the endings in the median eminence and posterior pituitary or neural lobe originate from the perikarya in the preoptic region as well as the tuber cinereum. Double immunostaining of sections shows that the SRIF-immunopositive perikarya and fibers are separate and distinct from those containing vasotocin or mesotocin (Vandesande and Dierickx, 1980; Van Vossel-Daeninck *et al.*, 1981). Brain lesioning and tract sectioning studies on *R. temporaria* further demonstrated that SRIF fibers in the median eminence originate from perikarya in both the preoptic region and the tuber cinereum, whereas the fibers in the posterior pituitary originate only from the preoptic region (Fig. 4) (Dierickx *et al.*, 1981). This study is notable, because it is the first to demonstrate the origin of the SRIF fibers in the posterior pituitary and to show that SRIF perikarya in the tuber cinereum have fibers terminating in the median eminence.

SRIF-immunopositive perikarya with relatively short extensions contacting the cerebrospinal fluid were found in the preoptic region of *R. temporaria* (Vandesande and Dierickx, 1980), in the tuber cinereum of tadpoles of *X. laevis* (Blähser *et al.*, 1982), and in both regions in the newt *T. cristatus carnifex* (Fasolo and Gaudino, 1981). The function of these cerebrospinal-fluid-contacting neurons is not known; however, amphibia may be a particularly useful model for investigation of these cells, due to their frequency and relative ease of access in the brain. In tadpoles of the frog *A. obstetricians*, SRIF-immunopositive perikarya were not detectable unless the animals had

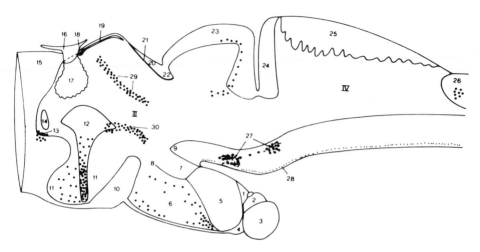

Fig. 3. Right half of the brain of *Rana temporaria*. The density of the large dots indicates the distribution of the somatostatinergic perikarya. 1, Neural lobe; 2, pars intermedia and 3, pars distalis of the hypophysis; 4, median eminence; 5, saccus infundibuli; 6, pars ventralis and 7, pars dorsalis of the tuber cinereum hypothalami; 8, sulcus lateralis infundibuli; 9, tuberculum posterius mesencephali; 10, regio chiasmatica; 11, recessus praeopticus; 12, magnocellular preoptic neurosecretory nucleus containing somatostatinergic perikarya; 13, subfornical organ; 14, foramen of Monro; 15, telencephalon; 16, paraphysis; 17, choroid plexus of third ventricle; 18, commissura habenularum; 19, epiphysis; 20, pars intercalaris; 21, subcommissural organ; 22, commissura posterior; 23, tectum opticum (paramedian sagittal section); 24, cerebellum; 25, choroid plexus of the fourth ventricle; 26, roof at the caudal end of calamus scriptorius; 27, large dots: distribution of somatostatinergic perikarya in the interpeduncular nucleus, and grey area: distribution of somatostatinergic fibers; 28, small dots: tract of somatostatinergic fibers; 29, strand of somatostatinergic perikarya in dorsal part of thalamus; 30, strand of somatostatinergic perikarya in middle part of thalamus; III, third ventricle; IV, fourth ventricle. [From Vandesande and Dierickx, 1980. Reproduced by permission.]

been treated previously with injections of bovine growth hormone (Rémy and Dubois, 1978). This suggests that growth hormone may have a developmental influence on SRIF neurons; amphibians may be useful for further investigation of this phenomenon.

Among the bony fishes, the brain distribution of SRIF-immunoreactive perikarya and fibers has been investigated in several teleost species [*Salmo gairdneri* (Dubois *et al.*, 1979; Vigh-Teichman *et al.*, 1983); *Carassius auratus* (Kah *et al.*, 1982b; Olivereau *et al.*, 1984a); *Xiphophorus maculatus* (Margolis-Kazan *et al.*, 1983); *Anguilla anguilla* (Vigh-Teichmann *et al.*, 1983; Olivereau *et al.*, 1984b; *Phoxinus phoxinus* (Vigh-Teichmann *et al.*, 1983); *Cyprinus carpio* (Olivereau *et al.*, 1984a,b); *Salmo irideus, S. fario, S. salar, Oncorhynchus keta, Mugil namada, Myoxocephalus octodecimspinosus,* and *Colisa lalia* (Olivereau *et al.*, 1984b); *Oreochromis mossambicus, Gillichthys mirabilis,* and *Fundulus heteroclitus* (Grau *et al.*, 1985)]. SRIF-immunoreactive perikarya were found in most species in the nucleus preop-

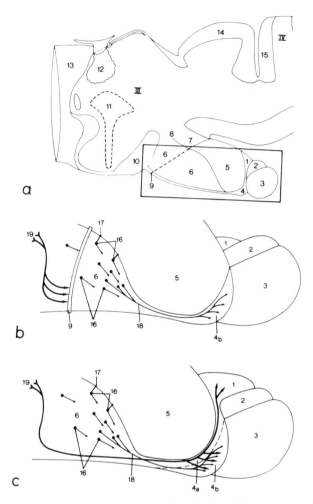

Fig. 4. Schematic drawings. (a) Right half of the brain of *Rana temporaria*. (b) and (c) Correspond to the region enclosed in the rectangle of a. (b) Represents a hypothalamic island (second group of animals operated on). (c) Same region in an intact animal. 1, Neural lobe; 2, pars intermedia and 3, pars distalis of the hypophysis; 4, median eminence: 4a, internal region, and 4b, external region; 5, saccus infundibuli; 6, pars ventralis and 7, pars dorsalis of the tuber cinereum; 8, sulcus lateralis infundibuli; 9, teflon barrier (location); 10, regio chiasmatica; 11, magnocellular neurosecretory preoptic nucleus (removed in the first group of animals operated on); 12, choroid plexus; 13, telencephalon; 14, tectum opticum, paramedian sagittal section; 15, cerebellum; 16, somatostatin perikarya located in the pars ventralis tuberis, sending dendrites (17) to the ventricle and axons (18) to the median eminence; 19, axons from somatostatin perikarya located outside the tuber cinereum, in c, running to the median eminence and neural lobe, and in b interrupted and ending at the teflon barrier; III, third ventricle; IV, fourth ventricle. [From Dierickx *et al.*, 1981. Reproduced by permission.]

ticus periventricularis [an anterior preoptic periventricular nucleus, see Peter and Gill (1975)], the nucleus lateralis tuberis in the ventrobasal hypothalamus (homologous to the arcuate or infundibular nucleus), and in a dorsal hypothalamic-ventromedial thalamic nucleus; SRIF perikarya were also demonstrated in other brain locations in some studies, but more thorough studies are needed in this regard. Vigh-Teichmann *et al.*, (1983) also observed SRIF-immunopositive perikarya in the magnocellular preoptic nucleus, continuous with the main group of cells in the nucleus preopticus periventricularis, in several species. Notably, Vigh-Teichmann *et al.* (1983) also found numerous SRIF-immunopositive perikarya with prominent extensions contacting the cerebrospinal fluid (Fig. 5). SRIF-immunopositive fibers have been observed in many hypothalamic and other brain areas. In the pituitary, fibers and endings have been found in only the neurohypophysial tissue invading the proximal pars distalis where growth hormone cells are located. Although additional studies are needed to confirm this selective distribution of SRIF fibers, teleost fish might be particularly useful for studying the brain origin(s) of the SRIF that is involved in regulation of growth hormone secretion.

D. Corticotropin-Releasing Factor (CRF)

1. The Nature of CRF

Ovine CRF is a 41-amino-acid peptide (Vale *et al.*, 1981, 1983). This peptide has potent effects on the release of corticotropin, β-endorphin, and melanotropin.

2. Mammals

A number of reviews have dealt with the actions and brain distribution of CRF in mammals (Yasuda *et al.*, 1982; Vale *et al.*, 1983; Fellmann *et al.*, 1984; Joseph *et al.*, 1985). Early studies on the hypothalamic distribution of CRF indicated the presence of CRF-immunopositive perikarya in the paraventricular nucleus, and fibers and endings in the external zone of the median eminence of the rat (Bugnon *et al.*, 1982a; Tramu and Pillez, 1982; Vigh *et al.*, 1982), guinea pig (Tramu and Pillez, 1982), sheep (Kolodziejczyk *et al.*, 1983), human (Pelletier *et al.*, 1983), dog, and cat (Clavequin *et al.*, 1983). In the squirrel monkey, CRF-immunopositive perikarya were demonstrated in the paraventricular and supraoptic nuclei (Paull *et al.*, 1984). A number of thorough studies on the brain distribution of CRF in the rat (e.g., Merchenthaler *et al.*, 1982; Joseph and Knigge, 1983; Swanson *et al.*, 1983; Joseph *et al.*, 1985; Skofitsch and Jacobowitz, 1985) demonstrated that in addition to the prominent accumulation of CRF perikarya in the anterior and medial parvocellular divisions of the paraventricular nuclei, CRF-immunopositive perikarya are also present in the supraoptic nucleus, the ventromedial nucleus, arcuate nucleus, medial and periventricular preoptic nuclei,

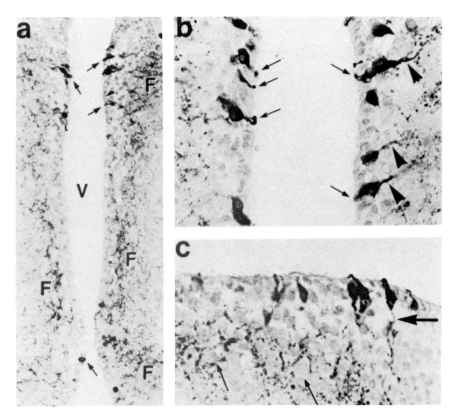

Fig. 5. Somatostatin immunoreactivity in the anterior periventricular nucleus of the European minnow, *Phoxinus phoxinus*. (a) Immunoreactive CSF-contacting neurons (arrows) are scattered in the dorsal and ventral divisions of this nucleus. F, strongly immunoreactive fiber plexus; V, third ventricle. (b) Strong immunoreaction in the CSF-contacting dendritic protrusions (arrows), the corresponding perikarya, and the basal processes (arrowheads) of the CSF-contacting neurons. (c) The basal processes (arrow) of the CSF-contacting neurons contribute to a somatostatin-positive fiber plexus, which encompasses nonimmunoreactive perikarya (small arrows) of the anterior periventricular nucleus. [From Vigh-Teichmann *et al.*, 1983. Reproduced by permission.]

and the mammillary and premammillary nuclei of the hypothalamus (Fig. 6). CRF-immunopositive perikarya were also found in a number of extrahypothalamic sites, some of the more prominent ones being along the diencephalic–mesencephalic border, the bed nuclei of the stria terminalis and anterior commissure, and the central nucleus of the amygdala; other less prominent accumulations are scattered widely in the brain (Fig. 6).

There are a number of well-defined CRF pathways in the brain (Fig. 6). Experimental studies to determine the major pathway from the paraventricular nucleus to the median eminence indicate that the fibers leave the paraventricular nucleus laterally, turn ventrally and then medially in the

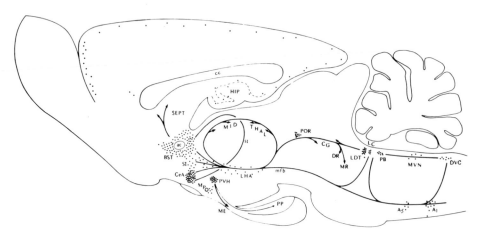

Fig. 6. Major CRF-stained cell groups (black dots) and fiber system illustrated schematically in sagittal view of the rat brain. Abbreviations as follows: A_1, A_5, noradrenergic cell groups; ac, anterior commissure; BST, bed nucleus of the stria terminalis (posterior part); CC, central canal; CeA, central nucleus (amygdala); CG, central gray; DR, dorsal raphe; DVC, dorsal vagal complex; HIP, hippocampus; LC, locus coeruleus; LDT, laterodorsal tegmental nucleus; LHA, lateral hypothalamic area; ME, median eminence; mfb, medial forebrain bundle; MID THAL, midline thalamic nuclei; MPO, medial preoptic nucleus; MR, median raphe; MVN, medial vestibular nucleus; PB, parabrachial nucleus; POR, perioculomotor nucleus; PP, peripeduncular nucleus; PVH, parventricular nucleus; SEPT, septal region; SI, substantia innominata; st, stria terminalis. [From Swanson *et al.*, 1983. Reproduced by permission of S. Karger AG, Basel.]

lateral hypothalamus, and then proceed to the median eminence in a coalesced pathway from the lateral retrochiasmatic area (Fig. 7) (Antoni *et al.*, 1983; Merchenthaler *et al.*, 1984b). In addition, a few fibers take a more direct periventricular route from the paraventricular nucleus to the median eminence (Fig. 7) (Merchenthaler *et al.*, 1984b).

CRF has been colocalized with oxytocin and oxytocin-neurophysin in some perikarya in the periventricular area of the anterior hypothalamus, in the supraoptic nucleus (Burlet *et al.*, 1983), and in some fibers of the internal zone of the median eminence (Dreyfuss *et al.*, 1984). Whereas Tramu *et al.*, (1983b) found no evidence for colocalization of immunoreactive oxytocin and CRF, they did find colocalization of CRF and vasopressin in adrenalectomized rats and guinea pigs, confirming earlier evidence for colocalization of these two peptides in the paraventricular, but not supraoptic, nucleus of rats (Roth *et al.*, 1982). Notably, some CRF paraventricular nucleus neurons also display colocalization with dynorphin (Roth *et al.*, 1983), enkephalin, and the peptide designated as PHI-27, which is a 27-amino-acid peptide with an N-terminal histidine and a C-terminal isoleucine amide (Hokfelt *et al.*, 1983). SRIF and gonadotropin-releasing hormone (GnRH) do not colocalize with CRF in the rat (Fellmann *et al.*, 1982). Additional research is necessary

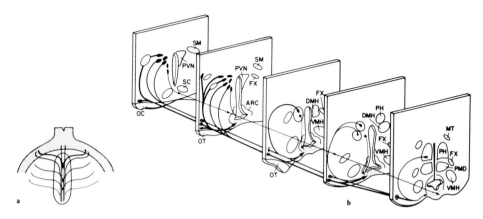

Fig. 7. (a) Schematic illustration of CRF afferents as they approach the median eminence (ME) at the base of the hypothalamus. The ME is represented by the oval area in the middle. CRF-containing fibers approach the ME from anterior and anterolateral directions. The majority reach the ME through the laterobasal retrochiasmatic area (thick lines), but scattered fibers can be seen caudally as far as the level of the pituitary stalk. The ipsilateral fibers form terminals in the entire length of the ME. Many of them cross the midline and give terminals on the contralateral side in regions close to the midline. (b) Schematic illustrations of the major sources and pathways of CRF fibers terminating in the ME. CRF afferents from the paraventricular nucleus (PVN) form a fan of fibers consisting of a medial, a more diffuse intermediate, and a lateral tract. Axons arising from the dorsomedial nucleus (DMH) and the perventricular hypothalamic nuclei join the curved fan of fibers formed by the paraventriculo–infundibular CRF system. Lines of different thickness reflect the differences in the relative number of fibers within the tracts. Other abbreviations: ARC, arcuate nucleus; DMH, dorsomedial nucleus; FX, fornix, MT, mammillothalamic tract; OC, optic chiasm; OT, optic tract; PMD, dorsal premammillary nucleus; PH, periventricular nucleus; PVN, paraventricular nucleus; SC, suprachiasmatic nucleus; SM, stria medullaris; VMH, ventromedial nucleus. [From Merchenthaler *et al.*, 1984b. Reproduced by permission of S. Karger AG, Basel.]

to determine the full range of colocalization of various peptides with CRF. CRF-immunoreactive fibers appear in the median eminence during days 16.5–17.5 of gestation in the rat (Daikoku *et al.*, 1984) and during the 17th week of human gestation (Bugnon *et al.*, 1982b); the ontogeny of colocalization remains to be investigated.

3. Nonmammalian Vertebrates

Ovine CRF-41 immunoreactive-like activity has been found in each of the major groups of vertebrates investigated. In the pigeon, *Columba liva domestica,* CRF-immunoreactive fibers were found in the median eminence, and perikarya were found in the paraventricular nucleus, subdivisions of the supraoptic nuclei, posterior hypothalamic nuclei, the dorsomedial and lateral thalamic nuclei, and a number of other brain regions (Péczely and Antoni, 1984; Bons *et al.*, 1985). In the chicken (Józsa *et al.*, 1984) and Japanese quail (Mikami and Yamada, 1984; Bons *et al.*, 1985; Yamada and

Mikami, 1985) CRF-immunopositive perikarya were found in the paraventricular, preoptic and mammillary nuclei of the hypothalamus, as well as in the septal area, nucleus accumbens, and some thalamic nuclei; the median eminence in both species and the OVLT in the chicken have dense localizations of CRF-immunopositive fibers. CRF-immunoreactive perikarya were detected in the paraventricular nuclei, and fibers were demonstrated in the median eminence of a blackbird (Fellmann *et al.*, 1984). In the turtle *Pseudemys scripta elegans,* a prominent group of CRF-immunopositive perikarya is located in the paraventricular organ, and fibers from these perikarya can be traced to the median eminence; other aspects of the distribution of CRF in the turtle show close homology to the situation in the rat (Fellmann *et al.*, 1984). In the frog *Rana esculenta,* the newt *Triturus alpestris* (Fellmann *et al.*, 1984), and the frog *R. ridibunda* (Verhaert *et al.*, 1984) CRF-immunopositive perikarya are present in the preoptic nucleus, the homologue of the supraoptic and paraventricular nuclei, and CRF-immunopositive fibers are abundant in the median eminence. Olivereau *et al.*, (1984c) found CRF perikarya somewhat more broadly distributed in the preoptic area and the interpeduncular nucleus of the frog *R. ridibunda,* the newt *Pleurodeles waltlii,* and the axolotl *Ambystoma mexicanum.* CRF-immunopositive perikarya in the ventral part of the preoptic nucleus of the *R. ridibunda* are distinct from perikarya immunopositive for arginine vasotocin and mesotocin (Tonon *et al.*, 1985); in addition to the median eminence, CRF fibers were found in a number of forebrain areas. In several species of teleosts, CRF-immunopositive perikarya have been demonstrated in the preoptic nucleus [*Salmo irideus, Carassius auratus, Tinca tinca, Scomber scombrus, Serranus cabrilla, Boops salpa* (Bugnon *et al.*, 1983a; Fellmann *et al.*, 1984); *Carassius auratus, Cyprinus carpio* (Olivereau *et al.*, 1984a)] and in the adjacent nucleus preopticus periventricularis of *Carassius auratus* and *Cyprinus carpio* (Olivereau *et al.*, 1984a). In these teleosts CRF-immunopositive fibers were generally found in the neurohypophysial tissue invading the rostral pars distalis, in the region of the corticotropin cells, and to a lesser extent in the neurointermediate lobe. The more general distribution of CRF in the brain of teleosts and amphibians remains to be described. However, it appears that the distribution of CRF-immunopositive perikarya and fibers is homologous throughout the vertebrates. Considering the relative ease of access to the preoptic region in teleosts and amphibians, these forms would be particularly useful models for investigating the regulation of CRF neuronal activity.

E. Thyrotropin-Releasing Hormone (TRH)

1. Mammals

Although the structure of the tripeptide TRH has been known for some time (for review, see Sandow and Konig, 1978), the study of its distribution in the brain by immunohistochemistry has proven difficult, presumably due to problems of fixation and retention of this small peptide in the tissue. However, by perfusion fixation with acrolein it has become possible to do such studies.

The distribution of TRH in the brain of mammals has been reviewed recently by Jackson and Lechan (1983). Early studies on the rat demonstrated TRH-immunopositive fibers in the median eminence and several hypothalamic nuclei; TRH-immunopositive perikarya were found in the paraventricular, periventricular, and dorsomedial nuclei; the lateral hypothalamus; and the perifornical region (see, e.g., Elde and Hokfelt, 1978; Johansson et al., 1980). More recently, the hypothalamic distribution of TRH-immunopositive perikarya has been confirmed and has been extended to show that the largest accumulation is in the anterior and medial parvocellular portions of the paraventricular nucleus and in the preoptic and arcuate nuclei (Fig. 8; Lechan and Jackson, 1982; Ishikawa et al., 1984a). In addition to the dense accumulation of TRH-immunopositive fibers in the median eminence, a high density of fibers also occurs in the paraventricular nucleus, the OVLT, the interstitial nucleus of the striae terminales, and the septum (Lechan and Jackson, 1982; Ishikawa et al., 1984a); the posterior pituitary also contains a relatively high density of TRH-immunopositive fibers. The source of the TRH fibers in the median eminence has not been investigated experimentally. However, on the basis of lesioning studies it has been shown that the primary origin of TRH for the regulation of thyrotropin secretion is the paraventricular nucleus (Aizawa and Greer, 1981; Ishikawa et al., 1984b).

Nakai et al., (1983) recently demonstrated colocalization of immunoreactive TRH and monoamines in nerve terminals in the median eminence, as determined by radioautography following injection of labeled dopamine, norepinephrine, and serotonin. Remarkably, in the rat brain, the distribution of perikarya and nerve fibers that show immunohistochemical cross-reactivity with human growth hormone appears to be identical with that of TRH (Lechan et al., 1983). Furthermore, sequential staining of the same tissue sections demonstrates colocalization of the human growth hormone–immunoreactive material and TRH in the same perikarya and fibers. The significance of these observations is not yet apparent.

2. Nonmammalian Vertebrates

The ontogeny of TRH-immunopositive perikarya and fibers in the chick embryo has been described by Thommes et al. (1985). The number of TRH perikarya in the supraoptic nucleus and lateral region of the preoptic nucleus gradually increased from day 6.5 through day 13.5 of embryonic development; TRH perikarya were first found at day 4.5 in the undifferentiated ependymal wall of the infundibulum. TRH-immunopositive fibers were first identified in the anterior median eminence on day 6.5, and on day 7.5 in the

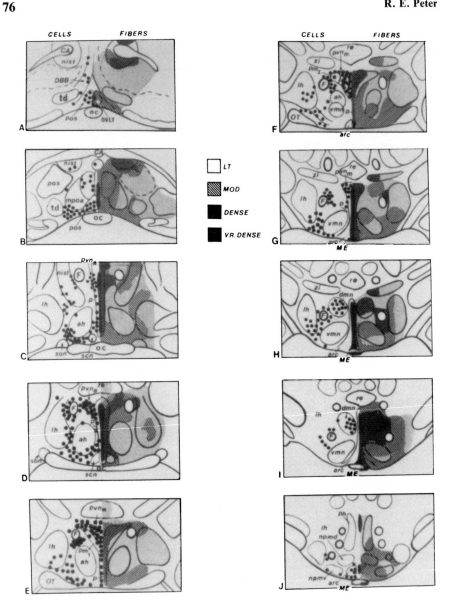

Fig. 8. Schematic diagram of the distribution of TRH-containing perikarya (dots) and fibers (textured areas) in coronal sections through the rat hypothalamus (panels A to J are progressively more caudal). Degrees of staining from light (LT) to very (VR.) dense areas are indicated by different textures. Abbreviations: ah, anterior hypothalamic nucleus; arc, arcuate nucleus; CA, anterior commissure, DBB, diagonal band of Broca; dmn, dorsomedial nucleus; F, fornix; lh, lateral hypothalamus; ME, median eminence; nist, interstitial nucleus of the striae terminalis; npmd, dorsal premammillary nucleus; npmv, ventral premammillary nucleus; OC, optic chiasm; OT, optic tract; OVLT, organum vasculosum of the lamina terminalis; p, periventricular nucleus; ph, posterior hypothalamic nucleus; pm_1, paraventricular nucleus, rostromedial

OVLT. At day 15.5, TRH perikarya and fibers were observed in other brain regions, including areas superior to the preoptic nucleus, and around the septomesencephalic tract and anterior commissure.

The distribution of immunoreactive TRH in the forebrain of the frog *Rana catesbeiana* was described by Seki *et al.* (1983). TRH-immunopositive perikarya were found in the anterior portion of the preoptic nucleus, the dorsal infundibular nucleus, and some locations in the telencephalon (Fig. 9); immunoreactive fibers and terminals were present in the lateral infundibulum, median eminence, posterior pituitary, and neurointermediate lobe, as well as in some locations in the telencephalon. The distribution of TRH in the frog appears to be homologous to that in the rat.

Additional studies on nonmammalian vertebrates would be informative.

F. Gonadotropin-Releasing Hormone (GnRH)

1. The Nature of GnRH

The decapeptide structure of LHRH has been known since 1971 (for review, see Krey and Silverman, 1983). Chromatographic and radioimmunoassay studies by King and Millar (1980) demonstrated that the GnRH(s) present in hypothalamic extracts from chicken, pigeon, a reptile (tortoise), a teleost (cichlid), and an elasmobranch (dogfish) are distinct from LHRH; however the GnRH in hypothalamic extracts from a toad and a frog were identical to LHRH. More recently, the structure of a GnRH in chum salmon, *Oncorhynchus keta,* was demonstrated to be [Trp^7,Leu^8]-LHRH; a second form, not yet characterized, is also present (Sherwood *et al.,* 1983). The structure [Trp^7,Leu^8]-LHRH is widely distributed in teleosts (tGnRH, "t" for teleost) Breton *et al.,* 1984; Sherwood *et al.,* 1984). The structures [Gln^8]-LHRH (King and Millar, 1982a,b; Miyamoto *et al.,* 1983) and [His^5,Trp^7,Tyr^8]-LHRH (Miyamoto *et al.,* 1984) have been established as the structures of GnRH in the domestic chicken. Clearly, there is a high degree of homology in this family of GnRH peptides, indicating conservation during evolution.

2. Mammals

The literature on the distribution of LHRH in the mammal brain has been reviewed on a number of occasions, and I refer to some such reviews as an aid in covering the large volume of early literature in this field (Hoffman *et*

magnocellular subdivision; pm$_2$, paraventricular nucleus, caudolateral magnocellular subdivision; mpoa, medial preoptic nucleus; pol, lateral preoptic nucleus; pos, suprachiasmatic preoptic nucleus; pvn$_a$, paraventricular nucleus, anterior parvocellular subdivision; pvn$_m$, paraventricular nucleus, medial parvocellular subdivision; re, nucleus reuniens; scn, suprachiasmatic nucleus; son, supraoptic nucleus; td, nucleus of the diagonal band of Broca; vmn, ventromedial nucleus; zi, zona incerta. [From Lechan and Jackson, 1982. Reproduced from *Endocrinology* **111,** 55–65, by permission from the Endocrine Society.]

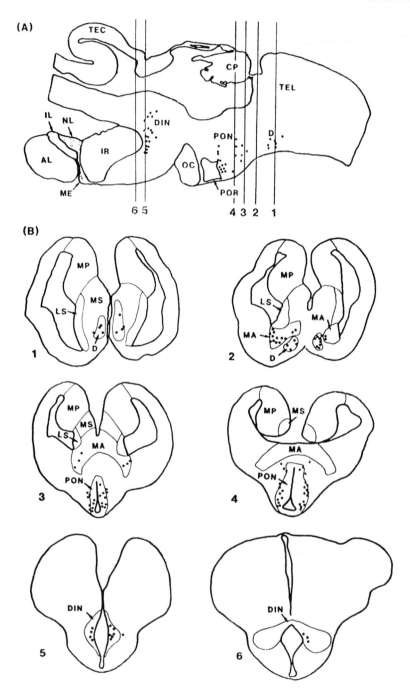

Fig. 9. Diagrams of parasagittal (A) and transverse (B) sections of bullfrog forebrain showing immunoreactive TRH perikarya (large dots) and nerve terminals (small dots). Transverse

al., 1978; Silverman and Zimmerman, 1978; Barry, 1979b; Silverman *et al.,* 1979; Krey and Silverman, 1983; Silverman and Pickard, 1983; King and Anthony, 1984).

The brain distribution of LHRH-immunopositive perikarya has been studied in a number of mammalian species, with the greatest number of studies published being on the rat. Notably, LHRH perikarya are not organized within the bounds of discrete nuclei; rather, there is a scattering of LHRH perikarya in a continuum from the olfactory bulb and nervus terminalis anteriorly through to the area of the premammilary nucleus. Although there has been much controversy (see, e.g., Silverman and Pickard, 1983), the arcuate nucleus of the rat is apparently devoid of LHRH perikarya (Kozlowski and Dees, 1984; Merchenthaler *et al.,* 1984a). Recent thorough studies on the rat described immunopositive perikarya in the olfactory bulb, the origin of the nervus terminalis, some regions of the hippocampus, vertical and horizontal limbs of the diagonal band of Broca, the medial septum, medial preoptic and suprachiasmatic areas, and the anterior and lateral hypothalamus (Fig. 10) (King *et al.,* 1982; Witkin *et al.,* 1982; Witkin and Silverman, 1983; Merchenthaler *et al.,* 1984a). The distribution of LHRH perikarya in a number of other mammals (guinea pig, hamster, mouse, rabbit, sheep, dog, and primates including humans) appears to be quite similar to that in the rat, except that LHRH perikarya are present in the arcuate nucleus in these species (for review, see Barry, 1979a,b; Krey and Silverman, 1983; King and Anthony, 1984). An additional difference between the rat and most other mammalian species, particularly primates, is that the rat apparently has fewer LHRH-immunopositive perikarya, particularly in the preoptic–anterior hypothalamic area. LHRH perikarya have been found in the midbrain of only a few species (Krey and Silverman, 1983); why the midbrain is an apparently inconsistent locality is not known.

In the rat, LHRH perikarya in the preoptic–anterior hypothalamic and septal areas project to the median eminence along four tracts, one with fibers coursing laterally, then ventrally and medially, another tract that courses medially along a periventricular route, another through the tractus infundibularis, and one that courses with the medial forebrain bundle to the posterior hypothalamus and then turns medially and ventrally to the median eminence (Bennett-Clarke and Joseph, 1982; Hoffman and Gibbs, 1982; Jew *et al.,* 1984); the lateral route appears to be principal (Fig. 11) (Jew *et al.,* 1984). Moreover, input from the medial preoptic nucleus and suprachiasmatic nuclei is necessary for the maintenance of spontaneous cyclic ovulation in the

---

sections numbered as indicated in parasagittal section. AL, anterior lobe; CP, choroid plexus; D, nucleus of diagonal bond of Broca; DIN, dorsal infundibular nucleus; IL, intermediate lobe; IR, infundibular recess; LS, lateral septal nucleus; MA, medial part of amygdala; ME, median eminence; MP, medial pallium; MS, medial septal nucleus; NL, neural lobe; OC, optic chiasma; PON, preoptic nucleus; POR, preoptic recess; TEC, tectum; TEL, telencephalon. [From Seki *et al.,* 1983. Reproduced by permission.]

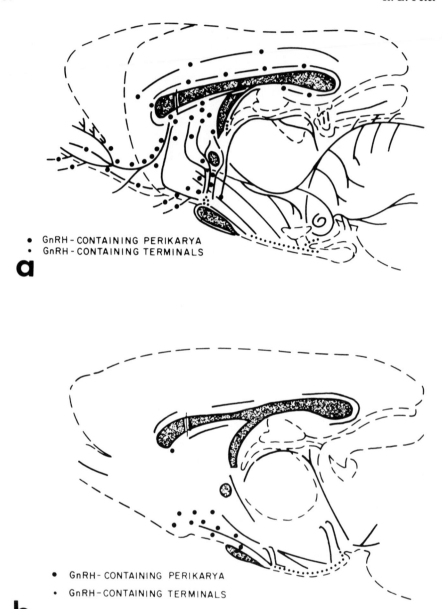

Fig. 10. Diagrams representing the distribution of LHRH neurons and their pathways in a 600-μm-thick median–sagittal segment (a), and a 500-μm-thick parasagittal segment (b) of the rat brain. [From Merchenthaler *et al.*, 1984a. Reproduced by permission.]

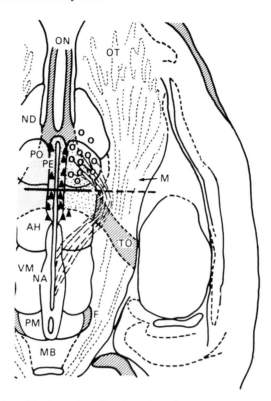

Fig. 11. Drawing of horizontal section through rat hypothalamus. A periventricular knife-cut (shown as a heavy line) or a bilateral lateral hypothalamic cut (shown only on one side as a heavy interrupted line) was made 8.8–9.0 mm anterior to the interaural line (1.0–1.2 mm caudal to the bregma or anterior commissure crossover level). Circles indicate locations of LHRH cell bodies, and triangles show locations of somatostatin cell bodies. Hypothesized LHRH-containing pathways to the median eminence which would be interrupted by the lateral cuts are indicated by interrupted lines. Abbreviations: AH, anterior hypothalamic nucleus; F, fornix; M, medial forebrain bundle; MB, mammillary body; NA, arcuate nucleus; ND, nucleus tractus diagonalis; PE, periventricular nuclei; PO, medial preoptic nucleus; ON, optic nerves; OT, olfactory tubercle; PM, premammillary nuclei; TO, optic tract; VM, ventromedial nucleus. [From Jew *et al.*, 1984. Reproduced by permission of S. Karger AG, Basel.]

rat (see, e.g., Wiegand *et al.,* 1980). However, some authors suggest that such lesions may only interrupt signals for the pulsatile release of LHRH and may not actually destroy LHRH perikarya (Terasawa and Davis, 1983). In female sheep, deafferentation of the preoptic region from the mediobasal hypothalamus causes complete disappearance of LHRH immunohisto-chemical-positive material from the median eminence, disappearance of leutinizing hormone cells, and cessation of estrous cycles (Polkowska, 1981), indicating that preoptic LHRH perikarya are the principal source of LHRH fibers ending in the median eminence; the relative importance of

LHRH perikarya in the acurate and ventromedial nuclei of sheep (Dees *et al.*, 1981) remains open. However, in several species in which LHRH-immunopositive perikarya are present in the mediobasal hypothalamus, LHRH-immunopositive fiber tracts can be traced from these cells to the median eminence [e.g., in the baboon (Marshall and Goldsmith, 1980) and in rhesus and pigtailed monkeys (Silverman *et al.*, 1982)]. Deafferentation of the mediobasal hypothalamus from preoptic input does not cause interruption of normal estrous cycles in the guinea pig (see, e.g., Terasawa and Wiegand, 1978) and rhesus monkey (Krey *et al.*, 1975), suggesting functional importance of the LHRH perikarya in the mediobasal hypothalamus in such species.

In humans, the rhesus and pigtailed monkeys, the ferret, and bats (*Myotis keenii* and *M. leibii*) but not in the rat, LHRH-immunopositive fibers were found to project to the posterior pituitary (Anthony *et al.*, 1984). This raises the possiblity that LHRH may reach the anterior pituitary by vascular flow from the posterior pituitary in these species.

The OVLT is a major site for termination of LHRH-immunopositive fibers from perikarya in the precommissural region to the area of the lamina terminalis in all mammalian species (for review, see Barry, 1979b; Silverman and Pickard, 1983). LHRH fibers also terminate in the subfornical organ, area postrema, and subcommissural organ, as well as being subependymal or projecting directly into the lumen of the third ventricle in the rat (Bennett-Clarke and Joseph, 1982; Witkin *et al.*, 1982; Merchenthaler *et al.*, 1984a). Whether these terminal areas have a neuroendocrine function is not known. LHRH-immunopositive fibers are found in numerous brain regions, including the olfactory tracts, stria terminalis, stria medullaris, medial forebrain bundles, and periaqueductal gray of the mesencephalon, to name a few (Witkin *et al.*, 1982; Merchenthaler *et al.*, 1984a). A peptide immunohistochemically related to adrenocorticotropin was colocalized with LHRH in nerve terminals in the median eminence of the guinea pig (Beauvillain *et al.*, 1981). The function of these fibers is not known.

In the rat, immunohistochemical-reactive LHRH first appears on gestation day 18.5 in the OVLT and on day 19.5 in the median eminence (Kawano *et al.*, 1980). LHRH-immunopositive perikarya are more numerous in neonatal male rats than in females, and estrogenization reduced the amount of immunoreactive material in neonatal males but not in females (Elkind-Hirsch *et al.*, 1981). This suggests an earlier functional development of LHRH neurons in males than occurs in females. Both rats (Polkowska and Jutisz, 1979) and squirrel monkeys (Barry, 1979a) show a depletion of immunoractive LHRH in the median eminence just prior to the time of ovulation. Castration of adult mice or rats of both sexes causes a depletion of immunoreactive LHRH in the median eminence; replacement therapy with sex steroids tends to replenish the stainability (Gross, 1980; Shivers *et al.*, 1983). On the other hand, aging and onset of persistent vaginal estrus is not associated with a change in immunocytochemical LHRH in the median eminence and perikarya in the preoptic region of the rat (Rubin *et al.*, 1984). Overall these studies demonstrate developmental and sexual effects on LHRH in the brain. Further studies coupling immunocytochemistry with measurements of tissue concentrations and turnover of LHRH are necessary to clarify the functional significance of these changes.

3. Nonmammalian Vertebrates

The distribution of immunohistochemical-positive LHRH perikarya and fibers in the brain has been investigated in the duck *Anas platyrhynchos* (McNeill *et al.*, 1976; Bons *et al.*, 1978); domestic chicken, *Gallus domesticus* (Hoffman *et al.*, 1978; Józsa and Mess, 1982; Sterling and Sharp, 1982); pheasant (Hoffman *et al.*, 1978); and Japanese quail, *Coturnix coturnix* (Mikami and Yamada, 1984). An overview was presented by Blähser (1984). Hoffman *et al.*, (1978) reported LHRH perikarya in the infundibular (=arcuate) nucleus, medial septal and medial preoptic nuclei, and olfactory bulb of both chicken and pheasant. Studies on the chicken by Józsa and Mess (1982) and Sterling and Sharp (1982) broadened the distribution of perikarya in the septal–preoptic region to include the periventricular region near the OVLT but did not confirm the presence of LHRH-immunopositive perikarya in the arcuate nucleus. In the Japanese quail, LHRH perikarya were found in the infundibular nucleus and the medial septal–preoptic region (Mikami and Yamada, 1984). In the duck, LHRH-immunopositive perikarya located in the preoptic region send a fiber tract to the median eminence (Fig. 12) (Bons *et al.*, 1978); no LHRH perikarya have been found in the arcuate nucleus. LHRH-immunopositive fibers have been found in a number of brain regions in the chicken (Józsa and Mess, 1982; Sterling and Sharp, 1982), including the olfactory bulbs, periventricular preoptic region, habenular nucleus, and posterior commissure. Thus, the situation in these few bird species suggests that the distribution of LHRH-immunopositive perikarya and fibers is homologous to the rat and other mammals.

In the snake *Elaphe climacophora*, LHRH-immunopositive perikarya were distributed in the medial-septal/medial-preoptic nuclei and the bed nucleus of the hippocampal commissure (Nozaki and Kobayashi, 1979); none were found in the ventrobasal hypothalamus. LHRH-immunopositive fiber tracts appeared to travel from the septal–preoptic region to the ventrobasal hypothalamus and median eminence. Doerr-Schott and Dubois (1978) noted a few scattered LHRH perikarya in the dorsal telencephalon of the lizard *Lacerta muralis*. Given the evolutionary position of reptiles in vertebrate phylogeny, it is unfortunate that more studies have not been done on this group.

The brain distribution of LHRH-immunopositive perikarya and fibers in amphibians has been reviewed by Ball (1981) and Peter (1983a). A group of LHRH-immunopositive perikarya was found in the medial-septal/anterome-

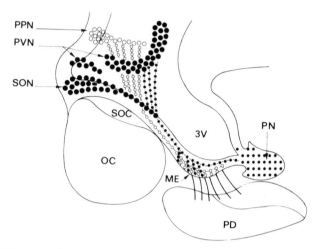

Fig. 12. Diagrammatic representation of neurosecretory systems in the hypothalamus of the duck; sagittal view. Nuclear areas containing LHRH (○) and vasotocin (●) perikarya. LHRH (○–○) and vasotocin (●–●) pathways. ME, median eminence; OC, optic chiasma; PD and PN, pars distalis and pars nervosa of the pituitary, respectively; PPN, periventricular preoptic nucleus; PVN, paraventricular nucleus; SOC, supraoptic commissure; SON, supraoptic nucleus; 3V, third ventricle. [From Bons *et al.*, 1978. Reproduced by permission.]

dial-preoptic region of all species of amphibians investigated and has been confirmed in recent studies on bullfrog tadpoles (*Rana catesbeiana*) undergoing metamorphosis (Crim, 1984a) and in sexually mature specimens of the bullfrog (Crim, 1985) and the toads (*Gastrophryne carolinensis* (Crim, 1984b) and *Bufo japonicus* (Jokura and Urano, 1985). Crim (1985) reported that juvenile bullfrogs, in contrast to the adults, had LHRH-immunopositive perikarya in only the anterior preoptic area and had few immunoreactive fibers in the median eminence, suggesting there is a developmental sequence in the maturation of LHRH perikarya in amphibians. However, Jokura and Urano (1985) reported a seasonal cycle in the number of immunopositive perikarya in the Japanese toad, with sexually regressed toads showing very few. Paired tracts emerging from the grouping of septal and preoptic region perikarya in adult amphibians were traced to the median eminence in several species. LHRH-immunopositive fibers first appear in the median eminence of bullfrogs during metamorphosis (Whalen and Crim, 1985). Although metamorphosis of bullfrogs is accelerated by treatment with thyroxine, the number of immunopositive fibers appears to be fewer and the gonads reduced in size in such animals (Whalen and Crim, 1985), suggesting that bullfrogs may be an excellent model for investigation of development of the LHRH neuronal system. LHRH-immunopositive fibers or tracts have also been variously described in the lateral telencephalon and olfactory bulbs, the thalamus and habenular nucleus, and the midbrain in adults of several amphibian species.

Early literature on brain distribution of LHRH-immunopositive perikarya and fibers in teleost fishes has been reviewed by Ball (1981), Peter (1983a,b) and Demski (1984). LHRH-immunopositive perikarya were found throughout the preoptic nucleus of the African catfish, *Clarias gariepinus* (Goos *et al.*, 1985). An LHRH-immunopositive fiber tract could be traced from the preoptic nucleus to the proximal pars distalis in the region of the gonadotropin cells in the African catfish. However, observations on the platyfish, *Xiphophorus* sp., and goldfish, *Carassius auratus*, demonstrated LHRH-immunopositive perikarya in the anterior preoptic region along the ventrolateral extensions of the preoptic recess, the nucleus lateralis tuberis in the ventrobasal hypothalamus, the ventral telencephalon bordering the olfactory bulbs and in the olfactory bulbs, and in the dorsal midbrain tegmentum posterior of the posterior commissure (Schreibman *et al.*, 1979; Münz *et al.*, 1981, 1982; Kah *et al.*, 1982a; Stell *et al.*, 1984; O. Kah, J. Dulka, and R. Peter, unpublished observations). Recent observations on the goldfish, using an enhanced sensitivity technique and antisera against [Trp[7],Leu[8]]-LHRH revealed tGnRH-immunopositive perikarya in a continuum from the olfactory bulbs through the ventral telencephalon into the anteroventral preoptic region, with reactive perikarya only rarely observed in the ventrobasal hypothalamus (O. Kah, J. Dulka, and R. Peter, unpublished observations). Whereas technical differences may explain the observation of a discontinuous distribution of LHRH-immunopositive perikarya in early studies versus a continuum of tGnRH-immunopositive perikarya through the telencephalon and into the preoptic region, this does not explain the scarcity of tGnRH-immunopositive perikarya in the ventrobasal hypothalamus. Perhaps there is a regional brain distribution of GnRH in the goldfish, and the second molecular form of GnRH may be localized in the ventrobasal hypothalamus? The description of the brain distribution of LHRH-immunopositive perikarya is relatively incomplete for other teleost species (see reviews by Ball, 1981; Peter, 1983a,b; Demski, 1984), although aspects of the observations reported are in agreement with the patterns exhibitied in platyfish and goldfish. A notable exception in this regard is the three-spined stickleback, *Gasterosteus aculeatus*, in which LHRH-immunopositive perikarya were found in only the dorsal midbrain tegmentum and in the magnocellular portion of the preoptic nucleus, continuous with the anterodorsal thalamic area (Borg *et al.*, 1982). This contrasts with the ventral localization in the preoptic region and in other brain areas in platyfish and goldfish. In any case, platyfish and goldfish apparently have a homologous distribution of LHRH- (tGnRH-) immunopositive perikarya compared to mammals.

LHRH-immunopositive fibers, originating from perikarya in the preoptic region appear to travel posteriorly along the ventral preoptic region and arch ventromedially through the anterior hypothalamus to the pituitary stalk in both platyfish (Schreibman *et al.*, 1979) and goldfish (Kah *et al.*, 1982a). LHRH-immunopositive fibers are widespread in the brain of platyfish (Münz *et al.*, 1981, 1982; Schreibman *et al.*, 1979, 1983, 1984) and goldfish (Kah *et al.,* 1982a; Stell *et al.,* 1984), including the dorsal telencephalon, olfactory bulbs and tracts, optic tracts, thalamus and epithalamus (particularly the habenular nuclei), posterior hypothalamus, cerebellum, medulla, and spinal cord. Notably, LHRH-immunopositive perikarya in the ventral telencephalon bordering the olfactory bulbs and in the olfactory bulbs constitute part of the nervus terminalis system and innervate the retina via the optic tracts (Münz *et al.,* 1982; Stell *et al.,* 1984).

In both early- and late-maturing strains of platyfish (excellent models for the investigation of the onset of puberty), there is a sequential appearance of LHRH-immunoreactive perikarya, first in the olfactory-bulbs/anterior-telencephalon, and then in the preoptic region and nucleus lateralis tuberis (Halpern-Sebold and Schreibman, 1983). Following hypophysectomy the staining intensity of perikarya decreases in both the preoptic and ventrobasal hypothalamus but increases in the olfactory-bulbs/anterior-telencephalon region (Schreibman *et al.,* 1983). This suggests that the perikarya in the preoptic and ventrobasal hypothalamic regions innervate the pars distalis. However, demonstration of the functional role of the three concentrations of perikarya in platyfish remains open.

LHRH-immunopositive perikarya are found in the preoptic nucleus of the Pacific lamprey, *Entosphenus tridentata* (Crim *et al.,* 1979a); western brook lamprey, *Lampetra richardsoni* (Crim *et al.,* 1979b); and Japanese lamprey, *E. japonica* (No.ʾaki and Kobayashi, 1979). LHRH-immunopositive fibers were found ventrally in the anterior hypothalamic region, presumably constituting a pathway from the perikarya in the preoptic region to the LHRH-immunopositive fibers ending in the neurohypophysis. This structural organization suggests that the native GnRH is released into the general circulation. The relationship of the apparent organization of the GnRH system in these chordates to the systems found in vertebrates remains open.

G. Other Peptides

Given the long list of peptides present in the nervous system, it is only reasonable to expect that several of these peptides, in addition to those classically recognized as neurohormones, will be released in the median eminence or posterior pituitary and will have a neurohormonal action. Table I summarizes a number of observations in which a given peptide was found in an anatomical situation, such as in the median eminence or posterior pituitary (=neurohypophysis), which would suggest a neurohormonal function.

H. Amines and Other Neurotransmitters

1. Mammals

It is well established that dopamine acts as a prolactin-release-inhibitory factor in mammals (Neill, 1980) and lower vertebrates (Ball, 1981). Dopamine may also have a neurohormonal role in the inhibition of thyrotropin secretion (Hershman and Pekary, 1985) and in the regulation of secretion by the pars intermedia (Smelik *et al.*, 1983). Although examples of putative neurohomronal actions of other amines and neurotransmitters could be cited, a review of such literature is not the intent. On the other hand, if a particular amine or neurotransmitter is located in nerve fibers in the median eminence, it is very likely that the anterior pituitary will be exposed to it via the hypothalamo-hypophysial portal system, and it could therefore have some influence on anterior pituitary activity. The purpose of this review is to discuss the brain origins of such fibers in the median eminence, or the posterior pituitary, without analyzing function.

Dopaminergic perikarya that project to the median eminence, and the pars intermedia via the posterior pituitary, are located in the arcuate and periventricular nuclei in the ventrobasal hypothalamus (Parent, 1979; Polenov *et al.*, 1981; Moore and Demarest, 1982; Lindvall and Bjorklund, 1983; Jennes *et al.*, 1984). The substantia nigra is also thought to contribute dopaminergic fibers to the median eminence (Parent, 1979; Polenov *et al.*, 1981). Norepinephrine-containing neurons project from the medulla to the median eminence (Polenov *et al.*, 1981; Lindvall and Bjorklund, 1983). Serotoninergic fibers in the median eminence are suggested to originate, at least in part, from the arcuate and ventromedial nuclei in the hypothalamus (Polenov *et al.*, 1981). However, other studies indicate that serotoninergic perikarya are absent from the telencephalon and diencephalon, and that serotoninergic fibers in the hypothalamus, median eminence, posterior pituitary, and pars intermedia originate from lower brain regions (Kawata *et al.*, 1984). Immunohistochemical studies on glutamic acid decarboxylase, combined with deafferentation studies, indicate that γ-aminobutyric acid (GABA) located in the median eminence originates from perikarya in the mediobasal hypothalamus (Racagni *et al.*, 1982; Jennes *et al.*, 1984). Finally, immunocytochemical localization of choline acetyltransferase in magnocellular preoptic nucleus neurons of the rat suggests that acetylcholine may also be present in the pituitary (Armstrong *et al.*, 1983).

2. *Nonmammalian Vertebrates*

The median eminence of birds appears to be much less densely innervated by catecholaminergic fibers than in mammals, reptiles, and amphibians (Parent, 1979). In addition, relatively few catecholaminergic perikarya are found in the hypothalamus of birds; experimental evidence suggests that catecholamine fibers in the mediobasal hypothalamus are largely extrahypothalamic in origin. Nevertheless, dopaminergic and noradrenergic perikarya, with projections to the median eminence, are indicated for the infundibular nucleus of birds (Polenov *et al.*, 1981). Furthermore, the hypothalamic paraventricular organ in birds, as well as in reptiles, amphibians, and teleosts and lower bony fishes, is thought to project noradrenergic and sertoninergic fibers to the median eminence and the pituitary; in amphibians

TABLE I

The Brain Origins of Some Brain Peptides Localized by Immunohistochemistry in the Median Eminence, Posterior Pituitary, or Neurohypophysial Tissue Innervating the Anterior Pituitary

Peptide	Localization suggesting neurohormonal function	Origin in brain	Species or class	Reference
Cholecystokinin	Median eminence	Paraventricular nucleus	Mammalia	Palkovits (1984)
	Posterior pituitary	Supraoptic and paraventricular nuclei	Rat and bovine	Vanderhaeghen et al., (1981)
Enkephalins	Median eminence	Origin?	Japanese quail	Mikami and Yamada (1984)
	Posterior pituitary	Arcuate, paraventricular, and supraoptic nuclei	Mammalia	Follénius and Dubois (1981)
				Goodman et al. (1983)
				DiFiglia and Aronin (1984)
				Palkovits (1984)
β-Endorphin	Median eminence	Ventrobasal hypothalamus	Mammalia	Palkovits (1984)
	Median eminence and posterior pituitary	Paraventricular nucleus	Rat	Kiss et al. (1984)
	Neurohypophysis innervating pars distalis	Ventrobasal hypothalamus	Teleostei	Follénius and Dubois (1981)
Neurotensin	Median eminence	Arcuate nucleus and other?	Mammalia	Aronin et al. (1983a)
				Ibata et al. (1984)
				Jennes et al. (1984)
				Palkovits (1984)
	Median eminence and posterior pituitary	Origin?	Japanese quail	Yamada and Mikami (1981)

Angiotensin II	Median eminence and posterior pituitary	Paraventricular nucleus	Mammalia	Ganong (1983) Lind et al. (1985)
Substance P	Median eminence and posterior pituitary	Several hypothalamic nuclei	Mammalia	Stoeckel et al. (1982) Aronin et al. (1983b) Palkovits (1984)
Vasoactive intestinal peptide	Median eminence	Ventral hypothalamus	Japanese quail	Mikami and Yamada (1984)
	Median eminence	Origin?	Mammalia	Mutt (1983) Palkovits (1984)
	Median eminence	Infundibular (=arcuate) nucleus	Japanese quail	Yamada and Mikami (1982) Mikami and Yamada (1984)
FMRF-amide-like	Posterior pituitary	Origin?	Rat	Williams and Dockray (1983)
Gastrin-like	Neurohypophysis innervating pars distalis	Nucleus lateralis tuberis (ventrobasal hypothalamus)	Rainbow trout	Notenboom et al. (1981)
Glucagon-like	Neurohypophysis innervating pars distalis and pars intermedia	Nucleus lateralis tuberis	Rainbow trout	Gielen and Terlou (1983)
	Median eminence	Infundibular nucleus	Japanese quail	Mikami and Yamada (1984)

and fishes the paraventricular organ also contains dopaminergic perikarya that project to the pituitary (Parent, 1979; Polenov et al., 1981; Kah and Chambolle, 1983; Ekström and Van Veen, 1984; Margolis-Kazan et al., 1985). Brain lesioning studies on goldfish demonstrate that the parventricular organ does indeed innervate the pituitary with catecholaminergic fibers (Fryer et al., 1985).

In reptiles and amphibians catecholaminergic perikarya are found in both the ventrobasal hypothalamus and the ventral preoptic region (Parent, 1979; Polenov et al., 1981). The perikarya in the preoptic region are part of the OVLT and the more anterior preoptic recess organ; the perikarya in these structures are cerebrospinal-fluid-contacting. In both amphibians (Polenov et al., 1981) and teleosts such as the goldfish (Kah et al., 1984), a grouping of dopaminergic perikarya in the anteroventral preoptic region projects to the median eminence and pituitary, respectively. The pathway may be unique in lower vertebrates; however, noradrenergic perikarya have been found in the anteroventral preoptic region of reptiles and birds. Finally, whereas early studies indicated the presence of catecholaminergic perikarya in the ventrobasal hypothalamus in teleosts and other bony fishes, more recent studies do not confirm this (Parent, 1979; Polenov et al., 1981). Perhaps these lower vertebrates are unique in not having catecholaminergic perikarya in this brain region.

IV. CONCLUDING REMARKS

It is notable that most neuropeptide systems appear to be highly conserved throughout vertebrate evolution. Between species that are relatively closely related (i.e., between family or order comparisons within the class Mammalia) there can be evolutionary changes to emphasize or de-emphasize the prominence of some particular neuropeptide in a specific brain region. Examples of changes that are as large or as small can be found with interclass comparisons in the vertebrates. The significance of this is that animal models from different classes of vertebrates may be suitable for investigation of some particular aspect of a neuropeptide system, provided the model has the desired emphasis or configuration.

Colocalization of various neuropeptides and neurotransmitters provides important variability for evolutionary change; through evolution it is possible that the relative amount of a given peptide or transmitter in a given brain region can change. If a neuron produces only one peptide or transmitter, the possible range of evolutionary change is more restricted. Colocalization has not yet been extensively investigated in nonmammalian vertebrates. Nevertheless, investigation of nonmammalian vertebrates will reveal the substrate of the mammalian neurohormonal system, thereby providing a rational basis for understanding some of the anomolous differences found between mammals in the relative prominence of a given neuropeptide in some particular brain region.

ACKNOWLEDGMENTS

Supported by grant A6371 from the Natural Sciences and Engineering Research Council of Canada. I wish to thank Jim Fryer, Kei-li Yu, and Olivier Kah for comments on the manuscript.

REFERENCES

Aizawa, T., and Greer, M. A. (1981). Delineation of the hypothalamic area controlling thyrotropin secretion in the rat. *Endocrinology (Baltimore)* **109**, 1731–1738.

Anthony, E. L. P., King, J. C., and Stopa, E. G. (1984). Immunocytochemical localization of LHRH in the median eminence, infundibular stalk and neurohypophysis. Evidence for multiple sites of releasing hormone secretion in humans and other mammals. *Cell Tissue Res.* **236**, 5–14.

Antoni, F. A., Palkovits, M., Markara, G. B., Linton, E. A., Lowry, P. J., and Kiss, J. Z. (1983). Immunoreactive corticotropin-releasing hormone in the hypothalamoinfundibular tract. *Neuroendocrinology* **36**, 415–423.

Armstrong, D. M., Saper, C. B., Levey, A. I., Wainer, B. H., and Terry, R. D. (1983). Distribution of cholinergic neurons in rat brain: Demonstrated by the immunocytochemical localization of choline acetyltransferase. *J. Comp. Neurol.* **216**, 53–68.

Aronin, N., Carraway, R. E., Difiglia, M., and Leeman, S. E. (1983a). Neurotensin. *In* "Brain Peptides" (D. T. Krieger, M. J. Brownstein, and J. B. Martin, eds.), pp. 753–781. Wiley, New York.

Aronin, N., Difiglia, M., and Leeman, S. E. (1983b). Substance P. *In* "Brain Peptides" (D. T. Krieger, M. J. Brownstein and J. B. Martin, eds.), pp. 783–803. Wiley, New York.

Baker, B. L., and Yu, Y.-Y. (1976). Distribution of growth hormone-release-inhibiting hormone (somatostatin) in the rat brain as observed with immunocytochemistry. *Anat. Rec.* **186**, 343–356.

Bakhit, C., Koda, L., Benoit, R., Morrison, J. H., and Bloom, F. E. (1984). Evidence for selective release of somatostatin-14 and somatostatin-28 (1-12) from rat hypothalamus. *J. Neurosci.* **4**, 411–419.

Ball, J. N. (1981). Hypothalamic control of the pars distalis in fishes, amphibians and reptiles. *Gen. Comp. Endocrinol.* **44**, 135–170.

Barry, J. (1979a). Immunofluorescence study of the preoptico-terminal LRH tract in the female squirrel monkey during the estrous cycle. *Cell Tissue Res.* **198**, 1–13.

Barry, J. (1979b). Immunohistochemistry of luteinizing hormone-releasing hormone-producing neurons in vertebrates. *Int. Rev. Cytol.* **60**, 179–221.

Bear, M. F., and Ebner, F. F. (1983). Somatostatin-like immunoreactivity in the forebrain of *Pseudemys* turtles. *Neuroscience* **9**, 297–307.

Beauvillain, J. C., Tramu, G., and Dubois, M. P. (1981). Ultrastructural immunocytochemical evidence of the presence of a peptide related to ACTH in granules of LHRH nerve terminals in the median eminence of the guinea pig. *Cell Tissue Res.* **218**, 1–6.

Bennett-Clarke, C., and Joseph, S. A. (1982). Immunocytochemical distribution of LHRH neurons and processes in the rat: Hypothalamic and extrahypothalamic locations. *Cell Tissue Res.* **221**, 493–504.

Bennett-Clarke, C., Romagnano, M. A., and Joseph, S. A. (1980). Distribution of somatostatin in the rat brain: Telencephalon and diencephalon. *Brain Res.* **188**, 473–486.

Benoit, R., Bohlen, P., Ling, N., Esch, F., Baird, A., Ying, S. Y., Wehrenberg, W. B., and Guillemin, R. (1985). Stomatostatin-28(1-12)-like peptides. *Adv. Exp. Med.* **188**, 89–107.

Berk, M. L., Reaves, T. A., Hayward, J. N., and Finkelstein, J. A. (1982). The localization of vasotocin and neurophysin neurons in the diencephalon of the pigeon. *Columba livia. J. Comp. Neurol.* **204**, 392–406.

Blähser, S. (1983). Topography, ontogeny, and functional aspects of immunoreactive neuropeptide systems in the domestic fowl. *In* "Avian Endocrinology" (S. Mikami, K. Honma, and M. Wada, eds.), pp. 11–24. Jpn. Sci. Soc. Press, Tokyo.

Blähser, S. (1984). Peptidergic pathways in the avian brain. *J. Exp. Zool.* **232**, 398–403.

Blähser, S., and Heinrichs, M. (1982). Immunoreactive neuropeptide systems in avian embryos (domestic mallard, domestic fowl, Japanese quail). *Cell Tissue Res.* **223**, 287–303.

Blähser, S., Fellmann, D., and Bugnon, C. (1978). Immunocytochemical demonstration of somatostatin-containing neurons in the hypothalamus of the domestic mallard. *Cell Tissue Res.* **195**, 183–187.

Blähser, S., Vigh-Teichmann, I., and Ueck, M. (1982). Cerebrospinal fluid-contacting neurons and other somatostatin-immunoreactive perikarya in brains of tadpoles of *Xenopus laevis. Cell Tissue Res.* **224**, 693–697.

Bloch, B., Brazeau, P., Bloom, F., and Ling, N. (1983a). Topographical study of the neurons containing hpGRF immunoreactivity in monkey hypothalamus. *Neurosci. Lett.* **37**, 23–38.

Bloch, B., Brazeau, P., Ling, N., Bohlen, P., Esch, F., Wehrenberg, W. B., Benoit, R., Bloom, F., and Guillemin, R. (1983b). Immunohistochemical detection of growth hormone-releasing factor in brain. *Nature (London)* **301**, 607–608.

Bloch, B., Gaillard, R. C., Brazeau, P., Lin, H. D., and Ling, N. (1984a). Topographical and ontogenetic study of the neurons producing growth hormone-releasing factor in human hypothalamus. *Regul. Pept.* **8**, 21–31.

Bloch, B., Ling, N., Benoit, R., Wehrenberg, W. B., and Guillemin, R. (1984b). Specific depletion of immunoreactive growth hormone-releasing factor by monosodium glutamate in rat median eminence. *Nature (London)* **307**, 272–273.

Bons, N. (1980). The topography of mesotocin and vasotocin systems in the brain of the domestic mallard and Japanese quail: Immunocytochemical identification. *Cell Tissue Res.* **213**, 37–51.

Bons, N., Kerdelhue, B., and Assenmacher, I. (1978). Immunocytochemical identification of an LHRH-producing system originating in the preoptic nucleus of the duck. *Cell Tissue Res.* **188**, 99–106.

Bons, N., Bouillé, C., Vaudry, H., and Guillaume, V. (1985). Localisation par immunofluorescence des neurones à corticolibérine dans l'encéphale de Pigeon. *C. R. Seances Acad. Sci., Ser. 3,* **300**, 49–52.

Borg, B., Goos, H. J. Th., and Terlou, M. (1982). LHRH-immunoreactive cells in the brain of the three-spined stickleback, *Gasterosteus aculeatus* L. (Gasterosteidae). *Cell Tissue Res.* **226**, 695–699.

Bresson, J.-L., Clavequin, M.-C., Fellmann, D., and Bugnon, C. (1984). Ontogeny of the neuroglandular system revealed with HPGRF 44 antibodies in human hypothalamus. *Neuroendocrinology* **39**, 68–73.

Breton, B., Motin, A., Kah, O., Lemenn, F., Geoffre, S., Precigoux, G., and Chambolle, P. (1984). Dosage radio-immunologique homologue d'un facteur hypothalamique de stimulation de la fonction gonadotrope hypophysaire de Saumon s-Gn-RH. *C. R. Seances Acad. Sci., Ser. 3* **299**, 383–388.

Bugnon, C., Fellmann, D., and Bloch, B. (1977). Immunocytochemical study of the ontogenesis of the hypothalamic somatostatin-containing neurons in the human fetus. *Cell Tissue Res.* **183**, 319–328.

Bugnon, C., Bloch, B., Lenys, D., and Fellmann, D. (1978). Cytoimmunological study of LH-RH neurons in humans during fetal life. *In* "Brain-Endocrine Interactions. (D. E. Scott, G. P. Kozlowski, and A. Weindl, eds.), Vol. 3, pp. 183–196. Karger, Basel.

Bugnon, C., Fellmann, D., Gouget, A., and Cardot, J. (1982a). Mise en évidence immunocyto-chimique du système neuroglandulaire à CRF dans l'hypothalamus du Rat. *C. R. Seances Acad. Sci., Ser. 3* **294**, 279–284.

Bugnon, C., Fellmann, D., Bresson, J.-L., and Clavequin, M. C. (1982b). Etude immunocyto-chimique de l'ontogénèse du système neuroglandudlaire à CRF chez l'Homme. *C. R. Seances Ada. Sci., Ser. 3* **294**, 491–496.

Bugnon, C., Cardot, J., Gouget, A., and Fellmann, D. (1983a). Mise en évidence d'un système neuronal peptidergique réactif à un immun-serum anti-CRF41, chez les téléostéens dulci-coles et marins. *C. R. Seances Acad. Sci., Ser. 3* **296**, 711–716.

Bugnon, C., Gouget, A., Fellmann, D., and Clavequin, M. C. (1983b). Immunocytochemical demonstration of a novel peptidergic neurone system in the cat brain with an anti-growth hormone-releasing factor serum. *Neurosci. Lett.* **38**, 131–137.

Buijs, R. M., De Vries, G. J., Van Leeuwen, F. W., and Swaab, D. F. (1983). Vasopressin and oxytocin: Distribution and putative functions in the brain. *Prog. Brain Res.* **60**, 115–122.

Burlet, A., Tonon, M.-C., Tankosic, P., Coy, D., and Vaudry, H. (1983). Comparative im-munocytochemical localization of corticotropin releasing factor (CRF-41) and neurohy-pophysial peptides in the brain of Brattleboro and Long-Evans rats. *Neuroendocrinology* **37**, 64–72.

Chronwall, B. M., Chase, T. N., and O'Donohue, T. L. (1984). Coexistence of neuropeptide Y and somatostatin in rat and human cortical and rat hypothalamic neurons. *Neurosci. Lett.* **52**, 213–217.

Clavequin, M.-C., Bresson, J.-L., Gouget, A., Fellmann, D., and Bugnon, C. (1983). Etude immunocytochimique comparée due système neuroglandulaire hypothalamique à CRF en période post-natale chez le Chien et le Chat. *C. R. Seances Soc. Biol. Ses Fil.* **177**, 28–36.

Crim, J. W. (1984a). Immunocytochemistry of luteinizing hormone-releasing hormone in brains of bullfrogs (*Rana catesbeiana*) during spontaneous metamorphosis. *J. Exp. Zool.* **229**, 327–337.

Crim, J. W. (1984b). Immunocytochemistry of luteinizing hormone-releasing hormone in brains of breeding eastern narrow-mouthed toads (*Gastrophryne carolinensis*). *Comp. Biochem. Physiol. A* **79A**, 283–287.

Crim, J. W. (1985). Immunocytochemistry of luteinizing hormone-releasing hormone and sex-ual maturation of the frog brain: Comparisons of juvenile and adult bullfrogs (*Rana cates-beiana*). *Gen. Comp. Endocrinol.* **59**, 424–433.

Crim, J. W., Urano, A., and Gorbman, A. (1979a). Immunocytochemical studies of luteinizing hormone-releasing hormone in brains of agnathan fishes. I. Comparisons of adult Pacific lamprey (*Entosphenus tridentata*) and the Pacific hagfish (*Eptatretus stouti*). *Gen. Comp. Endocrinol.* **37**, 294–305.

Crim, J. W., Urano, A., and Gorbman, A. (1979b). Immunocytochemical studies of luteinizing hormone-releasing hormone in brains of agnathan fishes. II. Patterns of immunoreactivity in larval and maturing western brook lamprey (*Lampetra richardsoni*). *Gen. Comp. Endo-crinol.* **38**, 290–299.

Cumming, R., Reaves, A. T., Jr., and Hayward, J. N. (1982). Ultrastructural immunocyto-chemical characterization of isotocin, vasotocin and neurophysin neurons in the magnocel-lular preoptic nucleus of the goldfish. *Cell Tissue Res.* **223**, 685–694.

Daikoku, S., Okamura, Y., Kawano, H., Tsuruo, Y., Maegawa, M., and Shibasaki, T. (1984). Immunohistochemical study on the development of CRF-containing neurons in the hypo-thalamus of the rat. *Cell Tissue Res.* **238**, 539–544.

Dees, W. L., Sorensen, A. M., Jr., Kemp, W. M., and McArthur, N. H. (1981). Immunohistochemical localization of gonadotropin-releasing hormone (GnRH) in the brain and infundibulum of the sheep. *Cell Tissue Res.* **215**, 181–191.

Demski, L. S. (1984). The evolution of neuroanatomical substrates of reproductive behavior: Sex steroid and LHRH-specific pathways including the terminal nerve. *Am. Zool.* **24**, 809–830.

Dierickx, K., Goossens, N., and Vandesande, F. (1981). The origin of somatostatin fibers in the median eminence and neural lobe of *Rana temporaria*. *Cell Tissue Res.* **215**, 41–45.

DiFiglia, M., and Aronin, N. (1984). Immunoreactive Leu-enkephalin in the monkey hypothalamus including observations on its ultrastructural localization in the paraventricular nucleus. *J. Comp. Neurol.* **225**, 313–326.

Doerr-Schott, J., and Dubois, M. P. (1977). Immunohistochemical demonstration of an SRIF-like system in the brain of the reptile: *Lacerta muralis* Laur. *Experientia* **33**, 947–949.

Doerr-Schott, J., and Dubois, M. P. (1978). Immunohistochemical localisation of different peptidergic substances in the brain of amphibians and reptiles. *In* "Comparative Endocrinology" (P. J. Gaillard and H. H. Boer, eds.), pp. 367–370. Elsevier/North-Holland, Biomedical Press, Amsterdam.

Dreyfuss, F., Burlet, A., Tonon, M. C., and Vaudry, H. (1984). Comparative immunoelectron microscopic localization of corticotropin-releasing factor (CRF-41) and oxytocin in the rat median eminence. *Neuroendocrinology* **39**, 284–287.

Dubois, M. P., Billard, R., Breton, B., and Peter, R. E. (1979). Comparative distribution of somatostatin, LH-RH, neurophysin, and α-endorphin in the rainbow trout: An immunocytological study. *Gen. Comp. Endocrinol.* **37**, 220–232.

Ekström, P., and Van Veen, T. (1984). Distribution of 5-hydroxytryptamine (serotonin) in the brain of the telost *Gasterosteus aculeatus* L. *J. Comp. Neurol.* **226**, 307–320.

Elde, R., and Hokfelt, T. (1978). Distribution of hypothalamic hormones and other peptides in the brain. *In* "Frontiers in Neuroendocrinology" (W. F. Ganong and L. Martini, eds.), Vol. 5, pp. 1–33. Raven Press, New York.

Elkind-Hirsch, K., King, J. C., Gerall, A. A., and Arimura, A. A. (1981). The luteinizing hormone-releasing hormone (LHRH) system in normal and estrogenized neonatal rats. *Brain Res. Bull.* **7**, 645–654.

Epelbaum, J., Arancibia, L. T., Herman, J. P., Kordon, C., and Palkovits, M. (1981). Topography of median eminence somatostatinergic innervation. *Brain Res.* **230**, 412–416.

Fasolo, A., and Gaudino, G. (1981). Somatostatin immunoreactive neurons and fibers in the hypothalamus of the newt. *Gen. Comp. Endocrinol.* **43**, 256–263.

Fasolo, A., and Gaudino, G. (1982). Immunohistochemical localization of somatostatin-like immunoreactivity in the hypothalamus of the lizard *Lacerta sicula*. *Gen. Comp. Endocrinol.* **48**, 205–212.

Fellmann, D., Bugnon, C., Gouget, A., and Cardot, J. (1982). Les neurones à corticoliberine (CRF) du cerveau de Rat. *C. R. Seances Soc. Biol., Ses Fil.* **176**, 511–516.

Fellmann, D., Gouget, A., and Bugnon, C. (1983). Mise en évidence d'un nouveau système neuronal peptidergique immunoreaactif à un immun-sérum anti-hpGRF 44 dans le cerveau du lérot (*Eliomys quercinus*). *C. R. Seances Acad. Sci., Ser. 3* **296**, 487–492.

Fellmann, D., Bugnon, C., Bresson, J. L., Gouget, A., Cardot, J., Clavequin, M. C., and Hadjiyiassemis, M. (1984). The CRF neuron: Immunocytochemical study. *Peptides* **5**, Suppl. 1, 19–33.

Fellmann, D., Bugnon, C., and Lavry, G. N. (1985). Immunohistochemical demonstration of a new neurone system in rat brain using antibodies against human growth hormone-releasing factor (1-37). *Neurosci. Lett.* **58**, 91–96.

Filby, A. B., and Gross, D. S. (1983). Distribution of immunoreactive somatostatin in the primate hypothalamus. *Cell Tissue Res.* **233**, 69–80.

Finley, J. C. W., Maderdrut, J. L., Roger, L. J., and Petrusz, P. (1981). The immunocytochemical localization of somatostatin-containing neurons in the rat central nervous system. *Neuroscience* **6**, 2173–2192.

Follènius, E., and Dubois, M. P. (1981). Contribution of endophinergic and enkephalinergic pathways to brain-pituitary relationships. *In* "Neurosecretion. Molecules, Cells, Systems" (D. S. Farner and K. Lederis, eds.), pp. 79–91. Plenum, New York.

Fryer, J. N., Boudreault-Chateauvert, C., and Kirby, R. P. (1985). Pituitary afferents originating in the paraventricular organ (PVO) of the goldfish hypothalamus. *J. Comp. Neurol.* **242**, 475–484.

Ganong, W. F. (1983). The brain renin-angiotensin system. *In* "Brain Peptides" (D. T. Krieger, M. J. Brownstein, and J. B. Martin, eds.), pp. 805–826. Wiley, New York.

Gielen, J. Th., and Terlou, M. (1983). Immunocytochemical evidence for hypophysial projections of the cells of the nucleus lateralis tuberis pars lateralis in the rainbow trout (*Salmo gairdneri*). *Cell Tissue Res.* **228**, 43–50.

Goodman, R. R., Fricker, L. D., and Synder, S. H. (1983). Enkephalins. *In* "Brain Peptides" (D. T. Krieger, M. J. Brownstein, and J. B. Martin, eds.), pp. 827–849). Wiley, New York.

Goos, H. J. Th., De Leeuw, R., de Zoeten-Kamp, C., Peute, J., and Blähser, S. (1985). Gonadotropin-releasing hormone-immunoreactive neuronal structures in the brain and pituitary of the African catfish, *Clarias gariepinus* (Burchell). *Cell Tissue Res.* **214**, 593–596.

Goossens, N., Blahser, S., Oksche, A., Vandesande, F., and Dierickx, K. (1977). Immunocytochemical investigation of the hypothalamo-neurohypophysial system in birds. *Cell Tissue Res.* **184**, 1–13.

Goossens, N., Dierickx, K., and Vandesande, F. (1980). Immunocytochemical localization of somatostatin in the brain of the lizard, *Ctenosauria pectinata*. *Cell Tissue Res.* **208**, 499–505.

Gorbman, A., Dickhoff, W. W., Vigna, S. R., Clark, N. B., and Ralph, C. L. (1983). "Comparative Endocrinology." Wiley, New York.

Grau, E. G., Nishioka, R. S., Young, G., and Bern, H. A. (1985). Somatostatin-like immunoreactivity in the pituitary and brain of three teleost fish species: Somatostatin as a potential regulator of prolactin cell function. *Gen. Comp. Endocrinol.* **59**, 350–357.

Gross, D. S. (1980). Effect of castration and steroid replacement on immunoreactive gonadotropin-releasing hormone in the hypothalamus and preoptic area. *Endocrinology (Baltimore)* **106**, 1442–1450.

Guillemin, R., Brazeau, P., Bohlen, P., Esch, F., Ling, N., Wehrenberg, W. B., Bloch, B., Mougin, C,., Zeytin, F., and Baird, A. (1984). Somatocrinin, the growth hormone releasing factor. *Recent Prog. Horm. Res.* **40**, 233–299.

Guy, J., Benoit, R., and Pelletier, G. (1985). Immunocytochemical localization of somatostatin-28 1-12 in the rat hypothalamus. *Brain Res.* **330**, 283–289.

Halpern-Sebold, L. R., and Schreibman, M. P. (1983). Ontogeny of centers containing luteinizing hormone-releasing hormone in the brain of platyfish (*Xiphophorus maculatus*) as determined by immunocytochemistry. *Cell Tissue Res.* **229**, 75–84.

Hershman, J. M., and Pekary, A. E. (1985). Regulation of thyrotropin secretion. *In* "The Pituitary Gland" (H. Imura, ed.), pp. 149–188. Raven Press, New York.

Hoffman, G. E., and Gibbs, F. P. (1982). LHRH pathways in the rat brain: 'Deafferentation' spares a sub-chiasmatic LHRH projection to the median eminence. *Neuroscience* **7**, 1979–1993.

Hoffman, G. E., and Hayes, T. A. (1979). Somatostatin neurons and their projections in dog diencephalon. *J. Comp. Neurol.* **186**, 371–392.

Hoffman, G. E., Melnyk, V., Hayes, T., Bennett-Clarke, C., and Fowler, E. (1978). Immunocytology of LHRH neurons. *In* "Brain-Endocrine Interaction" (D. E. Scott, G. P. Kozlowski, and A. Weindl, eds.), Vol. 3, pp. 67–82. Karger, Basel.

Hokfelt, T., Fahrenkrug, J., Tatemoto, K., Mutt, V., Werner, S., Hutting, A.-L., Terenius, L., and Chang, K. J. (1983). The PHI (PHI/27)/corticotropin-releasing factor/enkephalin immunoreactive hypothalamic neuron: Possible morphological basis for integrated control of prolactin, corticotropin, and growth hormone secretion. *Proc. Natl. Acad. Sci. U.S.A.* **80,** 895–898.

Ibata, Y., Fukui, K., Obata, H. L., Tanaka, M., Hisa, Y., Sano, Y., Ishigami, T., Imagawa, K., and Sin, S. (1982). Postnatal ontogeny of catecholamine and somatostatin neuron systems in the median eminence of the rat as revealed by a colocalization technique. *Brain Res. Bull.* **9,** 407–415.

Ibata, Y., Kawakami, F., Fukui, K., Okamura, H., Obata-Tsuto, H. L., Tsuto, T., and Terubayashi, H. (1984). Morphological survey of neurotensin-like immunoreactive neurons in the hypothalamus. *Peptides* **5,** Suppl. 1, 109–120.

Inagaki, S., Shiosaka, S., Takatsuki, K., Sakanaka, M., Takagi, H., Senba, E., Matsuzaki, T., and Tohyama, M. (1981). Distribution of somatostatin in the frog brain, *Rana catesbiana*, in relation to location of catecholamine-containing neuron systems. *J. Comp. Neurol.* **202,** 89–101.

Ishikawa, K., Inoue, K., Tosaka, H., Shimada, O., and Suzuki, M. (1984a). Immunohistochemical characterization of thyrotropin-releasing hormone-containing neurons in rat septum. *Neuroendocrinology* **39,** 448–452.

Ishikawa, K., Kakegawa, T., and Suzuki, M. (1984b). Role of the hypothalamic paraventricular nucleus in the secretion of thyrotropin under adrenergic and cold-stimulated conditions in the rat. *Endocrinology (Baltimore)* **114,** 352–358.

Iwanaga, T., Ito, S., Yamada, Y., and Fujita, T. (1983). Cytochemical demonstration of somatosstatin 28-(1-14)-like immunoreactivity in the rat hypothalamus and gastro-entero-pancreatic endocrine system. *Endocrinology (Baltimore)* **113,** 1839–1844.

Jackson, I. M. D., and Lechan, R. M. (1983). Thyrotropin releasing hormone (TRH). *In* "Brain Peptides" (D. T. Krieger, M. J. Brownstein, and J. B. Martin, eds.), pp. 661–685. Wiley, New York.

Jacobowitz, D. M., Schulte, H., Chrousos, G. P., and Loriaux, D. L. (1983). Localization of GRF-like immunoreactive neurons in the rat brain. *Peptides* **4,** 525–542.

Jennes, L., Stumpf, W. D., Bissette, G., and Nemeroff, C. B. (1984). Monosodium glutamate lesions in rat hypothalamus studied by immunohistochemistry for gonadotropin releasing hormone, neurotensin, tyrosine hydroxylase, and glutamic acid decarboxylase and by autoradiography for [^{3}H] estradiol. *Brain Res.* **308,** 245–253.

Jew, J. Y., Léranth, C., Arimura, A., and Palkovits, M. (1984). Preoptic LH-RH and somatostatin in the rat median eminence. *Neuroendocrinology* **38,** 169–175.

Johansson, O., Hokfelt, T., Jeffcoate, S. L., White, N., and Sternberger, L. A. (1980). Ultrastructural localization of TRH-like immunoreactivity. *Exp. Brain Res.* **38,** 1–10.

Jokura, Y., and Urano, A. (1985). An immunohistochemical study of seasonal changes in luteinizing hormone-releasing hormone and vasotocin in the forebrain and the neurohypophysis of the toad, *Bufo japonica*. *Gen. Comp. Endocrinol.* **59,** 238–245.

Joseph, S. A., and Knigge, K. M. (1983). Corticotropin releasing factor: Immunocytochemical localization in rat brain. *Neurosci. Lett.* **35,** 135–141.

Joseph, S. A., Pilcher, W. H., and Knigge, K. M. (1985). Anatomy of the corticotropin-releasing factor and opiomelanocortin systems of the brain. *Fed. Proc., Fed. Am. Soc. Exp. Biol.* **44,** 100–107.

Józsa, R., and Mess, B. (1982). Immunohistochemical localization of the luteinizing hormone releasing hormone (LHRH)-containing structures in the central nervous system of the domestic fowl. *Cell Tissue Res.* **227,** 451–458.

Józsa, R., Vigh, S., Schally, A. V., and Mess, B. (1984). Localization of corticotropin-releasing factor-containing neurons in the brain of the domestic fowl. An immunohistochemical study. *Cell Tissue Res.* **236,** 245–248.

Kah, O., and Chambolle, P. (1983). Serotonin in the brain of the goldfish, *Carassius auratus*. *Cell Tissue Res.* **234**, 319–333.

Kah, O., Chambolle, P., Dubourg, P., and Dubois, M. P. (1982a). Distribution of immunoreactive LH-RH in the brain of the goldfish. *Proc. Int. Symp. Reprod. Physio. Fish, 2nd, 1982*, p. 56.

Kah, O., Chambolle, P., Dubourg, P., and Dubois, M. P. (1982b). Localisation immunocytochimique de la somatostatine dans le cerveau antérieur et l'hypophyse de deux Téléostéens, le Cyprin (*Carassius auratus*) et *Gambusia* sp. *C. R. Seances Acad. Sci., Ser. 3* **294**, 519–524.

Kah, O., Chambolle, P., Thibault, J., and Geffard, M. (1984). Existence of dopaminergic neurons in the preoptic region of the goldfish. *Neurosci. Lett.* **48**, 293–298.

Kawano, H., Watanabe, Y. G., and Daikoku, S. (1980). Light and electron microscopic observation on the appearance of immunoreactive LHRH in perinatal rat hypothalamus. *Cell Tissue Res.* **213**, 465–474.

Kawata, M., Takeuchi, Y., Ueda, S., Matsuura, T., and Sano, Y. (1984). Immunohistochemical demonstration of serotonin-containing nerve fibers in the hypothalamus of the monkey, *Macaca fuscata. Cell Tissue Res.* **236**, 495–503.

King, J. A., and Millar, R. P. (1980). Comparative aspects of luteinizing hormone-releasing hormone structure and function in vertebrate phylogeny. *Endocrinology (Baltimore)* **106**, 707–717.

King, J. A., and Millar, R. P. (1982a). Structure of chicken hypothalamic luteinizing hormone-releasing hormone. I. Structural determination on partially purified material. *J. Biol. Chem.* **257**, 10722–10728.

King, J. A., and Millar, R. P. (1982b). Structure of chicken hypothalamic luteinizing hormone-releasing hormone. II. Isolation and characterizaiton. *J. Biol. Chem.* **257**, 10729–10732.

King, J. C., and Anthony, E. L. P. (1984). LHRH neurons and their projections in humans and other mammals: Species comparisons. *Peptides* **5**, Suppl. 1, 195–207.

King, J. C., Tobet, S. A., Snavely, F. L., and Arimura, A. A. (1982). LHRH immunopositive cells and their projections to the median eminence and organum vasculosum of the lamina terminalis. *J. Comp. Neurol.* **209**, 287–300.

Kiss, J. Z., Cassell, M. D., and Palkovits, M. (1984). Analysis of the ACTH/β-End/α-MSH-immunoreactive afferent input to the hypothalamic paraventricular nucleus of rat. *Brain Res.* **324**, 91–99.

Kolodziejczyk, E., Baertschii, A. J., and Tramu, G. (1983). Corticoliberin-immunoreactive cell bodies localized in two distinct areas of the sheep hypothalamus. *Neuroscience* **9**, 261–270.

Kozlowski, G. P., and Dees, W. L. (1984). Immunocytochemistry for LHRH neurons in the arcuate nucleus area of the rat: Fact or artifact? *J. Histochem. Cytochem.* **32**, 83–91.

Krey, L. C., and Silverman, A. J. (1983). Luteinizing hormone releasing hormone (LHRH). *In* "Brain Peptides" (D. T. Krieger, M. J. Brownstein, and J. B. Martin, eds.), pp. 687–709. Wiley, New York.

Krey, L. C., Butler, W. R., and Knobil, E. (1975). Surgical disconnection of the medial basal hypothalamus and pituitary function in the rhesus monkey. I. Gonadotropin secretion. *Endocrinology (Baltimore)* **96**, 1073–1087.

Krieger, D. T. (1983). Brain peptides: What, where and why? *Science* **222**, 975–985.

Krisch, B. (1981). Distribution of somatostatin and luliberin in the rat brain. *In* "Neurosecretion. Molecules, Cells, Systems" (D. S. Farner and K. Lederis, eds.), pp. 49–59. Plenum, New York.

Lechan, R. M., and Jackson, I. M. D. (1982). Immunohistochemical localization of thyrotropin-releasing hormone in the rat hypothalamus and pituitary. *Endocrinology (Baltimore)* **111**, 55–65.

Lechan, R. M., Molitch, M. E., and Jackson, I. M. D. (1983). Distribution of immunoreactive human growth hormone-like material and thyrotropin-releasing hormone in the rat central nervous system: Evidence for their coexistence in the same neurons. *Endocrinology (Baltimore)* **112**, 877–884.

Lechan, R. M., Lin, H. D., Ling, N., Jackson, I. M. D., Jacobson, S., and Reichlin, S. (1984). Distribution of immunoreactive growth hormone-releasing factor(1-44)NH_2 in the tuberoinfundibular system of the rhesus monkey. *Brain Res.* **309**, 55–61.

Lind, R. W., Swanson, L. W., and Ganten, D. (1985). Organization of angiotensin II immunoreactive cells and fibers in the rat central nervous system. *Neuroendocrinology* **40**, 2–24.

Lindvall, D., and Bjorklund, A. (1983). Dopamine- and norepinephrine-containing neuron systems: There anatomy in the rat brain. *In* "Chemical Neuroanatomy" (P. C. Emson, ed.), pp. 229–255. Raven Press, New York.

McNeill, T. H., Kozlowski, G. P., Abel, J. H., Jr., and Zimmerman, E. A. (1976). Neurosecretory pathways in the mallard duck (*Anas platyrhynchos*) brain: Localization by aldehyde fuchsin and immunoperoxidase techniques for neurophysin (NP) and gonadotropin releasing hormone (Gn-RH). *Endocrinology (Baltimore)* **99**, 1323–1332.

McQuillan, M. T. (1980). "Somatostatin." Eden Press, St. Albans, Vermont.

Margolis-Kazan, H., Schreibman, M. P., and Halpern-Sebold, L. (1983). Immunoreactive serotonin and somatostatin in the brain and pituitary of the platyfish. *Am. Zool.* **24**, 883.

Margolis-Kazan, H., Halpern-Sebold, L. R., and Schreibman, M. P. (1985). Immunocytochemical localization of serotonin in the brain and pituitary gland of the platyfish, *Xiphophorus maculatus*. *Cell Tissue Res.* **240**, 311–314.

Makara, G. B., Palkovits, M., Antoni, F. A., and Kiss, J. Z. (1983). Topography of the somatostatin-immunoreactive fibers to the stalk-median eminence of the rat. *Neuroendocrinology* **37**, 1–8.

Marshall, P. E., and Goldsmith, P. C. (1980). Neuroregulatory and neuroendocrine GnRH pathways in the hypothalamus and forebrain of the baboon. *Brain Res.* **193**, 353–372.

Martin, R., Geis, R., Holl, R., Schafer, M., and Voigt, K. H. (1983). Co-existence of unrelated peptides in oxytocin and vasopressin terminals of rat neurohypophyses: Immunoreactive methionine[5]-enkephalin-, leucine[5] -enkephalin- and cholecystokinin-like substances. *Neuroscience* **8**, 213–227.

Merchenthaler, I., Vigh, S., Petrusz, P., and Schally, A. V. (1982). Immunocytochemical localization of corticotropin-releasing factor (CRF) in the rat brain. *Am. J. Anat.* **165**, 385–396.

Merchenthaler, I., Gorcs, T., Sétáló, G., Petrusz, P., and Flerkó, B. (1984a). Gonadotropin-releasing hormone (GnRH) neurons and pathways in the rat brain. *Cell Tissue Res.* **237**, 15–29.

Merchenthaler, I., Hynes, M. A., Vigh, S., Schally, A. V., and Petrusz, P. (1984b). Corticotropin releasing factor (CRF): Origin and course of afferent pathways to the median eminence (ME) of the rat hypothalamus. *Neuroendocrinology* **39**, 296–306.

Merchenthaler, I., Vigh, S., Schally, A. V., and Petrusz, P. (1984c). Immunocytochemical localization of growth hormone-releasing factor in the rat hypothalamus. *Endocrinology (Baltimore)* **114**, 1082–1085.

Mikami, S.-I., and Yamada, S. (1984). Immunohistochemistry of the hypothalamic neuropeptides and anterior pituitary cells in the Japanese quail. *J. Exp. Zool.* **232**, 405–417.

Miyamoto, K., Hasegawa, Y., Igarashi, M., Chimo, N., Sakakibara, S., Kangawa, K., and Matsuo, H. (1983). Evidence that chicken hypothalamic luteinizing hormone-releasing hormone is [Gln^8]-LHRH. *Life Sci.* **32**, 1341–1347.

Miyamoto, K., Hasegawa, Y., Nomura, M., Igarashi, M., Kangawa, K., and Matsuo, H. (1984). Identification of the second gonadotropin-releasing hormone in chicken hypothalamus: Evidence that gonadotropin secretion is probably controlled by two distinct gonadotropin-releasing hormones in avian species. *Proc. Natl. Acad. Sci. U.S.A.* **81**, 3874–3878.

Moore, K. E., and Demarest, K. T. (1982). Tuberoinfundibular and tuberohypophyseal dopaminergic neurons. In "Frontiers in Neuroendocrinology" (W. F. Ganong and L. Martini, eds.), Vol. 7, pp. 161–190. Raven Press, New York.

Münz, H., Stumpf, W. E., and Jennes, L. (1981). LHRH systems in the brain of platyfish. Brain Res. **221**, 1–13.

Münz, H., Claas, B., Stumpf, W. E., and Jennes, L. (1982). Centrifugal innervation of the retina by luteinizing hormone releasing hormone (LHRH)-immunoreactive telencephalic neurons in teleostean fishes. Cell Tissue Res. **222**, 313–323.

Mutt, V. (1983). VIP, motilin and secretion. In "Brain Peptides" (D. T. Krieger, M. J. Brownstein, and J. B. Martin, eds.), pp. 871–901. Wiley, New York.

Nakai, Y., Shioda, S., Ochiai, H., Kudo, J., and Hashimoto, A. (1983). Ultrastructural relationship between monoamine- and TRH-containg axons in the rat median eminence as revealed by combined autoradiography and immunocytochemistry in the same tissue section. Cell Tissue Res. **230**, 1–14.

Neill, J. D. (1980). Neuroendocrine regulation of prolactin secretion. In "Frontiers in Neuroendocrinology" (L. Martini and W. F. Ganong, eds.), Vol. 5, pp. 129–155. Raven Press, New York.

Notenboom, C. D., Garaud, J. C., Doerr-Schott, J., and Terlou, M. (1981). Localization by immunofluorescence of a gastrin-like substance in the brain of the rainbow trout, Salmo gairdneri. Cell Tissue Res. **214**, 247–255.

Nozaki, M., and Kobayashi, H. (1979). Distribution of LHRH-likesubstance in the vertebrate brain as revealed by immunohistochemistry. Arch. Histol. Jpn. **42**, 201–219.

Ohtsuka, M., Hisano, S., and Daikoku, S. (1983). Electromicroscopic study of somatostatin-containing neurons in rat arcuate nucleus with special reference to neuronal regulation. Brain Res. **263**, 191–199.

Okamura, H., Murakami, S., Chihara, K., Nagatsu, I., and Ibata, Y. (1985). Coexistence of growth hormone releasing factor-like and tyrosine hydroxylase-like immunoreactivities in neurons of the rat arcuate nucleus. Neuroendocrinology **41**, 177–179.

Olivereau, M., Ollevier, F., Vandesande, F., and Verdonck, W. (1984a). Immunocytochemical identification of CRF-like and SRIF-like peptides in the brain and the pituitary of cyprinid fish. Cell Tissue Res. **237**, 379–382.

Olivereau, M., Ollevier, F., Vandesande, F., and Olivereau, J. (1984b). Somatostatin in the brain and the pituitary of some teleosts. Immunocytochemical identification and the effect of starvation. Cell Tissue Res. **238**, 289–296.

Olivereau, M., Vandesande, F., Boucique, E., Ollevier, F., and Olivereau, J.-M. (1984c). Mise en évidence immunocytochimique d'un système peptidergique de type CRF (corticotropin-releasing factor) dans le cerveau des Amphibiens. Comparaison avec la répartition du système à somatostatine. C. R. Seances Acad. Sci., Ser. 3 **299**, 871–876.

Palkovits, M. (1984). Neuropeptides in the hypothalamo-hypophyseal system: Lateral retrochiasmatic area as a common gate for neuronal fibers towards the median eminence. Peptides **5**, Suppl. 1, 35–39.

Pan, J. X., Lechan, R. M., Lin, H. D., and Jackson, I. M. D. (1985a). Immunoreactive neuronal pathways of growth hormone-releasing hormone (GRH) in the brain and pituitary of the teleost Gadus morhua. Cell Tissue Res. **241**, 487–493.

Pan, J. X., Lechan, R., Lin, H. D., Sohn, J., Reichlin, S., and Jackson, I. M. D. (1985b). Multiple forms of human pancreatic growth hormone releasing factor-like immunoreactivity in teleost brain and pituitary. Endocrinology (Baltimore) **116**, 1663–1665.

Parent, A. (1979). Anatomical organization of monoamine- and acetylcholinesterase-containing neuronal systems in the vertebrate hypothalamus. In "Handbook of the Hypothalamus" (P. J. Morgane and J. Panksepp, eds.) Vol. 1, pp. 511–554. Dekker, New York.

Paull, W. K., Phelix, C. F., Copeland, M., Palmiter, P., Gibbs, F. P., and Middleton, C. (1984).

Immunohistochemical localization of corticotropin releasing factor (CRF) in the hypothalamus of the squirrel monkey, *Saimiri sciureus. Peptides* **5** Suppl. 1, 45–51.

Péczely, P., and Antoni, F. A. (1984). Comparative localization of neurons containing ovine corticotropin releasing factor (CRF)-like and neurophysin-like immunoreactivity in the diencephalon of the pigeon (*Columba livia domestica*). *J. Comp. Neurol.* **228**, 69–80.

Pelletier, G., Désy, L., Coté, J., and Vaudry, H. (1983). Immunocytochemical localization of corticotropin-releasing factor-like immunoreactivity in the human hypothalamus. *Neurosci. Lett.* **41**, 259–263.

Peter, R. E. (1983a). Evolution of neurohormonal regulation of reproduction in lower vertebrates. *Am. Zool.* **23**, 685–695.

Peter, R. E. (1983b). The brain and neurohormones in teleost reproduction. *In* "Fish Physiology" (W. S. Hoar, D. J. Randall, and E. M. Donaldson, eds.), Vol. 9, pp. 97–135. Academic Press, New York.

Peter, R. E., and Gill, V. E. (1975). A sterotaxic atlas and technique for forebrain nuclei of the goldfish, *Carassius auratus. J. Comp. Neurol.* **159**, 69–102.

Polenov, A. L., Belenky, M. A., and Konstantinova, M. S. (1981). Morphological bases of the functional interaction of peptidergic and monoaminergic structures of the hypothalamo-hypophyseal complex. *In* "Neurosecretion. Molecules, Cells, Systems" (D. S. Farner and K. Lederis, eds.), pp. 105–116. Plenum, New York.

Polkowska, J. (1981). Immunocytochemistry of luteinizing hormone releasing hormone (LHRH) and gonadotropic hormones in the sheep after anterior deafferentations of the hypothalamus. *Cell Tissue Res.* **220**, 637–649.

Polkowska, J., and Jutisz, M. (1979). Local changes in immunoreactive gonadotropin releasing hormone in the rat median eminence during the estrous cycle. Correlation with the pituitary luteinizing hormone. *Neuroendocrinology* **28**, 281–288.

Racagni, G., Apud, J. A., Cocchi, D., Locatelli, V., and Muller, E. E. (1982). GABAergic control of anterior pituitary hormone secretion. *Life Sci.* **31**, 823–838.

Reaves, T. A., Jr., and Hayward, J. N. (1980). Functional and morphological studies of peptide-containing neuroendocrine cells in goldfish hypothalamus. *J. Comp. Neurol.* **193**, 777–788.

Reichlin, S. (1983). Somatostatin, *In* "Brain Peptides" (D. T. Krieger, M. J. Brownstein, and J. B. Martin, eds.), pp. 711–752. Wiley, New York.

Rémy, C., and Dubois, M. P. (1978). Immunofluorescence of somatostatin-producing sites in the hypothalamus of the tadpole, *Alytes obstetricans* Laur. *Cell Tissue Res.* **187**, 315–321.

Romagnano, M. A., Pilcher, W. H., Bennett-Clarke, C., Chafel, T. L., and Joseph, S. A. (1982). Distribution of somatostatin in the mouse brain: Effects of neonatal MSG treatment. *Brain Res.* **234**, 387–398.

Rorstad, O. P., Martin, J. B., and Terry, L. C. (1980). Somastatin and the nervous system. *In* "The Role of Peptides in Neuronal Function" (J. L. Barker and T. G. Smith, Jr., eds.), pp. 573–614. Dekker, New York.

Roth, K. A., Weber, E., and Barchas, J. D. (1982). Immunoreactive corticotropin releasing factor (CRF) and vasopressin are colocalized in a subpopulation of the immunoreactive vasopressin cells in the paraventricular nucleus of the hypothalamus. *Life Sci.* **31**, 1857–1860.

Roth, K. A., Weber, E., Barchas, J. D., Chang, D., and Chang, J.-K. (1983). Immunoreactive dynorphin-(1-8) and corticotropin-releasing factor in subpopulation of hypothalamic neurons. *Science* **219**, 189–191.

Rubin, B. S., King, J. C., and Bridges, R. S. (1984). Immunoreactive forms of luteinizing hormone-releasing hormone in the brains of aging rats exhibiting persistent vaginal estrus. *Biol. Reprod.* **31**, 343–351.

Sandow, J., and Konig, W. (1978). Chemistry of the hypothalamic hormones. *In* "The Endocrine Hypothalamus" (S. L. Jeffcoate and J. S. M. Hutchinson, eds.), pp. 149–211. Academic Press, New York.

Sawchenko, P. E., Swanson, L. W., Rivier, J., and Vale, W. W. (1985). The distribution of growth-hormone-releasing factor (GRF) immunoreactivity in the central nervous system of the rat: An immunohistochemical study using antisera directed against rat hypothalamic GRF. *J. Comp. Neurol.* **237**, 100–115.

Schreibman, M. P., and Halpern, L. R. (1980). The demonstration of neurophysin and arginine vasotocin by immunocytochemical methods in the brain and pituitary gland of the platyfish, *Xiphophorus maculatus. Gen. Comp. Endocrinol.* **40**, 1–7.

Schreibman, M. P., Halpern, L. R., Goos, H. J. Th., and Margolis-Kazan, H. (1979). Identification of luteinizing hormone-releasing hormone (LH-RH) in the brain and pituitary gland of a fish by immunocytochemistry. *J. Exp. Zool.* **210**, 153–160.

Schreibman, M. P., Halpern-Sebold, L., Ferin, M., Margolis-Kazan, H., and Goos, H. J. Th. (1983). The effect of hypophysectomy and gonadotropin administration on the distribution and quantity of LH-RH in the brains of platyfish: A combined immunocytochemistry and radioimmunoassay study. *Brain Res.* **267**, 293–300.

Schreibman, M. P., Margolis-Kazan, H., Halpern-Sebold, L., O'Neill, P. A., and Silverman, R. C. (1984). Structural and functional links between olfactory and reproductive systems: Puberty-related changes in olfactory epithelium. *Brain Res.* **302**, 180–183.

Seki, T., Nakai, Y., Shioda, S., Mitsuma, T., and Kikuyama, S. (1983). Distribution of immunoreactive thyrotropin-releasing hormone in the forebrain and hypophysis of the bullfrog, *Rana catesbeiana. Cell Tissue Res.* **233**, 507–516.

Sherwood, N. M., Eiden, L. E., Brownstein, M. J., Spiess, J., Rivier, J., and Vale, W. (1983). Characterization of a teleost gonadotropin-releasing hormone. *Proc. Natl. Acad. Sci. U.S.A.* **80**, 2794–2798.

Sherwood, N. M., Harvey, B., Brownstein, M. J., and Eiden, L. E. (1984). Gonadotropin-releasing hormone (Gn-RH) in striped mullet (*Mugil cephalus*), milkfish (*Chanos chanos*), and rainbow trout (*Salmo gairdneri*): Comparison with salmon Gn-RH. *Gen. Comp. Endocrinol.* **55**, 174–181.

Shiosaka, S., Takatsuki, K., Inagaki, S., Sakanaka, M., Takagi, H., Senba, E., Matsuzaki, T., and Tohyama, M. (1981). Topographic atlas of somatostatin-containing neuron system in the avian brain in relation to catecholamine-containing neuron system. II. Mesencephalon, rhombencephalon, and spinal cord. *J. Comp. Neurol.* **202**, 115–124.

Shiosaka, S., Takatsuki, K., Sakanaka, M., Inagaki, S., Takagi, H., Senba, E., Kawai, Y., Iida, H., Minagawa, H., Hara, Y., Matsuzaki, T., and Tohyama, M. (1982). Ontogeny of somatostatin-containing neuron system of the rat: Immunohistochemical analysis. II. Forebrain and diencephalon. *J. Comp. Neurol.* **204**, 211–224.

Shivers, B. D., Harlan, R. E., Morrell, J. I., and Pfaff, D. W. (1983). Immunocytochemical localization of luteinizing hormone-releasing hormone in male and female rat brains. *Neuroendocrinology* **36**, 1–12.

Silverman, A.-J., and Pickard, G. E. (1983). The hypothalamus. *In* "Chemical Neuroanatomy" (P. C. Emson, ed.), pp. 295–336. Raven Press, New York.

Silverman, A. J., and Zimmerman, E. A. (1978). Pathways containing luteinizing hormone-releasing hormone (LHRH) in the mammalian brain. *In* "Brain-Endocrine Interaction" (D. E. Scott, G. P. Kozlowski and A. Weindl, eds.), Vol. 3, pp. 83–96. Karger, Basel.

Silverman, A. J., Krey, L. C., and Zimmerman, E. A. (1979). A comparative study of the luteinizing hormone releasing hormone (LHRH) neuronal networks in mammals. *Biol. Reprod.* **20**, 98–110.

Silverman, A. J., Antunes, J. L., Abrams, G. M., Nilaver, G., Thau, R., Robinson, J. A., Ferin, M., and Krey, L. C. (1982). The luteinizing hormone-releasing hormone pathways in rhesus (*Macaca mulatta*) and pigtailed (*Macaca nemestrina*) monkeys: New observations on thick, unembedded sections. *J. Comp. Neurol.* **211**, 309–317.

Skofitsch, G., and Jacobowitz, D. M. (1985). Distribution of corticotropin releasing factor-like immunoreactivity in the rat brain by immunohistochemistry and radioimmunoassay: Comparison and characterization of ovine and rat/human CRF antisera. *Peptides* **6**, 319–336.

Smelik, P. G., Berkenbosch, F., Vermes, I., and Tilders, F. J. H. (1983). The role of catecholamines in the control of the secretion of pro-opiocortin-derived peptides from the anterior and intermediate lobes and its implications in the response to stress. *In* "The Anterior Pituitary Gland" (A. J. Bhatnagar, ed.), pp. 113–125. Raven Press, New York.

Sofroniew, M. V. (1983). Morphology of vasopressin and oxytocin neurons and their central and vascular projections. *Prog. Brain Res.* **60**, 101–114.

Sofroniew, M. V., Eckenstein, F., Schrell, U., and Cuello, A. C. (1984). Evidence for colocalization of neuroactive substances in hypothalamic neurons. *In* "Coexistence of Neuroactive Substances in Neurons" (V. Chan-Palay and S. L. Palay, eds.), pp. 73–90. Wiley, New York.

Stell, W. K., Walker, S. E., Chohan, K. S., and Ball, A. K. (1984). The goldfish nervus terminalis: An LHRH- and FMRFamide-immunoreactive olfactoretinal pathway. *Proc. Natl. Acad. Sci. U.S.A.* **81**, 940–944.

Sterling, R. J., and Sharp, P. J. (1982). The localisation of LH-RH neurones in the diencephalon of the domestic hen. *Cell Tissue Res.* **222**, 283–298.

Stoeckel, M. E., Porte, A., Klein, M. J., and Cuello, A. C. (1982). Immunocytochemical localization of substance P in the neurohypophysis and hypothalamus of the mouse compared with the distribution of other neuropeptides. *Cell Tissue Res.* **223**, 533–544.

Swanson, L. W., Sawchenko, P. E., Rivier, J., and Vale, W. W. (1983). Organization of ovine corticotropin-releasing factor immunoreactive cells and fibers in the rat brain: An immunohistochemical study. *Neuroendocrinology* **36**, 165–186.

Takatsuki, K., Shiosaka, S., Inagaki, S., Sakanaka, M., Takagi, H., Senba, E., Matsuzaki, T., and Tohyama, M. (1981). Topographic atlas of somatostatin-containing neuron system in the avian brain in relation to catecholamine-containing neuron system I. Telencephalon and diencephalon. *J. Comp. Neurol.* **202**, 103–113.

Tennyson, V. M., Hou-Yu, A., Nilaver, G., and Zimmerman, E. A. (1985). Immunocytochemical studies of vasotocin and mesotocin in the hypothalamo-hypophysial system of the chicken. *Cell Tissue Res.* **239**, 279–291.

Terasawa, E., and Davis, G. A. (1983). The LHRH neuronal system in female rats: Relation to the medial preoptic nucleus. *Endocrinol. Jpn.* **30**, 405–417.

Terasawa, E., and Wiegand, S. J. (1978). Effects of hypothalamic deafferentation on ovulation and estrous cyclicity in female guinea pig. *Neuroendocrinology* **26**, 229–248.

Thommes, R. C., Caliendo, J., and Woods, J. E. (1985). Hypothalamo-adenohypophyseal-thyroid interrelationships in the developing chick embryo. VII. Immunocytochemical demonstration of thyrotropin-releasing hormone. *Gen. Comp. Endocrinol.* **57**, 1–9.

Tonon, M.-C., Burlet, A., Lauber, M., Cuet, P., Jégou, S., Gouteux, L., Ling, N., and Vaudry, H. (1985). Immunohistochemical localization and radioimmunoassay of corticotropin-releasing factor in the forebrain and hypophysis of the frog *Rana ridibunda*. *Neuroendocrinology* **40**, 109–119.

Tramu, G., and Pillez, A. (1982). Localisation immunohistochimique des terminaisons nerveuses à corticolibérine (CRF) dans l'éminence du Cobaye et du Rat. *C. R. Seances Acad. Sci., Ser. 3* **294**, 107–114.

Tramu, G., Beauvillain, J. C., Pillez, A., and Mazzuca, M. (1983a). Présence d'une substance immunologiquement apparentée à la somatolibérine extraite d'une tumeur pancreéatique humaine (hpGRF) dans des neurones de l'air hypophysiotrope du Cobaye et du Rat. *C. R. Seances Acad. Ser. 3* **297**, 435–440.

Tramu, G., Croix, C., and Pillez, A. (1983b). Ability of the CRF immunoreactive neurons of the paraventricular nucleus to produce a vasopressin-like material. Immunohistochemical demonstration in adrenalectomized guinea pigs and rats. *Neuroendocrinology* **37**, 467–469.

Vale, W., Spiess, J., Rivier, C., and Rivier, J. (1981). Characterization of a 41-residue ovine hypothalamic peptide that stimulates secretion of corticotropin and β-endorphin. *Science* **213**, 1394–1397.

Vale, W., Rivier, C., Brown, M. R., Spiess, J., Koob, G., Swanson, L., Bilezikjian, L., Bloom, F., and Rivier, J. (1983). Chemical and biological characterization of corticotropin releasing factor. *Recent Prog. Horm. Res.* **39**, 245–270.

Vanderhaeghen, J. J., Lotstra, F., Vandesande, F., and Dierickx, K. (1981). Coexistence of cholecystokinin and oxytocin-neurophysin in some magnocellular hypothalamo-hypophyseal neurons. *Cell Tissue Res.* **221**, 227–231.

Vandesande, F., and Dierickx, K. (1980). Immunocytochemical localization of somatostatin-containing neurons in the brain of *Rana temporaria*. *Cell Tissue Res.* **205**, 43–53.

Van Vossel-Daeninck, J. V., Dierickx, K., Vandesande, F., and Van Vossel, A. (1981). Electron-microscopic immunocytochemical demonstration of separate vasotocinergic, mesotocinergic and somatostatinergic neurons in the hypothalamic magnocellular preoptic nucleus of the frog. *Cell Tissue Res.* **218**, 7–12.

Verbalis, J. G., and Robinson, A. G. (1985). Neurophysin and vasopressin: Newer concepts of secretion and regulation. *In* "The Pituitary Gland" (H. Imura, ed.), pp. 307–339. Raven Press, New York.

Verhaert, P., Marivoet, S., Vandesande, F., and De Loof, A. (1984). Localization of CRF immunoreactivity in the central nervous system of three vertebrate and one insect species. *Cell Tissue Res.* **238**, 49–53.

Vigh, S., Merchenthaler, I., Torres-Aleman, I., Suerias-Diaz, J., Coy, D. H., Carter, W. H., Petrusz, P., and Schally, A. V. (1982). Corticotropin releasing factor (CRF): Immunocytochemical localization and radioimmunoassay (RIA). *Life Sci.* **31**, 2441–2448.

Vigh-Teichmann, I., Vigh, B., Korf, H.-W., and Oksche, A. (1983). CSF-contacting and other somatostatin-immunoreactive neurons in the brains of *Anguilla anguilla, Phoxinus phoxinus,* and *Salmo gairdneri* (Teleostei). *Cell Tissue Res.* **233**, 319–334.

Weindl, A., Triepel, J., and Kuchling, G. (1984). Somatostatin in the brain of the turtle *Testudo hermanni* Gmelin. An immunohistochemical mapping study. *Peptides* **5**, Suppl. 1, 91–100.

Whalen, R., and Crim, J. W. (1985). Immunocytochemistry of luteinizing hormone-releasing hormone during spontaneous and thyroxine-induced metamorphosis of bullfrogs. *J. Exp. Zool.* **234**, 131–144.

Wiegand, S. J., and Price, J. L. (1980). Cells of origin of the afferent fibers to the median eminence in the rat. *J. Comp. Neurol.* **192**, 1–19.

Wiegand, S. J., Terasawa, E., Bridson, W. E., and Goy, R. W. (1980). Effects of discrete lesions of preoptic and suprachiasmatic structures in the female rat. *Neuroendocrinology* **31**, 147–157.

Williams, R. G., and Dockray, G. J. (1983). Immunohistochemical studies of FMRF-amide-like immunoreactivity in rat brain. *Brain Res.* **276**, 213–229.

Witkin, J. W., and Silverman, A.-J. (1983). Luteinizing hormone-releasing hormone (LHRH) in rat olfactory systems. *J. Comp. Neurol.* **218**, 426–432.

Witkin, J. W., Paden, C. M., and Silverman, A.-J. (1982). The luteinizing hormone-releasing hormone (LHRH) systems in the rat brain. *Neuroendocrinology* **35**, 429–438.

Yamada, S., and Mikami, S.-I. (1981). Immunocytochemical localization of neurotensin-containing neurons in the hypothalamus of the Japanese quail, *Coturnix coturnix japonica*. *Cell Tissue Res.* **218**, 29–39.

Yamada, S., and Mikami, S.-I. (1982). Immunohistochemical localization of vasoactive intestinal polypeptide (VIP)-containing neurons in the hypothalamus of the Japanese quail, *Coturnix coturnix*. *Cell Tissue Res.* **226**, 13–26.

Yamada, S., and Mikami, S.-I. (1985). Immunohistochemical localization of corticotropin-releasing factor (CRF)-containing neurons in the hypothalamus of the Japanese quail, *Coturnix coturnix*. *Cell Tissue Res.* **239**, 299–304.

Yasuda, N., Greer, M. A., and Aizawa, T. (1982). Corticotropin-releasing factor. *Endocr. Rev.* **3**, 123–140.

Zimmerman, E. A. (1983). Oxytocin, vasopressin, and neurophysins. *In* "Brain Peptides" (D. T. Krieger, M. J. Brownstein, and J. B. Martin, eds.), pp. 597–711. Wiley, New York.

4

The Pineal Organ

HORST-W. KORF AND ANDREAS OKSCHE

Department of Anatomy and Cytobiology
Justus Liebig University of Giessen
D-6300 Giessen, Federal Republic of Germany

I. INTRODUCTION

The pineal organ (epiphysis cerebri) is a derivative and an integral component of the vertebrate brain. During phylogeny, the structural and functional properties of the pineal complex have undergone considerable transformation. In poikilothermic vertebrates the pineal is primarily a sensory organ capable of perceiving light stimuli, which are transformed into an electrical response and subsequently propagated to the brain via prominent pinealofugal pathways. In addition, the pineal organ of poikilotherms is apparently endowed with neuroendocrine capacities (for review, see Vollrath, 1981; McNulty, 1984).

In mammals, the pineal body seems to exert its influence exclusively via neuroendocrine mechanisms. To date, the best investigated active principle of the pineal is melatonin, an indoleamine (Lerner *et al.*, 1958; Axelrod and Weissbach, 1960; Wurtman *et al.*, 1968). Numerous investigators have demonstrated the inhibitory influence of light on the indole metabolism in the mammalian pineal gland (cf. Wurtman *et al.*, 1968; Klein, 1982; Reiter, 1981). The control of the pineal indole metabolism in mammals depends on neural input from the sympathetic system (cf. Moore, 1978; Klein, 1982).

Thus, in evolving from a primarily sensory organ in poikilothermic vertebrates to a neuroendocrine organ in mammals, the pineal complex reflects a surprisingly high degree of evolutionary and adaptive plasticity (Ralph, 1970; Quay, 1974). The structural basis of this development will be reviewed in this chapter, with special reference to biomedical implications.

II. DEVELOPMENT AND GROSS ANATOMY

The pineal complex develops from the diencephalic roof in a circumscribed area between the habenular and posterior commissures. The cellular

105

VERTEBRATE ENDOCRINOLOGY:
FUNDAMENTALS AND BIOMEDICAL IMPLICATIONS
Volume 1

elements of the pineal complex are derivatives of the primitive neuroepithelium. As can be concluded from the presence of a parietal foramen in fossil skulls, the pineal complex apparently existed in certain Silurian and Devonian vertebrates, the ancestors of recent fishes, amphibians, and lacertilians (Edinger, 1955; Romer, 1966; Roth and Roth, 1980). In several species of recent vertebrates the pineal complex is divided into two distinct components.

In lampreys and most gnathostome fish the pineal complex consists of a pineal organ proper (epiphysis cerebri) and a parapineal organ, both located inside the cranium (Fig. 1). In several species of fish the tissue overlying the pineal complex displays distinct specializations forming a pineal window or fontanelle (cf. Vollrath, 1981, for references). The ventral aspect of the epiphysis cerebri is bordered by the dorsal sac. The parapineal organ is located to the left of the pineal stalk (Rüdeberg, 1969). The pineal organ proper is in anatomical continuity with the habenular and posterior commissures; the parapineal organ is connected to the left habenular ganglion. The lumen of the parapineal (very prominent in lampreys) is reduced to a capillary space in most other fishes, and the cytoarchitecture of the parapineal organ closely resembles that of the habenular nuclei. The pineal organ proper is a hollow structure in most species of fish, and its lumen is in open communication with the third ventricle.

The gross anatomy of the amphibian pineal complex exhibits striking variation in the three subclasses, i.e, anurans, urodeles, and Gymnophiona (Fig. 1). [For an outline of the ontogenetic development, see von Haffner (1952) and Kelly (1971).] The pineal complex of anurans is composed of a frontal organ located in an extracranial position in the skin and an intracranial pineal organ proper (Studnička, 1905; Oksche, 1955). The location of the frontal organ may be marked by a frontal spot devoid of pigment cells (Oksche, 1955). In several anuran species (and also individuals) the frontal organ may degenerate during ontogeny (von Haffner, 1952; Oksche, 1955; Korf et al., 1981). No frontal organ is found in Urodela and Gymnophiona (Kelly, 1963; Flight, 1973; Korf, 1976); in these animals the pineal complex is exclusively represented by a pineal organ proper (epiphysis). The amphibian pineal organ is a hollow structure with a more or less developed lumen communicating with the third ventricle; topographically it is related to the habenular and posterior commissures, as well as to the choroid plexus of the third ventricle.

The pineal complex of reptiles is highly variable. The pineal organ proper occurs fairly regularly in the typical location between the habenular and posterior commissures. It is claimed to be lacking in crocodilians (*Alligator mississippiensis, Crocodilus niloticus*). The distal component of the reptilian pineal complex, i.e., the parietal eye, is a consistent feature only of certain lizard species, in the adult stage (Fig. 1). The persistence of the parietal eye obviously depends on the latitudes the animals inhabit. Parietal eyes are predominantly present in the lizards at higher latitudes, whereas the species

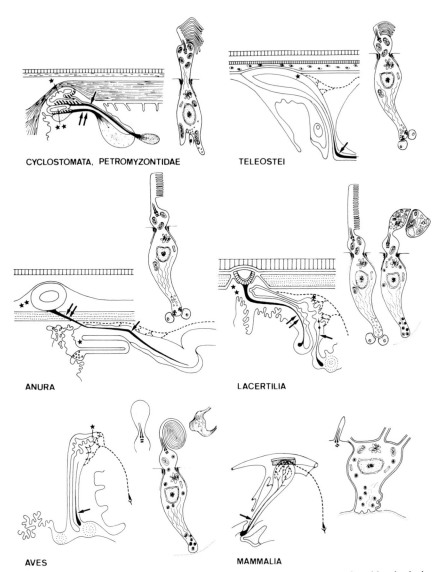

Fig. 1. Comparative representation of pineal complexes. Diagrammatic midsagittal views (after Studnička, 1905) in relation to the respective ultrastructural features of pinealocytes of the receptor line. Single star, pineal organ proper (epiphysis cerebri); double star, parapineal organ (cyclostomes, teleosts), frontal organ (anurans), and parietal eye (lacertilians); single arrow, central neural connections of the pineal organ proper; double arrows, central neural connections of the parapineal or frontal organ; dotted line, basal lamina; broken line, sympathetic nerve fibers. [Modified and extended after Oksche, 1983.]

at lower latitudes tend to lack a parietal eye (Gundy and Wurst, 1976; cf. Quay, 1979). If present, the parietal eye is highly differentiated, closely resembling the lateral eyes. The parietal eye is endowed with a cornea and lens; its parenchyma is arranged in a retinalike pattern. The parietal eye is located in the parietal foramen of the bony skull (Edinger, 1955; Kummer-Trost, 1956) and is connected to the left habenular ganglion via the parietal nerve (Kappers, 1967; Engbretson et al., 1981; Korf and Wagner, 1981). There is evidence that the parietal eye and the pineal organ proper develop from two thickenings lying one behind another (Trost, 1952; Petit, 1967; Viviens-Roels, 1976), although there are considerable species differences confusing the clarity of this picture (cf. Oksche, 1965).

In contrast to poikilothermic vertebrates, the pineal complex in birds is represented exclusively by the pineal organ proper (Fig. 1). The avian pineal apparently develops from a single protrusion of the diencephalic roof; however, there may be secondary and tertiary evaginations as well. In some species accessory pineal tissue is also found anterior to the habenular commissure (Quay and Renzoni, 1967). The avian pineal organ varies in size and shape (cf. Quay and Renzoni, 1967, for a detailed survey). In several species it extends from the intercommissural lamina (between the habenular and posterior commissures) to the surface of the brain, where it is found between the telencephalic hemispheres and the cerebellum directly beneath the confluens sinuum. Whereas the pineal organ of passerine birds is a hollow, saclike structure resembling the pineal organ of lacertilians, it displays a follicular shape in the pigeon and duck. In sexually mature galliform birds the epiphysis is compact and may resemble the mammalian pineal gland.

The mammalian pineal organ also exhibits conspicuous differences in size and shape. In the opossum, the pineal appears as a hollow evagination between the habenular and posterior commissures (Hofer et al., 1976). In other species (e.g., hedgehog and most primates, including humans) the pineal organ displays a solid deep portion found between the habenular and posterior commissures (part A according to Vollrath, 1981), and it lacks a stalk (part B according to Vollrath, 1981) or a superficial portion (part C according to Vollrath, 1981). Most of the rodent species possess a deep and superficial portion connected via a stalklike structure (Fig. 1). The stalk may be composed of pineal parenchyma, blood vessels, and nerve fibers. The superficial portion of the pineal is located between the occipital poles of the telencephalon and the cerebellum, directly beneath the confluens sinuum. The superficial portion and the stalk face the superior colliculi. In all mammalian species the pineal is in close topographical relationship to both the third ventricle and the subarachnoid space. The third ventricle may protrude rather deeply into the pineal tissue, forming the pineal recess [orangutan (Quay, 1970), *Macaca mulatta* (Hülsemann, 1967), golden hamster (Hewing, 1976)]. In this area the ependyma may be reduced, and the pineal parenchyma directly exposed to the cerebrospinal fluid (CSF) of the third ventricle (Hewing, 1978). Furthermore, in several mammalian species, a suprapineal recess of the third ventricle contacts the anterior portion of the pineal body (cf. Vollrath, 1981, for details and references).

A special feature of the mammalian pineal organ is the presence of calcareous concretions, which are typical of the pineal in humans and ungulates (Bargmann, 1943; Quay, 1974), but they are also found in the rhesus monkey (Lukaszyk and Reiter, 1975), the Mongolian gerbil (Japha *et al.,* 1976), and the rat (Diehl, 1978). In clinical medicine, these concretions serve as a landmark in X-ray analysis of the skull; they are composed of inorganic and organic material (Bargmann, 1943; Krstić, 1976). The functional significance of these concrements remains enigmatic. As shown in humans, the number of concretions increases with age (Kitay and Altschule, 1954). Whereas pineal calcifications were previously considered as a sign of gradual inactivation of the organ, more recent investigations indicate that the increase in calcification is not paralleled by a decrease in metabolic activity (Wurtman *et al.,* 1968). Reiter *et al.* (1976) discuss the possibility that pineal concretions may be related to secretory processes in the mammalian pineal, a concept that is backed by our own observations in humans.

As in all other vertebrate species, the primordium of the mammalian pineal organ is located between the habenular and posterior commissures (Gladstone and Wakeley, 1940; Bargmann, 1943; Vollrath, 1981). The human pineal organ appears in embryos of 6–8 mm total length (second month of gestation). In early embryonic stages an anterior lobe can be distinguished from a posterior one, the two being separated by a connective tissue septum (Hülsemann, 1971; Møller, 1974). These two lobes fuse during further development to form a solid mass, the pineal body.

Relatively few studies deal with the blood supply of the pineal complex in submammalian vertebrates (e.g., Mautner, 1965; see Vollrath, 1981, for further references). In mammals, the pineal organ receives its main arterial blood supply from branches of the posterior choroidal arteries originating from the posterior cerebral arteries (Bargmann, 1943; Quay, 1974). These branches give rise to a rich intrapineal capillary network; arterioles are only occasionally found in the pineal (e.g., in the trabeculae of human pineals). The capillaries are drained via trabecular and capsular venules into a vein that joins the vena cerebri magna of Galen. In humans, venous drainage may also be provided by internal cerebral veins. A vascular link has been shown to exist between the pineal organ and the choroid plexus in the frog (Mautner, 1965) and rat (Quay, 1973). A system of specialized portal vessels between the pineal organ and other neuroendocrine areas of the brain was, however, never observed.

III. HISTOLOGY, HISTOCHEMISTRY, AND ULTRASTRUCTURE

In comparative terms, pinealocytes, supporting (glial) cells, and neurons form the pineal parenchyma, which is separated from the adjacent connective tissue layer (capsule) and the capillaries by a basal lamina. Free cells, e.g., macrophages, occur in the perivascular spaces and sometimes also in the pineal lumen. It appears reasonable to divide pinealocytes into three main categories, since these cells display a conspicuous cytoevolution during phylogeny (Collin, 1971; Oksche, 1971): (1) pineal photoreceptor cells, (2) modified (rudimentary) pineal photoreceptor cells, and (3) pinealocytes *sensu stricto* (Collin, 1979; Collin and Oksche, 1981). All of these cell types represent elements of the "receptor line" (Collin, 1979), which can be readily distinguished from supportive (glial) elements and conventional neurons.

A. Pineal Photoreceptor Cells

Photoreceptor cells, the outer segment of which ultrastructurally resembles that of retinal cones, were demonstrated in pineal sense organs of petromyzontids, fishes, amphibians, and lacertilians (Eakin and Westfall, 1959; Oksche and von Harnack, 1963; Oksche and Kirschstein, 1967; Collin, 1971). This outer segment protrudes into the pineal lumen and consists of numerous disks produced by successive basoapical invaginations of the plasma membrane. The disks (20 to 30 nm in width) comprise membrane-limited spaces separated by thin layers of the outer segment cytoplasm and marginally remaining in communication with the extracellular space (i.e., the lumen of the pineal organ). The number of membranous disks varies between 10 and 300 in different species. The most highly differentiated outer segments were found in the parietal eye of lizards (Fig. 2).

Regular outer segments may be intermingled with irregularly lamellated ones, the number of which varies with the species and the age of the individual. The significance of these irregular images has not yet been definitely established. Apart from possible fixation artefacts they may (1) reflect genetically controlled rudimentation, or (2) indicate renewal mechanisms. The latter hypothesis gains support from recent investigations showing that photoreceptor cells in the epiphysis cerebri of the frog *Rana esculenta* cyclically lose all outer segment disks in response to photoperiodic stimuli. However, strong synchronization could be observed in the frog epiphysis only in summer under artificial long day (light/dark cycle 17L : 7D); it was never found in photoreceptors of the frontal organ (Hartwig, 1984).

As shown by microspectrographical recordings, the regularly lamellated outer segments of the frontal organ of *R. catesbeiana* contain a photolabile substance with an absorption maximum between 560 and 580 nm, and those in the pineal organ proper of *R. esculenta* contain a rhodopsinlike photopigment 502_1 (Hartwig and Baumann, 1974).

The outer segment of pineal photoreceptor cells is connected to the inner segment via a connecting piece containing a cilium of the $9+0 \times 2$ type. The cilium arises from one or two centrioles located in the apical portion of the inner segment. Ciliary rootlets extend from the basal body of the cilium to

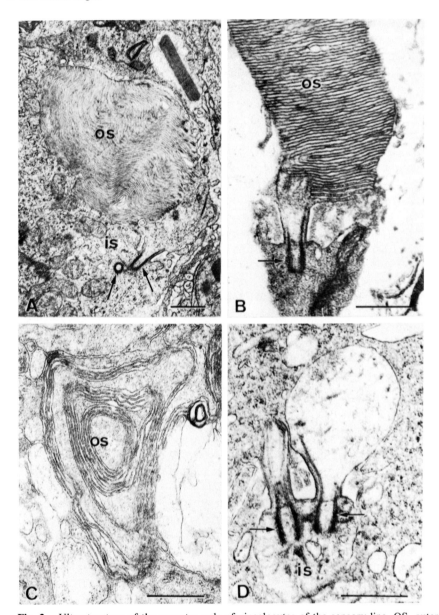

Fig. 2. Ultrastructure of the receptor pole of pinealocytes of the sensory line. OS, outer segments; IS, inner segments. Bar = 1 μm. (A) and (B) Regularly lamellated outer segments of typical pineal photoreceptor cells in the frontal organ of *Xenopus laevis* (A) and the parietal eye of *Trachydosaurus rugosus* (B). (C) and (D) Modified outer segment (C) and bulbous cilia (D) in the pineal organ of *Sturnus vulgaris*. [Part A from Korf *et al.*, 1981, courtesy of Springer Verlag. Part B from A. Oksche and B. T. Firth, unpublished results.]

the basal part of the inner segment. An accumulation of mitochondria forms the "ellipsoid" of the inner segment, which is supposed to be a source of energy required for the transduction processes in the outer segments. At the base of the inner segment a "paraboloid" is found, consisting of free ribosomes, granular endoplasmic reticulum, and a Golgi complex. The neck portion of the inner segments is attached to neighboring receptor or supportive cells via zonulae adhaerentes and occludentes. Abundant microvilli emerging from the inner segment are shown to encircle the outer segment (Flight, 1973; Korf, 1976). These microvilli may serve metabolic or mechanical functions (e.g., the stabilization of the position of the outer segments).

The basal processes of pineal photoreceptors originate from the perikaryon and contribute to conspicuous intrapineal neuropil formations (see p. 121). They are endowed with microfilaments and microtubules, and their enlarged terminals contain numerous electron-lucent synaptic vesicles (30–50 nm in diameter), intermingled with synaptic ribbons and scattered dense-core vesicles. The synaptic ribbons are polymorphic, frequently displaying a platelike appearance; they are often arranged perpendicular to the closely apposed presynaptic membrane (Fig. 3A).

Electrophysiological investigations have clearly shown that pineal photoreceptors are capable of transformation of photic stimuli into an electrical type of response that is transmitted synaptically to intrapineal second-order neurons (cf. Dodt, 1973; Hamasaki and Eder, 1977; Meissl and Dodt, 1981). The neurotransmitters involved in conveying signals of pineal photoreceptor cells have only partly been identified (for biochemical data, see Meissl et al., 1978).

It is still open to discussion whether pineal photoreceptors may convert photic information directly into an endocrine (hormonal) output. Serotonin, one of the precursors of melatonin, has been demonstrated in the pineal parenchyma of various poikilothermic vertebrates by means of the Falck-Hillarp technique (Hafeez and Quay, 1969; Owman and Rüdeberg, 1970). After administration of radioactively labeled 5-hydroxytryptophan and 5-hydroxytryptamine an increased labeling is found in the pineal parenchyma (Oguri et al., 1968; Hafeez and Zerihun, 1976). At the electron microscopic level, pineal photoreceptor cells of the pike are shown to contain melatoninlike immunoreactive material (Falcon, 1984). Pineal photoreceptors of the killifish (Fundulus heteroclitus) contain numerous dense-core vesicles (140–220 nm in diameter), the number of which is considerably increased in fish subjected to darkness (Omura and Ali, 1981). It must be emphasized, however, that several of the outer segments of the pineal sensory cells in the killifish exhibit irregularly arranged membrane disks and, thus, may represent modified photoreceptors (see p. 114). Pineal photoreceptor cells of the frontal organ of anurans and the parietal eye of lizards apparently lack the ability to metabolize indoleamines (Meiniel et al., 1973; Hartwig and Reinhold, 1981).

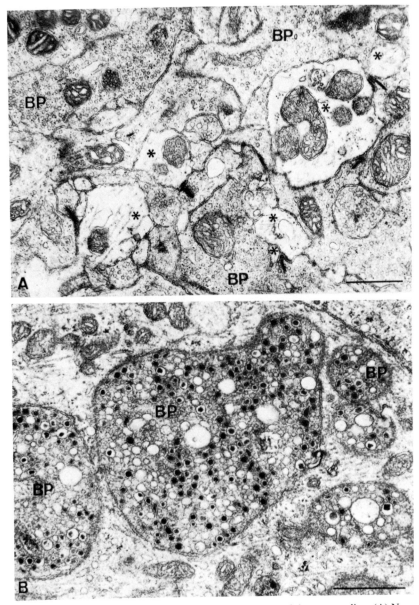

Fig. 3. Ultrastructure of the effector pole of pinealocytes of the sensory line. (A) Neuropil formation in the epiphysis cerebri of *Xenopus laevis*. Basal processes of pineal photoreceptors synaptically connected to dendrites of intrapineal neurons. (B) Basal processes of pinealocytes in the Mongolian gerbil. Note the unusually high number of dense-core vesicles. BP, basal processes of pinealocytes; asterisks, dendrites. Bar = 1 μm. [Part B, courtesy of M. Møller, Copenhagen.]

B. Modified (Rudimentary) Pineal Photoreceptors

Modified (rudimentary) pineal photoreceptor cells form the predominant cellular component in the pineal organ of chelonians, lacertilians, and birds, but they may also be intermingled with true photoreceptors in the pineal sense organs of anamniote species (Collin and Oksche, 1981; for a dualistic concept of cells of the receptor line, see Meiniel, 1981). The modified photoreceptor cells closely resemble characteristic pineal photoreceptors with respect to their segmental organization. As in pineal photoreceptors, an outer and inner segment, a perikaryon, and a basal process can be identified. However, modified photoreceptors display a conspicuous variety of their receptive and effective poles (Fig. 2C, D).

The outer segments of the modified photoreceptor cells are less regularly organized than they are in typical pineal photoreceptors and may contain polymorphic inclusions composed of tubules, clear and dense vesicles, vacuoles, dense and multivesicular bodies, and whorllike structures. Some cells have only a bulbous cilium lacking membrane disks (Fig. 2D).

The number of dense-core vesicles elaborated in the Golgi complex increases drastically parallel to these transformations of the outer segments (cf. Collin, 1979; Oksche, 1983). Dense-core vesicles (50–350 nm in diameter) aggregate in both the cell apex and the basal processes (Quay et al., 1968; Bischoff, 1969; Oksche et al., 1969; Ueck, 1973; Collin et al., 1976). The appearance of numerous dense-core vesicles is typical of secretory glandular elements and secretory neurons. These ultrastructural indications of a secretory (synthetic) capacity of modified photoreceptors conform to the demonstration of yellow serotonin fluorescence in the pineal parenchyma of sauropsids (Quay, 1968; Oksche et al., 1969; Wartenberg and Baumgarten, 1969; Collin, 1971; Meiniel et al., 1973; Ueck, 1973; Juillard et al., 1977). Autoradiographic and cytochemical investigations at the electron microscopic level, combined with pharmacological experiments, have shown that indoleamines are primarily localized within the dense-core vesicles (Collin and Meiniel, 1973; Juillard and Collin, 1978). In addition, the dense-core vesicles contain proteinaceous substances that have not yet been identified (cf. Collin, 1981). They may either represent enzymes involved in indole metabolism, or they may be a specific secretory product of the modified photoreceptors.

The active principles stored in the dense-core vesicles may be released by a diffusion (molecular dispersion) mechanism or exocytotic processes, which, however, have only rarely been observed (cf. Collin and Oksche, 1981). This may be due to the fact that like in secretory neurons, the number of classical exocytotic images may be extremely low and also species-dependent.

Synaptic ribbons, typical of pineal photoreceptor cells, are also found in the modified photoreceptors; they predominate in the terminal portion of the basal process. However, in contrast to pineal photoreceptor cells, modified photoreceptors have not been shown to contact dendrites or perikarya of intrapineal nerve cells. The basal processes of the modified photoreceptors may be located adjacent to the basal lamina separating the pineal parenchyma from the perivascular space or may be apposed to basal processes of other rudimentary photoreceptors.

C. Pinealocytes (*sensu stricto*)

Various studies have dealt with the structure, ultrastructure, and cytochemistry of pinealocytes which establish the main cellular component of the mammalian pineal organ. Pinealocytes *sensu stricto* may also occur in the solid parenchyma of the ophidian pineal organ.

In mammals, pinealocytes appear as mono-, bi-, and multipolar cells when stained with the silver method of Rio del Hortega (cf. Bargmann, 1943). The pinealocytes are considered as derivatives of pineal photoreceptor cells of poikilothermic vertebrates (Collin, 1971; Oksche, 1971). The segmental zonation characteristic of pineal photoreceptors and modified pineal photoreceptors is less conspicuous in pinealocytes, which lack regular or irregular outer segments as well as typical inner segments. However, cilia of the 9+0 type occurring in certain pinealocytes may represent equivalents or rudiments of outer segments and can be considered as the receptive poles of pinealocytes. Centrioles frequently display unusual features.

Synaptic ribbons, typical of pineal photoreceptor cells, may also occur in mammalian pinealocytes (Hopsu and Arstila, 1965; Vollrath, 1973). These organelles are polymorphous, varying in shape from synaptic ribbons to synaptic spherules (Leonhardt, 1967; Romijn, 1975; Hewing, 1978). Frequently, synaptic ribbons are located close to the cell membrane and face adjacent pinealocytes. Only occasionally are they spatially related to blood vessels or nerve fibers. Synaptic ribbons are often accompanied by clear vesicles (30–50 nm in diameter). Although the number of synaptic ribbons in mammalian pinealocytes displays characteristic changes under various physiological and experimental conditions (Vollrath, 1973; see Vollrath, 1981, for further references), they are still in need of a conclusive functional interpretation. These structures may be related to the release of pineal substances or they may be involved in intercellular communication. Intercellular communication may also be established via gap junctions, shown to be present in human fetuses (Møller, 1976) and also in rats (Taugner *et al.*, 1981).

Synaptic ribbons most frequently occur in the processes of pinealocytes. These processes contain microtubules and/or microfilaments, and they often possess bulbous terminals containing a varying number of clear and dense-core vesicles (Fig. 3B). Processes of pinealocytes predominantly terminate at the basal lamina of the perivascular space; only few studies indicate that they may contribute to classical neuropil formations (see, e.g., Vigh-Teich-mann and Vigh, 1979).

The perikarya and processes of pinealocytes contain serotonin, as shown by fluorescence microscopy and microspectrophotometry (Bertler *et al.,* 1964; Owman, 1964; Björklund *et al.,* 1972). These techniques reveal a widely varying concentration of serotonin in individual pinealocytes and different mammalian species. This histochemical variation conforms to the results from biological and chemical assays of pineal serotonin (Wurtman *et al.,* 1968; Quay, 1974). In accord with the conspicuous circadian variation in serotonin concentrations, the number of strongly fluorescent (serotonin-con-taining) pinealocytes decreases during the night (Smith *et al.,* 1972). In sev-eral mammalian species the number of strongly fluorescent pinealocytes located in the periphery of the organ (cortex) exceeds that of pinealocytes located in the center. This finding may indicate a functional zonation of the pineal organ in several mammalian species (see also p. 118; cf. Vollrath, 1981). The application of recently developed antibodies against serotonin confirms the presence of this indole in mammalian pinealocytes (Matsuura and Sano, 1983; Korf and Møller, 1985).

Only a few investigations have been concerned with the location of mela-tonin in pinealocytes (Bubenik *et al.,* 1974; Freund *et al.,* 1978; Vivien-Roels *et al.,* 1981). Virtually nothing is known about the precise intracellular locali-zation of the two enzymes most important for the regulation of the melatonin synthesis, i.e., *N*-acetyltransferase (NAT) and hydroxyindole-*O*-methyl-transferase (HIOMT).

Ultrastructural correlates of secretory activity are less conspicuous in pinealocytes than in most other endocrine cells. Pinealocytes contain forma-tions of rough and smooth endoplasmic reticulum, but the amount of these structures is widely variable among different species. A slight amount of endoplasmic reticulum is seen in pinealocytes of the guinea pig (Vollrath, 1981) and the hedgehog (Pévet and Saboureau, 1973), whereas the pinealocy-tes of the cat, rabbit, and hamster contain conspicuous arrays of this organ-elle (Rodin and Turner, 1965; Leonhardt, 1967; Sheridan and Reiter, 1970). In most mammalian species, the amount of smooth endoplasmic reticulum exceeds the granular type; occasionally the rough type is more abundant. In the mole, mole rat, and dormouse, the rough endoplasmic reticulum enve-lopes paracrystalline structures (Pévet, 1977; Roux *et al.,* 1977), which, by use of pronase, were shown to consist of proteinaceous material (see Collin, 1979; Pévet, 1979). It has not yet been clarified whether this proteinaceous material reflects a specific pineal protein or peptide later released from the pinealocytes. According to Pévet (1979) the Golgi apparatus is not involved in further processing of this material.

Dense-core vesicles have been demonstrated in pinealocytes of all mam-malian species, although their number is fairly small in comparison to other endocrine cells or neurons. These vesicles (40–350 nm in diameter) appear to originate from the Golgi complex; they are apparently transported into the basal process where they are found to be intermingled with clear vesicles of different size and shape. The number of dense-core vesicles undergoes circadian variation, as has been shown in the rabbit (Romijn et al., 1976) and mouse (Benson and Krasovich, 1977); it is high during the light phase of the photoperiod and reduced in darkness. Blinding of the animals leads to a strong increase in the number of dense-core vesicles (Clabough, 1971; Lin et al., 1975), whereas the amount of these vesicles decreases after sympathectomy (Lin et al., 1975; Benson and Krasovich, 1977). The core of the dense-core vesicles consists of a proteinaceous substance (cf. Collin, 1979, 1981). In addition, dense-core vesicles obviously contain indoleamines (Juillard et al., 1984).

In summary, pinealocytes represent neuroendocrine cells capable of synthesis of melatonin and/or related indoles. There is also evidence that pinealocytes, or at least a certain type of them, may synthesize and secrete(?) unidentified proteins or peptides. Moreover, as shown in the next paragraph, mammalian pinealocytes have retained several molecular markers characteristic of pineal and retinal photoreceptor cells.

D. Immunocytochemical Similarities between Pineal Photoreceptors, Modified Pineal Photoreceptors, Pinealocytes, and Retinal Photoreceptors

Opsin, the proteic component of the visual pigment rhodopsin, present in retinal rods, has been immunocytochemically demonstrated in the outer segments of pineal photoreceptors of several teleosts, amphibians, and chelonians (Vigh and Vigh-Teichmann, 1981; Vigh-Teichmann et al., 1982, 1983). Thus, pineal photoreceptor membranes resemble those of retinal rods in their molecular properties, but they are similar to retinal cones in their ultrastructural appearance (Vigh et al., 1983; Vigh-Teichmann et al., 1982, 1983). As shown in Rana esculenta, a population of pineal outer segments does not bind the opsin antibody (Vigh et al., 1985). Such segments belong to pineal photoreceptors endowed with a lipid droplet in the inner segment (see Oksche and von Harnack, 1963). These results speak in favor of—at least—two different types of pineal photoreceptor cells. Immunoreactive opsin also occurs in outer segments of certain modified photoreceptors in the avian pineal organ (Vigh and Vigh-Teichmann, 1981; Korf and Vigh-Teichmann, 1984). This is in accord with biochemical studies indicating that the modified photoreceptors of the avian pineal have not lost their photoreceptive capacity (Deguchi, 1981). Due to the apparent lack of synaptic contact of these cells, one might conclude that, in the avian pineal organ, photic signals are directly transformed into a neuroendocrine response. Irrespective of the conspicuous ultrastructural transformations during phylogeny, certain pinealocytes of mammals (mouse, cat) also display an opsin-positive immunoreaction (Korf et al., 1985a). The immunolabel is observed in the perikarya and processes of pinealocytes; it is very intense in the plasma membrane of these cells.

Furthermore, the S-antigen, first identified in the retina and apparently involved in the autoimmune disease uveoretinitis, was demonstrated in the pineal organs of all vertebrate species investigated to date (Kalsow and Wacker, 1977; Mirshahi et al., 1984; Korf et al., 1985b; 1986b,c). The functional significance of the S-antigen has been a matter of dispute. In recent studies Pfister et al. (1984) and Buzdygon et al. (1985) have identified the S-antigen as the "48K protein" of the retina. This protein is capable of reacting to light signals (1) by binding to rod outer segment membranes (Kühn, 1978, 1984; Pfister et al., 1984) and (2) by mediating rhodopsin-catalyzed ATP binding and quench of cyclic guanosine 5-phosphate-phosphodiesterase (GMP-PDE) activation (Buzdygon et al., 1985).

S-antigen-like immunoreactive material is selectively localized to retinal photoreceptor cells, pineal photoreceptor cells of poikilotherms, modified pineal photoreceptors of sauropsids, and pinealocytes of mammals; other cell types of neuroepithelial origin are immunonegative (Fig. 4) (Kalsow and Wacker, 1973, 1977; Mirshahi et al., 1984; Korf et al., 1985b, 1986b,c). The immunoreaction may occur in all compartments of the cell (i.e., outer and inner segments, perikarya, and basal processes). The number of labeled pinealocytes displays a conspicuous interspecific variation. Whereas in rodents and the cat, most of the pinealocytes are immunolabeled, in primates and humans only a limited number of pinealocytes contain S-antigen-like immunoreactive material. Moreover, the intensity of the S-antigen immunoreactivity varies considerably among individual pineal photoreceptors, modified pineal photoreceptors, and pinealocytes. Further studies are required to elucidate whether this variation reflects (1) the existence of different types of cells or (2) different activity stages of a single cell class. The topographical distribution of strongly immunoreactive pinealocytes in the periphery and weakly immunoreactive cells in the center of the pineal organs of rodents and the cat might point towards multiple types of pinealocytes. However, a definitive proof of this assumption can be expected only from the use of a wide spectrum of combined cytochemical techniques.

The comparative opsin- and S-antigen immunocytochemistry is of crucial importance for our knowledge of the functional significance of mammalian pinealocytes. The results reviewed strongly indicate that mammalian pinealocytes still possess characteristics of photoreceptor cells (for further discussion, see p. 134).

E. Supportive Cells

In all classes of vertebrates, pineal cells of the sensory line are intermingled with supportive or interstitial elements which bear characteristics of

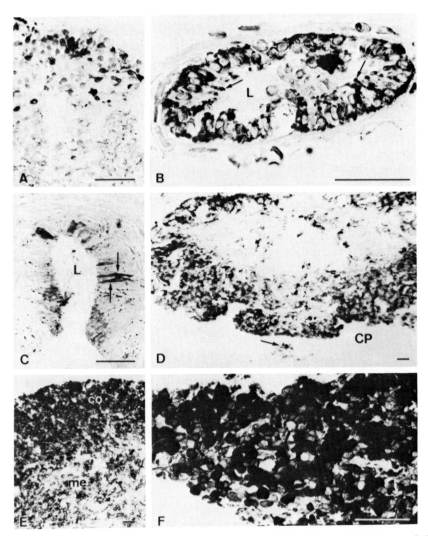

Fig. 4. Immunocytochemical demonstration of retinal S-antigen in pinealocytes of the sensory line. (A) and (B) Typical pineal photoreceptors in the pineal of the rainbow trout (A) and the frog *Rana esculenta* (B). L, pineal lumen; arrows, immunoreactive outer segments. (C) Immunoreactive modified pineal photoreceptors (arrows) in the pineal organ of *Tupinambis nigropunctatus*. L, pineal lumen. (D–F) Immunoreactive pinealocytes in the hedgehog (D), cat (E), and Mongolian gerbil (F). Arrow, immunoreactive pinealocytes scattered in the posterior commissure (CP). Note zonation in the pineal of the cat. co, Cortex; me, medulla. Bar = 1 μm. [Parts D, E, and F from Korf *et al.*, 1985b, courtesy Springer Verlag.]

glial cells (cf. Collin and Oksche, 1981; Vollrath, 1981; McNulty, 1984, for review and references). In hollow pineal organs these cells extend from the base of the pineal parenchyma to the lumen, and they ensheath pineal sensory cells and neuropil formations. Supportive cells are endowed with numerous microvilli and a few cilia of the 9 + 2 × 2 type protruding into the pineal lumen. The basal portion of these cells may form several interdigitating end feet or a single, broad end foot contacting the basal lamina. The cytoplasm of these cells contains conspicuous smooth endoplasmic reticulum. In some species (cf. McNulty, 1984) supportive cells contain a considerable amount of rough endoplasmic reticulum with dilated cisternae encompassing granular material. Dense-core vesicles (70–170 nm in diameter) are observed adjacent to the stacks of Golgi cisternae and, as accumulations, in the luminal and basal portion of the cells.

In lampreys, fishes, and amphibians (for anurans, see Oksche and von Harnack, 1963) the supportive cells display flattened stacks or packed arrays of membranes resembling myeloid bodies of the pigment epithelium of the retina. As shown by Flight and van Donselaar (1975), supportive cells incorporate ^{3}H-labeled vitamin A; the most heavily labeled organelles are the myeloid bodies. This finding might point toward an involvement of these cells in the metabolism of photopigments. Furthermore, supportive cells may serve the phagocytosis of outer segments during the action of renewal mechanisms.

In sauropsids and mammals, supportive cells apparently lack myeloid bodies. In the mammalian pineal the majority of supportive cells ultrastructurally resembles astrocytes, and as shown in the rat, supportive cells—like astrocytes—react with antibodies against two glial marker proteins (the S-100 protein and acidic glial fibrillary protein) (Møller et al., 1978a; see also Schachner et al., 1984). Fibrous processes of these cells form a dense glial network in the pineal organ. In addition, glial elements located outside the pineal in the hypendyma of the posterior commissure may contribute to these extended networks.

The functional significance of the supportive cells is poorly understood. Metabolically, these cells appear very active; they may serve the exchange of ions, nutrients, and metabolites between the vascular system and the pineal parenchyma; and, finally, they may be involved in the storage, transport, and inactivation of indoleamines produced by the cells of the sensory line.

F. Intrapineal Neurons and Related Neuropil Formations

Due to the small amount of granular endoplasmic reticulum, intrapineal neurons are difficult to detect by Nissl staining techniques. Also, silver impregnations of the Golgi type usually fail to reveal intrapineal neurons. Silver impregnations of the Bodian type and methylene blue staining have, in comparison, led to better (although not fully conclusive) results. The analysis of intrapineal neurons was promoted considerably by the use of histochemical and electron microscopic techniques. Neurons have been shown to be present in the pineal organ of all classes of vertebrates; their number, however, displayed wide interspecific variations. The functional significance of intrapineal neurons must be viewed in context with (1) the cytoevolution of pinealocytes and (2) the general morphologic pattern (*Bauplan*) of the epithalamic region.

In pineal sense organs of fishes and amphibians, complex neuropil formations are evident, connecting pineal photoreceptors to secondary intrapineal neurons (Fig. 3A). These neuropil formations do not form plexiform layers as in the retina of the lateral eyes but are unevenly distributed in the pineal parenchyma. The most conspicuous contact within these neuropil formations is the "ribbon synapse" (Eakin *et al.*, 1963; Oksche and von Harnack, 1963) involving (1) the basal process of photoreceptors as presynaptic element and (2) either dendrites (axodendritic synapses) or perikarya (axosomatic synapses) of intrapineal neurons as postsynaptic elements (Bayrhuber, 1972; Flight, 1973; Korf, 1976; Korf *et al.*, 1981; Omura, 1984). Most frequently, the basal process of a pineal photoreceptor is found in contact with a single dendrite which is occasionally invaginated into the basal process. This simple synaptic contact is typical of pineal sense organs and has never been observed in the outer plexiform layer of the retina, which contains synapses involving at least two postsynaptic elements of different origin (dyads, triads) (Werblin and Dowling, 1969). Occasionally, the basal process of pineal photoreceptors establishes synaptic contacts with two or three dendrites, thus forming dyads or triads of the retinal type.

The number of conventional synapses (presynaptic terminals lacking synaptic ribbons) within the neuropil formations of pineal sense organs is very limited. It is not yet clear whether these conventional presynaptic terminals belong to intrapineal neurons or to nerve cells located outside the pineal.

As can be expected from the fact that the number of fully differentiated pineal photoreceptors decreases during phylogeny, typical ribbon synapses are less frequent in the pineal organ of reptiles (Petit and Vivien-Roels, 1977) and birds (Ueck, 1973; Korf and Vigh-Teichmann, 1984). As far as mammals are concerned, ribbon synapses resembling those of poikilothermic vertebrates have been observed only in exceptional cases (cf. Vigh-Teichmann and Vigh, 1979).

Intrapineal neurons classified as bipolar, multipolar, and amacrinelike cells were clearly demonstrated in anurans by means of supravital staining with methylene blue (Paul *et al.*, 1971). With the histochemical acetylcholinesterase (AChE) method, intrapineal neurons are demonstrated in the pineal complex of several teleosts and amphibians (Wake, 1973; Wake *et al.*, 1974; Korf, 1974, 1976; Korf *et al.*, 1981; Herwig, 1981), reptiles (Meissl and Ueck, 1980), and birds (Ueck and Kobayashi, 1972; Sato and Wake, 1983).

The topographical distribution of AChE-positive nerve cells is not uniform. In teleosts, the number of AChE-positive nerve cells per unit area increases in a rostrocaudal direction, indicating that the convergence ratio of photoreceptor cells to secondary neurons is high in the end vesicle and lower in the pineal stalk. In ranid frogs, AChE-positive nerve cells located in the dorsal wall of the epiphysis outnumber those located in its ventral wall (ratio of 3 : 1). In reptiles and birds, AChE-positive neurons tend to aggregate in the proximal (stalk) portion of the pineal, whereas the distal portion (end vesicle) in some avian species is completely devoid of such elements (Sato and Wake, 1983). These regional differences in the innervation of the reptilian and avian pineals suggest that pinealocytes, ranging from more sensorylike to more secretorylike elements, are arranged in a mosaic pattern, as has been shown also for the pike (Falcon, 1979). Interestingly, in the quail and domestic fowl the number of AChE-positive neurons decreases during ontogenetic development (Sato and Wake, 1984), thus resembling the phylogenetic regression of the pineal sensory apparatus.

In all submammalian vertebrates studied to date, AChE-positive neurons contribute to the pinealofugal nervous pathways. However, the exact wiring diagram of the intrapineal neuronal network remains to be elucidated. AChE-positive neurons appear as large and small multipolar, (pseudo-) unipolar, and, to a lesser extent, bipolar elements, but the majority of AChE-positive intrapineal nerve cells could not be further characterized, due to a lack of the AChE reaction in their processes (Fig. 5B, D). A previous hypothesis that the unipolar nerve cells form the pinealofugal pathways and that the multipolar ones represent local interneurons (Wake et al., 1974) has been challenged by retrograde tracing studies with the frontal organ of the frog (Eldred and Nolte, 1981) and the pineal organ of the rainbow trout (Ekström and Korf, 1985). These studies provide clear-cut evidence that both unipolar and multipolar nerve cells contribute to the frontal nerve or the pineal tract, as was also inferred from the results with the methylene blue staining technique (Paul et al., 1971). However, this new evidence does not disprove the existence of interneurons. The studies of Wake et al. (1974) were restricted to AChE-positive nerve cells; methylene blue staining and retrograde labeling, in contrast, do not depend on the presence of a certain type of neurotransmitter or neurotransmitter-bound enzyme in the stained elements.

The tracing experiments in the trout (Fig. 5A, C) reveal an additional population of nerve cells not stained with the methylene blue or AChE technique. These cells, predominantly located in the pineal stalk, have a process directed towards the pineal lumen, and their axons can be traced into the pineal tract; they closely resemble CSF-contacting neurons (see Vigh-Teichmann and Vigh, 1983, for definition). Whether these cells are pineal photoreceptors or serve the perception of other sensory qualities is presently unknown.

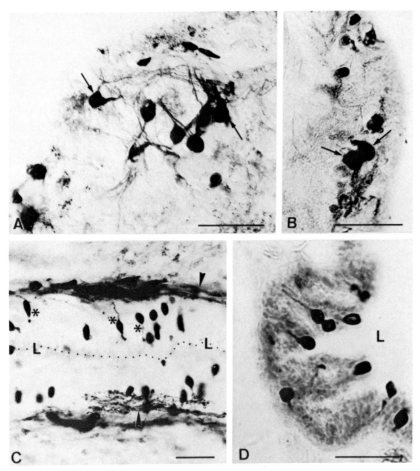

Fig. 5. Intrapineal neurons of the rainbow trout, as revealed by retrograde transport of HRP *in vitro* (A and C) and the AChE method (B and D). (A) and (B) Large multipolar neurons in the tip of the pineal organ (arrows). (C) and (D) Small neurons in the pineal stalk. Some of these cells (asterisks) display a labeled process directed toward the pineal lumen (L). Bar = 50 μm. [Parts A and C from P. Ekström and H. W. Korf, unpublished results.]

Neurons in the pineal organ of mammals must be viewed in light of these comparative data. As in submammalian vertebrates, the number of these neurons displays a conspicuous interspecific variation. AChE-positive nerve cells form aggregates in the pineal organ of the ferret (Trueman and Herbert, 1970; David and Herbert, 1973) and the rhesus monkey (David and Anand Kumar 1978) but are rare in the pineal of rodent species (Ueck, 1979; Guerillot *et al.*, 1979). At the ultrastructural level, neuronlike cells have been described in the opossum (McNulty and Hazlett, 1980) and the ground squirrel (Matsushima and Reiter, 1978). Neuroblasts endowed with axosomatic and axodendritic synapses occur in the pineal organ of the human fetus (Møller, 1978). In the pineal organ of the rabbit approximately 40 neurons are present (Romijn, 1973), which are assumed to be parasympathetic elements (Romijn, 1975). This assumption is in contrast with clear-cut evidence obtained from lesion experiments in the habenular nucleus of the ferret (David and Herbert, 1973) and the rhesus monkey (David *et al.*, 1975), demonstrating that in these species, the intrapineal neurons must be regarded as central nervous elements. The findings of David and Herbert (1973) and David *et al.* (1975) are in accord with the origin of the pineal organ from the embryonic neuroepithelium and conform to the general morphologic pattern of the epithalamic region. It is not yet clear whether neurons located in the mammalian pineal give rise to pinealofugal axonal projections or belong to pinealopetal systems (see Section III,G on central nervous connections). Neurons associated with the central neural connections of the pineal have been demonstrated in several poikilothermic and mammalian species (Wake *et al.*, 1974; Korf, 1974, 1976; Møllgård and Møller, 1973; cf. Korf and Møller, 1985) in the region of the subcommissural organ. These cells may participate in pinealopetal or pinealofugal projections.

In our opinion, the intrapineal neurons and those in the region of the subcommissural organ which are associated with the central pineal connections must be distinguished from nerve cells which, particularly in primates, form ganglionlike aggregates outside the pineal (e.g., Warburg's and Pastori's ganglion). These latter ganglia are considered to belong to the peripheral autonomic system; convincing proof of their physiological role is, however, still lacking (cf. Kenny, 1965).

G. Central Neural Connections of the Pineal Complex

Direct neural connections between the pineal complex and the central nervous system exist in all classes of vertebrates and pass through the habenular and/or the posterior commissures. There is, however, a major difference between submammalian and mammalian species in the direction and function of these nerve fibers. In poikilotherms, the central neural connections (pineal tract, frontal nerve, parietal nerve) are exclusively, or at least predominantly, established by pinealofugal axons originating from intrapineal neurons, whereas investigations of the central nervous connections of the mammalian pineal have revealed pinealopetal projections (Korf and Møller, 1984, 1985). In the lizard *Lacerta sicula* and the house sparrow, pinealopetal or parietopetal elements are intermingled with pinealofugal nerve fibers (Korf and Wagner, 1981; Korf *et al.*, 1982). The central neural pathways of the pineal complex, which may comprise as many as 3000 nerve fibers (cf. Collin and Oksche, 1981; Vollrath, 1981; McNulty, 1984, for review), are dominated by unmyelinated axons. Depending upon the species, up to 20% of the total of nerve fibers are myelinated.

The analysis of the target areas of pinealofugal axons within the central nervous system of several species of teleosts, anurans, and lacertilians and one avian species (*Passer domesticus*) (Paul *et al.*, 1971; Eldred *et al.*, 1980; Hafeez and Zerihun, 1974; Engbretson *et al.*, 1981; Korf and Wagner, 1981; Korf *et al.*, 1982; Ekström and van Veen, 1984; Ekström, 1984) indicates a basic pattern in these neuronal pathways, despite the conspicuous reduction of the sensory apparatus during phylogeny. Pinealofugal nerve fibers terminate within the reticular formation of the brainstem (central tegmental gray), pretectal area, habenular nuclei, several thalamic nuclei, periventricular gray of the hypothalamus (including the magnocellular system), preoptic region, and some archicortical areas.

Apparently, the projections from the extracranial components of the pineal complex are very similar to those of the pineal organ proper, as can be shown for the frontal organ and the epiphysis cerebri of the frog (Paul *et al.*, 1971; Eldred *et al.*, 1980), and the parietal eye and the epiphysis cerebri of the lizard *L. sicula* (Kappers, 1967; Korf and Wagner, 1981). However, it must be emphasized that the parietal nerve and, to a lesser extent, the frontal organ nerve display a striking asymmetry in the habenular region (Eldred *et al.*, 1980; Engbretson *et al.*, 1981; Korf and Wagner, 1981). Such an asymmetric arrangement is only occasionally observed in the projections of the pineal organ proper (pineal tract) (see Ekström, 1984).

The neural projections of the pineal sense organs in lower vertebrates serve the transmission of electrical signals generated in the pineal in response to light stimuli (chromatic and achromatic responses) to central systems within the brain. These systems are known to be involved in the neuroendocrine control of reproduction and reticulomotor activity. Thus, the analysis of the projections of pineal sense organs fully supports the concept that the pineal is an essential component of photoneuroendocrine systems (cf. Oksche and Hartwig, 1979). Several target areas of the pinealo-(parieto-)fugal projections also receive fiber inputs from the retina of the lateral eyes. This finding (Korf and Wagner, 1981; Ekström, 1984) favors the assumption that photoneuroendocrine mechanisms are controlled by at least two different photoreceptive systems, i.e., the retina and pineal organ. [A discussion of putative deep encephalic photoreceptors is beyond the scope of this article; for further details, see Hartwig and Oksche, (1982).] Such dual innervation may enhance the precision in conveying one of the biologically most important environmental signals, the photoperiod. It may also account for the partly ambiguous results obtained after pinealectomy or parietalectomy in lower vertebrates.

The functional significance of pinealofugal pathways in avian species is open to discussion. The avian pineal has not been shown to display ON/OFF responses to direct illumination (Ralph and Dawson, 1968), which are typical of pineal sense organs of lower vertebrates (Morita, 1966), and it is not yet clear whether the changes in the electrical activity of the pineal in response to light stimuli (Semm and Demaine, 1983) are neurally conveyed to the central nervous system or transformed into an endocrine type of response (see Deguchi, 1981). Transplantation experiments with the pineal of the house sparrow indicate that the pineal is involved in the regulation of circadian phenomena via its neuroendocrine output (Zimmerman and Menaker, 1979).

Also open to discussion is whether pinealofugal axons are present in mammalian species. During ontogeny, an unpaired midsagittal nerve is observed connecting the pineal organ with the most rostral part of the mesencephalic tectum in the human, rabbit, and sheep fetus (Møllgard and Møller, 1973; Møller, 1978; Møller et al., 1975). The presence of neuroblasts within the fetal nerve might point toward a sensory capacity of the mammalian pineal organ during embryonic development, and thus the fetal nerve may be a homologue of the pineal pathways of poikilothermic vertebrates. This assumption would conform to the demonstration of extraretinal photoreceptive elements in newborn rats (Zweig et al., 1966). However, the fetal nerve is no longer detectable at birth, and one might suggest that these connections undergo degeneration during ontogeny. On the other hand, pinealofugal projections may be incorporated into the system of nerve fibers connecting the mammalian pineal organ with the habenular and posterior commissures. It should be noted that the existence of pinealofugal connections may be postulated from electrophysiological experiments (Semm, 1983). Comparative data may provide a key for further consideration of these problems.

Central pinealopetal projections (axons originating from the brain and innervating the pineal complex) apparently increase in number and complexity during evolution, as has been shown by tracer studies (Korf and Wagner, 1980, 1981; Korf and Møller, 1984, 1985). The presence of such elements in pineal sense organs, especially of anurans, has been postulated on the basis of ultrastructural and electrophysiological observations (Morita, 1966; cf. Ueck, 1979, for further references); however, tracer studies conducted with the pineal organs of teleosts and anurans (Hafeez and Zerihun, 1974; Eldred et al., 1980; Ekström, 1984; Ekström and van Veen, 1984) have failed to demonstrate this type of connection. The parietal eye of the lizard *Lacerta sicula* receives a central pinealopetal (parietopetal) input from the hypothalamic paraventricular nucleus (Korf and Wagner, 1981). This nucleus, together with the medial and lateral habenular ganglia, also participates in the innervation of the pineal organ of the house sparrow (Korf et al., 1982).

In view of phylogenetic and functional considerations it was of special interest to determine whether (1) these nerve fibers also occur in mammals or (2) the mammalian pineal receives its neural input exclusively from the sympathetic nerve fibers (Kappers, 1960, 1965; Wurtman et al., 1968). With the use of the HRP method it has been shown that central pinealopetal projections (1) exist in mammalian species and (2) are apparently more complex than in submammalian vertebrates. In rodents, neurons innervating the pineal organ are scattered in the medial and lateral habenular complex, the hypothalamic paraventricular nucleus, and the nucleus of the posterior commissure (Korf and Wagner, 1980; Møller and Korf, 1983b; Korf and Møller, 1984, 1985). In the Mongolian gerbil, additional perikarya giving rise to pinealopetal projections are found in the lateral geniculate body adjacent to the pretectal area, and in the rat such perikarya are also found in the superior colliculi (Guerillot *et al.*, 1982).

Evidence for the complexity of the pineal innervation in mammals is also obtained from the electron microscopic analysis of terminal formations present in the pineal. On the basis of ultrastructural criteria two different types of terminals have been described, apart from the sympathetic nerve endings characterized by numerous small, dense-core vesicles (40–60 nm in diameter) (Fig. 6). The first type contains numerous small, clear transmitter vesicles (40–60 nm in diameter) intermingled with large, granular vesicles (approximately 100 nm in diameter). These boutons are shown to degenerate after habenular lesions in the monkey (David *et al.*, 1975), the ferret (David and Herbert, 1973), and the Mongolian gerbil (Møller and Korf, 1983a). Thus, they belong to central pinealopetal nerve elements. Most of these boutons end freely in the perivascular space or the pineal parenchyma; however, in the ferret and monkey they are shown to establish synaptic contacts with intrapineal neurons (cf. Korf and Møller, 1984, for review and references).

The other type of intrapineal nerve terminal contains numerous large, granular vesicles (approximately 100 nm in diameter) intermingled with small, clear vesicles. Due to the ultrastructure of these nerve terminals, they may represent peptidergic elements. These fibers end freely in the perivascular space or between pinealocytes.

The ultrastructural analysis of the aforementioned terminals indicates that (1) they influence the parenchymal elements of the pineal by release of their transmitter agents into the perivascular space or (2) they may act on single cells (pinealocytes or neurons) via direct apposition to the pinealocytic or neuronal membrane. The possible role of central pinealopetal fibers will be discussed after the description of the sympathetic innervation of the pineal organ.

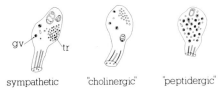

Fig. 6. Diagrammatic representation of nerve terminals found in the mammalian pineal organ. gv, Granular vesicles; tr, transmitter vesicles. [From Korf and Møller, 1984, courtesy Alan Liss.]

H. Sympathetic Innervation of the Pineal Organ

In structural and functional terms, the sympathetic innervation of the pineal organ is very conspicuous in mammals and less evident, or completely absent, in the pineal sense organs of lower vertebrates. Most frequently, investigations of the sympathetic nerve fibers have been performed by use of the histofluorescence method of Falck and Hillarp. In teleost fish, green fluorescent nerve fibers are located in the connective tissue surrounding the pineal organ, but these fibers do not enter the pineal parenchyma (Owman and Rüdeberg, 1970). A similar arrangement is found in ranid frogs. Green fluorescent nerve fibers are seen in the parenchyma and perivascular spaces of the reptilian pineal organ (Quay et al., 1968; Wartenberg and Baumgarten, 1969). In birds, the number and location of these fibers display considerable interspecific variation (cf. Ueck, 1979, for review and references). Microspectrophotometrically, the fluorescent nerve fibers were shown to contain noradrenaline and occasionally also serotonin (Ueck, 1973). As demonstrated by superior cervical ganglionectomy the green fluorescent fibers in the avian pineal originate from the superior cervical ganglion (Hedlund and Nalbandov, 1969).

Considering the functional significance of these fibers it must be emphasized that, in contrast to mammals, sympathetic denervation of the avian pineal does not abolish the effects of light and darkness on melatonin formation. This indicates a fundamental difference in the neural regulation of the neuroendocrine mechanisms in the pineal organ of birds and mammals (cf. Binkley, 1981).

In mammals, the origin of the pinealopetal sympathetic nerve fibers from the superior cervical ganglion was verified by Kappers in 1960 (cf. Owman, 1964, 1965). This finding was followed by numerous morphological, physiological, biochemical, and pharmacological studies elucidating the importance of the sympathetic input for the rhythmic biosynthesis of melatonin in mammals (cf. Axelrod et al., 1983). The sympathetic nerve fibers reach the pineal organ via either the vascular route or the nervus conarius, as has been described in several mammalian species including humans (Kappers, 1960; Kenny, 1961). Combined fluorescence and microspectrophotometric investigations have shown the presence of noradrenaline in these fibers (Bertler et al., 1964; Owman, 1964; Nielsen and Møller, 1978). In addition, sympathetic nerve fibers running within the pineal parenchyma may contain serotonin, probably due to an uptake of serotonin released from pinealocytes (for serotonin-immunoreactive nerve fibers not belonging to the sympathetic system, see p. 131). The sympathetic nerves of the mammalian pineal organ mainly follow the vascular system; however, axons also penetrate from the perivascular space into the pineal parenchyma. There is a wide range of interspecific variation from a rich parenchymal innervation in the rat, cat, and gerbil (Owman, 1965; Nielsen and Møller, 1978) to a sparse parenchymal representation of sympathetic nerves in the guinea pig (Owman, 1965).

Several electron microscopic investigations have shown that sympathetic nerve fibers form terminals endowed with small dense-core vesicles (40–60 nm in diameter), which occasionally are intermingled with larger granular vesicles (Arstila, 1967; Pellegrino de Iraldi and de Robertis, 1965; Jaim-Etcheverry and Zieher, 1983). Superior cervical ganglionectomy leads to a degeneration of this type of nerve terminal, reinforcing its sympathetic character (Møller and Korf, 1983a). Sympathetic nerve terminals end freely in the perivascular space or between pinealocytes. They are never seen to establish synapselike contacts with pinealocytes, interstitial cells, or neurons.

I. Sympathetic and Nonsympathetic Control Mechanisms in the Regulation of Neuroendocrine Functions of the Mammalian Pineal Organ

In mammals, the sympathetic innervation of the pineal organ mediates the inhibitory influence of light on melatonin formation (Wurtman et al., 1968; Reppert and Klein, 1980). Furthermore, the sympathetic nerve fibers are an essential prerequisite for the generation of the circadian rhythms in the pineal indole metabolism. The suprachiasmatic nucleus shown to be a central circadian pacemaker plays a crucial role in this regulation (Moore and Klein, 1974). Through the retinohypothalamic projection, this nucleus receives photoperiodic stimuli perceived in the retina of the lateral eyes (cf. Moore, 1978). The precise neuronal pathways conveying this information from the suprachiasmatic nucleus to the superior cervical ganglia have not yet been elucidated. Recently, evidence was obtained that the paraventricular nucleus apparently represents an important component in the neuronal chain, transmitting signals from the suprachiasmatic nucleus to the superior cervical ganglia (Klein et al., 1983). Interestingly, neurons of the paraventricular nucleus innervate the mammalian pineal organ also via direct (central pinealopetal) projections (Nürnberger and Korf, 1981; Korf and Møller, 1985).

In conclusion, in mammals the regulation of the pineal indole metabolism, depends on the sympathetic innervation of the organ. This apparently holds true also for humans (cf. Vaughan, 1984). However, there is growing evidence that pineal functions are not exclusively regulated by noradrenaline-containing sympathetic nerve fibers. Immunocytochemical investigations have revealed intrapineal nerve fibers and terminals immunoreactive with antisera against vasopressin and oxytocin (Buijs and Pévet, 1980; Nürnberger and Korf, 1981; Matsuura et al., 1983a) (Fig. 7B, C), neurophysin (Yulis and Rodríguez, 1982), vasoactive intestinal polypeptide (VIP) (Uddman et al., 1980; Møller et al., 1985), substance P (Rønnekleiv and Kelley, 1984; Korf and Møller, 1985), and serotonin (Matsuura et al., 1983b; Korf and Møller, 1985) (Fig. 7A).

In consideration of the diversity of neurotransmitters in pineal nerve fi-

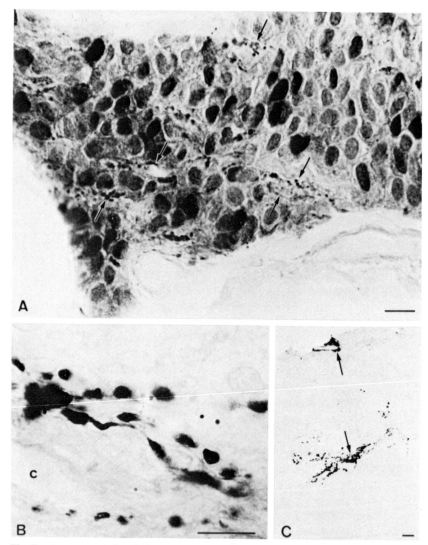

Fig. 7. Immunocytochemical characterization of nerve fibers in the mammalian pineal organ. (A) Serotonin-immunoreactive nerve fibers in the pineal organ of the Mongolian gerbil (arrows). [From Korf and Møller, 1985, courtesy Akadémiai Kiadó, Budapest/Elsevier.] (B) and (C) Vasopressin-immunoreactive (B) and oxytocin-immunoreactive (C) nerve fibers in the pineal of hibernating hedgehog (arrows). Bar = 10 μm. [From Nürnberger and Korf, 1981, courtesy Springer Verlag.]

bers, two important questions remain to be answered: (1) Is there a colocalization of neuropeptides and serotonin with noradrenaline in sympathetic nerve fibers? (2) Is there convincing evidence that peptidergic and serotoninergic nerve fibers may originate from extrasympathetic (central) sources? Serotonin, already shown to be present by uptake in sympathetic

nerve fibers, apparently is also the neurotransmitter of certain central pinealopetal projections, as may be inferred from the persistence of some
serotoninergic nerve fibers in the pineal after superior cervical ganglionectomy (Matsuura et al., 1983b). Substance-P-immunoreactive nerve fibers
were not affected by bilateral ablation of the superior cervical ganglia; they
belong to central pinealopetal projections (Rønnekleiv and Kelley, 1984). A
central origin may also be considered for the vasopressin-, oxytocin-, and
neurophysin-immunoreactive fibers due to their course via the stria medullaris and the habenular and posterior commissures. These fibers may originate from neurons of the paraventricular nucleus shown to innervate the
pineal organ in tracer experiments (see p. 127).

As far as the functional significance of peptidergic nerve fibers is concerned, the best results have been obtained with VIP that was shown to
stimulate the serotonin-N-acetyltransferase activity via a non-beta-receptor
mechanism (Yuwiler, 1983). This finding conforms to the biochemical and
autoradiographic demonstration of specific binding sites for VIP in the pineal
organ (Møller et al., 1985). Further studies are required to provide clear-cut
evidence for the functional role of the remaining peptidergic fibers in the
mammalian pineal.

The morphological observations on the central pinealopetal projections in
mammals are in accord with electrophysiological results (Dafny, 1980;
Semm, 1983). They may indicate that (1) photic stimuli are transmitted to the
pineal not only via sympathetic nerve fibers but also via more direct routes,
(2) the function of the pineal organ is regulated by environmental cues different from the photoperiod, and (3) the neuroendocrine apparatus of the hypothalamus participates in the control of pineal function via direct neural connections, which may have the character of neural short-loop feedback.

An additional feedback control may be exerted via endocrine (humoral)
pathways, as may be concluded from the demonstration of highly specific
binding sites for several reproductive hormones in the pineal (Cardinali,
1979).

All of these results indicate that our knowledge of the regulation of the
function of the pineal organ is still fragmentary.

J. Vascular Permeability (Blood–Brain Barrier) in the Pineal Organ

The vascular permeability in the pineal deserves particular attention with
respect to the neuroendocrine capacity of the pineal organ and the presumed
hormonal feedback control of pineal function.

Electron microscopic studies have revealed considerable interspecific variation in the ultrastructure of pineal capillaries. Whereas in some mammals
(mouse, rat, ground squirrel) fenestrated capillaries are frequently observed,
in a great number of other species the pineal contains exclusively or predominantly capillaries with nonfenestrated endothelial cells (Mongolian gerbil:

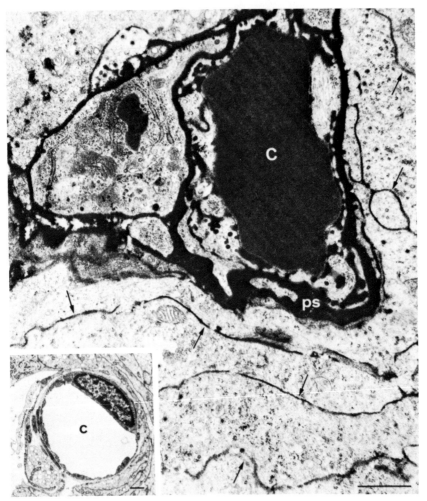

Fig. 8. Vascular permeability in the pineal organ of the mouse. Extravasation of intravenously applied HRP into the pineal parenchyma 5 min after injection. c, capillary; ps, perivascular space; arrows, intercellular spaces filled with HRP. Inset, fenestrated capillary in the hamster pineal. Bar = 1 μm. [Courtesy of M. Møller, Copenhagen.]

Welsh and Beitz, 1981). In certain rodents [golden hamster (Hewing and Bergmann, 1985)] fenestrated and nonfenestrated capillaries are arranged in a peculiar topographical pattern: the fenestrated capillaries predominate in the superficial portion; the nonfenestrated ones predominate in the deep portion of the pineal organ.

However, at this point it must be emphasized that the permeability of the blood–brain barrier cannot be inferred simply from the ultrastructure of the capillary endothelium. As shown by intravital tracer techniques, widely used for investigation of the blood–brain barrier, small hydrophilic molecules (e.g., lizamine green) or proteins (microperoxidase, HRP) readily extravasate from pineal capillaries (Fig. 8), irrespective of whether the latter belong to the fenestrated or nonfenestrated type (Møller et al., 1978b; Møller and van Veen, 1981; Welsh and Beitz, 1981). These findings provide convincing evidence that the pineal of the species investigated (hamster, mouse, Mongolian gerbil) is devoid of a blood–brain barrier for these substances. Considering the prolonged extravasation of Myofer (Dretzki, 1971; Wartenberg et al., 1972) it should not be overlooked that the permeability of the blood–brain barrier of the pineal organ may vary with the physicochemical properties of the tracer used.

With respect to the evolution of pinealocytes, investigations of the blood–brain barrier in submammalian vertebrates are of considerable interest. In the rainbow trout, systemically applied HRP and ferritin pass the fenestrated capillaries and are subsequently taken up by pineal supportive and sensory cells. By the use of tannic acid it has been shown that the junctional complexes between the neck portion of the pineal photoreceptor cells and the supportive cells form a structural barrier, preventing the intercellular passage of substances from the capillaries and perivascular compartments into the pineal lumen containing cerebrospinal fluid (Omura et al., 1985).

Further studies with the use of a wide spectrum of tracer techniques are required to elucidate whether the lack of a blood–brain barrier is a general feature in all vertebrate classes.

IV. CONCLUSIONS AND BIOMEDICAL IMPLICATIONS

As reviewed in this chapter, the pineal complex of vertebrates has undergone considerable changes during phylogeny. However, comparative studies have shown that, irrespective of these transformations, the pineal organ displays several characteristics that are common in all vertebrates: (1) The pineal organ is in close contact with the cerebrospinal fluid and belongs to the group of circumventricular organs, as defined by Bargmann and Hofer (Hofer, 1959, 1965; cf. Leonhardt, 1980). (2) The presence of daily rhythms in enzymatic activities and concentration of indoles, including melatonin, is a remarkable feature in all vertebrate classes (Quay, 1974). (3) The pineal is an important component of photoneuroendocrine systems (Oksche and Hartwig, 1979), which, according to the basic definition of Scharrer (1964), serve to integrate photoperiodic and intrinsic information for the control of autonomic and neuroendocrine functions. (4) All manifestations of pinealocytes (pineal photoreceptors of poikilotherms, modified pineal photoreceptors of sauropsids, pinealocytes of mammals) share several molecular markers with retinal photoreceptors (e.g., immunoreactive S-antigen; immunoreactive opsin).

These findings are of biomedical significance for several reasons.

1. They indicate that retinal photoreceptors and pinealocytes may develop from a common diencephalic primordium (see also von Frisch, 1911) and that mammalian pinealocytes have developed from pineal photoreceptor

cells of poikilotherms, as was suggested by Collin (1971) and Oksche (1971).

2. The apparent similarities in cell biology and function between the pineal organ and the retina relate to the phenomena of photoneuroendocrine regulation, circadian rhythms, and rhythm generators. There is growing evidence that disturbances of the circadian organization in humans are correlated with several affective and endocrine diseases (e.g., mania, depression, schizophrenia, insomnia) (see Wetterberg, 1979; Vaughan, 1984). However, the structural and functional fundament of these events is not yet established. Thus, continued basic research aimed at further elucidation of the phylogenetic development and the molecular, neuronal, and neuroendocrine mechanisms involved in the generation of biological rhythms may reveal data important for clinical medicine.

3. The different manifestations of pinealocytes may be considered as excellent models for studying molecular aspects of photoneuroendocrine regulation with a wide range of newly developed methods. In possessing both photoreceptive and neuroendocrine capacities, pinealocytes appear to be unique among endocrine cells (neurons). Results of comparative research lead to the suggestion that during the evolution of pineal systems, the pinealocytes have developed as *photoneuroendocrine cells* (Oksche, 1971, 1983; Oksche and Hartwig, 1979). It has been postulated that photoneuroendocrine cells are—at the cellular level—capable of perceiving photic stimuli and translating these stimuli into a neuroendocrine response. Experimental evidence for the existence of photoneuroendocrine cells is provided by biochemical studies with cultured pineal organs of the chicken (Deguchi, 1981). The concept of photoneuroendocrine cells may provide a useful guideline in further research. Experiments may be designed to clarify whether photoneuroendocrine cells are a common element in all vertebrate (and also invertebrate?) classes. In this context, it may be considered that pineal photoreceptors of poikilotherms display characteristics pointing towards their neuroendocrine capability, and that neuroendocrine mammalian pinealocytes contain proteins specifically involved in phototransduction. Thus, the function of the mammalian pinealocyte may resemble that of pinealocytes in lower vertebrates more closely than was previously surmised, and experiments with submammalian vertebrates may result in a more complete understanding of the cellular and molecular mechanisms in mammalian pinealocytes. These studies may also contribute to the problem of whether pinealocytes elaborate active principles different from melatonin and whether these substances participate in photoneuroendocrine regulation.

4. Finally, comparative investigations establishing similarities between the different manifestations of pinealocytes and retinal photoreceptors may increase our knowledge on the origin, development, molecular characteristics, and the differential diagnosis of certain brain tumors (retinoblastoma, pineocytoma, pineoblastoma, medulloblastoma) in humans. Preliminary results have shown that neoplastic cells of a pineocytoma removed from a

49-year-old patient display S-antigen-like immunoreaction (Korf *et al.*, 1986a), as well as opsin- and serotoninlike immunoreactions (H.-W. Korf, unpublished). Other tumors of the pineal region of different origin (glial tumors, teratomas, germinoblastomas) examined thus far are immunonegative.

Very strikingly, we have been able to demonstrate S-antigen- and opsinlike immunoreactions in certain medulloblastomas (H.-W. Korf, W. Schachenmayr, M. Czerwionka, I. Gery, and D. C. Klein, unpublished). Medulloblastomas are extremely malignant and preferentially occur in the midline of the cerebellum (vermis cerebelli) in children (8–12 years of age). Our immunocytochemical findings suggest that certain types of medulloblastomas may originate from a primordium which in normal development gives rise to photoreceptor or photoneuroendocrine cells. Thus, further comparative *in vivo* and *in vitro* studies on the phylogenetic and ontogenetic origin and differentiation of photoneuroendocrine pinealocytes may also provide a useful model helping to interpret the genesis of these tumors in humans.

ACKNOWLEDGMENTS

The experimental studies of the authors were supported by the Deutsche Forschungsgemeinschaft (Ok 1/25-3; Ko 758/2-2; 2-4). The authors are greatly indebted to Dr. Morten Møller, Institute of Medical Anatomy, Department B, University of Copenhagen, for providing previously unpublished electron micrographs. Excellent technical assistance by G. Döll, R. Liesner, and K. Michael is gratefully acknowledged.

REFERENCES

Arstila, A. U. (1967). Electron microscopic studies on the structure and histochemistry of the pineal gland of the rat. *Neuroendocrinology* **2**, Suppl. 1–101.

Axelrod, J., and Weissbach, H. (1960). Enzymatic O-methylation of N-acetylserotonin to melatonin. *Science* **131**, 1312.

Axelrod, J., Fraschini, F., and Velo, G. P., eds. (1983). "The Pineal Gland and its Endocrine Role." Plenum, New York.

Bargmann, W. (1943). Die Epiphysis cerebri. *In* "Handbuch der mikroskopischen Anatomie des Menschen" (W. von Möllendorff, ed.), Vol. VI, Part 4, pp. 309–502. Springer-Verlag, Berlin and New York.

Bayrhuber, H. (1972). Über die Synapsenformen und das Vorkommen von Acetylcholinesterase in der Epiphyse von *Bombina variegata* (L.). *Z. Zellforsch. Mikrosk. Anat.* **126**, 278–296.

Benson, B., and Krasovich, M. (1977). Circadian rhythm in the number of granulated vesicles in the pinealocytes of mice. Effects of sympathectomy and melatonin treatment. *Cell Tissue Res.* **184**, 499–506.

Bertler, A., Falck, B., and Owman, C. (1964). Studies on 5-hydroxytryptamine stores in pineal gland of the rat. *Acta Physiol. Scand.* **63**, Suppl. 239, 1–18.

Binkley, S. (1981). Pineal biochemistry: Comparative aspects and circadian rhythms. *In* "The Pineal Gland: Anatomy and Biochemistry" (R. J. Reiter, ed.), pp. 155–172. CRC Press, Boca Raton, Florida.

Bischoff, M. B. (1969). Photoreceptoral secretory structures in the avian pineal organ. *J. Ultrastruct. Res.* **28**, 16–26.

Björklund, A., Owman, C., and West, K. A. (1972). Peripheral sympathetic innervation and serotonin cells in the habenular region of the rat brain. *Z. Zellforsch. Mikrosk. Anat.* **127**, 570–579.

Bubenik, G. A., Brown, G. M., Uhlir, I., and Grota, L. J. (1974). Immunohistological localization of N-acetylindolealkylamines in pineal gland, retina and cerebellum. *Brain Res.* **81**, 233–242.

Buijs, R. M., and Pévet, P. (1980). Vasopressin- and oxytocin-containing nerve fibers in the pineal gland and subcommissural organ of the rat. *Cell Tissue Res.* **205**, 11–17.

Buzdygon, B., Philp, N., Zigler, J. S., Gery, I., and Zuckerman, R. (1985). Identity of S-antigen and the 48 kilodalton protein in retinal rod outer segments. *Invest. Ophthalmol. Visual Sci.* **26**, Suppl., 293.

Cardinali, D. P. (1979). Models in neuroendocrinology. Neurohumoral pathways of the pineal gland. *Trends in Neuroscience,* Oct., pp. 250–253.

Clabough, J. W. (1971). Ultrastructural features of the pineal gland in normal and light deprived golden hamsters. *Z. Zellforsch. Mikrosk. Anat.* **114**, 151–164.

Collin, J. P. (1971). Differentiation and regression of the cells of the sensory line in the epiphysis cerebri. *In* "The Pineal Gland" (G. E. W. Wolstenholme and J. Knight, eds.), pp. 79–125. Little, Brown, Boston, Massachusetts.

Collin, J. P. (1979). Recent advances in pineal cytochemistry. Evidence of the production of indoleamines and proteinaceous substances by rudimentary photoreceptor cells and pinealocytes of amniota. *In* "The Pineal Gland of Vertebrates Including Man" (J. A. Kappers and P. Pévet, eds.), pp. 271–296. Elsevier, Amsterdam.

Collin, J. P. (1981). New data and vistas on the mechanisms of secretion of proteins and indoles in the mammalian pinealocyte and its phylogenetic precursors; the pinealin hypothesis and preliminary comments on membrane traffic. *In* "The Pineal Organ: Photobiology-Biochronometry-Endocrinology" (A. Oksche and P. Pévet, eds.), pp. 187–210. Elsevier, Amsterdam.

Collin, J. P., and Meiniel, M. (1973). Métabolisme des indoléamines dans l'organe pinéal de Lacerta (Reptiles, Lacertiliens). II. L'activité MAO et l'incorporation de 5-HIP-³H, et de 5-HT-³H, dans les conditions normales et expérimentales. *Z. Zellforsch. Mikrosk. Anat.* **145**, 331–361.

Collin, J. P., and Oksche, A. (1981). Structural and functional relationships in the nonmammalian pineal gland. *In* "The Pineal Gland: Anatomy and Biochemistry" (R. J. Reiter, ed.), Vol. 1, pp. 27–67. CRC Press, Boca Raton, Florida.

Collin, J. P., Calas, A., and Juillard, M. T. (1976). The avian pineal organ. Distribution of exogenous indoleamines: a qualitative study of the rudimentary photoreceptor cells by electron microscopic radioautography. *Exp. Brain Res.* **25**, 15–33.

Dafny, N. (1980). Photic input to rat pineal gland conveyed by both sympathetic and central afferents. *J. Neural Transm.* **48**, 203–208.

David, G. F. X., and Anand Kumar, T. C. (1978). Histochemical localization of cholinesterase in neural tissue of pineal in rhesus monkey. *Experientia* **34**, 1067–1068.

David, G. F. X., and Herbert, J. (1973). Experimental evidence for a synaptic connection between habenula and pineal ganglion in the ferret. *Brain Res.* **64**, 327–343.

David, G. F. X., Umberkoman, B., Kumar, K., and Anand-Kumar, T. C. (1975). Neuroendocrine significance of the pineal. *In* "Brain-Endocrine Interaction" (K. M. Knigge, D. E. Scott, H. Kobayashi, and S. Ishii, eds.), Vol. 2, pp. 365–375. Karger, Basel.

Deguchi, T. (1981). Rhodopsin-like photosensitivity of the isolated chicken pineal gland. *Nature (London)* **290**, 706.

Diehl, B. J. M. (1978). Occurrence and regional distribution of calcareous concretions in the rat pineal gland. *Cell Tissue Res.* **195**, 359–366.

Dodt, E. (1973). The parietal eye (pineal and parietal organs) of lower vertebrates. In "Handbook of Sensory Physiology" (R. Jung, ed.), Vol. 7, Part 3B, pp. 113–140. Springer-Verlag, Berlin and New York.

Dretzki, J. (1971). Licht-und elektronenmikroskopische Untersuchungen zum Problem der Blut-Hirnschranke zirkumventrikulärer Organe der Ratte nach Behandlung mit Myofer. Z. Anat. Entwicklungsgesch. 134, 278–297.

Eakin, R. M., and Westfall, J. A. (1959). Fine structure of the retina in the reptilian third eye. J. Biophys. Biochem. Cytol. 6, 133–134.

Eakin, R. M., Quay, W. B., and Westfall, J. A. (1963). Cytological and cytochemical studies on the frontal and pineal organs of the tree frog, Hyla regilla. Z. Zellforsch. Mikrosk. Anat. 59, 663–683.

Edinger, T. (1955). The size of the parietal foramen and organ in reptiles. A rectification. Bull. Mus. Comp. Zool. 114, 1–34.

Ekström, P. (1984). Central neural connections of the pineal organ and retina in the teleost Gasterosteus aculeatus L. J. Comp. Neurol. 226, 321–336.

Ekström, P., and Korf, H. W. (1985). Morphology of pineal neurons projecting to the brain in the rainbow trout, Salmo gairdneri Richardson (Teleostei), as visualized by retrograde filling with horseradish peroxidase in vitro. Cell Tissue Res. 240, 693–700.

Ekström, P., and van Veen, Th. (1984). Pineal neural connections with the brain in two teleosts, the crucian carp and the european eel. J. Pineal Res. 1, 245–261.

Eldred, W. D., and Nolte, J. (1981). Multiple classes of photoreceptors and neurons in the frontal organ of Rana pipiens. J. Comp. Neurol. 203, 269–296.

Eldred, W. D., Finger, Th.E., and Nolte, J. (1980). Central projections from the frontal organ of Rana pipiens, as demonstrated by the anterograde transport of horseradish peroxidase. Cell Tissue Res. 211, 215–222.

Engbretson, G. A., Reiner, A., and Brecha, N. (1981). Habenular asymmetry and the central connections of the parietal eye of the lizard. J. Comp. Neurol. 198, 155–165.

Falcon, J. (1979). Unusual distribution of neurons in the pike pineal organ. In "The Pineal Gland of Vertebrates Including Man" (J. A. Kappers and P. Pévet, eds.), pp. 89–91. Elsevier, Amsterdam.

Falcon, J. (1984). Identification et propriétés des cellules photoneuroendocrines de l'organe pinéal. Thesis, University of Poitiers.

Flight, W. F. G. (1973). Observations on the pineal ultrastructure of the urodele, Diemictylus viridescens viridescens. Proc. Kn. Ned. Akad. Wet., Ser. C 76, 425–448.

Flight, W. F. G., and van Donselaar, E. (1975). Ultrastructural aspects of the incorporation of ³H-vitamin A in the pineal organ of the urodele, Diemictylus viridescens viridescens. Proc. Kn. Ned. Akad. Wet., Ser. C. 78, 130–142.

Freund, D., Arendt, J., and Vollrath, L. (1978). Tentative immunohistochemical demonstration of melatonin in the rat pineal gland. Cell Tissue Res. 181, 239–257.

Gladstone, R. J., and Wakeley, C. P. G. (1940). "The Pineal Gland." Baillière, London.

Guerillot, C., Leffray, P., Pfister, A., and Da Lage, C. (1979). Contribution to the study of the pineal stalk nerve fibers in the rat. In "The Pineal Gland of Vertebrates Including Man" (J. A. Kappers and P. Pévet, eds.), pp. 97–102. Elsevier, Amsterdam.

Guerillot, C., Pfister, A., Müller, J., and Da Lage, C. (1982). Recherche de l'origine des fibres nerveuses extraorthosympathiques innervant l'épiphyse du rat (étude du transport retrograde de la peroxydase de raifort). Reprod. Nutr. Dev. 22, 371–378.

Gundy, G. C., and Wurst, G. Z. (1976). The occurrence of parietal eyes in recent Lacertilia. J. Herpetol. 10, 113–121.

Hafeez, M. A., and Quay, W. B. (1969). Histochemical and experimental studies of 5-hydroxytryptamine in pineal organs of teleosts (Salmo gairdneri and Atherinopsis californiensis). Gen Comp. Endocrinol. 13, 211–217.

Hafeez, M. A., and Zerihun, L. (1974). Studies on central projections of the pineal nerve tract in rainbow trout, *Salmo gairdneri* Richardson, using cobalt chloride iontophoresis. *Cell Tissue Res.* **154**, 485–510.

Hafeez, M. A., and Zerihun, L. (1976). Autoradiographic localization of ³H-5-HTP and ³H-5-HT in the pineal organ and circumventricular areas in the rainbow trout, *Salmo gairdneri* Richardson. *Cell Tissue Res.* **170**, 61–76.

Hamasaki, D. I., and Eder, D. J. (1977). Adaptive radiation of the pineal system. *In* "Handbook of Sensory Physiology" (F. Crescitelli, ed.), Vol. 7, Part 5, pp. 497–548. Springer-Verlag, Berlin and New York.

Hartwig, H. G. (1984). Cyclic renewal of whole pineal photoreceptor outer segments. *Ophthalmic Res.* **16**, 102–106.

Hartwig, H. G., and Baumann, C. (1974). Evidence for photosensitive pigments in the pineal complex of the frog. *Vision Res.* **14**, 597–598.

Hartwig, H. G., and Oksche, A. (1982). Neurobiological aspects of extraretinal photoreceptive systems: Structure and function. *Experientia* **38**, 991–996.

Hartwig, H. G., and Reinhold, C. (1981). Microspectrofluorometry of biogenic monoamines in pineal systems. *In* "The Pineal Organ: Photobiology-Biochronometry-Endocrinology" (A. Oksche and P. Pévet, eds.), pp. 237–246. Elsevier, Amsterdam.

Hedlund, L., and Nalbandov, A. V. (1969). Innervation of the avian pineal body. *Am. Zool.* **9**, 1090.

Herwig, H. J. (1981). The pineal organ. An ultrastructural and biochemical study on the pineal organ of *Hemigrammus caudovittatus* and other closely related characid fish species with special reference to the Mexican blind cave fish *Astyanax mexicanus*. Thesis, University of Utrecht.

Hewing, M. (1976). Die postnatale Entwicklung der Epiphysis cerebri beim Goldhamster. *Verh. Anat. Ges.* **70**, 85–92.

Hewing, M. (1978). A liquor contacting area in the pineal recess of the golden hamster (*Mesocricetus auratus*). *Anat. Embryol.* **153**, 295–304.

Hewing, M., and Bergmann, M. (1985). Differential permeability of pineal capillaries to lanthanum ion in rat (*Rattus norvegicus*), gerbil (*Meriones unguiculatus*) and golden hamster (*Mesocricetus auratus*). *Cell Tissue Res.* **241**, 149–154.

Hofer, H. O. (1959). Zur Morphologie der circumventrikulären Organe des Ziwschenhirns der Säugetiere. *Zool. Anz.* **22**, 202–251.

Hofer, H. O. (1965). Circumventrikuläre Organe des Zwischenhirns. *In* "Primatologia" (H. O. Hofer, A. H. Schultz, and D. Starck, eds.), Vol. II, Part 2, No. 13, pp. 1–104. Karger, Basel.

Hofer, H. O., Merker, G., and Oksche, A. (1976). Atypische Formen des Pinealorgans der Säugetiere. *Verh. Anat. Ges.* **70**, 97–102.

Hopsu, V. K., and Arstila, A. V. (1965). An apparent somatosomatic structure in the pineal gland of the rat. *Exp. Cell Res.* **37**, 484–487.

Hülsemann, M. (1967). Vergleichende histologische Untersuchungen über das Vorkommen von Gliafasern in der Epiphysis cerebri von Säugetieren. *Acta Anat.* **66**, 249–278.

Hülsemann, M. (1971). Development of the innervation in the human pineal organ. Light and electron microscopic investigations. *Z. Zellforsch. Mikrosk. Anat.* **115**, 396–415.

Jaim-Etcheverry, G., and Zieher, L. M. (1983). Ultrastructural evidence for monoamine uptake by vesicles of pineal sympathetic nerves immediately after their stimulation. *Cell Tissue Res.* **233**, 463–470.

Japha, J. L., Eder, T. J., and Goldsmith, E. D. (1976). Calcified inclusions in the superficial pineal gland of the Mongolian gerbil, *Meriones unguiculatus*. *Acta Anat.* **94**, 533–544.

Juillard, M. T., and Collin, J. P. (1978). The avian pineal organ: Evidence for a proteinaceous component in the secretion granules of the rudimentary photoreceptor cells. An ultracytochemical and pharmacological study in the parakeet. *Biol. Cell.* **31**, 51–58.

Juillard, M. T., Hartwig, H. G., and Collin, J. P. (1977). The avian pineal organ. Distribution of endogenous monoamines; a fluorescence microscopic, microspectrofluorimetric and pharmacological study in the parakeet. *J. Neural Transm.* **40**, 269–287.

Juillard, M. T., Collin, J. P., Balemans, M. G. M., and Queau, A. (1984). In-vitro uptake and metabolism of ³H-5-hydroxytryptamine in the pineal glands of the rabbit, rat and hamster. *Cell Tissue Res.* **235**, 539–549.

Kalsow, C. M., and Wacker, W. B. (1973). Localization of a uveitogenic soluble retinal antigen in the normal guinea pig eye by an indirect fluorescent antibody technique. *Int. Arch. Allergy Appl. Immunol.* **44**, 11–20.

Kalsow, C. M., and Wacker, W. B. (1977). Pineal reactivity of anti-retina sera. *Invest. Ophthalmol. Visual Sci.* **16**, 181–184.

Kappers, J. A. (1960). The development, topographical relations and innervation of the epiphysis cerebri in the albino rat. *Z. Zellforsch. Mikrosk. Anat.* **52**, 163–215.

Kappers, J. A. (1965). Survey of the innervation of the epiphysis cerebri and the accessory pineal organs of vertebrates. *Prog. Brain Res.* **10**, 87–153.

Kappers, J. A. (1967). The sensory innervation of the pineal organ in the lizard, *Lacerta viridis*, with remarks on its position in the trend of pineal phylogenetic structural and functional evolution. *Z. Zellforsch. Mikrosk. Anat.* **81**, 581–618.

Kelly, D. E. (1963). The pineal organ of the newt. A developmental study. *Z. Zellforsch. Mikrosk. Anat.* **58**, 693–713.

Kelly, D. E. (1971). Developmental aspects of amphibian pineal systems. *In* "The Pineal Gland" (G. E. W. Wolstenholme and J. Knight, eds.), pp. 53–77. Little, Brown, Boston, Massachusetts.

Kenny, G. C. T. (1961). The "nervus conarii" of the monkey. (An experimental study.) *J. Neuropathol. Exp. Neurol.* **20**, 563–570.

Kenny, G. C. T. (1965). The innervation of the mammalian pineal body. (A comparative study.) *Proc. Aust. Assoc. Neurol.* **3**, 133–140.

Kitay, J. I., and Altschule, M. D. (1954). "The Pineal Gland." Harvard Univ. Press, Cambridge, Massachusetts.

Klein, D. C. (1982). "Melatonin Rhythm Generating System: Developmental Aspects." Karger, Basel.

Klein, D. C., Smoot, R., Weller, J. L., Higa, S., Markey, S. P., Creed, C. J., and Jacobwitz, D. M. (1983). Lesions of the paraventricular nucleus area of the hypothalamus disrupt the suprachiasmatic-spinal cord circuit in the melatonin rhythm generating system. *Brain Res. Bull.* **10**, 647–652.

Korf, H. W. (1974). Acetylcholinesterase-positive neurons in the pineal and parapineal organs of the rainbow trout, *Salmo gairdneri* (with special reference to the pineal tract). *Cell Tissue Res.* **155**, 475–489.

Korf, H. W. (1976). Histological, histochemical and electron microscopical studies on the nervous apparatus of the pineal organ in the tiger salamander, *Ambystoma tigrinum*. *Cell Tissue Res.* **174**, 475–497.

Korf, H. W., and Møller, M. (1984). The innervation of the mammalian pineal gland with special reference to central pinealopetal projections. *Pineal Res. Rev.* **2**, 41–86.

Korf, H. W., and Møller, M. (1985). The central innervation of the mammalian pineal organ. *In* "The Pineal Gland: Current Status of Pineal Research" (B. Mess, Cs.Ruzsas, L. Tima, and P. Pévet, eds.), pp. 47–69. Akadémiai Kiadó, Budapest.

Korf, H. W., and Vigh-Teichmann, I. (1984). Sensory and central nervous elements in the avian pineal organ. *Ophthalmic Res.* **16**, 96–101.

Korf, H. W., and Wagner, U. (1980). Evidence for a nervous connection between the brain and the pineal organ in the guinea pig. *Cell Tissue Res.* **209**, 505–510.

Korf, H. W., and Wagner, U. (1981). Nervous connections of the parietal eye in adult *Lacerta s. sicula* Rafinesque as demonstrated by anterograde and retrograde transport of horseradish peroxidase. *Cell Tissue Res.* **219**, 567–584.

Korf, H. W., Liesner, R., Meissl, H., and Kirk, A. (1981). Pineal complex of the clawed toad, *Xenopus laevis* Daud.: Structure and function. *Cell Tissue Res.* **216**, 113–130.

Korf, H. W., Zimmerman, N. H., and Oksche, A. (1982). Intrinsic neurons and neural connections of the pineal organ of the house sparrow, *Passer domesticus*, as revealed by anterograde and retrograde transport of horseradish peroxidase. *Cell Tissue Res.* **222**, 243–260.

Korf, H. W., Foster, R. G., Ekström, P., and Schalken, J. J. (1985a). Opsin-like immunoreaction in the retinae and pineal organs of four mammalian species. *Cell Tissue Res.* **242**, 645–648.

Korf, H. W., Møller, M., Gery, I., Zigler, J. S., and Klein, D. C. (1985b). Immunocytochemical demonstration of retinal S-antigen in the pineal organ of four mammalian species. *Cell Tissue Res.* **239**, 81–85.

Korf, H. W., Klein, D. C., Zigler, J. S., Gery, I. and Schachenmayr, W. (1986a). S-antigen-like immunoreactivity in a human pineocytoma. *Acta Neuropathol.* **69**, 165–167.

Korf, H. W., Oksche, A., Ekström, P., van Veen, T., Zigler, J. S., and Klein, D. C. (1986b). Pinealocyte projections into the mammalian brain revealed with S-antigen antiserum. *Science* **231**, 735–737.

Korf, H. W., Oksche, A., Ekström, P., van Veen, T., Zigler, J. S., Gery, I., Stein, P., and Klein, D. C. (1986c). S-antigen immunocytochemistry. *In* "Pineal-Retinal Relationships" (P. O'Brien and D. C. Klein, eds.) Academic Press, New York (in press).

Kristić, R. (1976). A combined scanning and transmission electron microscopic study and electron probe microanalysis of human pineal acervuli. *Cell Tissue Res.* **174**, 129–137.

Kühn, H. (1978). Light-regulated binding of rhodopsin kinase and other proteins to cattle photoreceptor membranes. *Biochemistry* **17**, 4389–4395.

Kühn, H. (1984). Interactions between photoexcited rhodopsin and light-activated enzymes in rods. *Prog. Retinal Res.* **3**, 123–156.

Kummer-Trost, E. (1956). Die Bildungen des Zwischenhirndaches der Agamidae nebst Bemerkungen über die Lagebeziehungen des Vorderhirns. *Morph. Jahrb.* **97**, 143–192.

Leonhardt, H. (1967). Über axonähnliche Fortsätze, Sekretbildung und Extrusion der hellen Pinealozyten des Kaninchens. *Z. Zellforsch. Mikrosk. Anat.* **82**, 307–320.

Leonhardt, H. (1980). Ependym und circumventrikuläre Organe. *In* "Handbuch der mikroskopischen Anatomie des Menschen" (A. Oksche and L. Vollrath, eds.), Vol. IV, Part 10, pp. 177–666. Springer-Verlag, Berlin and New York.

Lerner, A. B., Case, J. D., Takahashi, Y., Lee, Y., and Mori, W. (1958). Isolation of melatonin, the pineal gland factor that lightens melanocytes. *J. Am. Chem. Soc.* **80**, 2587.

Lin, H. S., Hwang, H. B., and Tseng, C. Y. (1975). Fine structural changes in the hamster pineal gland after blinding and superior cervical ganglionectomy. *Cell Tissue Res.* **158**, 285–299.

Lukaszyk, A., and Reiter, R. J. (1975). Histophysiological evidence for the secretion of polypeptides by the pineal gland. *Am. J. Anat.* **143**, 451–464.

McNulty, J. A. (1984). Functional morphology of the pineal complex in cyclostomes, elasmobranchs, and bony fishes. *Pineal Res. Rev.* **2**, 1–40.

McNulty, J. A., and Hazlett, J. C. (1980). The pineal region in the opossum *Didelphis virginiana*. I. Ultrastructural observations. *Cell Tissue Res.* **207**, 109–121.

Matsushima, S., and Reiter, R. J. (1978). Electron microscopic observations on neuron-like cells in the ground squirrel pineal gland. *J. Neural Transm.* **45**, 63–73.

Matsuura, T., and Sano, Y. (1983). Distribution of monoamine-containing nerve fibers in the pineal organ of untreated and sympathectomized dogs. *Cell Tissue Res.* **234**, 519–531.

Matsuura, T., Kawata, M., Yamada, H., Kojima, M., and Sano, Y. (1983a). Immunohistochemical studies on the peptidergic nerve fibers in the pineal organ of the dog. *Arch. Histol. Jpn.* **46**, 373–379.

Matsuura, T., Takeuchi, Y., and Sano, Y. (1983b). Immunohistochemical and electron microscopic studies on serotonin containing nerve fibers in the pineal organ of the rat. *Biomed. Res.* **4**, 261–270.

Mautner, W. (1965). Studien an der Epiphysis cerebri und am Subcommissuralorgan der Frösche. Mit Lebendbeobachtung des Epiphysenkreislaufes, Totalfärbung des Subcommissuralorgans und Durchtrennung des Reissnerschen Fadens. *Z. Zellforsch. Mikrosk. Anat.* **67,** 234–270.

Meiniel, A. (1981). New aspects of the phylogenetic evolution of sensory cell lines in the vertebrate pineal complex. *In* "The Pineal Organ: Photobiology-Biochronometry-Endocrinology" (A. Oksche and P. Pévet, eds.), pp. 27–48. Elsevier, Amsterdam.

Meiniel, A., Collin, J. P., and Hartwig, H. G. (1973). Pinéale et troisième oeil de *Lacerta vivipara* (J.), au cours de la vie embryonaire et postnatale. Étude cytophysiologique des monoamines en microscopie de fluorescence et an microspectrofluorimetrie. *Z. Zellforsch. Mikrosk. Anat.* **144,** 89–115.

Meissl, H., and Dodt, E. (1981). Comparative physiology of pineal photoreceptor organs. *In* "The Pineal Organ: Photobiology-Biochronometry-Endocrinology" (A. Oksche and P. Pévet, eds.), pp. 61–80. Elsevier, Amsterdam.

Meissl, H., and Ueck, M. (1980). Extraocular photoreceptor in the aquatic turtle, *Pseudemys scripta elegans. J. Comp. Physiol.* **140,** 173–179.

Meissl, H., Donley, C. S., and Wissler, J. H. (1978). Free amino acids and amines in the pineal organ of the rainbow trout (*Salmo qairdneri*). Influence of light and dark. *Comp. Biochem. Physiol.* **61C,** 401–405.

Mirshahi, M., Faure, J. P., Brisson, P., Falcon, J., Guerlotte, J., and Collin, J. P. (1984). S-antigen immunoreactivity in retinal rods and cones and pineal photosensitive cells. *Biol. Cell.* **52,** 195–198.

Møller, M. (1974). The ultrastructure of the human fetal pineal gland. I. Cell types and blood vessels. *Cell Tissue Res.* **152,** 13–30.

Møller, M. (1976). The ultrastructure of the human fetal pineal gland. II. Innervation and cell junctions. *Cell Tissue Res.* **169,** 7–21.

Møller, M. (1978). Presence of a pineal nerve (nervus pinealis) in the human fetus: A light and electron microscopical study of the innervation of the pineal gland. *Brain Res.* **154,** 1–12.

Møller, M., and Korf, H. W. (1983a). Central innervation of the pineal organ of the Mongolian gerbil. *Cell Tissue Res.* **230,** 259–272.

Møller, M., and Korf, H. W. (1983b). The origin of central pinealopetal nerve fibers in the Mongolian gerbil as demonstrated by the retrograde transport of horseradish peroxidase. *Cell Tissue Res.* **230,** 273–287.

Møller, M., and van Veen, Th. (1981). Vascular permeability in the pineal gland of the Mongolian gerbil and the Djungarian hamster. *In* "Human Reproduction" (K. Semm and L. Mettler, eds.), pp. 539–543. Elsevier, Amsterdam.

Møller, M., Møllgard, K., and Kimble, J. E. (1975). Presence of a pineal nerve in sheep and rabbit fetuses. *Cell Tissue Res.* **158,** 451–459.

Møller, M., Ingild, A., and Bock, E. (1978a). Immunohistochemical demonstration of S-100 protein and GFA protein in interstitial cells of rat pineal gland. *Brain Res.* **140,** 1–13.

Møller, M., van Deurs, B., and Westergaard, E. (1978b). Vascular permeability to proteins and peptides in the mouse pineal gland. *Cell Tissue Res.* **195,** 1–15.

Møller, M., Mikkelsen, J. D., Fahrenkrug, J. and Korf, H. W. (1985). The presence of vasoactive intestinal polypeptide (VIP)-like immunoreactive nerve fibers and VIP-receptors in the pineal gland of the Mongolian gerbil (*Meriones unguiculatus*). *Cell Tissue Res.* **241,** 333–340.

Møllgard, K., and Møller, M. (1973). On the innervation of the human fetal pineal gland. *Brain Res.* **52,** 428–432.

Moore, R. Y. (1978). Neural control of pineal function in mammals and birds. *J. Neural Transm., Suppl.* **13,** 47–58.

Moore, R. Y., and Klein, D. C. (1974). Visual pathways and the central neural control of circadian rhythm in pineal serotonin N-acetyltransferase activity. *Brain Res.* **71,** 17–33.

Morita, Y. (1966). Entladungsmuster pinealer Neurone der Regenbogenforelle (*Salmo irideus*) bei Beleuchtung des Zwischenhirns. *Pfluegers Arch. Gesamte Physiol. Menschen Tiere* **289**, 155–167.

Nielsen, J. T., and Møller, M. (1978). Innervation of the pineal gland in the Mongolian gerbil (*Meriones unguiculatus*). A fluorescence microscopical study. *Cell Tissue Res.* **187**, 235–250.

Nürnberger, F., and Korf, H. W. (1981). Oxytocin- and vasopressin-immunoreactive nerve fibers in the pineal gland of the hedgehog, *Erinaceus europaeus* L. *Cell Tissue Res.* **220**, 87–97.

Oguri, M., Omura, Y., and Hibiya, T. (1968). Uptake of ^{14}C-labelled 5-hydroxytryptophan into the pineal organ of the rainbow trout. *Bull. Jpn. Soc. Fish.* **34**, 687–690.

Oksche, A. (1955). Untersuchungen über die Nervenzellen und Nervenverbindungen des Stirnorgans, der Epiphyse und des Subcommissuralorgans bei anuren Amphibien. *Morphol. Jahrb.* **95**, 393–425.

Oksche, A. (1965). Survey of the development and comparative morphology of the pineal organ. *Prog. Brain Res.* **10**, 3–29.

Oksche, A. (1971). Sensory and glandular elements of the pineal organ. *In* "The Pineal Gland" (G. E. W. Wolstenholme and J. Knight, eds.), pp. 127–146. Little, Brown, Boston, Massachusetts.

Oksche, A. (1983). Aspects of evolution of the pineal organ. *In* "The Pineal Gland and its Endocrine Role" (J. Axelrod, F. Fraschini, and G. P. Velo, eds.), pp. 15–35. Plenum, New York.

Oksche, A., and Hartwig, H. G. (1979). Pineal sense organs–components of photoneuroendocrine systems. *In* "The Pineal Gland of Vertebrates Including Man" (J. A. Kappers and P. Pévet, eds.), pp. 113–130. Elsevier, Amsterdam.

Oksche, A., and Kirschstein, H. (1967). Die Ultrastruktur der Sinneszellen im Pinealorgan von *Phoxinus laevis*. *Z. Zellforsch. Mikrosk. Anat.* **78**, 151–166.

Oksche, A., and von Harnack, M. (1963). Elektronenmikroskopische Untersuchungen an der Epiphysis cerebri von *Rana esculenta*. *Z. Zellforsch. Mikrosk. Anat.* **59**, 582–614.

Oksche, A., Morita, Y., and Vaupel-von Harnack, M. (1969). Zur Feinstruktur und Funktion des Pinealorgans der Taube (*Columba livia*). *Z. Zellforsch. Mikrosk. Anat.* **102**, 1–30.

Omura, Y. (1984). Pattern of synaptic connections in the pineal organ of the ayu, *Plecoglossus altivelis* (Teleostei). *Cell Tissue Res.* **236**, 611–617.

Omura, Y., and Ali, M. A. (1981). Ultrastructure of the pineal organ of the killifish, *Fundulus heteroclitus*, with special reference to the secretory function. *Cell Tissue Res.* **219**, 355–369.

Omura, Y., Korf, H. W., and Oksche, A. (1985). Vascular permeability (blood-brain barrier) in the pineal organ of the rainbow trout, *Salmo gairdneri*. *Cell Tissue Res.* **239**, 599–610.

Owman, C. (1964). New aspects of the mammalian pineal gland. *Acta Physiol. Scand.* **63**, *Suppl. 240*, 1–40.

Owman, C. (1965). Localization of neuronal and parenchymal monoamines under normal and experimental conditions in the mammalian pineal gland. *Prog. Brain Res.* **10**, 423–453.

Owman, C., and Rüdeberg, C. (1970). Light, fluorescence, and electron microscopic studies on the pineal organ of the pike, *Esox lucius* L. with special regard to 5-hydroxytryptamine. *Z. Zellforsch. Mikrosk. Anat.* **107**, 522–550.

Paul, E., Hartwig, H. G., and Oksche, A. (1971). Neurone und zentralnervöse Verbindungen des Pinealorgans der Anuren. *Z. Zellforsch. Mikrosk. Anat.* **112**, 466–493.

Pellegrino de Iraldi, A., and de Robertis, E. (1965). Ultrastructure and function of catecholamine containing systems. *Int. Congr. Ser.—Excerpta Med.* **83**, 355–363.

Petit, A. (1967). Nouvelles observations sur la morphogénèse et l'histogénèse du complex épiphysaire des Lacertiliens. *Arch. Anat. Histol. Embryol.* **50**, 227–257.

Petit, A., and Viviens-Roels, B. (1977). Présence de contacts neurosensoriels et de synapse d'un type noveau dans l'épiphyse du Lezard des murailles (*Lacerta muralis*, Laurenti). *C. R. Hebd. Seances Ser. Acad. Sci.*, D **284**, 1911–1913.

Pévet, P. (1977). The pineal gland of the mole (*Talpa europaea* L.). IV. Effects of pronase on material present in cisternae of the granular endoplasmic reticulum of pinealocytes. *Cell Tissue Res.* **182**, 215–219.

Pévet, P. (1979). Secretory processes in the mammalian pinealocytes under natural and experimental conditions. *In* "The Pineal Gland of Vertebrates Including Man" (J. A. Kappers and P. Pévet, eds.), pp. 149–194. Elsevier, Amsterdam.

Pévet, P., and Saboureau, M. (1973). L'épiphyse du herrison (*Erinaceus europaeus* L.) male. I. Les pinéalocytes et leur variations ultrastructurales considerées au cour du cycle sexuel. *Z. Zellforsch. Mikrosk. Anat.* **143**, 367–385.

Pfister, C., Dorey, C., Vadot, R., Mirshahi, M., Deterre, P., Chabre, M., and Faure, J. P. (1984). Identité de la protein dite "48K" qui interagit avec la rhodopsine illuminée dans les batonnets retiniens et de l'"antigeñe S retinien" inducteur de l'uveo-retinite autoimmune expérimentale. *C. R. Hebd. Seances Acad. Sci.* **299**, 261–265.

Quay, W. B. (1968). Comparative physiology of serotonin and melatonin. *Adv. Pharmacol. Chemother.* **6**, Part A, 283–297.

Quay, W. B. (1970). Pineal structure and composition in the orangutan (*Pongo pygmaeus*). *Anat. Rec.* **168**, 93–104.

Quay, W. B. (1973). Retrograde perfusion of the pineal region and the question of pineal vascular routes to brain and choroid plexuses. *Am. J. Anat.* **137**, 387–402.

Quay, W. B. (1974). "Pineal Chemistry." Thomas, Springfield, Illinois.

Quay, W. B. (1979). The parietal eye-pineal complex. *In* "Biology of Reptiles" (C. Gans, ed.). Vol. 9, pp. 245–406. Academic Press, London.

Quay, W. B., and Renzoni, A. (1967). The diencephalic relations and variably bipartite structures of the avian pineal complex. *Riv. Biol.* **60**, 9–75.

Quay, W. B., Kappers, J. A., and Jongkind, J. F. (1968). Innervation and fluorescence histochemistry of monoamines in the pineal organ of a snake (*Natrix natrix*). *J. Neuro-visc. Relat.* **31**, 11–25.

Ralph, C. L. (1970). Structure and alleged functions of avian pineals. *Am. Zool.* **10**, 217–235.

Ralph, C. L., and Dawson, D. C. (1968). Failure of the pineal body of two species of birds (*Coturnix coturnix japonica* and *Passer domesticus*) to show electrical responses to illumination. *Experientia* **24**, 147–148.

Reiter, R. J. (1981). "The Pineal Gland," Vol. 1. CRC Press, Boca Raton, Florida.

Reiter, R. J., Welsh, M. G., and Vaughan, M. K. (1976). Age-related changes in the intact and sympathetically denervated gerbil pineal gland. *Am. J. Anat.* **146**, 427–432.

Reppert, S. M., and Klein, D. C. (1980). Mammalian pineal gland: Basic and clinical aspects. *In* "The Endocrine Functions of the Brain" (M. Motta, ed.), pp. 327–371. Raven Press, New York.

Rodin, A. E., and Turner, R. A. (1965). The relationships of intravesicular granules to the innervation of the pineal gland. *Lab. Invest.* **14**, 1644–1652.

Romer, A. S. (1966). "Vertebrate Paleontology." Univ. of Chicago Press, Chicago, Illinois.

Romijn, H. J. (1973). Parasympathetic innervation of the rabbit pineal gland. *Brain Res.* **55**, 431–436.

Romijn, H. J. (1975). Structure and innervation of the pineal gland of the rabbit, *Oryctolagus cuniculus* (L.). III. An electron microscopic investigation of the innervation. *Cell Tissue Res.* **157**, 25–51.

Romijn, H. J., Mud, M. T., and Wolters, P. S. (1976). Diurnal variations in number of Golgi-dense core vesicles in light pinealocytes of the rabbit. *J. Neural Trans.* **38**, 231–237.

Rønnekleiv, O. K., and Kelley, M. J. (1984). Distribution of substance P neurons in the epithalamus of the rat: An immunohistochemical investigation. *J. Pineal Res.* **1**, 355–370.

Roth, J. J., and Roth, E. C. (1980). The parietal-pineal complex among paleovertebrates. *In* "A Cold Look at the Warm-blooded Dinosaurs. AAAS Selected Symposium" (R. D. K. Thomas and E. C. Olson, eds.), pp. 189–231. Weshiew Press, Boulder, Colorado.

Roux, M., Richoux, J. P., and Cordonnier, J. L. (1977). Influence de la photopériod sur l'ultrastructure de l'épiphyse avant et pendant la phase génitale saissonaire chez la femelle de lerot (Eliomys quercinus). J. Neural Transm. 41, 209–223.

Rüdeberg, C. (1969). Structure of the parapineal organ of the adult rainbow trout, Salmo gairdneri Richardson. Z. Zellforsch. Mikrosk. Anat. 93, 282–304.

Sato, T., and Wake, K. (1983). Innervation of the avian pineal organ. A comparative study. Cell Tissue Res. 233, 237–264.

Sato, T., and Wake, K. (1984). Regressive post-hatching development of acetylcholinesterase-positive neurons in the pineal organs of Coturnix coturnix japonica and Gallus gallus. Cell Tissue Res. 237, 269–275.

Schachner, M., Huang, S. K., Ziegelmüller, P., Bizzini, B., and Taugner, R. (1984). Glial cells in the pineal gland of mice and rats. Cell Tissue Res. 237, 245–252.

Scharrer, E. (1964). Photo-neuro-endocrine systems: General concepts. Ann. N.Y. Acad. Sci. 117, 13–22.

Semm, P. (1983). Neurobiological investigations of the pineal organ and its hormone melatonin. In "The Pineal Gland and its Endocrine Role." (J. Axelrod, F. Fraschini, and G. P. Velo, eds.), pp. 437–465. Plenum, New York.

Semm, P., and Demaine, C. (1983). Electrical responses to direct and indirect photic stimulation of the pineal gland in the pigeon. J. Neural Transm. 58, 281–289.

Sheridan, M. N., and Reiter, R. J. (1970). Observations on the pineal system in the hamster. I. Relations of the superficial and deep pineal to the epithalamus. J. Morphol. 131, 153–161.

Smith, A. R., Kappers, J. A., and Jongkind, J. F. (1972). Alterations in the distribution of yellow fluorescing rabbit pinealocytes produced by p-chlorophenylalanine and different conditions of illumination. J. Neural Transm. 33, 91–111.

Studnička, F. K. (1905). Die Parietalorgane. In "Lehrbuch der vergleichenden mikroskopischen Anatomie" (A. Oppel, ed.), Vol. 5, pp. 1–254. Fischer, Jena.

Taugner, R., Schiller, A., and Rix, E. (1981). Gap junctions between pinealocytes. A freeze-fracture study of the pineal gland in rats. Cell Tissue Res. 218, 303–314.

Trost, E. (1952). Untersuchungen über die frühe Entwicklung des Parietalauges und der Epiphyse von Anguis fragilis, Chalcides ocellatus und Tropidonotus natrix. Zool. Anz. 148, 58–71.

Trueman, T., and Herbert, J. (1970). Monoamines and acetylcholinesterase in the pineal gland and habenula of the ferret. Z. Zellforsch. Mikrosk. Anat. 109, 83–100.

Uddman, R., Alumets, J., Håkanson, R., Lorén, I., and Sundler, F. (1980). Vasoactive intestinal peptide (VIP) occurs in nerves of the pineal gland. Experientia 36, 1119–1120.

Ueck, M (1973). Fluoreszenzmikroskopische und elektronenmikroskopische Untersuchungen am Pinealorgan verschiedener Vogelarten. Z. Zellforsch. Mikrosk. Anat. 137, 37–62.

Ueck, M. (1979). Innervation of the vertebrate pineal. In "The Pineal Gland of Vertebrates Including Man" (J. A. Kappers and P. Pévet, eds.), pp. 45–88. Elsevier, Amsterdam.

Ueck, M., and Kobayashi, H. (1972). Vergleichende Untersuchungen über acetylcholinesterasehaltige Neurone im Pinealorgan der Vögel. Z. Zellforsch. Mikrosk. Anat. 129, 140–160.

Vaughan, G. M. (1984). Melatonin in humans. Pineal Res. Rev. 2, 141–201.

Vigh, B., and Vigh-Teichmann, I. (1981). Light-and electron microscopic demonstration of immunoreactive opsin in the pinealocytes of various vertebrates. Cell Tissue Res. 221, 451–463.

Vigh, B., Vigh-Teichmann, I., Röhlich, P., and Oksche, A. (1983). Cerebrospinal fluid-contacting neurons, sensory pinealocytes and Landolt's clubs of the retina as revealed by means of electron microscopic immunoreaction against opsin. Cell Tissue Res. 233, 539–548.

Vigh, B., Vigh-Teichmann, I., Aros, B., and Oksche A. (1985). Sensory cells of the "rod"- and "cone-type" in the pineal organ of Rana esculenta, as revealed by immunoreaction against opsin and by the presence of an oil (lipid) droplet. Cell Tissue Res. 240, 143–148.

Vigh-Teichmann, I., and Vigh, B. (1979). Comparison of epithalamic, hypothalamic and spinal neurosecretory terminals. Acta Biol. Acad. Sci. Hung. 30, 1–39.

Vigh-Teichmann, I., and Vigh, B. (1983). The system of cerebrospinal fluid-contacting neurons. *Arch. Histol. Jpn.* **46**, 427–468.

Vigh-Teichmann, I., Korf, H. W., Oksche, A., and Vigh, B. (1982). Opsin-immunoreactive outer segments and acetylcholinesterase-positive neurons in the pineal complex of *Phoxinus phoxinus* (Teleostei, Cyprinidae). *Cell Tissue Res.* **227**, 351–369.

Vigh-Teichmann, I., Korf, H. W., Nürnberger, F., Oksche, A., Vigh, B., and Olsson, R. (1983). Opsin-immunoreactive outer segments in the pineal and parapineal organs of the lamprey (*Lampetra fluviatilis*), the eel (*Anguilla anguilla*), and the rainbow trout (*Salmo gairdneri*). *Cell Tissue Res.* **230**, 289–307.

Vivien-Roels, B. (1976). L'épiphyse des Cheloniens. Étude embryologique, structurale, ultrastructurale; analyse qualitative et quantitative de la serotonine dans des conditions normales et expérimentales. Thesis, Strasbourg.

Viviens-Roels, B., Pévet, P., Dubois, M. P., Arendt, J., and Brown, G. M. (1981). Immunohistochemical evidence for the presence of melatonin in the pineal gland, the retina and the Harderian gland. *Cell Tissue Res.* **217**, 105–115.

Vollrath, L. (1973). Synaptic ribbons of a mammalian pineal gland. Circadian changes. *Z. Zellforsch. Mikrosk. Anat.* **145**, 171–183.

Vollrath, L. (1981). The pineal organ. *In* "Handbuch der mikroskopischen Anatomie des Menschen" (A. Oksche and L. Vollrath, eds.), Vol. IV, Part 10, pp. 1–665. Springer-Verlag, Berlin and New York.

von Frisch, K. (1911). Beiträge zur Physiologie der Pigmentzellen in der Fischhaut. *Pfluegers Arch. Gesamte Physiol. Menschen Tiere* **138**, 319–387.

von Haffner, K. (1952). Die Pinealblase (Stirnorgan, Parietalauge) von *Xenopus laevis* Daud, und ihre Entwicklung, Verlagerung und Degeneration. *Zool. Jahrb. Abt. Anat. Ontog. Tiere* **71**, 375–412.

Wake, K. (1973). Acetylcholinesterase-containing nerve cells and their distribution in the pineal organ of the goldfish, *Carassius auratus*. *Z. Zellforsch. Mikrosk. Anat.* **145**, 287–298.

Wake, K., Ueck, M., and Oksche, A. (1974). Acetylcholinesterase-containing nerve cells in the pineal complex and subcommissural area of the frogs, *Rana ridibunda* and *Rana esculenta*. *Cell Tissue Res.* **154**, 423–442.

Wartenberg, H., and Baumgarten, H. G. (1969). Untersuchungen zur fluoreszenz- und elektronenmikroskopischen Darstellung von 5-Hydroxytryptamin (5-HT) im Pinealorgan von *Lacerta viridis* und *Lacerta muralis*. *Z. Anat. Entwicklungsgesch.* **128**, 185–210.

Wartenberg, H., Hadziselimovic, F., and Seguchi, H. (1972). Experimentelle Untersuchungen über die Passage der Blut-Gewebs-Schranke in den zirkumventrikulären Organen des Meerschweinchengehirns. *Verh. Anat. Ges.* **66**, 345–355.

Welsh, M. G., and Beitz, A. J. (1981). Modes of protein and peptide uptake in the pineal gland of the Mongolian gerbil: An ultrastructural study. *Am. J. Anat.* **162**, 343–355.

Werblin, F. S., and Dowling, J. E. (1969). Organization of the retina of the mudpuppy, *Necturus maculosus*. *J. Neurophysiol.* **32**, 315–338.

Wetterberg, L. (1979). Clinical importance of melatonin. *In* "The Pineal Gland of Vertebrates Including Man" (J. A. Kappers and P. Pévet, eds.), pp. 539–547. Elsevier, Amsterdam.

Wurtman, R. J., Axelrod, J., and Kelly, D. E. (1968). "The Pineal." Academic Press, New York.

Yulis, C. R., and Rodríguez, E. M. (1982). Neurophysin pathways in the normal and hypophysectomized rat. *Cell Tissue Res.* **227**, 93–112.

Yuwiler, A. (1983). Vasoactive intestinal peptide stimulation of pineal serotonin N-acetyltransferase activity: General characteristics. *J. Neurochem.* **41**, 146–153.

Zimmerman, N. H., and Menaker, M. (1979). The pineal: A pacemaker within the circadian system of the house sparrow. *Proc. Natl. Acad. Sci. U.S.A.* **76**, 999.

Zweig, M., Snyder, S. H., and Axelrod, J. (1966). Evidence for nonretinal pathway of light to pineal gland in newborn rats. *Proc. Natl. Acad. Sci. U.S.A.* **56**, 515–520.

5

The Caudal Neurosecretory System in Fishes

**HIDESHI KOBAYASHI, KYOKO OWADA,
CHIFUMI YAMADA, AND YUJI OKAWARA**

Department of Biology
Faculty of Science
Toho University
Funabashi, Chiba 274, Japan

I. INTRODUCTION

In 1914 Dahlgren first described the presence of unusually large cells in the caudal region of the spinal cord in 11 species of skates. He noted that these cells had characteristics of secretory cells and supposed that they had some relation to the electric apparatus of the skates. However, Speidel (1919) pointed out that these large cells had no relation to the electric organs, because of the differences in their distributions. Further, Speidel (1922) confirmed the presence of large, caudal, glandular cells in 30 species, including teleosts, elasmobranchs, and ganoids, and he designated these cells as Dahlgren cells. Since then little attention was paid to the Dahlgren cells until Scharrer and Scharrer (1937, 1945) pointed out that these cells had been interpreted by Speidel (1919, 1922) as glandlike nerve cells. Speidel was the first to describe the secretion of nerve cells on the basis of a large number of observations.

The enlargement of the caudal spinal cord was first observed by Arsaky (1813). Between then and 1900 more than 15 biologists anatomically described this enlargement (see Favaro, 1925, 1926). Verne (1914) observed that eosinophilic grains were densely located around the blood vessels in the caudal enlargements of 11 teleostean species. He considered that the grains were formed by glial cells therein. Thus, he pointed out a similarity between the caudal enlargement of the spinal cord and the neural lobe of the pituitary in higher vertebrates, since, at that time, the pituicytes were considered to secrete hormones in the neural lobe. Favaro (1925) reported that most teleosts, except muraenoids and lophobranchs, have a caudal swelling of the

147

VERTEBRATE ENDOCRINOLOGY:
FUNDAMENTALS AND BIOMEDICAL IMPLICATIONS
Volume 1

spinal cord, and he considered it to be an endocrine organ. Little attention was paid to the caudal enlargement of the spinal cord in the interval between Favaro's works (1925, 1926) and the 1950s.

Thirty years later, Enami (1955) found that in the Japanese eel, *Anguilla japonica,* the Dahlgren cells in the spinal cord are neurosecretory cells and send their fibers to the caudal enlargment. Enami's contribution was to link the caudal neurosecretory cells to the storage–release organ of neurosecretory material in teleosts. He designated this system as the caudal neurosecretory system (Enami, 1955) (Fig. 1). For the terminal enlargement, many terms have been suggested, but the term urophysis has been generally accepted (see Holmgren, 1959; Sano, 1961; Fridberg, 1962a; Arvy, 1966; Fridberg and Bern, 1968).

Enami's concept of the caudal neurosecretory system was soon confirmed by the studies of Sano (1958a,b) and Holmgren (1958, 1959). During the 1960s, the caudal neurosecretory system was the subject of a number of reviews mostly concerned with the anatomy and histology of this system (see Arvy, 1966; Gabe, 1966; Fridberg and Bern, 1968; Bern, 1969; Kriebel, 1980). During the 1970s, most investigations were pharmacological, physiological, or chemical (see Lederis *et al.,* 1974; Lederis, 1974, 1977; Bern and Lederis, 1978; Loretz *et al.,* 1982). Between 1980 and 1984, urophysial hormones, urotensin I (UI) and urotensin II (UII) were purified, and their primary structures were determined (Pearson *et al.,* 1980; Lederis *et al.,* 1982, 1983; Ichikawa *et al.,* 1982, 1984; McMaster and Lederis, 1983). The availability of synthetic urotensins has facilitated physiological and immunocytochemical studies of the caudal neurosecretory system. Biochemical data of UI and UII are reviewed by Ichikawa *et al.* (1986).

II. ANATOMY AND HISTOLOGY

A. Occurrence of the Caudal Neurosecretory System

In the cephalochordate *Branchiostoma belcheri,* there are no caudal neurosecretory cells and no urophysis (Sano, 1965). In cyclostomes, Speidel

Fig. 1. Carp caudal neurosecretory system revealed by immunocytochemistry using antiserum to UII. L, large cell; M, medium-sized cell; S, small cell; T, axonal tract; U, urophysis. Scale bar = 500 μm.

(1922) could not find the Dahlgren cells in the lamprey *Petromyzon*. No elements of the caudal neurosecretory system were identified in the hagfish *Eptatretus burgeri* and the lampreys *Entosphenus reissneri* and *Entosphenus japonicus* (Sano, 1965). Dorsal cells found by Sterba (1962, 1972) and Peyrot (1964) in the posterior spinal cord of lampreys may not be comparable to the caudal neurosecretory cells of elasmobranchs and teleosts, despite Sterba's suggestion. Sterba's postulate becomes unlikely in the light of data from a pharmacological analysis of the caudal spinal cord tissue of the cyclostomes, which showed no UII activity (Bern *et al.*, 1973); furthermore, immunoreactivity to antisera of UII and corticotropin-releasing factor (CRF) which cross-reacts with UI was not detected in the lamprey dorsal cells (Owada *et al.*, 1985b).

The sharks and skates possess caudal neurosecretory cells (Fig. 2, Speidel, 1919), but they do not have a urophysis (see Fridberg and Bern, 1968; Bern, 1969, 1972). For the holocephalans, it has been suggested that in *Hydrolagus colliei* the caudal neurosecretory elements might not be present (Bern and Takasugi, 1962). However, further detailed studies are needed.

Lower actinopterygians, *Acipenser medirostris* (Hamana, 1962; Sano, 1965) and *A. güldenstädti* (Saenko, 1970, 1978) possess Dahlgren cells, but, as in the elasmobranchs, they lack a urophysis. The Dahlgren cells were found in *Polypterus* (Fridberg, 1962a) and in the garpike, *Lepidosteus osteus* (Speidel, 1922), but it is not certain whether these forms possess a urophysis.

To date, more than 350 species of teleosts have been appropriately studied by a number of investigators (see Bern and Takasugi, 1962; Sano, 1965; Fridberg, 1962a; Hamana, 1962; Enami and Imai, 1955, 1956a,b; Arvy, 1966), and it was found that only the syngnathids (Fridberg, 1962a; Hamana, 1962) and the molas (Hamana, 1962) do not possess any of the elements of a caudal neurosecretory sytem. Favaro (1926) had already reported earlier that fishes of the families Molidae, Syngnathidae, and Muraenidae do not possess the system. However, Hamana (1962) found that the system exists in *Gymnothorax kidako* and *G. hepatica* (Muraenidae), although in these species the caudal enlargement is flattened.

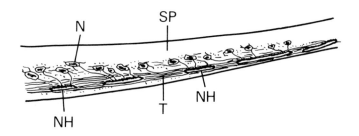

Fig. 2. Diagram of caudal region of the spinal cord (SP), indicating neurosecretory cells (N), axonal tract (T), and neurohemal areas (NH) in the banded dogfish, *Triakis scyllia*.

In 33 teleostean species studied, urophyses ranged in weight from 0.04 to 0.54 mg per 100 g of body weight. The fishes of Cypriniformes have heavier urophyses, and the fishes of Anguilliformes and Scorpaeniformes have lighter urophyses, relative to body weight (Kobayashi *et al.*, 1986a).

In dipnoans, Bern and Takasugi (1962) and Bern (1969) observed some large neurons in the caudal spinal cord of *Protopterus aethiopicus,* which may or may not be equivalent to Dahlgren cells. Mimura (1978) could not identify either Dahlgren cells or a urophysis in *Lepidosiren paradoxa.* Bern *et al.* (1973) pharmacologically showed that UII was not present in spinal cord tissue of dipnoans and that its presence was questionable in *Hydrolagus,* a holocephalan. There is no evidence for the presence of caudal neurosecretory cells in tetrapod spinal cords (Bern, 1969). Speidel (1922) could not find any cells homologous to the Dahlgren cells in the spinal cords of the amphibians *Diemyctylus* and *Necturus* or in the ventral nerve cord of two invertebrates *Homarus* and *Limulus.*

B. Distribution and Types of Caudal Neurosecretory Cells

In teleosts, the perikarya of the neurosecretory cells are usually located ventrolateral to the central canal and are distributed anteriorly as far as the fifth to eighth preterminal vertebrae. The cell somata are often distributed caudally in the filum terminale (Sano, 1958a; Fridberg, 1962a; Jaiswal and Belsare, 1973; Kriebel *et al.*, 1979; Kriebel, 1980). In elasmobranchs, the neurosecretory cell perikarya are huge (Section V) and occur extensively along the caudal spinal cord (Fig. 2). For instance, the adult skate, *Raia ocellata,* has about 120 vertebrae, and about 600 Dahlgren cell somata are found in the spinal cord from the level of the 64th vertebra to the tip of the tail (Speidel, 1919).

Caudal neurosecretory cells are usually large and polymorphic, and their nuclei are often lobated (Section V, Figs. 8–11). Although caudal neurosecretory cells may generally be stained with acidic dyes, a small number of basophilic cells and chromophobe cells are encountered (Fridberg and Bern, 1968). Details of the cytology of the neurosecretory cells have been reported in a number of fishes (see Fridberg, 1962a; Sano, 1965; Arvy, 1966; Fridberg and Bern, 1968; Feustel and Luppa, 1968; Bern, 1969; Kriebel *et al.*, 1979; Kriebel, 1980). Detailed histoenzymology of the caudal neurosecretory system has also been carried out (see Luppa *et al.*, 1968; Uemura, 1965; Kobayashi *et al.*, 1979).

On the basis of cell size, two or three types of cells have been noted in several species of teleosts (Figs. 1,8) (Enami, 1955, 1956; Sano, 1958a; Fridberg, 1962a; Fridberg *et al.*, 1966a; Fisher *et al.*, 1984; Owada *et al.*, 1985a; Yamada *et al.*, 1986b). In elasmobranchs, too, the caudal neurosecretory cells decrease in size caudally (Fridberg, 1962b; Owada *et al.*, 1985b). It has not been elucidated whether these different types of neurosecretory cells have different functions (Section V).

C. Neurosecretory Tract and Urophysis

In teleosts, most of the neurosecretory cells send their axons, which are generally unmyelinated, caudally to the urophysis. With the light microscope, Herring bodies are often seen in the axonal tracts. The neurosecretory cell processes are often associated with the central canal (Fridberg *et al.*, 1966a,b; Fridberg and Nishioka, 1966; Kriebel *et al.*, 1979; Owada *et al.*, 1985a) and also with capillaries abundant in the regions densely populated with neurosecretory cells (Sano and Kawamoto, 1960; Kriebel *et al.*, 1976) (Section V, Fig. 9). The urophysis is richly vascularized, and the neurosecretory axons terminate at the capillaries (Fig. 3). Glial cells are relatively few.

In the elasmobranch *Squalus acanthias,* axonlike neurosecretory processes run ventrolaterally to terminate at the vascular bed of the ventrolateral meninx sheath in the neurohemal area (Fridberg, 1962b; see Fridberg and Bern, 1968) (Fig. 4A). In *Raia batis* and *R. radiata,* most neurosecretory processes proceed ventrolaterally, as seen in *Squalus,* and some run ventrocaudally to terminate in more distal neurohemal areas. However, in *Trygon violacea* and *Torpedo ocellata,* the neurosecretory tract penetrates the meninx of the ventromedian part of the spinal cord to reach a vascular reticulum (Fig. 4B) (Fridberg, 1962b; Fridberg and Bern, 1968). Fridberg (1962b) considered that the *Squalus* type of neurohemal area is more primi-

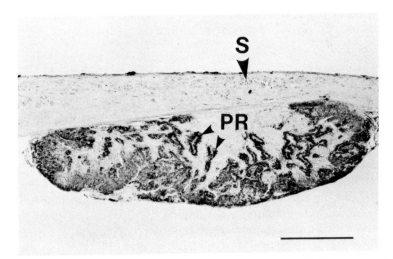

Fig. 3. Carp urophysis revealed by immunocytochemistry using CRF antiserum. PR, immunoreactive products around capillaries; S, small neurons immunoreactive with CRF antiserum. Scale bar = 500 μm.

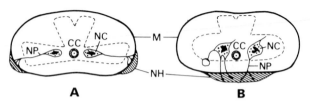

Fig. 4. Diagrams representing transverse sections through the caudal spinal cord of *Squalus* (A) and *Torpedo* (B) to show relations of neurosecretory cell processes (NP) to neurohemal area (NH). CC, central canal; M, meninx; NC, neurosecretory cell. [From Fridberg, 1962b]

tive than the *Torpedo* type. These two types appear sequentially during early stages of development of the caudal neurosecretory system in the teleost *Esox lucius* (Fridberg, 1962b). Acipenserids (Osteichthyes) have the primitve *Squalus* type of caudal neurosecretory structure (Hamana, 1962; Sano, 1965; Saenko, 1970, 1978). From these findings it seems likely that the elasmobranch type is a primitive form of a storage–release organ (urophysis) of the teleost. Fridberg and Bern (1968) presented the idea of a possible evolutionary sequence of the caudal neurosecretory system from the elasmobranch type to the generalized teleost type through the isospondylous teleosts. However, it is also possible that the elasmobranch type and the teleostean type have evolved independently.

The neurosecretory cells of the dogfish, *Triakis scyllia,* extend their long axons ventrocaudally to form loose tracts as in teleosts, and the fibers in the tracts terminate in more distal neurohemal areas distributed ventrolaterally to and along the caudal spinal cord (Figs. 2 and 11B) (Owada *et al.*, 1985b). By contrast, some cells extend short axons straight to the nearest neurohemal areas located on the ventrolateral surface of the spinal cord (Figs. 2 and 11A) (Owada *et al.*, 1985b).

D. Vascular Supply

In the char, *Salvelinus leucomaenis pluvius,* the caudal artery gives off a pair of branches, the dorsal segmental arteries, at the level of the penultimate vertebra (Fig. 5). The segmental arteries branch again, and the anterior branches supply the urophysis to form a vascular plexus. The capillaries join relatively thick veins, and the blood finally drains into the segmental vein above the fourth from last vertebra. The blood of the segmental vein flows into the caudal vein (Honma and Tamura, 1967) and then into the renal portal vein. Similar observations were reported by Nag (1967), Chan (1971), and Jaiswal and Belsare (1973) in different species. The blood in the portal vein drains into the kidney (including interrenal and chromaffin tissues, Stannius corpuscles, and juxtaglomerular apparatus). In addition, drainage from the caudal vein into the liver, bladder, and swim bladder has been suggested (see Fridberg and Bern, 1968; Bern *et al.*, 1985).

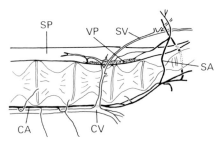

Fig. 5. Schematic illustration of caudal vascular links in *Salvelinus leucomaenis pluvius.* CA, caudal artery; CV, caudal vein; SA, segmental artery; SP, spinal cord; SV, segmental vein; VP, vascular plexus. [From Honma and Tamura, 1967.]

III. DEVELOPMENT AND REGENERATION OF THE CAUDAL NEUROSECRETORY SYSTEM

Speidel (1919) described that the Dahlgren cells in the skate, *Raia punctata,* are derived from the same group of neuroblasts which also give rise to the motor nerve cells of the anterior horn. Sano and Kawamoto (1959), Sano *et al.* (1962), and Sano (1965) described that Dahlgren cells are derived from special hypochromatic ependymal cells in the region dorsal to the developing urophysis in *Lebistes reticulatus* and *Salmo irideus.* Fridberg (1962a) stated that the Dahlgren cells in the anterior part of the neurosecretory system of the roach are derived from neuroblasts of the same appearance as those which give rise to motor neurons and that the small Dahlgren cells located dorsally to the urophysis seemed to be derived from cells that resemble asteroblasts.

Several investigators have studied the development of the urophysis in different species of fish (Favaro, 1926; Holmgren, 1959; Sano and Kawamoto, 1959; Sano *et al.,* 1962; see Fridberg, 1962a). The time at which the urophysis develops differs among species, but generally, it differentiates at later larval stages. For instance, in *L. reticulatus,* a viviparous teleost, the urophysis starts to develop at the time of birth, taking an adult form about 20 days after birth (Sano and Kawamoto, 1959). In *S. irideus,* the urophysis begins to develop around 50 days after hatching, taking an adult form 6 months after hatching (Sano *et al.,* 1962). Sano and his colleagues stated that pituitary development had already been completed and that neurosecretory material had been stored in the neural lobe at the beginning of urophysial development in *Lebistes* and *Salmo.* Similar observations were reported in *Salvelinus leucomaenis pluvius* (Honma and Tamura, 1967).

When the neurosecretory tract is cut proximal to the urophysis, the neurohemal area regenerates at the site of transection, and the neurosecretory cells proximal to the cut survive (Enami, 1956; Sano and Hartmann, 1959; Inoue, 1959; Imai, 1965). It was also observed that in several teleostean species ependymal cells transform into Dahlgren cells on regeneration after total extirpation of the caudal neurosecretory system (Imai, 1965; Fridberg *et al.*, 1966b; Jaiswal and Belsare, 1974). However, Tsuneki and Kobayashi (1979) could not observe any transformation from ependymal cells to new Dahlgren cells as late as 3 months after the total removal of the caudal neurosecretory system in the adult loach, *Misgurnus anguillicaudatus,* and in the adult medaka, *Oryzias latipes.* The discrepancy between their data and ours may be due to age or species differences in the fishes used.

IV. ELECTRON MICROSCOPY

Electron microscopy of the neurosecretory system was initially studied in the Japanese eel (Enami and Imai, 1958) and later in *Tinca vulgaris* (Sano and Knoop, 1959) and *Fundulus heteroclitus* (Holmgren and Chapman, 1960). Since then, many investigators have described the ultrastructural morphology of the caudal neurosecretory cells, which is essentially the same as that characteristic of neurosecretory cells in general (see Bern and Knowles, 1966; Fridberg and Bern, 1968; Chevalier, 1976; Kriebel, 1980). The neurosecretory granules emanate from the Golgi apparatus, mature in the cytoplasm (Fig. 6), and are transported to the neurohemal organ through the axons (Fig. 7). Microtubules and mitochondria are usually observed in the axons (Fig. 7).

The axon terminals in the urophysis contain electron-dense neurosecretory granules (around 200 nm) and small synaptic vesiclelike structures (around 60 nm) (Fig. 7). The number of the granules changes according to the degree of the environmental salinity in teleosts (Fridberg *et al.*, 1966a; Chevalier, 1976; Kriebel, 1980; Gauthier *et al.*, 1983). There seem to be at least two types of secretory granules, larger and smaller ones (Holmgren and Chapman, 1960; Oota, 1963a; Sano *et al.*, 1966), suggesting the possible presence of at least two different active substances. It was found that the granules obtained by ultracentrifugation contain urophysial hormones (Lederis *et al.*, 1971, 1974). Ichikawa and Kobayashi (1978) fractionated one layer containing mainly small vesicles (50–80 nm) from the carp urophysis, although the layer was contaminated with some large granules (150–200 nm), and they detected therein the highest acetylcholine (ACh) concentration of all the fractions. Since the granules contain urophysial hormones (Lederis *et al.*, 1974), the small vesicles must be, at least partly, the carriers of ACh. It has been shown by bioassay, and chemically, that the urophysis contains a high concentration of ACh (Kobayashi *et al.*, 1963; Ichikawa and Kobayashi, 1978; Ichikawa, 1978). However, there is the possibility that some of the small vesicles are derived from vesiculation of the plasma membrane after exocytosis of secretory granules, although exocytosis is as rare in neurohemal areas of the urophysis as it is in the neural lobe of the pituitary.

Pale or electron-lucent vesicles equal in size to the electron-dense secretory granules are often found in the axon terminals (Fig. 7), suggesting that

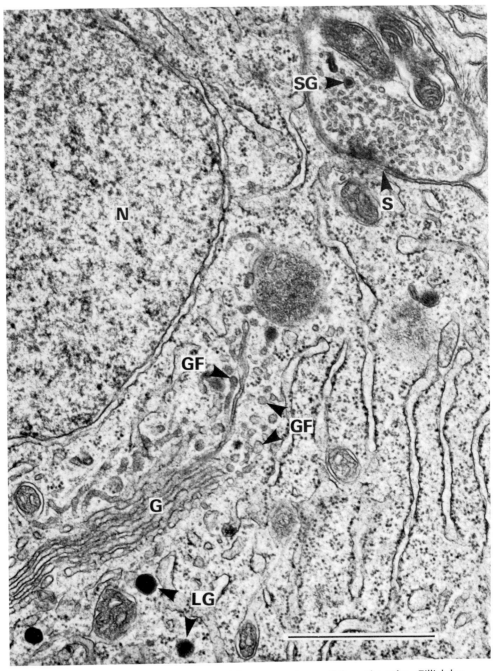

Fig. 6. Electron micrograph of a caudal neurosecretory cell from the goby, *Gillichthys mirabilis*. Granule formation (GF) is seen in Golgi area (G). Synapse (S) is seen between probable monoaminergic axon and neurosecretory perikaryon. LG, neurosecretory granule; N, nucleus; SG, small granules probably containing monoamine. Scale bar = 1 μm. [Courtesy of R. S. Nishioka and H. A. Bern.]

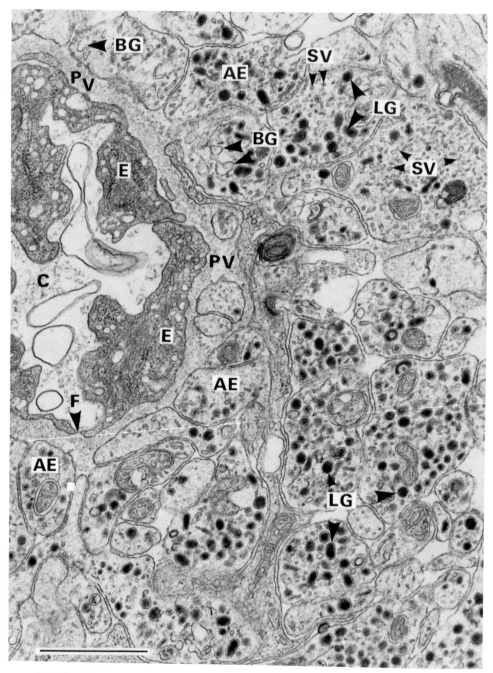

Fig. 7. Electron micrograph of urophysis of *Gillichthys mirabilis*. Capillary (C) is surrounded by neurosecretory axon endings (AE). BG, broken granule; E, endothelial cell; F, fenestration of endothelial cell; LG, neurosecretory granule; PV, perivascular space; SV, small vesicle. Scale bar = 1 μm. [Courtesy of R. S. Nishioka and H. A. Bern.]

the mode of secretion into the capillaries may involve diffusion rather than exocytosis (Fridberg, 1963; Oota, 1963a; Fridberg *et al.*, 1966a).

Glial cells seem to be relatively few in the urophysis (Fridberg *et al.*, 1966a). Kriebel *et al.* (1979) examined the glial cells electron microscopically in the blueback herring, *Pomolobus aestivalis*, and pointed out similarities to the pituicyte in the neural lobe. At present, it is only possible to speculate on the function of both glial and pituicyte cells.

V. IMMUNOCYTOCHEMISTRY

Urotensin I (UI) has amino acid sequences homologous with cortico-tropin-releasing factor (CRF) and sauvagine (see Lederis *et al.*, 1982; Ichi-kawa *et al.*, 1982), and UII has amino acid sequences partially homologous with somatostatin (see Pearson *et al.*, 1980; Ichikawa *et al.*, 1984). The availability of synthetic preparations of these peptides has facilitated im-munocytochemical studies of the caudal neurosecretory system.

We have demonstrated in the carp that immunoreactivity to carp UI anti-serum could be localized in caudal neurosecretory cells of three different size classes (Fig. 8; Yamada *et al.*, 1986b). However, neurosecretory cells which were not immunoreactive with UI antiserum were also encountered in each cell class (Fig. 8). Fisher *et al.* (1984) observed that all the caudal neurosecretory cells and the urophysis showed immunoreactivity to UI anti-serum in the sucker, *Catostomus commersoni*. They believe that the hypoth-esis of two populations of neurons, producing UI and UII, respectively, is untenable. They suggested further that the UI-immunoreactive small cells dorsal to the urophysis project their fibers proximally to feed back on the large caudal neurosecretory cells. Renda *et al.* (1981, 1982) demonstrated sauvaginelike immunoreactive substances in the urophysis of *Tinca tinca* and *Salmo gairdneri*. Immunoreactivity to ovine CRF antiserum was de-tected in most caudal neurosecretory cells and the urophysis in several

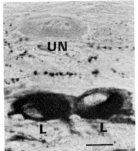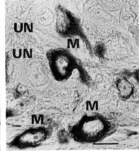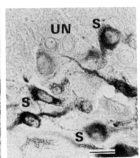

Fig. 8. Immunoreactive large (L), medium-sized (M), and small (S) caudal neurosecretory cells and nonimmunoreactive cells (UN) to UI antiserum in the carp. Scale bar = 20 μm.

species of teleosts (Onstott and Elde, 1984; Yamada *et al.*, 1985), although there were also nonimmunoreactive large cells.

Most of the caudal neurosecretory cells, processes, and urophyses were reactive to UII antiserum in 12 species of teleosts (Figs. 1 and 9) (Owada *et al.*, 1985a). In 5 species of teleosts examined, neurosecretory cells reacting to UII antiserum differentiated into three types by the size (Fig. 1), as well as by those reacting to UI antiserum. It seems, therefore, that the neurosecretory cells produce the same hormones, irrespective of their size. However, some unstained cells were also seen among the populations of these cell groups (Fig. 9).

By applying UI or UII antiserum to alternate sections of the carp caudal spinal cord, we observed at least three types of cells: UI cells, UII cells, and cells reactive to both UI and UII antisera (Fig. 10). However, Bern *et al.*, (1985), using a double-immunofluorescence technique on single sections, showed two populations of immunoreactive neurons in *Gillichthys*: CRF/UI cells and cells immunoreactive to both CRF/UI and UII antisera. Ishimura and Lederis (1983) observed colocalization of immunoreactive UI and UII in the same cells in the sucker. It is possible, however, that all the neurosecretory cells produce both UI and UII, and that cells which show immunoreactivity to only either UI or UII antiserum may contain insufficient amounts of

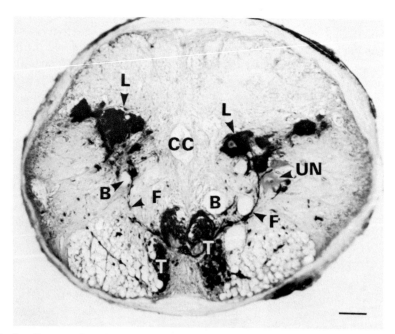

Fig. 9. Cross section of the carp spinal cord showing large, UII-immunoreactive cells (L). B, blood vessel; CC, central canal; F, UII-reactive fiber; T, UII-reactive fiber tract; UN, nonimmunoreactive large cell. Scale bar = 50 μm.

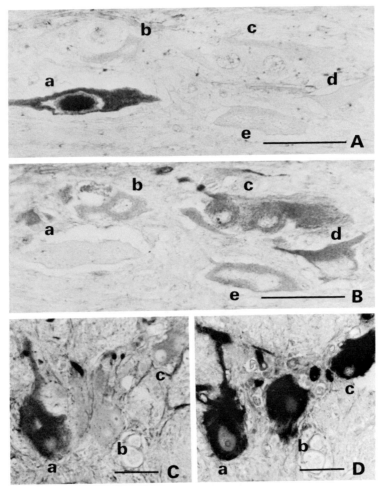

Fig. 10. Carp caudal neurosecretory cells on the alternate section (A and B; C and D). In A and B, cell a reacted to UI antiserum, but did not react to UII antiserum. Cells b, c, d, and e did not react to UI antiserum but did react to UII antiserum. In C and D, cell a reacted to both UI and UII antisera. Cells b and c did not react to UI antiserum but did react to UII antiserum. A and B, 4 μm thick; C and D, 10 μm thick. Scale bar = 50 μm.

the other urotensin type to be demonstrated by immunocytochemical methods. The resolution of this problem will need further studies, with different fixatives, different techniques for preparing sections, and antiserum specific for UI or UII.

Immunoreactive UI was detected with radioimmunoassay in different regions of the brain, spinal cord, pituitary, and plasma of *Catostomus commersoni* (Suess *et al.*, 1985). Immunoassayable UII was also detected in different parts of the brain, spinal cord, and blood of the same species (Kobayashi

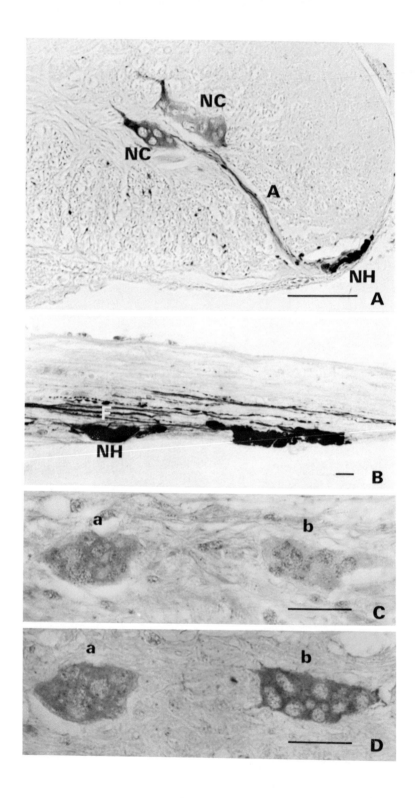

et al., 1986b). However, the brains of the carp, goldfish, and goby, *Acantho-gobius flavimanus*, did not contain neurons immunoreactive to UII antiserum (Owada *et al.*, 1985a). Radioimmunoassay failed to detect UII in the anterior spinal cord and brain of *Gillichthys mirabilis* (Bern *et al.*, 1985). The discrepancy may arise from the differences in the antibodies used in these laboratories or from differences in sensitivity between radioimmunoassay and immunocytochemistry.

In elasmobranchs, the caudal neurosecretory system of the cat shark, *Heterodontus japonicus*, and the swell shark, *Cephaloscyllium umbratile*, showed immunoreactivities with carp UI and goby UII antisera (Yamada *et al.*, 1986a). The caudal neurosecretory system of the tiger shark, *Scyliorhinus torazame*, and the stingray, *Dasyatis akajei*, showed immunoreactivity with UII antiserum (Owada *et al.*, 1985b). In these species, UI immunoreactivity was not examined. In the dogfish, *Triakis scyllia*, the caudal neurosecretory neurons were immunoreactive to human CRF (1–20) [hCRF(1–20)] and UII antisera (Fig. 11A, B). (Owada *et al.*, 1985b). In the swell shark, two consecutive sections were stained with carp UI or goby UII antiserum, and two types of neurons were identified: those immunoreactive to both UI and UII antisera and those immunoreactive to only UII antiserum (Fig. 11C, D). The number of neurons of the former type was greater than that of the latter one (Yamada *et al.*, 1986a). These results were similar to those obtained in *Triakis scyllia* using CRF and UII antisera.

The dorsal cells in the spinal cord of the lamprey *Lampetra japonica*, which have been considered to be possible forerunners of Dahlgren cells (Sterba, 1972), did not react to either UII or hCRF(1–20) antiserum (Owada *et al.*, 1985b).

The presence of arginine vasotocin (AVT) in the urophysis was suggested by bioassay and radioimmunoassay in several teleostean species (Lacanilao, 1972a,b; Lacanilao and Bern, 1972; Holder *et al.*, 1979). However, Goossens (1976), Gill *et al.* (1977), and Holder *et al.* (1979) could not detect AVT with an immunocytochemical or immunofluorescent technique in the urophysis. Whether AVT is identified in the urophysis may depend on the use of different techniques or on species differences. Oxytocin, vasopressin, neurotensin, somatostatin, gastrin-releasing peptide, vasoactive intestinal polypeptide, and substance P could not be detected by immunocytochemical methods in the carp urophysis (Yamada *et al.*, 1986b).

Fig. 11. (A) Cross section of the spinal cord of *Triakis scyllia*. NC, UII-immunoreactive neurosecretory cells; A, UII-immunoreactive fiber proceeding to neurohemal area (NH). Scale bar = 200 μm. (B) Longitudinal section of the spinal cord of *Triakis scyllia*. F, UII-immunoreactive fibers forming a loose tract; NH, neurohemal area. Scale bar = 200 μm. (C) Cell a, UI-immunoreactive cell; cell b, UI-nonreactive cell. (D) Cells a and b, UII-immunoreactive cell. Cell a was reactive to both UI and UII antisera, but cell b was immunoreactive to only UII antiserum in *Cephaloscyllium umbratile*. Scale bar = 50 μm.

VI. NERVOUS CONTROL OF THE CAUDAL
NEUROSECRETORY SYSTEM

Electrical stimulation of the spinal cord anterior to the caudal neurosecretory cells induces postsynaptic potentials in the neurosecretory cells in teleosts (Morita *et al.*, 1961; Ishibashi, 1962; Yagi and Bern, 1965) and in elasmobranchs (Bennett and Fox, 1962). Bouton-type synapses were morphologically found by use of Bodian's technique on the neurosecretory perikarya of the Japanese eel (Ishibashi, 1962), and axosomatic synapses were identified electron microscopically in teleosts and elasmobranchs (Fridberg, 1963; Fridberg *et al.*, 1966a; Sano *et al.*, 1966; Baumgarten *et al.*, 1970; Lederis *et al.*, 1974; Kobayashi *et al.*, 1979; Kriebel, 1980).

It is evident from studies with fluorescence microscopy and electron microscopy that some of the monoaminergic axons terminate at some of the neurosecretory perikarya (Baumgarten *et al.*, 1970; Wilén and Fridberg, 1973; Swanson *et al.*, 1975; Kobayashi *et al.*, 1980; Audet and Chevalier, 1981). These fluorescent fibers are possibly noradrenergic, because noradrenaline was found fluorometrically (0.19 ± 0.01 μg/g of tissue, $n = 6$) in a portion of the carp spinal cord that includes neurosecretory cells, and adrenaline and dopamine were not detected therein (Kobayashi *et al.*, 1980). Monoamine oxidase was found in axons in close contact with the caudal neurosecretory perikarya of the carp (Luppa *et al.*, 1968; Kobayashi *et al.*, 1979). Audet and Chevalier (1981) found two types of aminergic cells among neurosecretory cells and green fluorescent CSF-contacting neurons in the posterior spinal cord of the brook trout. They suggest possible functional links between these three types of aminergic cells and caudal neurosecretory cells. That cholinergic neurons may innervate caudal neurosecretory cells has also been suggested from the presence of small vesicles in the endings forming synapses with these neurosecretory cells (Baumgarten *et al.*, 1970; Kobayashi *et al.*, 1979). This is supported by the presence of acetylcholinesterase (AChE) in the neurosecretory cells (Uemura, 1965; Luppa *et al.*, 1968; Kobayashi *et al.*, 1979). On a morphological basis, O'Brien and Kriebel (1983) have suggested the presence of monoaminergic, cholinergic, and peptidergic nerve terminals forming synaptic contacts with neurosecretory cells in *Poecilia sphenops*.

In the urophysis, monoamine fluorescence was not observed in *Gillichthys mirabilis* (Lederis *et al.*, 1974; Swanson *et al.*, 1975) or in the brook trout (Audet and Chevalier, 1981). However, Kobayashi *et al.* (1980) found dense, green fluorescent varicosities over the entire urophysis in the carp. It is evident that the urophysis is innervated by monoaminergic neurons in at least some species. Chemical determination demonstrated that the concentration of noradrenaline was 0.1 μg/g in the pike urophysis (Baumgarten *et al.*, 1970) and 2.22 ± 0.33 μg/g ($n = 5$) in the carp urophysis (Kobayashi *et al.,* 1980), and that neither adrenaline nor dopamine were detectable in either species (Baumgarten *et al.,* 1970; Kobayashi *et al.,* 1980).

Cholinergic innervation in the urophysis has not been demonstrated. Acetylcholinesterase activity was detected in the urophysis in some teleostean species (Uemura *et al.,* 1963; Kobayashi *et al.,* 1979) but not in others (Uemura, 1965; Luppa *et al.,* 1968). High concentrations of ACh have been found in urophyses of several species (Kobayashi *et al.,* 1963; Uemura *et al.,* 1963; Ichikawa and Kobayashi, 1978; Ichikawa, 1978). The source of this ACh is not known. Thus, these findings support the presence of cholinergic innervation in the urophysis. It should be noted here that in the carp the small vesicles (25–65 nm) in the neurosecretory endings are aggregated like synaptic vesicles in which the plasma membrane is in contact with the perivascular space (Oota, 1963b). It is postulated that if ACh is in the vesicles they may play a part in the release of hormones into capillaries from the axon endings.

The presence of cholinergic and monoaminergic innervations in the caudal neurosecretory cells is further suggested by observations that adrenaline, noradrenaline, and acetylcholine stimulate release of UI *in vitro* from a piece of the carp spinal cord containing the entire caudal neurosecretory system. Acetylcholine was more potent than noradrenaline, and both dopamine and 5-hydroxytryptamine were ineffective (Ichikawa and Kobayashi, 1978; Ichikawa, 1979).

As to the origins of the innervation of the caudal neurosecretory system, Yagi and Bern (1965) have suggested, based on electrophysiological studies, that caudal neurosecretory cells are controlled synaptically by a center probably located in the brain. O'Brien and Kriebel (1982) identified the descending systems which coordinate the link between the caudal neurosecretory system and cranial centers (Fig. 12). Using the horseradish peroxidase technique, they found that in *Poecilia sphenops,* the neurons in the reticular nuclei of the medulla and the neurons in the nuclei in the mesencephalic dorsal tegmentum send projections to caudal neurosecretory cells. Further, they found that the medullary projections originate from two groups of neurons: (1) cells in the ventromedial division of the reticular formation, which are cholinergic, and (2) small neurons forming a rostrocaudal longitudinal column in the dorsolateral aspects of the medulla. These neurons seem to be catecholaminergic. Both medullary nuclear groups have direct or indirect connections with various nerves and brain centers: the lower rhombencephalic cranial nerves, visceral sensory nuclei, and hypothalamic regions (Kriebel *et al.,* 1985).

The neurons in the mesencephalon, which have projections to the caudal neurosecretory perikarya, are found in the dorsal tegmental magnocellular nucleus (DTMN) and in the nucleus of the medial longitudinal fascicle (FLM). The cells of DTMN seem to be peptidergic, since they contain large

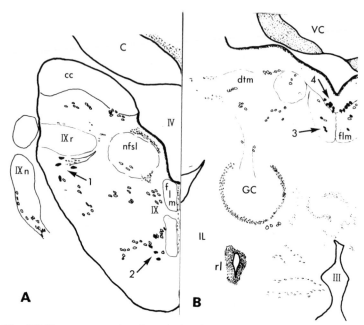

Fig. 12. (A) Transverse section through rhombencephalon at the level of the glossopharyngeal nerve and root (IXn and IXr, respectively) and nucleus (IX) in *Poecilia sphenops*. Neurons which project to the caudal neurosecretory system are present in visceral sensory nuclei (arrow 1) and in the medial reticular zone (arrow 2). C, cerebellum; cc, crista cerebellaris; nfsl, nucleus and fasciculus solitarii; flm, medial longitudinal fascicle; IV, fourth ventricle. (B) Transverse section through mesencephalon just caudal to posterior commisure in *Poecilia sphenops*. Arrow 3 shows projection cells in the nucleus of the medial longitudinal fascicle (flm). Peptidergic neurons are found in the dorsal tegmentum (arrow 4). dtm, dorsal tegmentum of mesencephalon; VC, valvula cerebelli; GC, glomerulosus nuclear complex; IL, inferior lobe; rl, nucleus recessus lateralis; III, third ventricle. [From Kriebel *et al.*, 1985.]

cytoplasmic granules, which may contain an immunoreactive leuteinizing-hormone-releasing hormone (LHRH) substance. Interestingly, LHRH-like staining of fibers and terminals has been seen in caudal neurosecretory system (Kriebel *et al.*, 1985). The peptidergic input from the DTMN is suggested as providing a photoneuroendocrine afferent connection to the caudal neurosecretory cells. The small cells in the FLM are cholinergic and seem to receive a variety of afferent inputs, including general visceral sensory information (Kriebel *et al.*, 1985).

In additon to DTMN peptidergic input to neurosecretory cells, peptidergic terminals appear to arise from collaterals of the caudal neurosecretory system itself (Kriebel *et al.*, 1985). Further, catecholaminergic neurons are present among caudal neurosecretory neurons (Audet and Chevalier, 1981), suggesting functional links between both types of neurons.

Thus, activity of the caudal neurosecretory system is coordinated with brain centers by descending cholinergic, catecholaminergic, and peptidergic projections and further seems to be controlled by local cathecholaminergic and peptidergic neurons.

There is the additional possibility, however, that the caudal neurosecretory cells can receive information directly from the blood vessels or cerebrospinal fluid, since, at least in some species, the cells are often in contact with blood vessels or extend their processes to the central canal (Fig. 9).

VII. SUMMARY

1. No caudal neurosecretory system exists in hagfish, lampreys, holocephalans, and dipnoans. Dahlgren cells, but not urophyses, are present in selachians, garpike, and acipenserids among lower actinopterygians. In the teleosts, syngnathids and molas do not have any elements of the caudal neurosecretory system.

2. On the basis of cell size, there seem to be at least two or three types of caudal neurosecretory cells in teleosts and elasmobranchs. It is not known whether cells of different sizes have different functions. There is immunocytological evidence, however, that they produce the same hormones, irrespective of differing size.

3. The caudal neurosecretory system develops at the late larval stage and later than the diencephalic neurosecretory system. After total extirpation of the system, the ependymal cells apparently transform into neurosecretory cells, and the system regenerates. However, there are species in which the system does not regenerate.

4. Electron microscopy reveals that axon endings terminating at capillaries in the urophysis contain large neurosecretory granules and small vesicles. The granules contain urophysial hormones, and the small vesicles, at least partly, are the carriers of acetylcholine.

5. Most neurosecretory cells in teleosts react to antisera of corticotropin-releasing factor (CRF), sauvagine, urotensin I (UI), and urotensin II (UII). A number of cells are immunoreactive to both UI and UII antisera, but some react only to antiserum of UI or UII. UI-, CRF-, and UII-like substances are detected in the caudal neurosecretory cells in several species of elasmobranchs. Most of these neurosecretory cells are immunoreactive to both UI and UII, but some are immunoreactive only to the antiserum of UI or UII. The dorsal cells of the lamprey are immunoreactive to neither CRF nor UII antiserum.

6. Projections from four brain centers to the caudal neurosecretory perikarya have been demonstrated: (1) cholinergic neurons in the ventromedial division of the reticular formation, (2) small catecholaminergic neurons

forming a rostrocaudal longitudinal column in dorsolateral aspects of the medulla, (3) peptidergic neurons in the dorsal tegmental magnocellular nucleus, and (4) small cholinergic neurons in the nuleus of the medial longitudinal fascicle. In addition, the caudal neurosecretory cells seem to be innervated by collaterals of caudal neurosecretory nerves and by catecholaminergic neurons located in the posterior spinal cord.

7. The urophysis contains monoaminergic fibers in some fishes but not in others. It also shows a high concentration of acetylcholine, suggesting the presence of cholinergic fibers. It is not yet fully understood whether neural mechanisms control hormone release at the level of the urophysis.

VIII. COMPARATIVE CONTRIBUTIONS AND BIOMEDICAL IMPLICATIONS

The evolution of the endocrine system has been conservative, with most endocrine glands present in fishes retained during the evolutionary process through to mammals. In some cases, however, modifications in the form of changes in the anatomical locations and physiological functions of these glands have occurred. The understanding of these evolutionary alterations is of great biomedical significance. By understanding the original function of an endocrine gland, novel interpretations may be developed for the complicated endocrine problems and, furthermore, may reveal new functions of the endocrine glands, other than the functions known to exist in higher vertebrates.

Some endocrine systems in fishes have, however, been lost with the evolution of fishes to amphibians. In these cases, it may be that the functions of these glands have been taken over by some other endocrine organs already present in tetrapods. Some examples will be described, principally based on morphological data.

1. The teleost caudal neurosecretory system, which secretes UI and UII, is missing in tetrapods. However, in higher vertebrates CRF shares several amino acid sequences with UI, and somatostatin exhibits homologous amino acid sequences with UII. Furthermore, some aspects of the physiological actions of CRF and somatostatin are similar to those of UI and UII, respectively (see Bern *et al.*, 1985). It is possible, therefore, that some of the functions of the urophysial hormones now known in fishes may come to be identified as new, but as yet unknown, functions of some other endocrine gland in higher vertebrates (Section V).

2. Speidel (1919, 1922) interpreted the Dahlgren cells as being glandlike nerve cells. This was the first formulation of the concept that nerve cells function as gland cells, as later pointed out by Scharrer and Scharrer (1937, 1945). Scharrer (1928) discovered secretory nerve cells in the hypothalamus

of the teleost *Phoxinus laevis*. This was the most important and significant finding for the establishment of the concept of neurosecretion. Thus, the concept of neurosecretion originated in fish, and this implies that findings obtained in lower vertebrates often lead investigators to formulate new concepts general in vertebrates (see Section I, Introduction).

3. Similarities in anatomical structure are seen between the caudal neurosecretory system and the hypothalamo-hypophysial neurosecretory system. However, the former has more simple structures than the latter; the caudal neurosecretory cells send axons only into the urophysis, but the hypothalamic neurosecretory system has two cell groups, preoptic and paraventricular nuclei, which send their axons to the median eminence and the neural lobe. Furthermore, each nucleus is surrounded by several other nonneurosecretory nuclei. Because of its structural simplicity, the caudal neurosecretory system may serve as a useful, novel experimental model to investigate fundamental questions about neurosecretion in general. For instance, a piece of the caudal spinal cord containing the entire caudal neurosecretory system can easily be removed and cultured in physiological saline solution for physiological studies (Ichikawa, 1978).

4. Electron microscopical profiles of the caudal neurosecretory neurons are similar to the hypothalamic neurosecretory neurons. The axonal endings of both caudal and hypothalamic neurosecretory cells contain secretory granules and synaptic vesiclelike structures. The vesicles are, at least partly, the carriers of ACh (Section IV), a finding which has significant biomedical implications. It is assumed from this that the vesicles in the neurosecretory axon endings in the median eminence and the neural lobe may be, in part, the carriers of ACh and may therefore be involved in the release of hypothalamic hormones from these endings.

5. After total extirpation of the caudal neurosecretory system, by cutting off the posterior part of the tail of the eel, loach, and tench, the ependymal cells near the cut transform into Dahlgren cells, and the caudal neurosecretory system completely regenerates (Section III). This process, which is revealed by Gomori's aldehyde fuchsin staining, may provide fundamental information for studies on the regeneration of the hypothalamic neurosecretory system, or even of ordinary neuronal tissues in higher vertebrates. Information obtained in lower vertebrates would also contribute to the understanding of higher vertebrates in this respect.

6. Because of the simple organization of the caudal neurosecretory system and the large size of the cells, this neurosecretory system in teleosts has been subjected to electrophysiological studies several times (Section VI). Recently, it has been elucidated in *Poecilia sphenops* that cholinergic, catecholaminergic, and peptidergic neurons located in four brain centers project their axons to the caudal neurosecretory cells (Section VI). This pattern of innervation may provide a good model for studying the nervous control of the secretion of hypothalamic neurohormones.

The caudal neurosecretory cells are not aggregated but are diffusely distributed in the caudal spinal cord. It is, therefore, expected that the caudal neurosecretory cells may send their axons to other caudal neurosecretory cells in order to transfer information. Neurosecretory cells producing adenohypophysial hormone releasing (inhibiting) hormones are either diffusely distributed or occur as separate cell groups in the brain of higher vertebrates. The caudal neurosecretory system may be a good model for studying mechanisms controlling the release of the neurohormones from diffusely distributed neurosecretory cells.

7. As mentioned previously, CRF with amino acid sequences, homologous with UI, and somatostatin, homologous with UII, are present in the mammalian brain. It has recently been revealed by immunocytochemistry that neuropeptides found in mammalian nervous tissues are frequently detected in invertebrates, and vice versa (Falkmer *et al.*, 1984). The fact that neuropeptides are found in lower vertebrates or invertebrates has significant biomedical importance. The physiological actions of these peptides in lower animals may contribute to the understanding of physiological actions in higher vertebrates, and vice versa. New functions of the peptides may be found in both vertebrates and invertebrates.

Each animal species has its own long evolutionary history of body structure and functions. Each animal has its ontogenetic history. When we know more of the phylogenetic and ontogenetic histories of each species, we shall be able to reach a much deeper understanding of animal phenomena.

ACKNOWLEDGMENTS

We would like to express our sincere appreciation to Professor H. A. Bern and Mr. R. S. Nishioka for providing beautiful unpublished electron micrographs and to Dr. B. A. Larson for helpful comments on the manuscript. We thank Professors S. Mikami and Y. Sano for helping with the immunohistochemistry of neuropeptides and Professors G. Fridberg, Y. Honma, and R. M. Kriebel for permission to use their figures. The work was supported by a Grant-in-Aid for Scientific Research and for Cooperative Research from the Ministry of Education, Science and Culture of Japan and by grants from the Ito Foundation for Promotion of Ichthyology and the Suzuken Memorial Foundation.

REFERENCES

Arsaky, A. (1813). De piscium cerebro et medulla spinali. Dissertatio inaugeralis, Halae.
Arvy, L. (1966). Le système neurosécréteur spinal des poissons ou 'système neurosécréteur caudal' d'Enami (1955). *Bull. Soc. Zool. Fr.* **91,** 217–249.
Audet, C., and Chevalier, G. (1981). Monoaminergic innervation of the caudal neurosecretory system of the brook trout *Salvelinus fontinalis* in relation to osmotic stimulation. *Gen. Comp. Endocrinol.* **45,** 189–203.
Baumgarten, H. G., Falck, B., and Wartenberg, H. (1970). Adrenergic neurons in the spinal

cord of the pike (*Esox lucius*) and their relation to the caudal neurosecretory system. *Z. Zellforsch. Mikrosk. Anat.* **107**, 479–498.

Bennett, M. V. L., and Fox, S. (1962). Electrophysiology of caudal neurosecretory cells in the skate and fluke. *Gen. Comp. Endocrinol.* **2**, 77–95.

Bern, H. A. (1969). Urophysis and caudal neurosecretory system. *In* "Fish Physiology" (W. S. Hoar and D. J. Randall, eds.), Vol. 2, pp. 399–418. Academic Press, New York.

Bern, H. A. (1972). Some questions on the nature and function of cranial and caudal neurosecretory systems in nonmammalian vertebrates. *Prog. Brain Res.* **38**, 85–96.

Bern, H. A., and Knowles, F. G. W. (1966). Neurosecretion. *In* "Neuroendocrinology" (L. Martini and W. F. Ganong, eds.), Vol. 1, pp. 139–186. Academic Press, New York.

Bern, H. A., and Lederis, K. (1978). The caudal neurosecretory system of fishes in 1976. *In* "Neurosecretion and Neuroendocrine Activity: Evolution, Structure and Function" (W. Bargmann, A. Oksche, A. Polenov, and B. Scharrer, eds.), pp. 341–349. Springer-Verlag, Berlin and New York.

Bern, H. A., and Takasugi, N. (1962). The caudal neurosecretory system of fishes. *Gen. Comp. Endocrinol.* **2**, 96–110.

Bern, H. A., Gunther, R., Johnson, D. W., and Nishioka, R. S. (1973). Occurrence of urotensin II (bladder-contracting activity) in the caudal spinal cord of anamniote vertebrates. *Acta Zool. (Stockholm)* **54**, 15–19.

Bern, H. A., Pearson, D., Larson, B. A., and Nishioka, R. S. (1985). Neurohormones from fish tails—The caudal neurosecretory system. I. "Urophysiology" and the caudal neurosecretory system of fish. *Recent Prog. Horm. Res.* **41**, 533–552.

Chan, D. K. O. (1971). The urophysis and the caudal circulation of teleost fish. *Mem. Soc. Endocrinol.* **19**, 391–412.

Chevalier, G. (1976). Ultrastructural changes in the caudal neurosecretory cells of the trout *Salvelinus fontinalis* in relation to external salinity. *Gen. Comp. Endocrinol.* **29**, 441–454.

Dahlgren, U. (1914). The electric motor nerve centers in the skates (*Rajidae*). *Science* **40**, 862–863.

Enami, M. (1955). Studies in neurosecretion. II. Caudal neurosecretory system in the eel (*Auguilla japonica*). *Gunma J. Med. Sci.* **4**, 23–36.

Enami, M. (1956). Studies in neurosecretion. VIII. Changes in the caudal neurosecretory system of the loach (*Misgurnus anguillicaudatus*) in response to osmotic stimuli. *Proc. Jpn. Acad.* **32**, 759–764.

Enami, M., and Imai, K. (1955). Studies in neurosecretion. V. Caudal neurosecretory system in several freshwater teleosts. *Endocrinol. Jpn.* **2**, 107–116.

Enami, M., and Imai, K. (1956a). Studies in neurosecretion. VI. Neurohypophysis-like organization near the caudal extremity of the spinal cord in several estuarine species of teleosts. *Proc. Jpn. Acad.* **32**, 197–200.

Enami, M., and Imai, K. (1956b). Studies in neurosecretion. VII. Further observations on the caudal neurosecretory system and neurohypophysis spinalis (Urohypophysis) in marine teleosts. *Proc. Jpn. Acad.* **32**, 633–638.

Enami, M., and Imai, K. (1958). Studies in neurosecretion. XII. Electron microscopy of the secrete granules in the caudal neurosecretory system of the eel. *Proc. Jpn. Acad.* **34**, 164–168.

Falkmer, S., El-Salhy, M., and Titlbach, M. (1984). Evolution of the neuroendocrine system in vertebrates. *In* "Evolution and Tumour Pathology of the Neuroendocrine System" (S. Falkmer, R. Häkanson, and F. Sundler, eds.), pp. 59–87. Elsevier, Amsterdam.

Favaro, G. (1925). Contributi allo studio morfologico dell' ipofisi caudale (rigonfiamento caudale della midolla spinale) dei teleostei. *Atti Accad. Naz. Lincei, Cl. Sci. Fis., Mat. Nat., Mem.* **1**, 30–72.

Favaro, G. (1926). Contribution à l'étude morphologique de l'hypophyse caudale (renflement caudal de la moelle èpiniére) des téléostéens. *Arch. Ital. Biol.* **75**, 164–170.

Feustel, G., and Luppa, H. (1968). Bausteinhistochemische Untersuchungen am caudalen neurosekretorischen System von *Cyprinus carpio* L. *Acta Histochem.* **31,** 358–380.

Fisher, A. W. F., Wong, K., Gill, V., and Lederis, K. (1984). Immunocytochemical localization of urotensin I neurons in the caudal neurosecretory system of the white sucker (*Catostomus commersoni*). *Cell Tissue Res.* **235,** 19–23.

Fridberg, G. (1962a). Studies on the caudal neurosecretory system in teleosts. *Acta Zool. (Stockholm)* **43,** 1–77.

Fridberg, G. (1962b). The caudal neurosecretory system in some elasmobranchs. *Gen. Comp. Endocrinol.* **2,** 249–265.

Fridberg, G. (1963). "Morphological Studies on the Caudal Neurosecretory System in Teleosts and Elasmobranchs," pp. 1–20. Åke Nyblom & Co., Boktryckeri AB, Stockholm.

Fridberg, G., and Bern, H. A. (1968). The urophysis and the caudal neurosecretory system of fishes. *Biol. Rev. Cambridge Philos. Soc.* **43,** 175–199.

Fridberg, G., and Nishioka, R. S. (1966). Secretion into the cerebrospinal fluid by caudal neurosecretory neurons. *Science* **152,** 90–91.

Fridberg, G., Bern, H. A., and Nishioka, R. S. (1966a). The caudal neurosecretory system of the isospondylous teleost, *Albula vulpes,* from different habitats. *Gen. Comp. Endocrinol.* **6,** 195–212.

Fridberg, G., Nishioka, R. S., Bern, H. A., and Fleming, W. R. (1966b). Regeneration of the caudal neurosecretory system in the cichlid teleost *Tilapia mossambica. J. Exp. Zool.* **162,** 311–335.

Gabe, M. A. (1966). The caudal neurosecretory pathway. *In* "Neurosecretion," pp. 658–674. Macmillan (Pergamon), New York.

Gauthier, L., Audet, C., and Chevalier, G. (1983). Régulations aminergique et cholinergique du système caudal neurosécréteur de l'omble de fontaine, *Salvelinus fontinalis,* en relation avec l'osmo-iono-regulation. *Can. J. Zool.* **61,** 2856–2867.

Gill, V. E., Burford, G. D., Lederis, K., and Zimmerman, E. A. (1977). An immunocytochemical investigation for arginine vasotocin and neurophysin in the pituitary gland and the caudal neurosecretory system of *Catostomus commersoni. Gen Comp. Endocrinol.* **32,** 505–511.

Goossens, N. (1976). Immunohistochemical evidence against the presence of vasotocin in the trout urophysis. *Gen. Comp. Endocrinol.* **30,** 231–233.

Hamana, K. (1962). Über die Neurophysis spinalis caudalis bei Fischen. *J. Kyoto Prefect. Univ. Med.* **71,** 478–490 (in Japanese, German abstract).

Holder, F. C., Schroeder, M. D., Guerne, J. M., and Vivien-Roels, B. (1979). A preliminary comparative immunohistochemical, radioimmunological, and biological study of arginine vasotocin (AVT) in the pineal gland and urophysis of some teleostei. *Gen. Comp. Endocrinol.* **37,** 15–25.

Holmgren, U. (1958). On the caudal neurosecretory system of the teleost fish *Fundulus heteroclitus* L. *Anat. Rec.* **132,** 454–455.

Holmgren, U. (1959). On the caudal neurosecretory system of the eel, *Anguilla rostrata. Anat. Rec.* **135,** 51–59.

Holmgren, U., and Chapman, G. B. (1960). The fine structure of the urophysis spinalis of the teleost fish *Fundulus heteroclitus. J. Ultrastruct. Res.* **4,** 15–25.

Honma, Y., and Tamura, E. (1967). Studies on Japanese chars of the genus *Salvelinus.* IV. The caudal neurosecretory system of the Nikkô-iwana, *Salvelinus leucomaenis pluvius* (Hilgendorf). *Gen. Comp. Endocrinol.* **9,** 1–9.

Ichikawa, T. (1978). Acetylcholine in the urophysis of several species of teleosts. *Gen. Comp. Endocrinol.* **35,** 226–233.

Ichikawa, T. (1979). Release of urotensin I by neurotransmitters and ultrastructural changes in the carp urophysis *in vitro. Gunma Symp. Endocrinol.* **16,** 87–95.

Ichikawa, T., and Kobayashi, H. (1978). Acetylcholine in the urophysis and release of urophysial hormones by neurotransmitters *in vitro. In* "Neurosecretion and Neuroendocrine Activity: Evolution, Structure and Function'' (W. Bargmann, A. Oksche, A. Polenov, and B. Scharrer, eds.), pp. 350–352. Springer-Verlag, Berlin and New York.

Ichikawa, T., McMaster, D., Lederis, K., and Kobayashi, H. (1982). Isolation and amino acid sequence of urotensin I, a vasoactive and ACTH-releasing neuropeptide, from the carp (*Cyprinus carpio*) urophysis. *Peptides* **3**, 859–867.

Ichikawa, T., Lederis, K., and Kobayashi, H. (1984). Primary structures of multiple forms of urotensin II in the urophysis of the carp, *Cyprinus carpio*. *Gen. Comp. Endocrinol.* **55**, 133–141.

Ichikawa, J., Pearson, D., Yamada, C., and Kobayashi, H. (1986). The caudal neurosecretory system of fishes. *Zool. Sci.* **3**, (in press).

Imai, K. (1965). Malformed caudal neurosecretory system in the eel, *Anguilla japonica*. *Embryologia* **9**, 78–97.

Inoue, S. (1959). Morphological changes in the caudal neurosecretory system of eel produced by the surgical transection. *Gunma J. Med. Sci.* **8**, 263–280.

Ishibashi, T. (1962). Electrical activity of the caudal neurosecretory cells in the eel *Anguilla japonica* with special reference to synaptic transmission. *Gen. Comp. Endocrinol.* **2**, 415–424.

Ishimura, K., and Lederis, K. (1983). Immunohistochemistry of the caudal neurosecretory system in the white sucker, *Catostomus commersoni*. *Folia Endocrinol. Jpn.* **59**, 4.

Jaiswal, A. G., and Belsare, D. K. (1973). Comparative anatomy and histology of the caudal neurosecretory system in teleosts. *Z. Mikrosk.-Anat. Forsch.* **87**, 589–609.

Jaiswal, A. G., and Belsare, D. K. (1974). Regeneration of the caudal neurosecretory system in the catfish, *Clarias batrachus* L. *Z. Mikrosk.-Anat. Forsch.* **88**, 987–996.

Kobayashi, H., Uemura, H., Oota, Y., and Ishii, S. (1963). Cholinergic substance in the caudal neurosecretory storage organ of fish. *Science* **141**, 714–716.

Kobayashi, Y., Ichikawa, T., and Kobayashi, H. (1979). Innervation of the caudal neurosecretory system of the teleost. *Gunma Symp. Endocrinol.* **16**, 81–86.

Kobayashi, Y., Kobayashi, H., Ohshiro, S., Osumi, Y., and Fujiwara, M. (1980). Monoaminergic innervation of the caudal neurosecretory system of the carp. *Cyprinus carpio*. *Zentralbl. Veterinaermed., Reihe C* **9**, 65–72.

Kobayashi, Y., Ichikawa, T., Okawara, Y., and Kobayashi, H. (1986a). Weight of the urophysis of teleosts. *J. Fac. Sci., Univ. Tokyo, Sect. 4* (in press).

Kobayashi, Y., Lederis, K., Rivier, J., Ko, D., Mcmaster, D., and Poulin, P. (1986b). Radioimmunoassays for fish tail neuropeptides. II. Development of a specific and sensitive assay for and the occurrence of immunoreactive urotensin II in the central nervous system and blood of *Catostomus commersoni*. *J. Pharmacol. Methods* **15**, 321–334.

Kriebel, R. M. (1980). The caudal neurosecretory system of *Poecilia sphenops*. *J. Morphol.* **165**, 157–165.

Kriebel, R. M., Meetz, G. D., and Burke, J. D. (1976). The relation between capillaries and neurons in the caudal neurosecretory system of *Pomatomus saltatrix*. *Experientia* **32**, 70.

Kriebel, R. M., Burke, J. D., and Meetz, G. D. (1979). Morphologic features of the caudal neurosecretory system in the blueback herring, *Pomolobus aestivalis*. *Anat. Rec.* **195**, 553–571.

Kriebel, R. M., Parsons, R. L., and Miller, K. E. (1985). Innervation of caudal neurosecretory cells. *In* ''Neurosecretion and the Biology of Neuropeptides'' (H. Kobayashi, H. A. Bern, and A. Urano, eds.), pp. 205–211. Jpn. Sci. Soc. Press, Tokyo.

Lacanilao, F. (1972a). The urophysial hydrosmotic factor of fishes. I. Characteristics and similarity to neurohypophysial hormones. *Gen. Comp. Endocrinol.* **19**, 405–412.

Lacanilao, F. (1972b). The urophysial hydrosmotic factor of fishes. II. Chromatographic and pharmacologic indications of similarity to arginine vasotocin. *Gen. Comp. Endocrinol.* **19**, 413–420.

Lacanilao, F., and Bern, H. A. (1972). The urophysial hydrosmotic factor of fishes. III. Survey of fish caudal spinal cord regions for hydrosmotic activity. *Proc. Soc. Exp. Biol. Med.* **140**, 1252–1253.

Lederis, K. (1974). Chemical and pharmacological properties of urotensin I, a long acting mammalian hypotensive peptide. *Acta Physiol. Lat. Am.* **24**, 481–483.

Lederis, K. (1977). Chemical properties and the physiological and pharmacological actions of urophysial peptides. *Am. Zool.* **17**, 823–832.

Lederis, K., Bern, H. A., Nishioka, R. S., and Geschwind, I. I. (1971). Some observations on biological and chemical properties and subcellular localization of urophysial active principles. *Mem. Soc. Endocrinol.* **19**, 413–433.

Lederis, K., Bern, H. A., Medakovic, M., Chan, D. K. O., Nishioka, R. S., Letter, A., Swanson, D., Gunther, R., Tesanovic, M., and Horne, B. (1974). Recent functional studies on the caudal neurosecretory system of teleost fishes. *In* "Neurosecretin—The Final Neuroendocrine Pathway" (F. Knowles and L. Vollrath, eds.), pp. 94–103. Springer-Verlag, Berlin and New York.

Lederis, K., Letter, A., McMaster, D., Moore, G., and Schlesinger, D. (1982). Complete amino acid sequence of urotensin I, a hypotensive and corticotropin-releasing neuropeptide from *Catostomus commersoni*. *Science* **218**, 162–164.

Lederis, K., Letter, A., McMaster, D., Ichikawa, T., MacCannell, K. L., Kobayashi, Y., Rivier, J., Rivier, C., Vale, W., and Fryer, J. (1983). Isolation, analysis of structure, synthesis and biological actions of urotensin I neuropeptides. *Can. J. Biochem. Cell Biol.* **61**, 602–614.

Loretz, C. A., Bern, H. A., Foskett, J. K., and Mainoya, J. R. (1982). The caudal neurosecretory system and osmoregulation in fish. *In* "Neurosecretion: Molecules, Cells, Systems" (D. S. Farner and K. Lederis, eds.), pp. 319–328. Plenum, New York.

Luppa, H., Weiss, J., and Feustel, G. (1968). Histochemische Untersuchungen zur Lokalisation von Acetylcholin-esterase, Monoaminoxydase und Monoaminen im kaudalen neurosekretorischen System von *Cyprinus carpio*. *Z. Zellforsch. Mikrosk. Anat.* **89**, 499–508.

McMaster, D., and Lederis, K. (1983). Isolation and amino acid sequence of two urotensin II peptides from *Catostomus commersoni* urophyses. *Peptides* **4**, 367–374.

Mimura, O. M. (1978). Ocorréncia do sistema neurossecretor caudal em *Lepidosiren paradoxa* (Peixe Dipnóico) em fase estival. *Bol. Fisiol. Anim., Univ. Sao Paulo* **2**, 43–48.

Morita, H., Ishibashi, T., and Yamashita, S. (1961). Synaptic transmission in neurosecretory cells. *Nature (London)* **191**, 183.

Nag, A. C. (1967). Functional morphology of the caudal region of certain clupeiform and perciform fishes with reference to the taxonomy. *J. Morphol.* **123**, 529–558.

O'Brien, J. P., and Kriebel, R. M. (1982). Brain stem innervation of the caudal neurosecretory system. *Cell Tissue Res.* **227**, 153–160.

O'Brien, J. P., and Kriebel, R. M. (1983). Caudal neurosecretory system synaptic morphology following deafferentation: An electron microscopic degeneration study. *Brain Res. Bull.* **10**, 89–96.

Onstott, D., and Elde, R. (1984). Immunohistochemical localization of urotensin I/Corticotropin-releasing factor immunoreactivity in neurosecretory neurons in the caudal spinal cord of fish. *Neuroendocrinology* **39**, 503–509.

Oota, Y. (1963a). Fine structure of the caudal neurosecretory system of the carp. *Cyprinus carpio*. *J. Fac. Sci., Univ. Tokyo, Sect. 4* **10**, 129–141.

Oota, Y. (1963b). On the synaptic vesicles in the neurosecretory organs of the carp, bullfrog, pigeon and mouse. *Annot. Zool. Jpn.* **36**, 167–172.

Owada, K., Kawata, M., Akaji, K., Takagi, A., Moriga, M., and Kobayashi, H. (1985a). Urotensin II-immunoreactive neurons in the caudal neurosecretory system of freshwater and seawater fish. *Cell Tissue Res.* **239**, 349–354.

Owada, K., Yamada, C., and Kobayashi, H. (1985b). Immunoreactivity to antisera of urotensin II and corticotropin-releasing factor in the caudal neurosecretory system of the elasmobranchs and the dorsal cells of the lamprey. *Cell Tissue Res.* **242**, 527–530.

Pearson, D., Shivery, J. E., Clark, B. R., Geschwind, I. I., Barkley, M., Nishioka, R. S., and Bern, H. A. (1980). Urotensin II: A somatostatin-like peptide in the caudal neurosecretory system of fishes. *Proc. Natl. Acad. Sci. U.S.A.* **77,** 5021–5024.

Peyrot, A. (1964). Il sistema neurosecernente caudale degli ittiopsidi: Osservazioni sulla lampreda di ruscello (*Lampetra zanandrei* Vladikov). *Boll. Soc. Ital. Biol. Sper.* **40,** 207–211.

Renda, T., D'Este, L., Negri, L., and Lomanto, D. (1981). Evidenziazione immunoistochimica di un nervo peptide (sauvagina) Nellúrifisi di alcuni teleostei. *Basic Appl. Histochem.* **25,** *Suppl.* 83.

Renda, T., D'Este, L., Negri, L., and Lomanto, D. (1982). Sauvagine-like immunoreactivity in the bony fish urophysis and caudal neurosecretory system. *Basic Appl. Histochem.* **26,** 89–98.

Saenko, I. I. (1970). Caudal neurosecretory system in sturgeons. *Dokl. Akad. Nauk. SSSR* **194,** 218–221.

Saenko, I. I. (1978). Caudal neurosecretory system in Acipenseridae and some aspects of its evolution. *In* "Neurosecretion and Neuroendocrine Activity: Evolution, Structure and Function" (W. Bargmann, A. Oksche, A. Polenov, and B. Scharrer, eds.), pp. 353–356. Springer-Verlag, Berlin and New York.

Sano, Y. (1958a). Über die Neurophysis (sog. Kaudalhypophyse, 'Urohypophyse') des Teleostiers *Tinca vulgaris. Z. Zellforsch. Mikrosk. Anat.* **47,** 481–497.

Sano, Y. (1958b). Weitere Untersuchungen über den Feinbau der Neurophysis spinalis caudalis. *Z. Zellforsch. Mikrosk. Anat.* **48,** 236–260.

Sano, Y. (1961). Das caudale neurosekretorische System bei Fischen. *Ergeb. Biol.* **24,** 191–212.

Sano, Y. (1965). The caudal neurosecretory system. *In* "Central Regulation of Internal Secretion" (S. Katsuki, ed.), pp. 469–522. Igaku Shoin, Tokyo (in Japanese, German abstract).

Sano, Y., and Hartmann, F. (1959). Über durchschneidungsversuche am kaudalen neurosekretorischen System von *Tinca vulgaris* (mit berücksichtigung des reissnerschen Fadens). *Z. Zellforsch. Mikrosk. Anat.* **50,** 415–424.

Sano, Y., and Kawamoto, M. (1959). Entwicklungsgeschichtliche Beobachtungen an der Neurophysis spinalis caudalis von *Lebistes recticulatus* Peters. *Z. Zellforsch. Mikrosk. Anat.* **51,** 56–64.

Sano, Y., and Kawamoto, M. (1960). Histologische Untersuchungen endozellulärer kapillaren neurosekretorischer Zelen. *Z. Zellforsch. Mikrosk. Anat.* **51,** 152–156.

Sano, Y., and Knoop, A. (1959). Elektronenmikroskopische Untersuchungen am kaudalen neurosekretorischen System von *Tinca vulgaris. Z. Zellforsch. Mikrosk. Anat.* **49,** 464–492.

Sano, Y., Kawamoto, M., and Hamana, K. (1962). Entwicklungsgeschichtliche Untersuchungen am kaudalen neurosekretorishen System von *Salmo irideus. Acta Anat. Nippon.* **37,** 117–125.

Sano, Y., Iida, T., and Taketomo, S. (1966). Weitere elektronenmikroskopische Untersuchungen am kaudalen neurosekretorischen System von Fischen. *Z. Zellforsch. Mikrosk. Anat.* **75,** 328–338.

Scharrer, E. (1928). Die Lichtempfindlichkeit blinder Elritzen (Unterzuchungen über das Zwischenhirn der Fische I). *Z. Vergl. Physiol.* **7,** 1–38.

Scharrer, E., and Scharrer, B. (1937). Über Drüsen-Nervenzellen und neurosekretorische Organe bei wirbellosen und Wirbeltieren. *Biol. Rev. Cambridge Philos. Soc.* **12,** 185–216.

Scharrer, E., and Scharrer, B. (1945). Neurosecretion. *Physiol. Rev.* **25,** 171–181.

Speidel, C. C. (1919). Gland-cells of internal secretion in the spinal cord of the skates. *Pap. Dep. Mar. Biol., Carnegie Inst. Washington* **13,** 1–31.

Speidel, C. C. (1922). Further comparative studies in other fishes of cells that are homologous to the large irregular glandular cells in the spinal cord of the skates. *J. Comp. Neurol.* **34,** 303–317.

Sterba, G. (1962). Distribution of nerve cells with secretory-like granules in Petromyzontes. *Nature (London)* **193,** 400–401.

Sterba, G. (1972). Neuro- and gliasecretion. *In* "The Biology of Lampreys" (M. W. Hardisty and I. C. Potter, eds.), Vol. 2, pp. 69–89. Academic Press, New York.

Suess, U., Lawrence, J., Ko, D., and Lederis, K. (1986). Radioimmunoassays for fish tail neuropeptides. I. Development of assay and measurement of immunoreactive urotensin I in *Catostomus commersoni* brain, pituitary and plasma. *J. Pharmacol. Methods* (in press)

Swanson, D. D., Nishioka, R. S., and Bern, H. A. (1975). Aminergic innervation of the cranial and caudal neurosecretory systems in the teleost *Gillichthys mirabilis*. *Acta Zool (Stockholm)* **56,** 225–237.

Tsuneki, K., and Kobayashi, H. (1979). Regeneration of the caudal neurosecretory cells in the teleosts, *Misgurnus anguillicaudatus* and *Oryzias latipes*. *Gunma Symp. Endocrinol.* **16,** 69–79.

Uemura, H. (1965). Histochemical studies on the distribution of cholinesterase and alkaline phosphatase in the vertebrate neurosecretory system. *Annot. Zool. Jpn.* **38,** 79–96.

Uemura, H., Kobayashi, H., and Ishii, S. (1963). Cholinergic substances in the neurosecretory storage release organs. *Zool. Mag.* **72,** 204–212.

Verne, J. (1914). A l'étude des cellules névrogliques. *Arch. Anat. Microsc. Morphol. Exp.* **16,** 149–192.

Wilén, P. E., and Fridberg, G. (1973). Ultrastructural studies on the ontogenesis of the caudal neurosecretory system in the roach, *Leuciscus rutilus*. *Z. Anat. Entwicklungsgesch.* **139,** 207–216.

Yagi, K., and Bern, H. A. (1965). Electrophysiologic analysis of the response of the caudal neurosecretory system of *Tilapia mossambica* to osmotic manipulations. *Gen. Comp. Endocrinol.* **5,** 509–526.

Yamada, C., Owada, K., and Kobayashi, H. (1985). Colocalization of corticotropin-releasing factor/urotensin I and urotensin II in the caudal neurosecretory neurons in the carp, *Cyprinus carpio*. *Zool. Sci.* **2,** 813–816.

Yamada, C., Owada, K., Ichikawa, T., Iwanaga, T., and Kobayashi, H. (1986a). Immunohistochemical localization of urotensin I and II in the caudal neurosecretory neurons of the carp *Cyprinus carpio* and the sharks *Heterodontus japonicus* and *Cephaloscyllium umbratile*. *Arch. Histol. Jpn.* **49,** 39–44.

Yamada, C., Ichikawa, T., Owada, K., Yamada, S., Iwanaga, T., and Kobayashi, H. (1986b). Immunohistochemical localization of urotensin I and several other neuropeptides in the caudal neurosecretory system of the carp, *Cyprinus carpio* and the shark, *Heterodontus japonicus* and *Cephaloscyllium umbratile*. *Cell Tissue Res.* (in press).

6

The Thyroid Gland

JAMES DENT
Department of Biology
University of Virginia
Charlottesville, Virginia 22901

I. HISTORICAL EVENTS

The historical details of the discovery of the thyroid gland and its functions have been succinctly documented by Rolleston (1936) and Werner (1978). The thyroid was given a casual description by Galen in his *De Voce* (cited by Rolleston, 1936) but was first described in detail by A. Vesalius in 1543 (cited by Rolleston, 1936). Its endocrine status was anticipated by G. Casserio, who in 1601 noted that it was lacking a duct (cited by Rolleston, 1936). It was referred to merely as a "gland of the laryngeal region" until 1656 when Thomas Wharton (cited by Rolleston, 1936) called it the thyroid gland. The gland was so named because in man it is positioned on the larynx and applied to the thyroid cartilage which is shield-shaped. The term "thyroid" is from Greek roots that signify "oblong shield."

Galen considered that the thyroid gland gave forth a lubricant which facilitated movement of the laryngeal cartilages. Other early conjectures concerning its function ranged from the speculation that it was there to round out the angularity of the larynx and to thus beautify the neck (T. Wharton, 1656, cited by Rolleston, 1936) to the idea that it served as a receptacle for worms which extruded ova that passed by ducts into the esophagus (Cowper, 1698).

Although it has been claimed that Chinese physicians as early as 3000 B.C. treated enlargement of the thyroid with iodine-containing marine plants (Trotter, 1964), it was not until the end of the past century that Baumann (1895) discovered large concentrations of iodine to be a consistent feature of the thyroid. The growth-promoting function of the thyroid was recognized in 1526 when the cretinism syndrome was described clinically as resulting from thyroidal deficiency (Norris, 1980).

Surgical removal of the thyroid gland was first reported by Astley Cooper in 1836 (cited by Rolleston, 1936). Over a period of years the experiment was repeated by numerous workers, but the results obtained were generally con-

175

VERTEBRATE ENDOCRINOLOGY:
FUNDAMENTALS AND BIOMEDICAL IMPLICATIONS
Volume 1

fusing, particularly since the parathyroid glands were often removed along with the thyroid, until the topographical association of the parathyroids and the thyroid was emphasized by E. Gley in 1891 (cited by Rolleston, 1936). The action of the thyroid in maintaining mammalian metabolism at its proper level was established in 1895 by Magnus-Levy (cited by Rolleston, 1936).

Comparative studies of thyroidal function may be said to have begun in 1912 with the discovery by Gudernatsch that tadpoles that were fed with bits of thyroid from a horse underwent precocious metamorphosis. These findings were expanded by Allen (1916, 1929), who reported that extirpation of the thyroid prevented metamorphosis. Among the lower vertebrates, thyroidal activity is directed primarily toward developmental changes and alterations in integumentary structures, although it is often found to serve a permissive role in the actions of other hormones.

The dual nature of the mammalian thyroid gland was foretold in 1876 when Baber reported that in addition to cells of the follicular epithelium in the dog, cells of a second type could be identified. He called those cells "parenchymatous cells" to distinguish them from the predominant or "principal cells" of the gland. Other investigators found them to be a consistent feature of the thyroids of other mammals (Stoeckel and Porte, 1970). In general, they were found to stain more lightly than the epithelial cells but to contain granules that gave evidence of secretory function (Nonidez, 1932). It was suggested by Born in 1883 that they were introduced not from the thyroidal anlage but from the fourth pharyngeal or ultimobranchial pouch. This view became firmly established by a series of cytological, histochemical, and ultrastructural studies (Stoeckel and Porte, 1970). The ultimobranchial glands of submammalian vertebrates persist as separate entities and do not fuse with the thyroid gland (see Chapter 8).

The function of this secondary cell type remained a mystery until the early 1960s when it was discovered to be the source of the hypocalcemic hormone calcitonin. In consequence they are now referred to as "C cells."

II. DEVELOPMENT OF THE MAMMALIAN THYROID

The comparative aspects of the development, both phylogenetic and ontogenetic, of the thyroid gland will be discussed in the following sections of this chapter. It is convenient, however, in the beginning to describe briefly the developmental anatomy of the mammalian gland and to thus provide a basis for the comparisons that will be made later. A similar approach will be used as we move to considerations of microanatomy and ultrastructural features. Among vertebrates, in general, the thyroidal anlage makes its appearance early in the differentiation of the pharynx. The mammalian primordium is first seen as an identation in the floor of the presumptive pharynx at a stage when the neural folds are fusing to form the neural tube and when the tubular heart is beginning to elongate and to fold upon itself (Fig. 1A–C). Within the indentation, cells become rounded and evaginate to form the thyroidal anlage, a protuberance that elongates ventrad and caudad between the hyomandibular pouches. The anlage is initially an outpocketing, but as its development proceeds its lumen usually collapses, except at the proximal end, where it persists as a depression called the *foramen caecum* (Norris, 1918). As development continues the anlage bifurcates distally. The cells of the distal portions proliferate laterally and posteriorly, but the unbranched proximal portion ceases to grow. It remains for a time as the thyroglossal cord, a connection between the expanding distal portions and the pharyngeal floor at the base of the differentiating tongue. In rare anomalous instances, the duct remains visible in the adult; however, it typically degenerates while the paired distal masses continue to increase in size and to differentiate as the lateral lobes of the thyroid.

With their continued growth these distal masses lose their compact character, and their cells become more loosely distributed. They are invaded by mesenchymatous mesodermal cells which will give rise to stromal connective tissue and to the abundance of venules and capillaries that will eventually carry blood to and from the developing gland. Later, during the seventh gestational week in man (Arey, 1974), a second invasion of mesenchymatous cells stems from the most posterior of the pharyngeal pouches, the ultimobranchial pouches, against which the paired expanding thyroidal anlagen begin to press. Those cells will differentiate to form the C cells of the definitive gland.

The loosely arranged endodermal cells of the paired anlagen next begin to come together to form labyrinthian cords that are separated by spaces in which the mesodermal cells begin to form blood vessels.

Still later, during the eighth week in man (Arey, 1974), cavities arise within swollen parts of the cords. These cavities soon fill with colloid and become functional follicles. While the follicles differentiate, most of the presumptive C cells separate themselves from the follicular walls and come

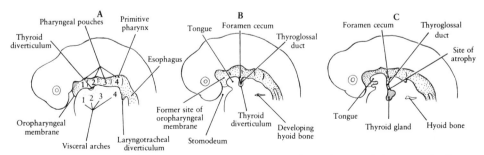

Fig. 1. Diagrams of sagittal sections of the human head showing successive stages in the development of the thyroid gland at 4, 5, and 6 weeks (A, B, and C, respectively). [From Hopper and Hart, 1985.]

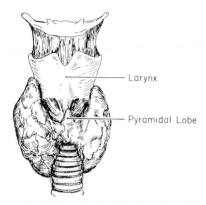

— Larynx

— Pyramidal Lobe

Fig. 2. The human thyroid gland in its definitive form. [From Hadley, 1984.]

together in loose clusters within the interfollicular spaces, although a few remain interspersed among the principal cells as constituent parts of the follicular epithelium (Nonidez, 1932). After the regression of the thyroglossal cord the lateral lobes retain a medial attachment which becomes the isthmus. In about 30% of human thyroids a small median pyramidal lobe extends anteriorly from the isthmus (Fig. 2). In mammals and other tetrapods a capsule of connective tissue differentiates from adjacent mesenchyme to surround the definitive gland.

III. MICROANATOMY

The follicle is the basic functional unit of the thyroid gland. Fluid-filled cavities are occasionally seen within the pars distalis of the pituitary gland, but except for the ultimobranchial gland of the urodele, the thyroid gland is the only consistently follicular vertebrate endocrine organ.

The follicles are usually spherical or ovoidal in shape. They consist of an external layer of simple basophilic epithelial cells which confines a central mass of coagulable fluid referred to as the colloid (Fig. 3).

The colloid is clear and homogeneous in appearance, except for rare desquamated epithelial cells and even more rare macrophages. It stains vividly with either basic or acidic aniline dyes, and with trichrome stain, it may be both acidophilic and basophilic in the same follicle (Bloom and Fawcett, 1975). No physiological significance is ascribed to these tinctorial responses, but the colloid also stains intensely with the periodic acid-Schiff (PAS) reagent, giving evidence of the glycoproteinacious nature of the thyroglobulin (Tg) which, aside from water, is the major constituent of the colloid.

The follicles are encompassed by thin basal lamellae which, in turn, are overlaid with plexi of capillaries and small venules (Fig. 4), lymphatic ves-

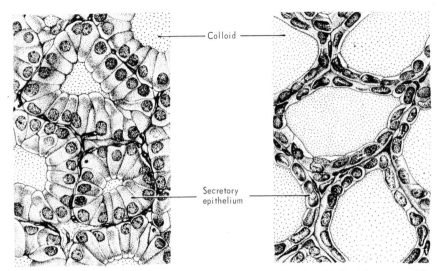

Fig. 3. Diagram showing, on the left, a section of an active thyroid gland from a rat. Note the high columnar cells of the secretory epithelium and the reduced colloidal space. On the right, a section from an inactive gland with low cuboidal epithelium and abundant colloid. [From Turner and Bagnara, 1976.]

sels, and sympathetic nerve fibers. Although exceptional responses are sometimes encountered (Dickhoff and Darling, 1983), the character of the colloid, the thickness (or height) of the epithelium, and the diameter of the blood vessels tend to vary with the functional state of the gland. In its active phase, the epithelium of the follicle is thickened. Its cells are columnar in form and the lumen is correspondingly reduced (Fig. 3). Large numbers of small, nonstaining vacuoles (reabsorption lacunae) appear in the colloid adjacent to the epithelium, and droplets of colloid are often present within the epithelial cytoplasm. The blood vessels of the active gland become distended and engorged with blood.

The epithelium of the inactive follicle is squamous in character. Vacuoles are rarely seen in the colloid, and the blood vascularity is not excessive. There is usually considerable variability among the responses of the follicles of a given gland. Although at any time many of the follicles will be in an active, inactive, or intermediate condition, some are usually out of phase with the majority, giving evidence of greater or lesser activity.

The C cells stain less deeply with routine histological procedures than do the principal cells. With the silver nitrate method of Cajál they are distinguished from the principal cells by the presence of black or brown cytoplasmic granules which, with trichrome stain, exhibit an affinity for aniline blue. The C cells that are intercalated in the follicular epithelium never border directly on the lumen, being separated from it by processes of the neighboring principal cells (Fig. 5) (Bloom and Fawcett, 1975).

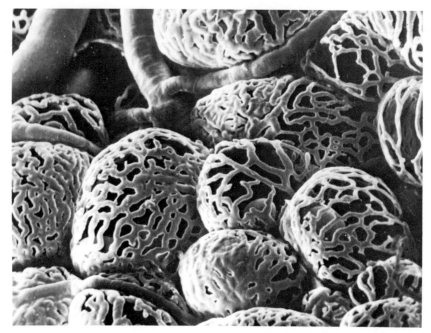

Fig. 4. Scanning electron micrograph demonstrating the pattern of blood vessels within the thyroid gland of a monkey. The vessels had been injected with a liquid plastic material which hardened and remained after the tissues were digested away. The persistent casts thus produced show blood vessel spaces of an arteriole and a venule in the upper left and the spaces of the many capillaries that lay on the rounded surfaces of follicles that are no longer present. [Provided by Professor Hisao Funita.]

IV. EPITHELIAL ORGANELLES

As explained earlier, various aspects of mammalian glands will be described initially to provide bases for comparison with the glands of other vertebrates. That approach is particularly useful in the treatment of ultrastructural elements in which the wealth of detailed knowledge of the mammalian gland completely outbalances the information regarding the glands of other vertebrate classes. The follicular epithelium is a "two-way street." In the very earliest cytological studies of the gland (R. J. Ludford and W. Cramer, 1928, cited by Rolleston, 1936), one finds suggestions that hormone is produced in the epithelial cells and is stored in the colloid to pass back through the cells before being discharged into the bloodstream. These early suggestions were soon confirmed, but the details of hormonal synthesis, storage, recovery, and release were not revealed until techniques employing radioactive iodine, electron microscopy, and immunochemistry became available.

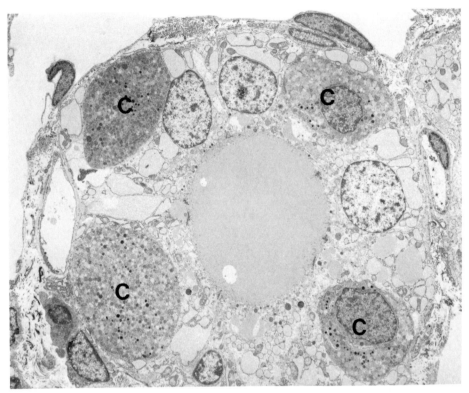

Fig. 5.   In this electron micrograph the epithelium of a follicle from the thyroid gland of an opossum contains intercalated C cells (C) of the ultimobranchial gland. ×3000. [Provided by Dr. James A. Fortney.]

The current understanding of the functional events that take place in the mammalian thyroid follicle is presented schematically in Figure 6. It is immediately apparent from this figure that the plasma membrane of the epithelial cell is a truly important structure. First of all, it contains the receptors which link with molecules of thyroid-stimulating hormone (TSH) brought to the cell by the bloodstream. The hormone–receptor complex activates molecules of adenylate cyclase, which induce the production of cyclic adenosine monophosphate (cAMP). All of the subsequent follicular activities appear to be mediated by cAMP-stimulated phosphorylation of substrate proteins, since cAMP analogs can mimic most actions of TSH (Van Herle *et al.*, 1979).

Glucose, the source of energy for those actions, passes through the membrane, as does the iodide, which is the critical constituent of the hormonal molecule. It was Baumann who in 1895 discovered a particularly high concentration of iodine in the thyroid gland. The concentration of iodide may

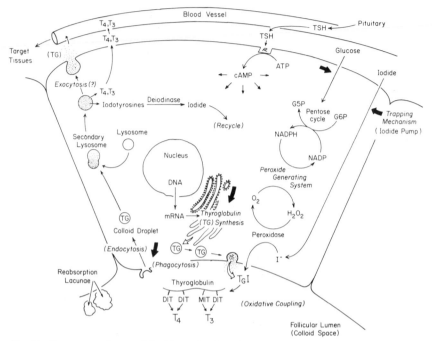

Fig. 6. Summary scheme of the biosynthesis and secretion of thyroid hormone. The actual oxidative coupling of iodinated tyrosine residues probably takes place at the apical surface of the follicular cell rather than in the lumen. [From Hadley, 1984.]

range from 25 to 500 times the level of the iodide circulating in the blood stream. In man, thyroidal iodine constitutes 90% of that found in the whole body. Such concentrations of iodine obviously are not achieved by passive diffusion. It is clear that the basal plasma membrane contains a powerful pump which transports iodide into and across the cell against an electrical gradient (Chow *et al.*, 1982).

Peroxidase-generating systems produce hydrogen peroxide from an unknown source (Ekholm and Björkman, 1985). Perioxidase converts iodide to an oxidized species. The oxidation of iodide is essential to its binding to tyrosyl groups of thyroglobulin molecules to form monoiodotyrosine (MIT) and diiodotyrosine (DIT) residues (Ekholm and Wollman, 1975). Within the Tg molecule, the same peroxidase may be responsible for the oxidative coupling of the iodinated tyrosines to produce some triiodothyronine (T_3) and larger amounts of tetraiodothyronine (T_4) (Taurog, 1974, 1979).

Thyroglobulin is a large, globular glycoprotein with a molecular weight of 600,000 and a sedimentation coefficient of 19 S (Edelhoch, 1965). Its synthesis begins in the nucleus where, as reviewed by Van Herle *et al.* (1979), its mRNA precursor is transcribed and undergoes polyadenylation, and proba-

bly capping as well, before being processed and transported to the cytoplasm. The nuclear events of the scheme are accepted largely on faith, but
the translational events are somewhat better understood. For example, a 33
S mRNA molecule with some 8600 bases has been isolated from the cow
(Vassart *et al.*, 1975, 1977). Upon its introduction into the oocyte of a frog by
the method of Gurdon *et al.* (1971), the molecule has been shown to translate
into a 300,000-Da polypeptide, immunologically and chemically related to
the Tg of the cow (Vassert *et al.*, 1975).

Rough endoplasmic reticulum (RER) occupies most of the basal and paranuclear regions of the principal cells (Fig. 7) (Ekholm, 1979). It has been
shown by ultrastructural autoradiography that the molecular cores of Tg

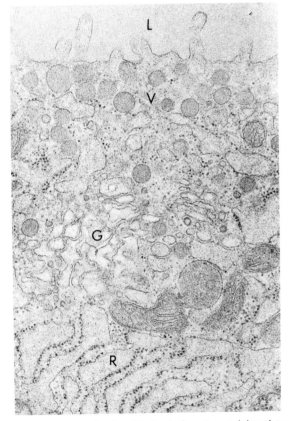

Fig. 7.   An epithelial cell from the thyroid gland of a rat containing cisternae of the rough
endoplasmic reticulum (R) associated with Golgi profiles (G) which, in turn, relate to exocytotic
vesicles (V) that are distributed between the Golgi apparatus and the apical cell membrane
adjoining the follicular lumen (L). ×21,000. [By permission from Ragnar Ekholm (1979). Anatomy and Development. *In* "Endocrinology" (L. J. De Groot, ed) Vol. 1, Grune and Stratton,
Inc., New York.

form on the polysomes of the RER and pass into its cisternae where glycosylation begins (Feeney and Wissig, 1972). Carbohydrate chains are synthesized and transferred to the core Tg molecules (Waechter and Lennarz, 1976). Eventually, sialic acid and additional carbohydrate residues are added to each molecule (Spiro and Spiro, 1985).

Small vesicles leave the endoplasmic reticulum to join the forming face of the Golgi apparatus where glycosylation is completed (Whür et al., 1969). A wide variety of vesicles are found in the apical cytoplasm. Originally, it was not clear which were transporting Tg from the Golgi apparatus to the luminal surface. It was discovered, however, that if the thyrotropic stimulation were removed by treatment with thyroid hormone, although synthesis of Tg was slowed (Melander and Rerup, 1968; Ericson, 1981b), removal of colloid from the lumen was, for a time, essentially halted (Björkman et al., 1974). By this means it was then shown in T_4-treated rats that small, 150-nm, exocytotic vesicles carrying newly synthesized (and suitably labeled) Tg moved through the apical cytoplasm to discharge their contents into the follicular lumen (Fig. 7). Furthermore, peroxidase was also shown to be carried in those same vesicles (Björkman et al., 1976). A second type of vesicle, which is somewhat larger, carries Tg but not peroxidase (Fig. 8; Novikoff et al., 1974). Destruction of microtubules with vinblastine inhibits the appearance of newly formed protein in the colloid, suggesting their involvement in movement of exocytotic vesicles (Ericson, 1980). Actin filaments, such as have been shown to affect secretion by exocytosis in other cells (Allison and Davies, 1974), are present in the apical cytosol of the principal cells, but their effect on the exocytosis of Tg is unknown (Ericson, 1981b).

Although both peroxidase and newly synthesized Tg occur together in the same vesicles, there is no evidence that iodination takes place either within them or within any other part of the follicular cell (Ericson, 1981b). On the other hand, it has been shown by light and electron microscopic studies that within short time intervals (seconds to minutes) following injection of ^{125}I, iodinated Tg appears at the luminal surface of the follicular epithelium (Fig. 9) (Wollman and Ekholm, 1981). Peroxidase is abundant on the apical surfaces of the epithelial cells, except on the projecting pseudopods, which will be discussed later (Tice and Wollman, 1974). As pointed out previously, the peroxidase (with H_2O_2 serving as an electron receptor) is essential both for the activation of iodide to permit the iodination of the tyrosyls and, apparently, for the coupling of the tyrosyls to form T_3 and T_4 (Taurog, 1974, 1979).

Colloid is removed from the follicular lumen by pinocytosis, also called endocytosis. Relatively large increments of colloid are taken in by macropinocytosis, which is essentially identical with phagocytosis (Ericson, 1981b). In this process, pseudopods protrude from the surface of the apical epithelium, increase in length and complexity, and then come together distally to engulf droplets of colloid up to 3 μm in diameter (Fig. 10). This response is seen only occasionally in moderately active cells but accelerates

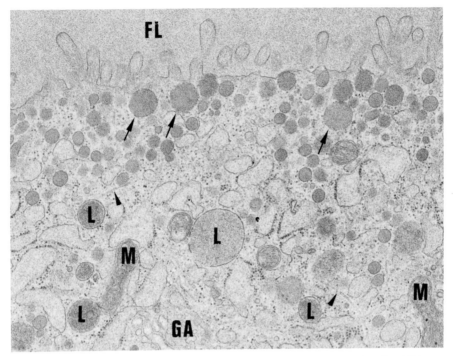

Fig. 8. Apical cytoplasm from a follicular cell of a t_4-treated rat showing, in addition to typical exocytotic vesicles with a diameter of about 150 nm, some larger vesicles (arrows) 300–400 nm in diameter. Also shown: lysosomes (L), mitochondria (M), Golgi area (GA), Folicular lumen (FL), microtubules (arrowheads). ×21,000. [From Ericson, 1981a.]

within minutes and is widespread after the administration of TSH (Ketelbant-Balasse *et al.*, 1976). The portions of the apical plasmalemma that are given over to macropinocytosis bear no peroxidase, and none enters the droplets that are formed (Tice and Wollman, 1974).

Occasional microtubules and microfilaments are found within the cytoplasm of the pseudopods. The microtubules perform no obvious function. Yet, although phagocytosis continues in other cell types after disruption of microtubules, macropinocytosis is halted in the follicular epithelum when microtubules are destroyed by treatment with colchicine or vinblastine (Ekholm *et al.*, 1974). A contractive role might be anticipated for actin filaments in the pinching off of colloid droplets from the luminal mass. In support of this suggestion are the findings that disassembly of filaments with cytochalasin B halts macropinocytosis (Wolff and Williams, 1973) and that administration of TSH induces a redistribution of actin into forming pseudopods (Gabrion *et al.*, 1980).

Most of the process for intake of colloid appears to be accomplished by micropinocytosis in which small vesicles form by involution of the apical

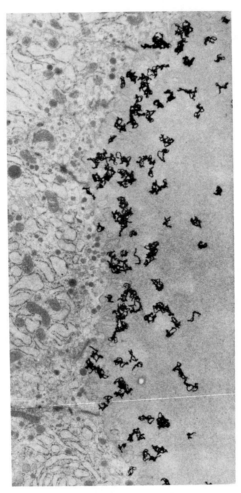

Fig. 9.   Electron microscopic autoradiogram prepared from the thyroid gland of a rat fixed
30–40 sec following administration of ^{125}I-labeled iodine. Autoradiographic tracks are concentrated over the peripheral zone of the follicular lumen; no tracks are located over the apical
cytoplasm of the follicular cell. ×7000. [By permission from Ragnar Ekholm (1979). Anatomy
and Development. *In* "Endocrinology" (L. J. De Groot, ed.) Vol. 1, Grune and Stratton, Inc.,
New York.

plasma membrane (Ericson, 1981b). These vesicles range in diameter from
about 50 nm up to about 800 nm, at which they begin to be indistinguishable
from small macropinocytotic droplets (Fig. 10). Some are smooth and some
are bristle-coated (Ekholm *et al.,* 1975). They are, in general, less dense than
macropinocytotic droplets. Both macro- and micropinocytotic vesicles take
up ferritin and thorotrast when those tracers are introduced into the follicular lumen by microinjection (Seljelid *et al.,* 1970). Microinjected ^{125}I initially

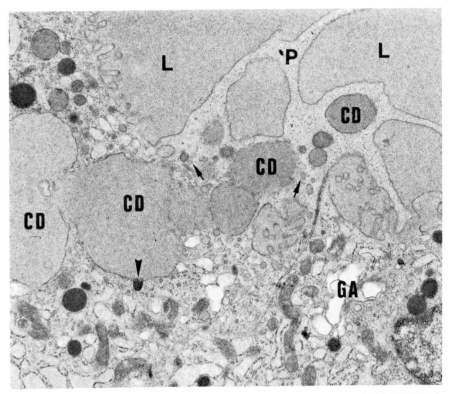

Fig. 10.   Apical portion of two follicular cells from a T_4-treated rat injected with TSH 20 min before perfusion fixation with glutaraldehyde. Few exocytotic vesicles remain. A pseudopod (P) protrudes into the follicular lumen (L). Colloid droplets (CD) are present within the pseudopod and in the adjacent cytoplasm. Small vesicles (arrows), morphologically different from exocytotic vesicles, lie in the cytoplasm and in the basal portion of the pseudopod. A small lysosome (arrowhead) appears to fuse with a colloid droplet. A Golgi area (GA) is also evident. ×13,300. [From Ericson, 1981a.]

accumulates in the apical region, presumably in the micropinocytotic vesicles, but is later concentrated in the larger droplets, indicating that the vesicles fuse with the droplets (Seljelid *et al.*, 1970); however, the relative importance of macro- and micropinocytosis in thyroidal secretion remains unresolved.

The studies which demonstrated the fusion of micropinocytotic vesicles with droplets also showed a movement of droplets toward the basal portion of the principal cell. In cytochemical preparations it was found that small (0.1–0.2 μm in diameter), spherical, membrane-bound dense bodies occupy the cytoplasm of many parts of the follicular cell and that these dense bodies, or lysosomes, contain hydrolytic enzymes capable of degrading Tg (Wollman *et al.*, 1964; Ekholm and Smeds, 1966; Seljelid, 1967). Indeed, under the stimulation of TSH these lysosomes show a tendency to move apically and to fuse with basally progressing colloid droplets (Ekholm, 1979) (Fig. 10). A mixing of the contents appears to result (Seljelid, 1967), and it is assumed that degradation of Tg occurs within the fused entities (Wollman, 1969). The thyroid hormones are released during degradation and are passed (presumably by diffusion) into the cytoplasm and on through the basal follicular membranes out into the bloodstream (Wollman, 1969). Some Tg is a normal secretory product of the thyroid gland, and although there is little supportive evidence, one must conclude that exocytosis can take place through the basal, as well as the apical, plasmalemma (Van Herle *et al.,* 1979). Thyroid hormones may pass from the cell by this route as well as, or even instead of, by diffusion.

It is generally assumed that the cytophysiological events just described for the mammalian gland apply broadly to all the jawed vertebrates, but few data to support that assumption are available for the fishes (Eales, 1979), reptiles (Lynn, 1970), or birds (Astier, 1980). The amphibian thyroid has been studied somewhat more extensively, as will be pointed out later on.

V. MORPHOLOGICAL VARIATION

It is apparent from Sections III and IV that at ultrastructural, histological, and even gross anatomical levels, the morphology of the thyroid can undergo great variation to reflect changes in functional state. The primary regulator of the gnathostomal thyroid gland is thyroid-stimulating hormone (TSH), which is secreted by the thyrotropic cells in the pars distalis of the adenohypophysis. Hypophyseal regulation of the agnathan thyroid has not been demonstrated clearly, although some experimental evidence points in that direction (Dickhoff and Darling, 1983). The thyroids of the jawed fishes and the tetrapods respond to TSH with increases in epithelial cell height and vascularity and with reductions in colloidal volume. They give all the classic cytological indications of increased functional acitivity seen in mammalian glands (Fig. 3). Correspondingly, those effects are reversed by a reduction in levels of TSH.

During the life cycles of various vertebrates, changes in circulating levels of thyroid hormones and their receptors are usually accompanied by changes in thyroidal structure and implicate the thyroid as a factor in the control of growth development and reproduction (Dickhoff and Darling, 1983), as well as in such specialized processes as hibernation, metamorphosis, and migration. Ordinarily a nice balance between the output of thyroid hormone and TSH is maintained (Sterling and Lazarus, 1977), but chronic reduction of thyroid hormone levels can be brought about by a variety of factors resulting in increased levels of TSH which, in turn, induce hyperplastic enlargement of the thyroid gland, the condition known as goiter. The most common naturally occurring cause of goiter is a deficiency of iodine in drinking water. Also, excessive dietary consumption of plants which contain cyanogenic glucosides or thioglucosides can cause goiter, particularly in regions where the water content of iodine is low. Thiocyanates and some other monovalent cations reduce hormonal output by inhibiting the transport of iodine into the gland (Hadley, 1984). A number of synthetic drugs, including thiourea and thiouracil and other thiocarbamide derivatives, often referred to as goitrogens, are used to inhibit synthesis of thyroid hormone and to thus produce goitrous conditions for experimental study.

VI. COMPARATIVE MORPHOLOGY

A. Evolutionary Origins

In Section IV, entitled "Epithelial Organelles," the complex sequence of events that takes place in the synthesis, storage, and release of T_4 and T_3 was outlined. Some of those events have been traced to beginnings that lie far back in the sequence of evolutionary succession. For example, the iodination of tyrosine occurs in a number of invertebrate and even algal species (Berg et al., 1959; Dickhoff and Darling, 1983). The coupling of tyrosyls to form iodotryosines is also accomplished by some invertebrates (Berg et al., 1959).

The immediate vertebrate ancestors, the protochordates, are filter-feeding organisms provided with a ciliated groove in the floor of the pharynx, called the endostyle. The endostyle is equipped with cells that secrete mucoid substances to which food particles adhere and are transported into the intestine by the cilia. Other endostylar cells, however, bind iodine. In this ragard, the endostyle and the thyroid are clearly related structures, and the search for the beginnings of thyroidal function in the endostyles of the protochordates has been, and is, an intriguing one.

Among the tunicates, which constitute one major group of protochordates, protein-bound, iodinated tyrosyl residues and T_4 were reported to be present in Ciona intestinalis (Barrington and Thorpe, 1965). In Mogula manhattensis, M. occidentalis, and Styela plicata, however, only MIT and DIT have been detected (Dunn, 1975). Other endostylar features of these organisms are perhaps anticipatory of thyroidal function. Iodine was found to be bound within cells only in zone 7 of the endostyle (Fig. 11) (Dunn, 1974). In those cells, bound iodine was demonstrated in Golgi profiles and in multivesicular bodies which appear to arise from the Golgi. In these organelles, peroxidase activity was also found (Figs. 11 and 12) (Dunn, 1974). It is further interest that the Golgi and multivesicular bodies contain carbohydrate and that the drug methimazole, which inhibits the binding of iodine by the thyroid, also prevents the binding of iodine in the tunicate endostyle (Dunn, 1974).

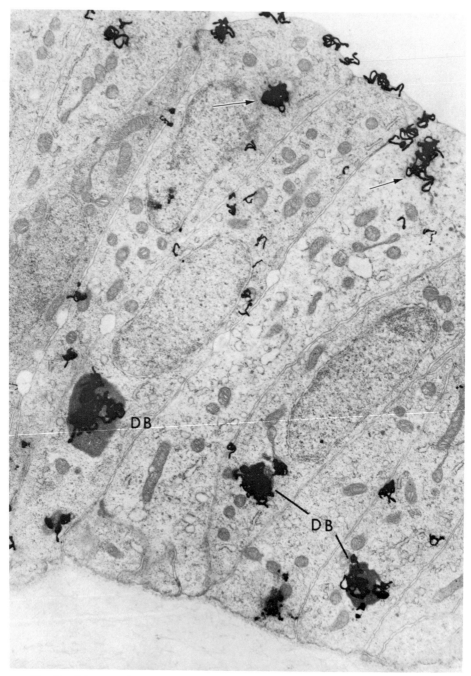

Fig. 11. Electron microscopic autoradiogram of cells from zone 7 of the endostyle of the ascidian *Molgula manhattensis*, fixed after exposure to 1.5 mCi of ^{125}I per liter for 24 hr. Clumps of grains are prominent along the apical cell membrane and over dense bodies (DB). Clumps in the apical portion of the cell (arrows) are probably associated with multivesicular bodies. ×16,000. [From Dunn, 1974.]

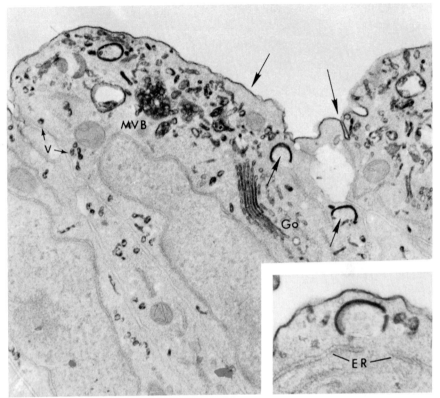

Fig. 12. Section through cells of zone 7 from the endostyle of the ascidian *Molgula manhattensis*, incubated in diaminobenzidine medium for the demonstration of peroxidase activity. A reaction product is evident along the apical cell membrane (arrows) within multivesicular bodies (MVB), the Golgi zone (Go), and in individual vesicles (v). The limina of saclike structures (arrows) show a particularly heavy reaction. ×27,000. (Inset) The endoplasmic reticulum (ER) does not react. ×43,000. [From Dunn, 1974.]

Surprisingly, in *M. manhattensis,* significant incorporation of iodine takes place in the tunic as well as in the endostyle. However, whereas iodination in the endostyle appears to be mediated by an enzymatic mechanism similar to that of the thyroid, iodination in the tunic has more nearly the character of quinone tanning, the process by means of which scleroprotein is iodinated in invertebrates (Dunn, 1975).

Several of the organ systems of the other major group of protochordates, the Cephalochordata, as represented by amphioxus (*Branchiostoma lanceolatum*), strongly resemble what might be expected of an idealized archetypal vertebrate ancestor. The adult binds iodine to produce MIT, DIT, T_3, and T_4 (Covelli *et al.,* 1960; Tong *et al.,* 1962). In the larva, an endostylar zone, corresponding to the paired iodinating zones of the adult, also binds iodine and may synthesize and release thyroid hormones (Fredrickson *et al.*, 1984). The endostyle of the adult, although lacking the follicular form, has been shown to produce a molecule similar to the 19 S form of Tg which is characteristic of most vertebrates, and, further, T_3 and T_4 are synthesized on this molecule at the same rate as on control mammalian Tg (Monaco *et al.*, 1981).

B. Cyclostomes

The phylogenetic link between the protochordates and the vertebrates became quite obvious when it was discovered that the ammocoetes larva of the lamprey is provided with an endostyle which is transformed into a thyroid gland during the metamorphosis of the larva into an adult organism. The lampreys and hagfishes are jawless animals that make up the Cyclostomata, the most primitive class of vertebrates. The lampreys and hagfishes are often studied in the hope of learning about the evolutionary beginnings of the vertebrate organ systems, but both are often very specialized and it is usually difficult to decide whether a structure which differs from those of the higher vertebrates is truly primitive or merely degenerate or specialized.

The endostyle of the larval lamprey begins development as a ciliated groove quite similar in form to that of protochordates, but as the larva develops the groove closes over and becomes subdivided to form two blind sacs which extend anteriorly and three which extend posteriorly from a small duct which is the persistent attachment of the endostyle to the pharyngeal floor (Fig. 13) (Barrington and Sage, 1972). The duct may be thought of as the forerunner of the thyroglossal cord. The two anterior sacs and the two lateral posterior sacs constitute together two structures referred to as the lateral tubes. The glandular cells of this endostyle (which may now be called the subpharyngeal gland) are gathered together in tracts that form ridges which extend into, and occupy most of the epithelial-lined lumina of the tubes (Fig. 14) (Wright and Youson, 1976). The epithelial cells that cover the

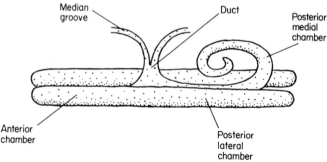

Fig. 13. Diagrammatic lateral view of the endostyle of the larval lamprey. [After Barrington and Sage, 1972.]

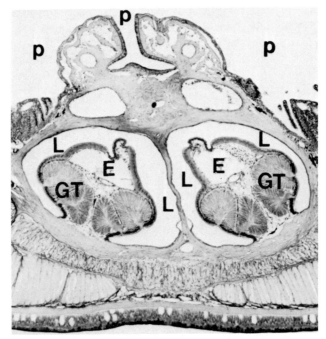

Fig. 14. Transverse section through the anterior portion of an ammocoete endostyle which lies medially beneath the pharynx (p) and consists of a pair of epithelial lined tubes into the lumina (L) of which protrude epithelial ridges (E) that are triangular in section and contain prominent glandular tracts (GT). ×64. [From Wright and Youson, 1976.]

ridges are still ciliated, but obviously the covered endostyle no longer enters directly into the process of food gathering.

With the advent of radioactive iodine, the endostyle of the larval lamprey was shown to metabolize and to store iodine and to synthesize thyroid hormones (Gorbman and Creaser, 1942; Monaco and Dominici, 1985). The iodine is bound by certain of the epithelial cells in the ridges. Throughout the sequential changes of metamorphosis, as described by Wright and Youson (1976), cells of the transforming endostyle continue to bind iodine. During metamorphosis the endostyle becomes increasingly compressed, and the lumen disappears. Cells are lost, although mitoses continue to be detected among epithelial cells. Eventually, paired disorganized masses of cells remain. Within these masses of cells, follicles begin to be apparent. It is not clear whether the lumina of these follicles are remnants of the original endostylar lumen or whether they form by cavitation within clumps of cells. The newly formed follicles do not contain PAS-positive (PAS⁺) colloid, but the colloid of the adult gland is PAS⁺.

The thyroid gland of the adult lamprey consists of a mass of sparsely vascularized, discrete follicles embedded in connective tissue ventral to the pharynx (Fig. 15) (Barrington and Sage, 1972). From the cyclostomes up through the mammals, there have been some variations in size and occasional irregularities in shape, but there has been no alteration in the basic form, and perhaps no alteration in the basic functional mechanisms, of the follicles that make up the thyroid gland. It is of interest that cells ciliated like those of the endostyle appear occasionally in the definitive thyroid glands of members of all the higher vertebrate classes.

The thyroglobulin of the andromous sea lamprey is very similar in composition to the Tg of mammals (Suzuki *et al.*, 1975). It cross-reacts against antibodies made against bovine Tg, and by microscopic immunocytochemical techniques it has been shown to occupy essentially the same locations in the follicle of the lamprey as in that of the rat (Wright *et al.*, 1978).

The thyroids of the hagfishes develop directly from embryonic anlagen without passing through an endostylar stage (Dickhoff and Darling, 1983). Observations made on the thyroid of the Pacific hagfish give evidence that at a very early stage in the evolution of the vertebrates similarity in structure

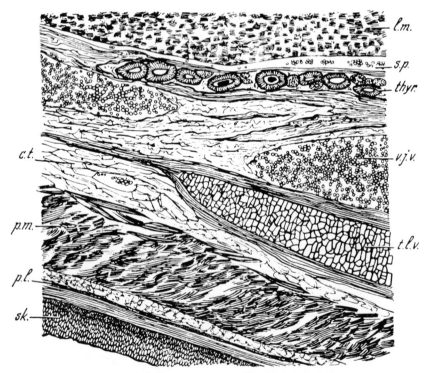

Fig. 15.   Longitudinal section through the ventral part of the branchial region of the head of an adult river lamprey. c.t., Connective tissue; l.m., lingual musculature; sk., skin; s.p., sinus perimandibularis; thyr., thyroid follicle; t.l.v., taenia longitudinalis ventralis (cartilage of the branchial basket); v.j.v., ventral jugular vein. ×60. [From Barrington and Sage, 1972.]

and function to the thyroid of the rat has already been achieved (Henderson and Gorbman, 1971). With the electron microscope one sees that macropinocytosis (Fig. 16) and micropinocytosis (Fig. 17) occur. Dense granules, probably lysosomal in nature, appear to be fusing with colloid droplets (Fig. 17), and inclusion bodies in which the degradation of Tg may be taking place are seen (Fig. 18). Precise distinction between endocytosis and exocytosis may be possible if cyclostomes are studied by the technique employed in the rat in which endocytosis is halted by treatment with T_4 while exocytosis is stimulated with TSH (Ericson, 1981b); however, it is already evident that the Golgi elements of the cyclostome follicle produce vesicles that could well be carrying Tg to the apical plasmalemma (Fig. 16).

Despite the marked similarities between the thyroids of the cyclostomes and those of the higher vertebrates, there has been no convincing demonstration of an endocrine function of thyroid hormone among the cyclostomes.

C. Jawed Fishes

Beginning with the jawed fishes, the embryonic development of the thyroid glands is initiated with an outpocketing of the midventral pharynx just

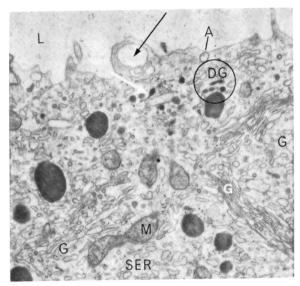

Fig. 16.   Apical area of a cell from the follicular epithelim of the thyroid gland of a hagfish. A colloid droplet (black arrow) appears to have been recently removed from the follicular lumen (L) and to be still contained within a pseudopod. The white arrow points to a site where continuity may occur between a lysomal dense granule (DG) and a segment of smooth endoplasmic reticulum (SER). A = Acanthosome; G, golgi profile; M, mitochondrian. ×16,800. [From Henderson and Gorbman, 1971.]

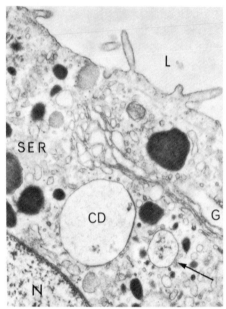

Fig. 17. Apical area of a cell from the follicular epithelium of the thyroid gland of a hagfish showing a presumed newly ingested colloid droplet (CD) that lacks an aggregation of satellite bodies (lysosomal dense granules) and two others (arrows) which are surrounded by satellite bodies. Compare with Fig. 10. SER, smooth endoplasmic reticulum; N, nucleus; G, Golgi profile; L, follicular lumen. ×18,500. [From Henderson and Gorbman, 1971.]

posterior to the level of the first pair of pharyngeal pouches. In the jawed fishes the evagination very quickly becomes a solid rod of cells (see review in Wabuke-Bunoti and Firling, 1983). The cells proliferate in a posterior direction, become loosely associated, and then group together to form follicles. In the fathead minnow (*Pimephales promales*), iodine is concentrated by thyroidal cells before colloid is demonstrated histochemically, and colloid is synthesized prior to the formation of follicles (Wabuke-Bunoti and Firling, 1983). The follicles of teleost fishes are most commonly not encapsulated and, like those of cyclostomes, are scattered singly or in small groups within loose connective tissue ventral to the pharynx. It is not uncommon for the nonencapsulated follicles to migrate to a variety of ectopic locations, most notably to the well-vascularized head kidney (Baker-Cohen, 1959), providing a means of increasing functional thyroidal mass under conditions of iodide deficiency (Eales, 1979). In some exceptional species (Bermuda parrotfish, swordfish, tuna), the follicles are massed together in single (or tandemly arranged, double) median lobes (Fig. 19) (Gorbman and Bern, 1962). In general, the thyroid glands of teleosts exhibit great heterogeneity with respect to follicular size and to cytophysiological state of individual follicular cells (Eales, 1979).

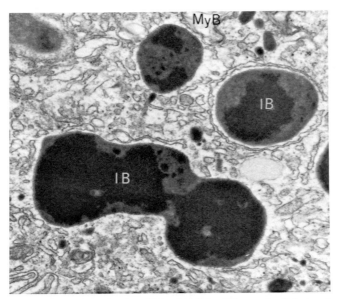

Fig. 18. Paranuclear region of a thyroidal epithelial cell from a hagfish showing inclusion bodies (IB) that appear to be formed through fusion of lysosomal dense granules and colloid droplets and which may be sites of hormone release. MyB, myloid body. ×20,000. [From Henderson and Gorbman, 1971.]

In the Chondrichthyes the thyroid gland is a discrete median structure gland (Fig. 19) and is encapsulated by a thin layer of connective tissue, as are the unpaired median glands of the dipnoan lungfishes and the Coelacanth, *Latimeria,* which is considered to be at least a close relative of the Crosspterygian line from which the tetrapods originated (Chavin, 1976).

D. Amphibians

The thyroidal anlagen of the amphibians, like those of the jawed fishes, arises typically as a thickening of the midventral pharyngeal endoderm at a level just porterior to the first pair of pharyngeal pouches. The thickening evaginates slightly to form the foramen caecum and then proliferates posteriorly as a solid rod of cells which constitutes the tractus thyreoglossus. As proliferation continues the tract bifurcates to give rise to two ovoid masses of cells which represent the anlagen of the two lobes of the amphibian gland (Dent, 1942; Hanoaka *et al.,* 1973). The bilobular nature of this tetrapod gland is perhaps anticipated by the posterior bifurcation seen in the unpaired thyroid glands of the South American and African lungfishes (Chavin, 1976). Later, the tract degenerates and the lobular anlagen separate and move laterally. The lateral progression is brief in the Anura, in which the lobes remain in close contact with hyoid cartilage, but in the urodeles the lobes are

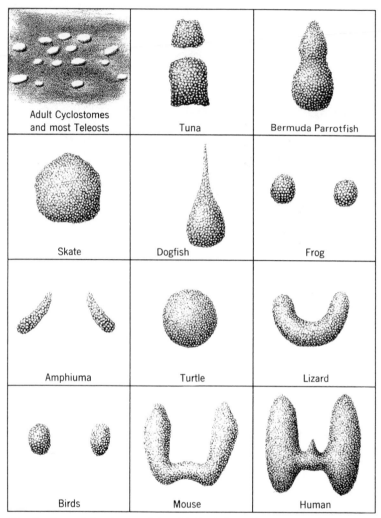

Fig. 19. Outlines of the shapes of thyroid glands in various groups of vertebrates [From Gorbman and Bern, 1962.]

more widely separated to ventrolateral positions and lie immediately internal to the procoracoid cartilages. In their definitive positions the cells of the lobular anlagen become loosely arranged and then come together in small masses which differentiate to become follicles while capillaries are forming among them. Iodine is incorporated in the presumptive thyroidal cells during the formation of the initial pharyngeal evagination, and synthesis of T_3 and T_4 precedes follicular differentiation (Hanoaka *et al.,* 1973).

The primary metamorphosis of an amphibian larva, particularly that of an anuran tadpole, is a dramatic and striking event (Dent, 1968). The discovery

by Gudernatsch (1912) that the thyroid gland induces the marked morphological, physiological, and behavioral changes that make up metamorphosis was
a finding of such magnitude that it can be considered to mark the beginning
of the field of comparative endocrinology. As a consequence, more attention
has been given to the amphibian thyroid than to the glands of others of the
lower vertebrates. Physiological studies show that the thyroidal uptake of
iodine and the levels of TH in the blood are low in the larva, increase
progressively during premetamorphosis and prometamorphosis, peak during
metamorphic climax, and return to a low level which persists during postcli-
max and in the postmetamorphic animal (Dodd and Dodd, 1976; Mondou
and Kaltenbach, 1979; Norris, 1983). Careful cytophysiological observations
relate changes in epithelial cell height, rough endoplasmic reticulum, distribution of peroxidase, distribution of microvesicles and colloid droplets to
the physiological findings (Fig. 20) (Regard, 1978). These observations also
strongly support the proposition that the functional model of the mammalian
gland, presented in the preceding sections of this chapter, is generally applicable to the lower vertebrates as well.

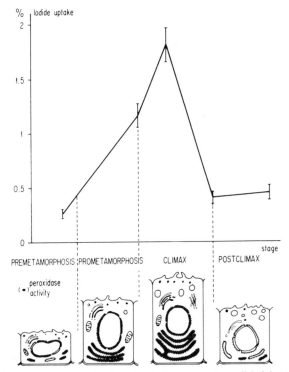

Fig. 20.   Relations of uptake of radioactive iodide, follicular cell height, and intracellular
localization of peroxidase activity in thyroid glands from larvae of *Xenopus* at four developmental stages after 72 hr of incubation in a bath containing ^{125}I (16 μCi/ml). [From Larras-Regard,
1978.]

E. Reptiles

The embryonic origin and early development of the reptilian thyroid show no unusual features (Lynn, 1960). The morphology of the adult gland has been cataloged by Lynn (1970). He reported that in turtles and snakes the thyroid is an unpaired, roughly spherical structure (Fig. 19) lying ventral to the trachea and just anterior to the heart. In the Crocodilia, mammalian anatomy is anticipated in a gland with two well-defined lobes connected by a narrow isthmus that stretches across ventral to the trachea. Considerable variation in form is encountered among the glands of the lizards. Some appear as in Fig. 19. Among other families some have a single compact median gland, some a bilobed gland with an isthmus, and still others have completely separate paired lateral lobes. The thyroid of *Sphenodon* is a single, narrow, transversely elongate body.

Although the ultimobranchial glands of reptiles are basically separate bodies distinct from the thyroid, parafolicular cells, which presumably are C cells, have been reported within the thyroids of some species (Gabe, 1961; Saint Girons and Duguy, 1962). Also, at the branching points of arteries in some snakes, Yamamoto (1960) found funnel-shaped sphincter apparatuses which resembled similar structures seen in mammals. In general, the micro-anatomy of the reptilian gland is essentially the same as that of the mammal.

F. Birds

The embryogeny of the avian thyroid gland is of the usual vertebrate pattern, except that the original evagination is a more well defined sac than most others (Hamilton, 1952). From the anlage two lobes differentiate (Fig. 19) and become displaced laterally to lie eventually at the base of the neck, usually in positions very close to the common carotid artery and the jugular vein (Astier, 1980).

A number of ultrastructural studies have been carried out on the thyroid glands of birds (Fujita, 1963; French and Hodges, 1977). The same complement of organelles seen in the mammalian gland have been found to be present, and although cytophysiological studies are lacking, it is generally assumed that they perform the same functions as those of the mammal (Astier, 1980).

VII. OVERVIEW AND BIOMEDICAL IMPLICATIONS

In the foregoing sections we have seen that the thyroid gland is character-ized by its evolutionary stability. The structure of the follicle remains un-changed from its first appearance in the cyclostomes upward throughout the whole vertebrate line. Beginning with the jawed fishes, it responds to thyro-tropic hormone, and where detailed studies have been made, evidence points to consistency in the form of organelles and in the nature of biochemical mechanisms.

In the mammal, the cells of the ultimobranchial gland are dispersed among the follicles and even within the follicular epithelium. In some mammalian species the parathyroid glands are embedded completely within the lobes of the thyroid. In others they are partially embedded or are at least closely applied to the thyroid. Since the thyroid glands of submammalian vertebrates are free of these anatomical complications, it is surprising that more use is not made of the lower species in investigations of thyroidal function. Even so, comparative studies have produced enlightening results of a general nature.

The anatomical discreteness of the thyroid and the ultimobranchial gland in the shark and the chicken were utilized by Copp and his collaborators (1967) in demonstrating clearly for the first time that calcitonin is produced by the C cells rather than by the principal cells of the follicular epithelium.

The results of two studies which have been useful in understanding the nature of goiter point to the consistency of thyroidal function among the vertebrates. In the first of these Marine and Lenhart (1910) investigated the abnormal and excessive growth of thyroid tissue which sometimes occurred in carnivorous game fish, such as salmon and trout, that were being reared under crowded conditions in captivity. They demonstrated that the stricken fish suffered from nothing more than endemic goiter, which could be prevented by the addition of iodine to the water that passed through the tanks in which the fish were kept. Some years later Dodd and Callan (1955) found that sexually mature neotenic specimens of the newt *Triturus helveticus* in a pond near the Scottish coast had thyroid glands that were goitrous and extremely enlarged. That condition could hardly be attributed to a lack of iodine in the environment, since the pond was situated within a few miles of the North Sea, and the iodine content in the drinking water of the nearby town of Crail was 5 g per liter. Dobb and Callan noted that kale and turnips were cultivated in the vicinity of the pond. The goitrogenic effects of plants of the genus *Brassica* are well known. It became clear that rabbits frequented the area, consumed the kale and turnips, and deposited feces on the slopes from which water drained into the pond. It was concluded that the drainage water accumulated sufficient "brassic factor" to induce goiter in larvae that were overwintering in the pond.

Interrelations between T_4 and T_3 cell in thyroidal action are far from clear. The physiological effects of T_3 are many times greater than those of T_4, yet T_4 is thought to exert a negative feedback action the hypothalamus. In general, extrathyroidal 5'-monodeiodination of T_4 appears to be the major source of T_3. Indeed, it has been suggested that T_3 is the physiologically relevant hormone and that T_4 is merely the progenetor of T_3. As pointed out earlier, although the thyroid follicles of most bony fishes are loosely distributed, those of some are arranged in compact masses. The thyroid gland of one of the latter, the Hawaiian parrotfish, has been suggested as an *in vitro* model system for the study of interrelations between T_4 and T_3. The thyroid of that fish has been shown to release significant amounts of T_4, but no detectable T_3, in culture, whereas *in vivo* T_3 is found in concentrations that are among the highest yet measured in teleost fishes (Grau *et al.*, 1986). This provides the first direct evidence that in bony fishes, plasma T_3 may be derived principally, perhaps exclusively, from the peripherial mono-deiodination of T_4.

REFERENCES

Allen, B. M. (1916). The results of extirpation of the anterior lobe of the hypophysis and of the thyroid of *Rana pipiens* larvae. *Science* **44**, 755–757.

Allen, B. M. (1929). The influence of thyroid gland and hypophysis upon growth and development of amphibian larvae. *Q. Rev. Biol.* **4**, 325–352.

Allison, A. C., and Davies, P. (1974). Interactions of membranes, microfilaments, and microtubules in endocytosis and exocytosis. *Adv. Cytopharmacol.* **2**, 237–248.

Arey, L. B. (1974). "Developmental Anatomy," Rev. 7th ed. Saunders, Philadelphia, Pennsylvania.

Astier, H. (1980). Thyroid gland in birds: structure and function. *In* "Avian Endocrinology" (A. Epple and M. H. Stetson, eds.), pp. 167–189. Academic Press, New York.

Baber, E. C. (1876). Contribution to the minute anatomy of the thyroid gland of the dog. *Philos. Trans. R. Soc. London* **166**, 557–568.

Baker-Cohen, K. R. (1959). Renal and other heterotopic thyroid tissue in fishes. *In* "Comparative Endocrinology" (A. Gorbman, ed.), pp. 283–301.

Barrington, E. J. W., and Sage, M. (1972). The endostyle and thyroid. *In* "The Biology of Lampreys" (M. W. Hardisty and I. C. Potter, eds.), Vol. 2, pp. 105–134. Academic Press, New York.

Barrington, E. J. W., and Thorpe, A. (1965). The identification of monoiodotyrosine, diiodotyrosine and thyroxine in extracts of the endostyle of the ascidian, *Ciona intestinalis* L. *Proc. R. Soc. London, Ser. B* **163**, 136–149.

Baumann, E. (1895). Uber das normale Vorkommen von Jod im Thierkörper. *Hoppe-Seyler's Z. Physiol. Chem.* **21**, 319–327.

Berg, O., Gorbmann, A., and Kobayashi, H. (1959). The thyroid hormones in invertebrates and lower vertebrates. *In* "Comparative Endocrinology" (A. Gorbman, ed.), pp. 302–319. Wiley, New York.

Björkman, U., Ekholm, R., Elmqvist, L.-G., Melander, A., and Smeds, S. (1974). Induced unidirectional transport of protein into the thyroid follicular lumen. *Endocrinology (Baltimore)* **95**, 1506–1517.

Björkman, U., Ekholm, R., Ericson, L. E., and Öfverholm, T. (1976). Transport of thyroglobulin and peroxidase in the thyroid follicle cell. *Mol. Cell. Endocrinol.* **5**, 3–17.

Bloom, W., and Fawcett, D. W. (1975). "A Textbook of Histology." Saunders, Philadelphia, Pennsylvania.

Born, G. (1883). Uber die Derivate der embryonalen Schlundspalten bei Säugetieren. *Arch. Mikrosk. Anat.* **22**, 271–318.

Chavin, W. (1976). Thyroid of sarcopterygian fishes (Dipnoi and Crossopterygii) and origin of tetrapod thyroid. *Gen. Comp. Endocrinol.* **50**, 167–171.

Chow, N. Y., Yen-Chow, Y. C., and Woodbury, D. M. (1982). Effects of thyrotropin, acetazolamide, 4-acetamido-4'-isothiocyanostilbene-2, 2'-disulfonic acid, perchlorate, and ouabain on the distribution of iodide ion in cells and luminal fluid of turtle thyroid. *Endocrinology (Baltimore)* **110**, 121–125.

Copp, D. H., Cockroft, D. W., and Kueh, Y. (1967). Calcitonin from ultimobranchial glands of dogfish and chickens. *Science* **158**, 924–925.

Covelli, I., Salvatore, G., Sena, L., and Roche, J. (1960). Sûr la formation de hormones thyroidiennes et de leurs precurseurs par *Branchiostoma lanceolatum*. *C.R. Seances Soc. Biol. Ses. Fil.* **154**, 1165–1169.

Cowper, W. (1698). "The Anatomy of Humane Bodies." Oxford Univ. Press, London and New York.

Dent, J. N. (1942). The embryonic development of *Plethodon cinereus* as correlated with the differentiation and functioning of the thyroid gland. *J. Morphol.* **71**, 577–601.

Dent, J. N. (1968). Survey of amphibian metamorphosis. *In* "Metamorphosis: A Problem in Developmental Biology" (W. Etkin and L. J. Gilbert, eds.), pp. 371–311. Appleton, New York.

Dickhoff, W. W., and Darling, D. S. (1983). Evolution of thyroid function and its control in lower vertebrates. *Am. Zool.* **23**, 697–707.

Dodd, J. M., and Callan, H. G. (1955). Neoteny with goitre in *Triturus helveticus*. *Q. J. Microsc. Sci.* **96**, 121–131.

Dodd, M. H. I., and Dodd, J. M. (1976). The biology of metamorphosis. *In* "Physiology of the Amphibia" (B. Lofts, ed.), Vol. 3, pp. 467–599. Academic Press, New York.

Dunn, A. D. (1974). Ultrastructural autoradiography and cytochemistry of the iodine-binding cells in the ascidian endostyle. *J. Exp. Zool.* **188**, 103–123.

Dunn, A. D. (1975). Iodine metabolism in the Ascidian, *Molgula manhattensis*. *Gen. Comp. Endocrinol.* **25**, 83–95.

Eales, J. G. (1979). Thyroid functions in cyclostomes and fishes. *In* "Hormones and Evolution" (E. J. W. Barrington, ed.), Vol. 1, pp. 341–436. Academic Press, New York.

Edelhoch, H. (1965). The structure of thyroglobulin and its role in iodination. *Recent Prog. Horm. Res.* **21**, 1–31.

Ekholm, R. (1979). Anatomy and development. *In* "Endocrinology" (L. J. DeGroot, ed.), Vol. I, pp. 305–309. Grune & Stratton, New York.

Ekholm, R., and Björkman, U. (1985). Generation of hydrogen peroxide, a factor in thyroglobulin iodination. *In* "Thyroglobulin—The Prothyroid Hormone (Progress in Endocrine Research and Therapy." (M. C. Eggo and G. N. Burrows, eds.), Vol. 2, pp. 153–158. Raven Press, New York.

Ekholm, R., and Smeds, S. (1966). On dense bodies and droplets in the follicular cells of the guinea pig thyroid. *J. Ultrastruct. Res.* **16**, 71–82.

Ekholm, R., and Wollman, S. H. (1975). Site of iodination in the rat thyroid gland deduced from electron microscopic autoradiographs. *Endocrinology (Baltimore)* **97**, 1432–1444.

Ekholm, R., Ericson, L. E., Josefsson, J.-O., and Melander, A. (1974). *In vivo* action of vinblastine on thyroid ultrastructure and hormone secretion. *Endocrinology (Baltimore)* **94**, 641–649.

Ekholm, R., Engström, G., Ericson, L. E., and Melander, A. (1975). Exocytosis of protein into the thyroid follicle lumen: An early effect of TSH. *Endocrinology (Baltimore)* **97**, 337–346.

Ericson, L. E. (1980). Vinblastine-induced inhibition of protein transport in the mouse thyroid *in vivo*. *Endocrinology (Baltimore)* **106**, 833–841.

Ericson, L. E. (1981a). Exocytosis and endocytosis in the thyroid follicle cell. *Mol. Cell. Endocrinol.* **20**, 87–89.

Ericson, L. E. (1981b). Exocytosis and endocytosis in the thyroid cell. *Mol. Cell. Endocrinol.* **22**, 1–24.

Feeney, L., and Wissig, S. L. (1972). A biochemical and radioautographic analysis of protein secretion by thyroid lobes incubated *in vitro*. *J. Cell Biol.* **53**, 510–522.

Fredrickson, G., Ericson, L. E., and Olsson, R. (1984). Iodine binding in the endostyle of larval *Branchiostoma lanceolatum* (Cephalochordata). *Gen. Comp. Endocrinol.* **56**, 177–184.

French, E. I., and Hodges, R. D. (1977). Fine structural studies on the thyroid gland of the normal domestic fowl. *Cell Tissue Res.* **78**, 397–410.

Fujita, H. (1963). Electron microscopic studies on the thyroid gland of domestic fowl, with special reference to the mode of secretion and the occurrence of a central flagellum in the follicular cell. *Z. Zellforsh. Mikrosk. Anat.* **60**, 615–632.

Gabe, M. (1961). Donnees histologiques sur les macrothyreocytes (cellules parafolliculaires) de quelques sauropsides et anammiotes. *Acta Anat.* **47**, 34–54.

Gabrion, J., Travers, F., Benyamin, Y., Sentein, P., and Van Thoai, N. (1980). Characterization of actin microfilaments at the apical pole of thyroid cells. *Cell Biol. Int. Rep.* **4**, 59–67.

Gorbman, A., and Bern, H. A. (1962). "A Textbook of Comparative Endocrinology." Wiley, New York.

Gorbman, A., and Creaser, C. W. (1942). Accumulation of radio-active iodine by the endostyle of larval lampreys and the problem of homology of the thyroid. *J. Exp. Zool.* **89**, 391–406.

Grau, E. G., Helms, L. M. H., Shimoda, S. K., Ford, C., LeGrand, J., and Yamauchi, K. (1986). The thyroid gland of the Hawaiian parrot fish and its use as an *in vitro* model system. *Gen. Comp. Endocrinol.* **61**, 100–108.

Gudernatsch, J. F. (1912). Feeding experiments on tadpoles. I. The influence of specific organs given as food on growth and differentiation. A contribution to the knowledge of organs with internal secretion. *Arch. Entwicklungsmech. Org.* **35**, 457–481.

Gurdon, J. B., Lane, C. D., Woodland, H. R., and Marbaix, G. (1971). Use of frog eggs and oocytes for the study of messenger RNA and its translation in living cells. *Nature (London)* **233**, 177–182.

Hadley, M. E. (1984). "Endocrinology." Prentice-Hall, Englewood Cliffs, New Jersey.

Hamilton, H. L. (1952). "Lillie's Development of the Chick." Holt, New York.

Hanaoka, Y., Koya, S. M., Kondo, Y., Kobayashi, Y., and Yamamoto, K. (1973). Morphological and functional maturation of the thyroid during early development of anuran larvae. *Gen. Comp. Endocrinol.* **21**, 410–423.

Henderson, N. E., and Gorbman, A. (1971). Fine structure of the thyroid follicle of the Pacific hagfish, *Eptatretus stouti. Gen. Comp. Endocrinol.* **16**, 409–429.

Hopper, A. F., and Hart, N. H. (1985). "Foundations of Animal Development," 2nd ed. Oxford Univ. Press, London and New York.

Ketelbant-Balasse, P., Van Sande, J., Neve, P., and Dumont, J. R. (1976). Time sequence of 3', 5'-cyclic AmP accumulation and ultrastructural changes in dog thyroid slices after acute stimulation by TSH. *Horm. Metab. Res.* **8**, 212–215.

Laras-Regard, E. (1978). Cytophysiology of the amphibian thyroid gland through larval development and metamorphosis. *Int. Rev. Cytol.* **52**, 81–118.

Lynn, W. G. (1960). Structure and functions of the thyroid gland in reptiles. *Am. Midl. Nat.* **64**, 309–326.

Lynn, W. G. (1970). The biology of the reptiles. *In* "The Thyroid" (A. d'A. Bellairs, C. Ganes, and E. E. Williams, eds.), pp. 201–234. Academic Press, New York.

Marine, D., and Lenhart, C. H. (1910). Observations and experiments on the so-called thyroid carcinoma of *S. fontinalis* and its relation to ordinary goitre. *J. Exp. Med.* **12**, 311–337.

Melander, A., and Rerup, C. (1968). Studies on the thyroid activity in the mouse. I. The effects of thyroxine, corticosteroids, hypophysectomy and changes in environmental temperature. *Acta Endocrinol. (Copenhagen)* **58**, 202–214.

Monaco. F., and Dominici, R. (1985). Role of carbohydrates in the evolutionary pathways of thyroglobulin biosynthesis and secretion (short and long loop). *In* "Thyroglobulin–The Prothyroid Hormone (Progress in Endocrine Research and Therapy) (M. C. Eggo and G. N. Burrow, eds.), Vol. 2, pp. 115–126. Raven Press, New York.

Monaco, F., Dominici, R., Andreoli, M., DePirro, R., and Roche, J. (1981). Thyroid formation in thyroglobulin synthesized in the amphioxus *Branchiostoma lanceolatum. Comp. Biochem. Physiol. B.* **B70**, 341–343.

Mondou, P. M., and Kaltenbach, J. C. (1979). Thyroxine concentrations in blood serum and pericardial fluid of metamorphosing tadpoles and adult frogs. *Gen. Comp. Endrocrinol.* **39**, 343–349.

Nonidez, J. F. (1932). The origin of the parafolicular cells, a second epithelial component of the thyroid gland of the dog. *Am. J. Anat.* **49**, 479–495.

Norris, D. O. (1978). Hormonal and environmental factors involved in the determination of neoteny in urodeles. *In* "Comparative Endocrinology" (P. J. Gaillard and H. H. Boer, eds.), pp. 109–112. Elsevier/North-Holland Biomedical Press, Amsterdam.

Norris, D. O. (1980). "Vertebrate Endocrinology." Lea & Febiger, Philadelphia, Pennsylvania.

Norris, D. O. (1983). Evolution of endocrine regulation of metamorphosis in lower vertebrates. *Am. Zool.* **23**, 709–718.

Norris, E. H. (1918). The early morphogenesis of the human thyroid gland. *Am. J. Anat.* **24**, 443–466.

Novikoff, A. B., Novikoff, P. M., Ma, H., Shin, W.-Y., and Nelson, A. (1974). Cytochemical studies of secretory and other granules associated with the endoplasmic reticulum in rat thyroid epithelial cells. *Adv. Cytopharmacol.* **2**, 349–368.

Regard, E. (1978). Cytophysiology of the amphibian thyroid through larval development and metamorphosis. *Int. Rev. Cytol.* **52**, 81–118.

Rolleston, H. D. (1936). "The Endocrine Organs in Health and Disease." Oxford Univ. Press, London.

Saint Girons, H., and Duguy, R. (1962). Donnees histologiques sur le cycle annuel de la glande thyroïde chez les viperes. *Gen. Comp. Endocrinol.* **2**, 337–346.

Seljelid, R. (1967). Endocytosis in thyroid follicle cells. V. On the redistribution of cytosomes following stimulation with thyrotropic hormone. *J. Ultrastruct. Res.* **18**, 479–488.

Selijelid, R., Reith, A., and Nakken, K. F. (1970). The early phase of endocytosis in rat thyroid follicle cells. *Lab. Invest.* **23**, 595–605.

Spiro, M. J., and Spiro, R. G. (1985). Synthesis and processing of thyroglobulin carbohydrate units. *In* "Thyroglobulin—The Prothyroid Hormone (Progress in Endocrine Research and Therapy." (M. C. Eggo and G. N. Burrows, eds.), Vol. 2, pp. 103–113. Raven Press, New York.

Sterling, K., and Lazarus, J. H. (1977). The thyroid and its control. *Annu. Rev. Physiol.* **39**, 349–372.

Stoeckel, M. E., and Porte, A. (1970). Origine embryonnaire et differenciation secretoire des cellules a calcitonine (cellules C) dans la thyroïde foetale du rat. *Z. Zellforsch. Mikrosk. Anat.* **106**, 251–268.

Suzuki, S., Gorbham, A., Rolland, M., Montford, M. F., and Lissitzky, S. (1975). Thyroglobulin of cyclostomes and an elasmobranch. *Gen. Comp. Endocrinol.* **26**, 59–69.

Taurog, A. (1974). Biosynthesis of iodoamino acids. *In* "Handbook of Physiology," (M. A. Greer and D. H. Solomon, eds.), Sect. 7, Vol. III, pp. 101–133. Williams & Wilkins, Baltimore, Maryland.

Taurog, A. (1979). Hormone synthesis. *In* "Endocrinology" (L. J. DeGrott, ed.), Vol. 1, pp. 331–334. Grune & Stratton, New York.

Tice, L. W., and Wollman, S. H. (1974). Ultrastructural localization of peroxidase on pseudopods and other structures of the typical thyroid epithelial cell. *Endocrinology (Baltimore)* **94**, 1555–1567.

Tong, W., Kerkof, P., and Chiakoff, I. L. (1962). Identification of labelled thyroxine and triiodothyronine in amphioxus treated with [131]I. *Biochim. Biophys. Acta* **561**, 326–331.

Trotter, W. R. (1964). Historical introduction. *In* "The Thyroid Gland" (R. Pitt-Rivers and W. R. Trotter, eds.). Vol. 1, pp. 1–8. Butterworth, London.

Turner, C. D., and Bagnara, J. T. (1976). "General Endocrinology," 6th ed. Saunders, Philadelphia, Pennsylvania.

Van Herle, A. J., Vassari, G., and Dumont, J. E. (1979). Control of thyroglobulin synthesis and secretion. *N. Engl. J. Med.* **301**, 239–249, 307–314.

Vassart, G., Brocas, H., Lecocq, R., and Dumont, J. (1975). Thyroglobulin messenger RNA:

Translation of a 33-S mRNA into a peptide immunologically related to thyroglobulin. *Eur. J. Biochem.* **55,** 15–22.

Vassart, G., Verstreken, L., and Dinsart, C. (1977). Molecular weight of thyroglobulin 33 S messenger RNA as determined by polyacrylamide gel electrophoresis in the presence of formamide. *FEBS Lett.* **79,** 15–18.

Wabuke-Bunoti, M. A. N., and Firling, C. E. (1983). The prehatching development of the thyroid gland of the fathead minnow, *Pimephales promelas* (Rafinesque). *Gen. Comp. Endocrinol.* **49,** 320–331.

Waechter, C. H., and Lennarz, W. J. (1976). The role of polyprenol-linked sugars in glycoprotein synthesis. *Annu. Rev. Biochem.* **45,** 95–112.

Werner, S. C. (1978). Historical resume. *In* "The Thyroid" (S. C. Werner and S. H. Ingbar, eds.), pp. 5–8. Harper (Hoeber), New York.

Whür, P., Herscovics, A., and Leblond, C. P. (1969). Radioautographic visualization of the incorporation of galactose-^{3}H and mannose-^{3}H by rat thyroids in vitro in relation to the stages of thyroglobulin synthesis. *J. Cell Biol.* **43,** 289–311.

Wolff, J., and Williams, J. A. (1973). The role of microtubules and microfilaments in thyroid secretion. *Recent Prog. Horm. Res.* **29,** 229–278.

Wollman, S. H. (1969). Secretion of thyroid hormones. *In* "Lysosomes in Biology and Pathology" (J. T. Dingle and H. B. Fell, eds.), pp. 483–512. North-Holland Publ., Amsterdam.

Wollman, S. H., and Ekholm, R. (1981). Site of iodination in hyperplastic thyroid glands deduced from autoradiographs. *Endocrinology (Baltimore)* **108,** 2082–2085.

Wollman, S. H., Spicer, S. S., and Burstone, M. S. (1964). Localization of esterase and acid phosphatase in granules and colloid droplets in rat thyroid epithelium. *J. Cell Biol.* **21,** 191–201.

Wright, G. M., and Youson, J. H. (1976). Transformation of the endostyle of the anadromous sea lamprey, *Petromyzon marinus* L, during metamorphosis. 1, Light microscopy and autoradiography with ^{121}I. *Gen. Comp. Endocrinol.* **30,** 243–257.

Wright, G. M., Filosa, M. F., and Youson, J. H. (1978). Light and electron microscopic immunocytochemical localization of thyroglobulin in the thyroid gland of the anadromous sea lamprey, *Petromyzon marinus* L., during its upstream migration. *Cell Tissue Res.* **187,** 473–478.

Yamamoto, Y. (1960). Comparative histological studies of the thyroid gland of lower vertebrates. *Folia Anat. Jpn.* **37,** 353–387.

7

The Parathyroid Glands

NANCY B. CLARK,* KAREN KAUL,†
AND SANFORD I. ROTH†

* Department of Physiology and Neurobiology
University of Connecticut
Storrs, Connecticut 06268

† Department of Pathology
Northwestern University School of Medicine
Chicago, Illinois 60611

I. INTRODUCTION

The parathyroid glands are present in all terrestrial vertebrates and perform an important function in calcium homeostasis. The embryological derivation of these glands has been determined in some species and is assumed to be similar in others. Parathyroid III and IV refer to glands derived from the third and fourth branchial pouches, respectively. The general pattern is two pairs of parathyroid glands, although in a number of species only one pair (usually parathyroid III) remains in adults. In mammals, the upper parathyroids (parathyroid IV) or the single pair are often located in close association with, or embedded within, the capsule of the thyroid gland, a fact that has given them their name, "parathyroid." In nonmammalian vertebrates the glands are usually present in the region of the thyroid and thymus but are not intimately related to the thyroid. Thus it is not necessary to perform thyroparathyroidectomy in nonmammalian vertebrates in order to assure complete removal of the parathyroids.

The major parathyroid cells, the chief cells, produce the hypercalcemic parathyroid hormone. These cells are similar in appearance throughout the vertebrates. In most species they form clusters or cords of cells, among which run blood vessels and nerves supported by connective tissue fibers. The rich blood supply assures adequate nutrition of the cells and provides a route for a rapid response of the gland to hypocalcemia. The function of parathyroid innervation is not clear. In most cases, the nerves supply only the blood vessels and thus regulate blood flow to the gland. Innervation of the chief cells has been described in a few studies; this might influence the

207

secretion of parathyroid hormone, but no definitive evidence of a neural response has been established.

The chief cells are often described as existing in an ultrastructurally light or dark phase. These phases have been hypothesized to indicate the state of activity of the cells. Generally, the "light" chief cells have less well developed cytoplasmic organelles and are considered to be in an inactive state. However, in many investigations in laboratory animals, these differences in cytoplasmic density may simply reflect the method of fixation for light and electron microscopy. The dark and light chief cell variants are seen most often in parathyroids immersed in fixative but not in those in which the fixative is perfused through the gland.

Though the differences in cytoplasmic density may be fixation artifacts, there is abundant ultrastructural evidence in a number of mammalian species for a cyclical change in activity of parathyroid chief cells. The resting stage is characterized by dispersed endoplasmic reticulum, poorly developed Golgi regions, few secretory or prosecretory granules, and abundant glycogen and lipid. During the synthetic phase, the granular endoplasmic reticulum aggregates into the juxtanuclear body. The secretory phase is characterized by well-developed, large Golgi membranes, many prosecretory and secretory granules, and little glycogen and lipid. Finally, during the involutionary phase, the cells regress to the resting level. Each chief cell seems able to respond independently to the level of plasma calcium.

The other cell type seen occasionally in parathyroid glands is the oxyphil. The ultrastructure of these cells appears to be virtually identical in all species. The cytoplasm is tightly packed with mitochondria, and it is difficult to discern any other organelles. These cells are most often seen in older specimens and may be an age-related variant of chief cells. Their function is unknown.

This chapter attempts to bring together the large body of literature regarding the anatomy, embryology, and structure of the parathyroid glands of vertebrates. Much of the early work in this field and the human studies have been described in detail elsewhere (Roth and Schiller, 1976). Thus we will provide here an overview, emphasizing those areas in which extensive new data are available.

II. CLASS MAMMALIA

A. Subclass Eutheria

1. Order Primata—Monkeys, Apes, Baboons, Humans, and Others

It is assumed that parathyroid glands arise from the third and fourth branchial pouches, although in fact no embryological studies of primates other than humans have been reported (Norris, 1937; Bargmann, 1939).

There is considerable variation in the number and position of the parathyroid glands among the various primate species. For example, a single pair of parathyroids (probably parathyroid III) is reported in macaques (*Macaque* sp.) and baboons (*Papio* sp.) (Pischinger, 1937). In contrast, vervet monkeys and relatives (*Cercopithecus* sp.), mangabeys (*Cercocebus* sp.), and chimpanzees (*Pan* sp.) have two pairs of parathyroid glands. The outer pair (parathyroid III) generally is located lateral to the thyroid or near the thymus, while the inner glands (parathyroid IV) are embedded within the thyroid gland (Roth and Schiller, 1976).

The parathyroids of primates are composed of compact cords of cells with varying amounts of vascularity, connective tissue, and fat (Coleman *et al.*, 1980). Four cell types have been described in macaque parathyroids, including a chief cell, pale and dark oxyphil cells, and a water-clear cell. Chief and oxyphil cells are present in baboons. Three chief cell variants have been described that are assumed to represent the resting cell (light), active cell (dark), and an intermediate stage. Active cells have prominent endoplasmic reticulum and Golgi membranes and show evidence of hormone synthesis and secretion. Abundant membrane-bound secretory granules measuring 200–500 nm in diameter appear to have developed from the Golgi (Nakagami, 1965; Trier, 1958; Coleman *et al.*, 1980). The chief cells of the macaque respond to experimental hypercalcemia similarly to other species (Swarup *et al.*, 1979). Administration of exogenous calcium or vitamin D_2 causes an increase in serum-calcium levels and inactivity and degenerative changes in the parathyroid chief cells. Hyperadrenocortism in the baboon results in functional parathyroid hyperactivity, with few ultrastructural alterations (Coleman *et al.*, 1980).

2. Order Carnivora—Dogs and Cats.

There are four parathyroid glands in the dog (*Canis familiaris*). The parathyroids are surrounded by dense, fibrous, connective tissue capsules containing varying amounts of fat. Light and dark chief cells are arranged in sheets, cords, and clusters, with a slight tendency to form follicles (Bergdahl and Boquist, 1973). Oxyphil cells are rare or absent in young animals and increase with age. Other cells described in the dog parathyroid include the water-clear cell (Weymouth and Baker, 1954) and occasional multinucleated cells or syncytial cells (Oksanen, 1980). The latter increase with age and hyperparathyroidism and contain several small, chromatin-rich nuclei. Atwal (1981) has described mixed bundles of myelinated and unmyelinated nerve fibers running along with the parathyroid blood vessels.

Light and dark chief cells, representing the resting and active cell phases, respectively, and intermediate forms have been described (Nakagami, 1965). The cells appear to undergo a secretory cycle. There are two classes of membrane-bound granules, dark ones measuring 500–800 nm and light ones measuring 100–200 nm. The light cells in neonatal dogs closely resemble the intermediate or transitional cell types seen in older animals (Wild, 1980).

Chief cells from neonatal dogs show a relatively higher volume of mitochondria, rough endoplasmic reticulum, and lysosomes than in adults, suggesting an enhanced secretory activity in the neonatal animal (Wild and Manser, 1980). The parathyroids of senile dogs contain more syncytial cells, oxyphil cells, and "colloid follicles," changes which appear to be the result of aging (Setoguti, 1977; Oksanen, 1980; Meuten *et al.*, 1984). Numerous chief cell inclusions, such as cilia and axons, have been described (Nakagami, 1965; Yeghiayan *et al.*, 1972).

Canine chief cells show the same responses to alterations in serum calcium as do other species. The calcium-chelating agent EGTA [ethyleneglycol-bis-(β-aminoethylether)-N,N'-tetraacetic acid] induces hypocalcemia, which in turn increases serum parathyroid hormone concentration. Within minutes, the Golgi regions become more prominent, the cisternae of the rough endoplasmic reticulum narrow, and the cell membrane becomes more tortuous, all changes that suggest enhanced secretory activity and hormone synthesis (Wild and Becker, 1980). Administration of calcium chloride to dogs causes ultrastructural alterations consistent with decreased synthetic and secretory activity. Nuñez and co-workers (1974) noted that the aforementioned changes are transitory and that with chronic feeding of a high-calcium diet, the chief cells eventually resume a normal ultrastructural morphology.

Other canine species are not well studied. Tewari and Swarup (1979) examined the effects of experimental hypercalcemia on Indian jackal puppies (*Canis aureus*) and noted an increase in connective tissue and parathyroid cell atrophy and shrinkage after 30 days of treatment.

The domestic cat (*Felis catus*) has two pairs of parathyroid glands, located in or near the thyroid gland. Histologically, feline parathyroid glands resemble those of other mammals, consisting of a connective tissue capsule overlying a parenchyma composed of cords of chief cells separated by vascularized connective tissue septae. Both active and inactive chief cells have been described. Membrane-bound secretory granules measure up to 250 nm in diameter and have central, electron-dense cores. The chief cells of adults contain prominent Golgi complexes and aggregations of rough endoplasmic reticulum, whereas those of kittens are characterized by numerous secretory granules, large amounts of glycogen, dispersed endoplasmic reticulum, and small Golgi complexes (Capen and Rowland, 1968,a,b; Stoeckel and Porte, 1969).

Descriptions of the parathyroid glands of other feline species and other carnivores are few and come from Forsyth's (1908) work, as cited in the review by Pischinger (1937).

3. Order Chiroptera—Bats

One or two pairs of parathyroids are present in bats, located on or near the thyroid glands. Ultrastructural studies by Nuñez *et al.* (1972) in two species of bats (*Myotis lucifugus; Pipistrellus pipistrellus*) reveal a secretory cycle of the chief cells similar to that of other mammals. The cells contain numerous secretory granules. The gland appears to be activated at the onset of hibernation and less active during arousal from hibernation.

4. Order Proboscidea—Elephants

Elephants appear to have only a single pair of parathyroid glands of typical cell cord structure. Both chief cells and oxyphil cells have been described, but detailed studies are not available (Fujita and Kamiya, 1963; Roth and Schiller, 1976).

5. Order Artiodactyla—Swine, Cows, Sheep, Deer, and Others

There is a single pair of parathyroid glands in pigs, located adjacent to or embedded within the thymus gland, rather than the thyroid. Both light and dark chief cells have been described. The light cells, which are the most common, are considered to be inactive. They contain few cytoplasmic organelles and occasional membrane-bound secretory granules, with a diameter of about 240 nm. The dark or active cells contain more prominent aggregations of granular endoplasmic reticulum, large Golgi complexes, and numerous prosecretory granules. Secretory granules are distributed throughout the cytoplasm, including regions within cytoplasmic projections of the plasma membrane (Capen, 1971; Oldham *et al.*, 1971). The number of secretory granules decreases in pigs fed a low-calcium diet, suggesting an increase in the secretory activity of the parathyroid (Nuñez *et al.*, 1976).

Four parathyroids are observed in the cow; the inner pair (parathyroid IV) is located on or within the thyroid glands, and the outer pair (parathyroid III) near the common carotid artery. The gland consists of both chief cells and oxyphils, arranged in typical cell cord structure. Both light and dark chief cells have been observed (Capen *et al.*, 1965a,b; Roth and Schiller, 1976).

Jacobwitz and Brown (1980) have investigated the innervation of the bovine parathyroid glands and have reported on the relative absence of noradrenergic nerve endings, which is in contrast to the direct catecholamine innervation seen in human parathyroids. Large numbers of dopamine-containing mast cells are noted, however, to be scattered throughout the interstitium.

The inactive, or light, chief cell, which contains few cytoplasmic organelles or secretory granules, predominates in the nonlactating, nonpregnant cow. Dark-phase chief cells, characterized by prominent Golgi and endoplasmic reticulum membranes and numerous secretory granules, are increased in pregnant, lactating cows. Two types of secretory granules, smaller ones measuring 100–200 nm in diameter, and larger, lysosomelike granules containing acid phosphatase and measuring over 400 nm in diameter, have been observed (Capen *et al.*, 1965a,b; Shannon and Roth, 1971).

Habener and co-workers (1977) have used pulse–chase labeling and electron microscopic autoradiographic techniques to follow the path of parathyroid hormone precursors in the bovine parathyroid gland from the endoplasmic reticulum to the Golgi and immature granules and finally into mature secretory granules, which then migrated to the cell membrane. The secretory granules have been seen in cytoplasmic projections and the extracellular space and capillary endothelium (Capen, 1971; Capen *et al.*, 1965a,b; Capen and Young, 1967; Capen and Roth, 1973; Shannon and Roth, 1971).

Synthesis and secretion of bovine parathyroid hormone increases in response to vitamin A, cytochalasin B, and decreased calcium values (Chertow *et al.*, 1975). Stimulated cells show increased numbers of secretory granules, increased complexity of the plasma membrane and the endoplasmic reticulum, and dilation of the Golgi membranes. High calcium decreases parathyroid hormone secretion and increases the number of lipid vacuoles and phagolysosomes. Vinblastine also reduces hormone secretion and causes decreased numbers of microtubules and less stacking of the endoplasmic reticulum. Vinblastine and hydrocortisone block the stimulatory effects and ultrastructural changes caused by vitamin A. Decreased circulating parathyroid hormone levels and ultrastructural changes consistent with inactivity are observed in parathyroids of pregnant cows fed a high-calcium diet. Conversely, cows fed a diet normal in calcium show evidence of parathyroid gland stimulation prepartum (Black *et al.*, 1973).

Sheep and goats have two pairs of parathyroid glands, which are similar in position and histology to bovine parathyroids. Follicles of chief cells have been observed scattered throughout the glandular parenchyma of the sheep parathyroid (Tsuchiya and Tamate, 1981).

The ultrastructure of the sheep parathyroids is similar to that of other mammals, especially primates and rats. Secretory granules are localized primarily near the plasma membrane of chief cells. Adenylate cyclase activity is localized in the plasma membrane, and thus a role for adenylate cyclase in the regulation of parathyroid hormone secretion has been postulated (Tsuchiya and Tamate, 1979).

The chief cells in Virginia deer (*Odocoileus virginianus*) are arranged in sheets and cords separated by connective tissue and vessels. Mature secretory granules (100–400 nm diameter) appear to originate from the immature secretory granules in the Golgi region (Munger and Roth, 1963; Roth and Munger, 1963).

6. Order Perissodactyla—Horses and Rhinoceroses

The parathyroid glands of the horse (*Equus caballus*) have been well studied. Horses have two pairs of parathyroids, but their embryologic origin has not been established (Roth and Schiller, 1976). Intraglandular canals and cysts are seen in the majority of the parathyroids, while a minority of the glands contain ectopic thyroid or thymus tissue (Krook and Lowe, 1964).

Microscopically, dense connective tissue and fat cells may be seen between the cell cords of the equine parathyroid; neither component varies with the age of the animal. Four cell types have been described: chief cells with light and dark phases, oxyphil cells, and water-clear cells (Fujimoto *et al.*, 1967). Oxyphil cells are rare and are usually seen only in older animals. Water-clear cells are also rare but are found in both young and adults. These latter cells may represent chief cells modified in response to disease (Roth and Schiller, 1976).

Light chief cells of the horse parathyroid gland have prominent Golgi complexes containing numerous prosecretory and secretory granules and stacks of rough endoplasmic reticulum, indicative of an actively secreting cell (Fujimoto *et al.*, 1967). The dark cells, in contrast, show little evidence of secretory activity. There is no evidence of a parathyroid secretory cycle in the horse.

The histology of the parathyroid glands of the Indian rhinoceros (*Rhinoceros unicornis*) is similar to that of the horse (Cave and Aumonier, 1966).

7. Order Lagomorpha—Rabbits

The parathyroid glands of rabbits are usually located caudal to the thyroid along the carotid (parathyroid III) and are embedded within the thyroid (parathyroid IV). Accessory glands are numerous, and parathyroid III derivatives may be located within the thymus. The chief cells are arranged in cords and sheets; occasionally, oxyphil cells and colloid-filled follicles are present. Mature secretory granules measure 200–400 nm and exit from the gland by exocytosis. The rabbit parathyroid is innervated by parasympathetic fibers originating in the ipsilateral superior cervical ganglion and the dorsal nucleus of the vagus (Shoumura *et al.*, 1983).

This innervation may normally inhibit parathyroid hormone synthesis and secretion, as vagotomy causes increased numbers of free ribosomes, Golgi regions, and prosecretory granules, and decreased numbers of secretory granules in the rabbit parathyroid gland (Isono and Shoumura, 1980). Administration of L-asparaginase, which causes hypocalcemic tetany, leads to degranulation and eventual degeneration of chief cells (Young *et al.*, 1973). Fluoride has no significant effect on either serum calcium, parathyroid histology, or ultrastructure in rabbits (Rosenquist and Boquist, 1973).

8. Order Rodentia—Rats, Mice, Hamsters, and Others

The rat (*Rattus* sp.) has a single pair of parathyroid glands (parathyroid III), located dorsolaterally on the superior pole of the thyroid, often embedded within the thyroid tissue (Fig. 1) (Pour *et al.*, 1983b). Accessory parathyroid glands are commonly located between the thymus and the primary parathyroid gland or within the thymus. Parathyroid chief cells exist in light and dark phases; the light cell comprises 75–95% of the chief cell population and is considered to be in an inactive state (Rosof, 1934; Weymouth and Baker, 1954). Rat parathyroid glands are richly innervated. Most of the

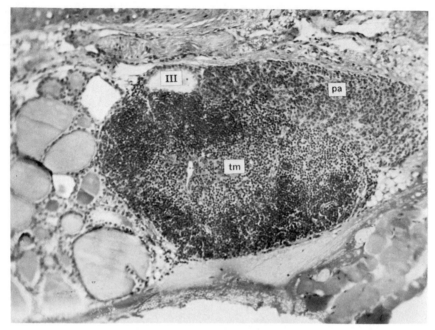

Fig. 1. Parathyroid gland (pa) histology in the mouse, in which the embryological relationship of the gland with pharyngeal pouch III (III) and thymus (tm) has been retained. [From Gorbman and Bern, 1962, "A Textbook of Comparative Endocrinology." Copyright © 1962. Reprinted by permission of John Wiley & Sons, Inc.]

nerves supply the blood vessels that run between the cell cords. Adrenergic innervation of the chief cells has also been reported (Raybuck, 1952).

Setoguti *et al.* (1980) reported the highest alkaline phosphatase and adenosine triphosphatase activities at the plasma membranes adjacent to the pericapillary space, possibly reflecting the energy requirements for active transport of substances across the plasma membrane.

Ultrastructural studies of rat parathyroid glands are numerous. Ravazzola and Orci (1977) have found evidence of both tight and gap junctions between rat chief cells using freeze–fracture techniques. A secretory cycle of the parathyroid gland has been described by a number of investigators (Roth and Raisz, 1964, 1966). In addition to infrequent, large storage granules (750 nm), abundant, small (100–250 nm), vesicular prosecretory granules have been described. Morphometric studies of stimulated and suppressed rat parathyroid glands indicate that the membranes of the secretory granules fuse with the cell membranes at the time of secretion. In EGTA-stimulated cells, there is a membrane shift from the Golgi complex and secretory granules to the plasma membrane within 1 hr. The plasma membrane appears to be retrieved by endocytosis between 1 and 3 hr (Wild *et al.*, 1982, 1984). The time course for the production and release of secretory granules is 45–60 min (Nakagami *et al.*, 1971).

The number of large storage granules in rats decreases in response to EDTA-(ethylenediaminetetraacetic acid disodium salt)-induced hypocalcemia. Large granules on stalks bud from the Golgi cisternae and square cylindrical crystals appear in the endoplasmic reticulum (Soji et al., 1974; Setoguti et al., 1981; Setoguti and Inoue, 1983). Depressions and elevations of the plasma membrane are thought to represent the sites and endo- and exocytosis. Treatment of rats with epinephrine or growth hormone leads to a rapid decrease in the amount of rough endoplasmic reticulum and of the mean number of storage granules in parathyroid chief cells, along with an increase in serum calcium and phosphate values (Altenähr and Kampf, 1976; Setoguti et al., 1984). Parathyroid glands from uremic nephrectomized rats show evidence of increased secretory activity, which can be partially prevented by the administration of vitamin D (Weisbrode and Capen, 1974). Vitamin D administration results in atrophy of the chief cells, with an accumulation of prosecretory granules. Cystic changes in the parenchyma and widened intracellular spaces have also been noted (Koyama et al., 1984). Large storage granules increase in number in animals administered calcium chloride and vitamin D_2 (Setoguti et al., 1981). Autoradiographic studies have shown that $1,25\text{-}(OH)_2D_3$ is taken up by rat parathyroid chief cells and is concentrated in the nucleus, consistent with the pattern seen for steroid hormones (Stumpf et al., 1981). Reaven and Reaven (1975) noted a decrease in the number of microtubules along with an increase in secretory granules in colchicine-treated chief cells, suggesting a role of microtubules in the release of parathyroid hormone.

In the mouse (Mus sp.), a single pair of encapsulated parathyroid glands is located on the lateroventral or caudolateral surface of the thyroid. Although the parathyroids may be embedded in the thyroid surface, they are generally more superficial than in the rat (Pour et al., 1983a). Mouse parathyroids also show less prominent trabeculation and vascularization than rat parathyroids. Accessory or ectopic glands have been reported, often in the caudal tracheal region (Frith and Fetters, 1983). Cordier and Haumont (1983) described the embryological development of the parathyroid glands in two strains of mice.

At both the light and ultrastructural level, mouse parathyroid chief cells are similar to those of the rat; 50–90% of the cells are normally in the light phase, depending upon the secretory activity of the gland (Pour et al., 1983a). Stoeckel and Porte (1966) attribute the variation in cytoplasmic density to fixation technique, but Nakagami (1967) has described a mouse parathyroid secretory cycle that is similar to that reported for the rat and other species.

Ultrastructural studies have demonstrated desmosomes and interdigitating plasma membranes between mouse parathyroid chief cells (Wild et al., 1982). Golgi complexes and rough endoplasmic reticulum are large and well developed, although few organelles are seen in the cytoplasm at birth (Stoeckel and Porte, 1969; Isono et al., 1977). The most common class of secretory granule is spherical and measures 150–200 nm in diameter. Less frequent ellipsoidal bodies measuring 270–410 nm in diameter also have been described (Hara and Nagatsu-Ishibashi, 1964). The granules originate in the Golgi apparatus and travel to the apical and lateral regions of the cell where exocytosis occurs (Stoeckel and Porte, 1969).

The ultrastructure of the mouse chief cells can be altered experimentally. The carbonic anhydrase inhibitor acetazolamide appears to inhibit the secretory activity of the mouse parathyroid, resulting in a poorly developed Golgi complex, few free ribosomes, and few secretory granules (Isono *et al.*, 1980). In contrast, the glucocorticoid triamcinolone lowers plasma calcium by interfering with intestinal calcium uptake and causes a stimulation of the mouse parathyroid gland (Coleman and Silbermann, 1978). Cadmium also has a stimulatory effect on the parathyroid, leading to an increase in the number of secretory and prosecretory granules along with more prominent Golgi complexes and more numerous ribosomes (Isono *et al.*, 1979).

Fewer studies have been published on the parathyroids of other rodents. Two pairs of glands are generally seen in the guinea pig, whereas only parathyroid III is present in hamsters. In the Syrian hamster, *Cricetus aurata,* the parathyroids are generally found at the lateral lower margins of the thyroid gland and are difficult to identify grossly. They may be surrounded by a fibrous capsule and may thus be separated from the thyroid gland or may project into the thyroid tissue. Fat cells occasionally occur in the connective tissue septa. Ectopic or accessory parathyroid glands have also been reported (Pour, 1983).

Seasonal variation occurs in the numbers of secretory granules in the hamster (*C. cricetus*), suggesting increased secretory activity of the glands during the winter months (Kayser *et al.*, 1961). A secretory cycle similar to that proposed for other species has been described (Stoeckel and Porte, 1969).

Boquist and Fahraeus (1975) have described a single pair of parathyroid glands placed anterolaterally to the superior poles of the thyroid gland of the gerbil (*Meriones unguiculatus*). A three-phase secretory cycle was suggested by Kapur (1977), who noted light-, dark-, and intermediate-phase cells in gerbil parathyroid glands. Recently, however, Larsson *et al.* (1984) proposed that the light- and dark-cell phases are an artifact of fixation and did not reflect a functional cycle, based on their observations that light chief cells are not conspicuous in glands that had been fixed by perfusion rather than immersion. Boquist and Lundgren (1975) also concluded that gerbil parathyroid glands lack a secretory cycle.

Gerbil parathyroids respond to the administration of calcium, magnesium, exogenous parathyroid hormone, and vitamin D metabolites in a manner similar to other species (Boquist, 1975a, 1977; Boquist *et al.*, 1979). Culture of parathyroid glands in high-calcium medium causes chief cell suppression and the appearance of oxyphil cells, which are not normally seen in gerbils (Boquist, 1975b). Degenerative changes also occur in suppressed chief cells and are associated with the cytoplasmic and intramitochondrial accumulation of calcium.

The parathyroid chief cells of the woodchuck, *Marmota monax,* are arranged in clusters around capillaries (Frink *et al.,* 1978). Increased numbers of secretory granules are observed during the fall and winter, whereas evidence of increased secretory activity is seen during spring and summer. Seasonal changes in the histology of the parathyroid glands of the Indian palm squirrel *Funambulus pennanti* have also been reported. Males exhibit the lowest level of secretory activity during the period of the breeding season, January to April. A progressive increase in secretory activity is observed during the remainder of the year, characterized by degranulation of the chief cells and a concomitant increase in serum calcium levels. Females show the highest secretory activity from January to June, coincidental with the period of breeding and pregnancy. Additionally, chronic vitamin D_2-induced hypercalcemia results in atrophy of the parathyroids (Swarup and Tewari, 1978).

9. Order Insectivora—Moles, Shrews, Hedgehogs, and Others

Studies of insectivore parathyroids are generally limited to descriptions of the number and location of the glands. Ultrastructural studies are lacking. Chronic hypercalcemia and parathyroid atrophy was induced in the house shrew, *Suncus murinus,* by administration of exogenous parathyroid hormone (Srivastav and Swarup, 1979).

B. Subclass Metatheria

Order Marsupalia—Kangaroos, Opossums, and Others

Two pairs of parathyroid glands have been noted in marsupials, with parathyroid III larger than IV. Parathyroid III is located in the vicinity of the common carotid bifurcation, whereas parathyroid IV is found in the thorax with the thymus gland. In some species, parathyroid IV may be fragmented or absent. The marsupial parathyroids are not associated with the thyroid gland (Adams, 1955; Kraus and Cutts, 1983). In general, marsupial parathyroid glands are densely cellular. In the opossum *Didelphis virginiana,* a homogenous population of chief cells devoid of oxyphil cells has been described. Ultrastructural studies reveal rod-shaped mitochondria, granular endoplasmic reticulum, free ribosomes, and small, perinuclear Golgi complexes in polyhedral cells with centrally placed nuclei. Small, electron-dense secretory granules surrounded by a delicate membrane are associated with either the Golgi apparatus or the basal cell membrane (Krause and Cutts, 1983).

C. Subclass Prototheria

Order Monotremata—Echidnas and Platypus

The four parathyroids of the echidna, *Tachyglossus aculeatus,* are derived from the third and fourth branchial pouches. In adults, parathyroid IV is located at the carotid bifurcation, whereas parathyroid III is embedded within or adjacent to the thymus gland. A pair of parathyroid glands is found in the duckbilled platypus, *Ornithorhynchus anatinus,* positioned at the origin of the trachea. Additionally, a separate glandular body, the "parathymus gland," is located near the thymus and may represent additional parathyroid tissue. Histological and ultrastructural studies of the parathyroid glands of the Prototheria are lacking (Roth and Schiller, 1976). Clearly, further investigation of these glands in monotremes is warranted.

III. CLASS AVES

The anatomy and micromorphology of the parathyroid glands of many species of birds have been described by Forsyth (1908). There are two or four parathyroids, and these originate from the third and fourth pharyngeal pouches. Occasionally parathyroids III and IV share the same connective tissue capsule. The yellowish parathyroids are separate from the larger and reddish-colored thyroid, a fact that makes parathyroidectomy much easier to accomplish in birds than in mammals without interfering with other endocrine systems. The parathyroid glands lie along the jugular vein between the thyroid and ultimobranchial glands. It is sometimes necessary to dissect the interclavicular air sacs in order to see the glands. Accessory parathyroid tissue is commonly found in the adjacent area, including within the thyroid, thymus, and ultimobranchial gland.

Avian parathyroid glands are richly innervated, by both cholinergic (vagal; Butler and Osborne, 1975; Wideman, 1980) and adrenergic fibers (Bennett, 1971). It has been suggested that the nerves regulate the blood flow to the gland, since parathyroid arterioles are well innervated. Additionally, some of the fibers innervate the parathyroid cells directly and thus may affect parathyroid hormone release. In many species of birds the parathyroids are located adjacent to, or partially envelope, the carotid body and share its vascular supply (Fig. 2). Surprisingly little has been done to clarify a functional relationship between the parathyroids and carotid body or to determine the function of the parathyroid gland innervation. A possible neurohemal association between the parathyroid and carotid body has been suggested in the starling. Vagotomy has not led to changes in ultrastructural characteristics of the starling parathyroid, as it does in rabbits (Wideman, 1980).

The parathyroid glands of birds consist of epithelial chief cells arranged in

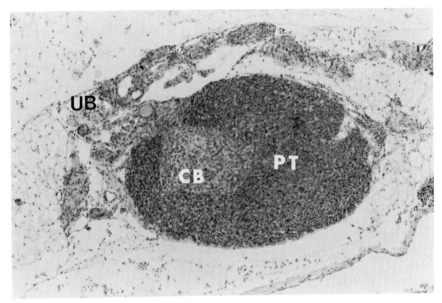

Fig. 2. Parathyroid gland (PT) of the starling (*Sturnus vulgaris*), which encloses the carotid body (CB). Ultimobranchial follicles (UB) are nearby. [Figure kindly provided by Dr. Robert F. Wideman, Pennsylvania State University.]

cords. Among the cords run blood vessels and connective tissue. The ultrastructure of the parathyroid has been described for several avian species (Narbaitz, 1972; Youshak and Capen, 1970; Nevalainen, 1969; Gould and Hodges, 1971; Stoeckel and Porte, 1973; Wideman, 1980); it is generally similar to that of mammals, except for a lack of oxyphil cells and a paucity of mature secretory granules in the cytoplasm. There are both dark and light variants of chief cells in chickens (Chan, 1977) but not in starlings; this may simply relate to the mode of fixation. Avian parathyroids contain many prosecretory granules which can be seen budding from the Golgi apparatus. Coated vesicles are also present. Granular endoplasmic reticulum is generally located in the region of the Golgi. It is suggested that in birds, as in mammals, most parathyroid hormone is not stored but is secreted without being packaged into granules. This may be the result of a constant need for parathyroid hormone for bone remodeling during growth and egg production.

Hypertrophy of the parathyroid glands of chickens occurs in response to deprivation of ultraviolet light, dietary vitamin D, or calcium (Mueller *et al.*, 1970). Electron microscopic studies of parathyroid glands of chicks maintained on a low-calcium diet show increased amounts of endoplasmic reticulum and Golgi membranes, indicating increased protein synthetic activity (Chan, 1976). Similarly, chick embryos grown apart from their shells (shell-

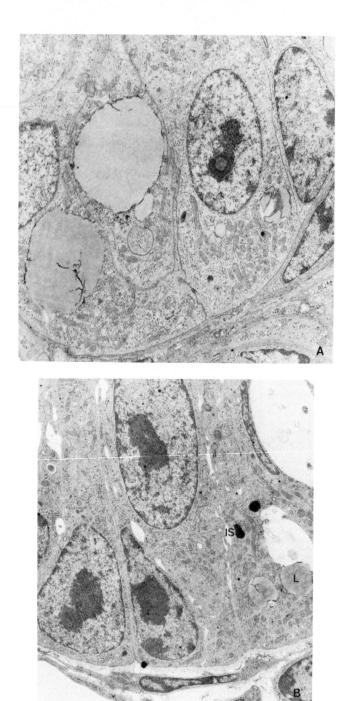

Fig. 3. Seasonal changes in the parathyroid gland of the starling. (A) Typical ultrastructural appearance in November, with large lipid droplets and no intracellular spaces. ×12,000. (B) Typical ultrastructural appearance in February, with enlarged intracellular spaces (IS) and decreased amounts of lipid (L). ×11,000. [Figures kindly provided by Dr. Judith A. Cole, University of Missouri School of Medicine.]

less culture), which become progressively hypocalcemic during development, demonstrate a progressive increase in amounts of granular endoplasmic reticulum and Golgi lamellae in their parathyroids relative to normal *in ovo* embryos of the same age (Narbaitz, 1972). In 15-day, shell-less embryos, mature secretory granules are seen, and the endoplasmic reticulum is very conspicuous and arranged in lamellar stacks (N. B. Clark and B. E. Dunn, 1986). All of these morphological changes suggest increased parathyroid hormone production in the hypocalcemic shell-less embryos. This system enables easy manipulation of parathyroid gland structure and function; such experiments would be much more difficult to accomplish using mammalian embryos.

A number of studies have shown that no major seasonal changes occur in the structure of the avian parathyroid gland. However, lipid inclusions seem to be more common in the winter (Stoeckel and Porte, 1973; R. F. Wideman, unpublished). Dilated intercellular spaces are seen in the parathyroid of starlings in the spring, which may reflect hypersecretion (R. F. Wideman, unpublished; J. A. Cole, unpublished) (Fig. 3). A similar structural change has been noted in hens maintained for a long time on a low-calcium diet (Gould and Hodges, 1971).

IV. CLASS REPTILIA

The anatomy and morphology of reptilian parathyroids has been reviewed by Clark (1970). Reptiles are subdivided into a number of distinct groups (turtles, crocodiles, snakes, lizards) in which the location of the parathyroids differs considerably. In most forms the parathyroid glands are located just anterior to the heart along the major arteries. In snakes, one of the pairs is far from this location, near the angle of the jaw (see Clark, 1970). The structure of the parathyroid glands in reptiles is similar to that of birds and mammals, consisting of epithelial cords that are richly vascularized and surrounded by a connective tissue capsule. Additionally, in the parathyroids of some species of turtles, lizards, and crocodiles, there are conspicuous follicles that contain colloidlike inclusions. Large cysts that comprise much of the parathyroid gland have also been reported (Clark, 1965; Singh and Kar, 1983a).

Reptiles have one or two pairs of parathyroid glands, presumed to be derivatives of pharyngeal pouches 3 and 4, although this has not been determined embryologically in all cases. The presence of additional, smaller parathyroid glands has been reported occasionally, and it is presumed that these represent accessory tissue (Adams, 1939; Peters, 1941; Oguro and Sasayama, 1976; Singh and Kar, 1983a). The reptilian parathyroid is most frequently reported to be composed of a single cell type, the chief cell. Additional cell types reported, such as the "clear," "dark," and "epithelial" cells described in several species of lizard (Rogers, 1963), have been suggested by others (Sidky, 1965) to be variants of the chief cell. However, oxyphil cells are found occasionally in some species of freshwater turtles (Clark and Khairallah, 1972). Additionally, dark, irregularly shaped stellate cells have been described at the periphery of cell cords of the iguana parathyroid (Anderson and Capen, 1976).

Most studies of reptilian parathyroid structure have not been concerned with seasonal changes in the histology of the glands. However, in a few cases seasonal changes have been noted; in these, signs of glandular activation occur during the breeding season (summer), and degenerative changes are observed during the winter (Peters, 1941; Singh and Kar, 1983a). In a few other studies it has been specifically noted that seasonal changes in parathyroid histology do not occur (Sidky, 1965; Anderson and Capen, 1976; Singh and Kar, 1983b). Thus the extent of seasonal variation in parathyroid histology of reptiles is at present unknown.

There have been only a few studies of the ultrastructure of reptilian parathyroid glands. In freshwater turtles (*Pseudemys scripta, Chrysemys picta, Clemmys japonica*) both light and dark variants of the chief cell have

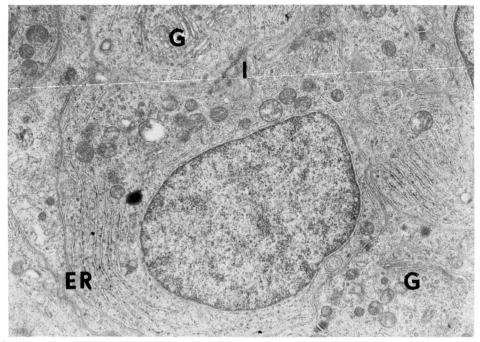

Fig. 4. Ultrastructure of an activated chief cell in the lizard, *Iguana iguana*, fed a low-calcium diet for 16 months. The Golgi (G) and endoplasmic reticulum (ER) membranes are large and conspicuous. Numerous prosecretory granules are in the region of the Golgi and the plasma membranes have complex interdigitations (I). ×4700. [From Anderson and Capen, 1976.]

been noted (Clark and Khairallah, 1972; Chin, 1974). Additionally, in *Pseudemys* and *Chrysemys,* oxyphils were seen (Clark and Khairallah, 1972). In lizards (*Iguana iguana*) the chief cells all have light or moderately electron-dense cytoplasm. A second cell type which is irregular in shape and has electron-dense cytoplasm is located at the periphery of the cords. These stellate cells do not respond to changes in calcium concentration but may instead have a supportive function (Anderson and Capen, 1976). In the snake (*Thamnophis sirtalis*) parathyroid, only chief cells are present, and all seem to have about the same electron density (Roth and Schiller, 1976). In the five species just mentioned, the ultrastructure of the chief cell is similar, containing all of the organelles involved in protein synthesis, including prominent rough endoplasmic reticulum, Golgi lamellae, and prosecretory granules. In turtles and snakes but not in lizards, there are large lipid inclusions in the cytoplasm.

Alteration of parathyroid ultrastructure in response to decreased calcium availability has been studied in lizards. Iguanas begin to show signs of increased chief cell activity by 1 month after being placed on a low-calcium diet. These changes, including increased prominence of the endoplasmic reticulum and Golgi regions and increased numbers of secretory granules, are noted as long as the animals remain calcium deficient (Fig. 4). In some calcium-deficient iguanas stacks of rough endoplasmic reticulum fill the cytoplasm (Anderson and Capen, 1976).

V. CLASS AMPHIBIA

The greatest variation in parathyroid form and function occurs in the amphibians. In urodeles (salamanders) the glands appear to be similar in structure to those of mammals, birds, and reptiles, consisting of cell cords invested with a rich vascular supply, nerves, and connective tissue. In anurans (frogs, toads), on the other hand, the glands have a distinctly different appearance, consisting of tightly packed epithelial cells. The cell cord structure is absent. It is difficult at present to categorize the parathyroid glands of the caecilians (limbless amphibians), since there have been very few studies.

A. Order Caecilia

The structure of the parathyroid gland has been described for only one species of caecilian, *Chthonerpeton indistinctum* (Welsch and Schubert, 1975). At the light microscopic level the gland appears to be thyroidlike, consisting entirely of follicles. If this is indeed parathyroid tissue it has a different structure from that of all other vertebrates, but the authors do not remark upon this point. The ultramicrographs, however, display cells that do resemble other parathyroid cells, and there is no evidence of a follicular structure. The cells are of a single type. The cytoplasm contains poorly developed endoplasmic reticulum, small Golgi bodies, and few lysosomes. Lipid droplets, secretory granules, and glycogen particles are found in all cells, which are bound together with extensive interdigitations. No nerves or blood vessels are seen between the cells. It is not clear how many parathyroid glands are present in this species.

B. Order Urodela

In some neotenic salamanders, such as the mudpuppy (*Necturus*), the parathyroid glands never develop. Most salamandrid species, however, do have parathyroids, at least during their adult life. The structure of the urodele parathyroid resembles that of other vertebrate groups, both at the light and electron microscopic level. The glands are composed of cell cords, among which runs a rich supply of nerves and blood vessels (Scholz, 1933). The innervation of the parathyroid glands has not been studied in detail for any amphibian species.

Urodeles generally have a single pair of parathyroid glands located near the carotid and systemic arches anterior to the heart. Accessory parathyroid glands are not uncommon (Oguro, 1969, 1973; Whittle and Dent, 1979). A single cell type is described in urodele parathyroids studied at the light microscopic level. However, in an extensive series of ultrastructural studies of the parathyroid of the newt *Triturus pyrrhogaster,* two cell types have been described (Isono *et al.,* 1971; Isono and Shoumura, 1973; Isono and Shoumura-Sakurai, 1973; Setoguti *et al.,* 1970a,b). These include basal cells, which do not respond to hormonal stimuli or to alterations in plasma calcium and are thought to play a role in support, and suprabasal cells, which are considered to be the equivalent of chief cells.

During natural hibernation the suprabasal cells in newts contain well-developed Golgi bodies, numerous prosecretory granules, numerous lysosomes, and several types of dense bodies that are believed to be precursors of secretory granules or lysosomes. The glands are stimulated at the time of emergence from hibernation in the spring (Setoguti *et al.,* 1970a,b) and by administration of phosphate salts (Isono *et al.,* 1971) or EDTA (Isono and Shoumura, 1973). These changes include increased numbers or amount of mitochondria, granular endoplasmic reticulum, and Golgi bodies. Additionally, there are increased numbers of large, dense bodies and small, dense granules, which the authors suggest to be the precursors of mature secretory granules. Autoradiographic studies of the newt parathyroid gland demonstrate that tritiated leucine first localizes over the granular endoplasmic reticulum and subsequently moves to the Golgi region and finally to secretory granules and intercellular spaces, demonstrating the expected events characteristic of protein synthesis (Isono and Shoumura, 1973). Conversely, injecting the newts with parathyroid hormone depresses protein synthesis, as indicated by decreased amounts of endoplasmic reticulum, Golgi cisternae, and secretory granules (Isono and Shoumura-Sakurai, 1973).

Gross seasonal changes in parathyroid gland structure, especially vacuolation and degeneration of cells such as have been reported in many species of anurans, have not been reported in urodeles. Whittle and Dent (1979) noted occasional vacuolated cells in the parathyroids of the newt *Notophthalmus viridescens*, but the appearance of the parathyroids was not associated with any particular area of the gland or season. Newts, maintained for up to 4 months at either 4 or 23°C, show no gross changes in their parathyroid structure. Scholz (1933) reported structural changes in the parathyroid glands of several species of urodeles that had been in captivity for long periods of time, but not in freshly caught specimens.

C. Order Anura

The parathyroid glands of anuran amphibians, generally two pairs, are located anterior to the heart near the ventral branchial body and carotid artery. In almost all species the glands are composed of closely packed epithelial cells surrounded by a connective tissue capsule. In many cases the cells are arranged in whorls; this is more common in frogs than in toads (Fig. 5). Blood vessels and nerves are supplied to the connective tissue capsule region, and in most species the blood vessels do not penetrate into the center of the gland (Boschwitz, 1961; Lange and von Brehm, 1965; Coleman, 1969). Instead, intercellular channels are described which may be the means by which the cells receive nutrition and the route by which parathyroid hormone is transported to the periphery (Hara *et al.*, 1959; Isono, 1960; Lange and von Brehm, 1965; Cortelyou and McWhinnie, 1967; Coleman, 1969). In a few species, blood vessels have been described that pass into the parathyroid gland (Waggener, 1929; von Brehm, 1963; Rogers, 1965). The vascularity of the parathyroid gland in *Xenopus* appears to be correlated with its more "urodelelike" cell cord structure (von Brehm, 1963).

The anuran parathyroid glands consist of a single type of cell, the chief cell, which often has light and dark variants. These differences in electron density of cytoplasm and nucleus may simply be fixation artifacts, since they appear most frequently in glands that are immersed rather than perfused with fixative. The central parathyroid cells in many anuran species are smaller, of more elongate shape, and have a denser nucleus and cytoplasm (Sasayama and Oguro, 1974). It has been suggested (Lange and von Brehm, 1965) that the central and peripheral cells in avascular anuran parathyroids are in different nutritional states, owing to their differential distance from the blood supply. Both the light and the dark cell have been interpreted to be the active form of the chief cell. Regardless of the electron density of the cytoplasm, active parathyroid cells are interpreted to be those in which there are abundant mitochondria, secretory granules, Golgi, and endoplasmic reticulum, whereas in hypofunctional cells the cytoplasmic inclusions including secretory granules are less abundant. Much lipid is often present in the cytoplasm of inactive cells.

Increased activity of the parathyroid cells in several species of anurans is stimulated by injection of phosphate or EDTA. Conversely, activity of the glands is depressed by treatment with calcium salts (Montskó *et al.*, 1963; von Brehm, 1964). Cytoplasmic and nuclear crystalline inclusions are found in active parathyroid gland cells of *Rana temporaria* (Coleman and Phillips, 1972; Lange and von Brehm, 1965). Such cytoplasmic inclusions are found close to the Golgi; crystalline inclusions in either the nucleus or cytoplasm are most frequently found in glands of winter frogs or animals kept at cold temperature. Whether this relates to formation and storage of parathyroid hormone or its secretion is uncertain. A search for crystalline inclusions in several other species of anurans was unsuccessful (Coleman and Phillips, 1972).

Seasonal changes in the histological appearance of the parathyroid glands are a common finding in anurans. The glands undergo a seasonal cytolysis in winter (Fig. 5), followed by a regeneration of the normal epithelial structure in the spring. The vacuolation of the gland, which preceeds degeneration, has been variously interpreted to be a sign of degeneration (Cortelyou *et al.*, 1960), secretory exhaustion (Yoshida and Talmage, 1962), or glandular activation (Isono, 1960; Hara and Yamada, 1965; Boschwitz, 1965). Degeneration of the gland is consistent with decreased normal blood calcium values. Also, the impact of parathyroidectomy or of administration of parathyroid hormone is least in winter and highest in summer (Robertson, 1977). Von Brehm (1964) has shown that exposure of frogs and toads to cold temperatures for a week or longer mimicked the natural seasonal change in parathyroid structure.

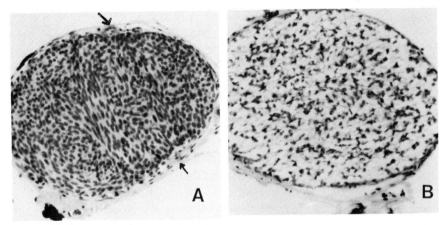

Fig. 5. Parathyroid gland histology in the frog *Rana pipiens*. (A) Normal appearance of the frog parathyroid, showing its whorllike arrangement of epithelial cells. A dense capillary network (arrows) is located just below the connective tissue capsule. ×165. (B) Degenerated appearance of the parathyroid gland in a winter frog. ×165. [From Cortelyou *et al.*, 1960.]

Parathyroid glands have not been found in fish, despite numerous investigations. One or more pituitary hormones may enable fish to raise their blood calcium values (Pang *et al.*, 1978; Parsons *et al.*, 1978). Additionally, a parathyroid hormone-like factor has been suggested to be secreted by the corpuscles of Stannius (Lopez *et al.*, 1984).

VI. BIOMEDICAL IMPLICATIONS AND CONCLUSIONS

Knowledge of the location and morphology of vertebrate parathyroid glands can provide scientists with information needed to choose ideal systems for studying parathyroid function. In laboratory mammals, it is often necessary to perform thyroidectomy in order to assure removal of all parathyroid tissue. The anatomical separation of thyroid and parathyroid in nonmammalian vertebrates provides animal models in which the parathyroids can be removed without damage to thyroid tissue, blood vessels, or nerves and without inadvertent release of thyroid hormone and calcitonin. Studies of the function of parathyroid innervation on blood flow or parathyroid hormone secretion can likewise be more easily studied in species in which the parathyroid is anatomically separate from other surrounding glands.

Special structural features of parathyroid glands of some nonmammalian species may enable researchers to learn more about basic cellular processes. For example, the study of the formation and resorption of parathyroid "colloid" material may provide information about mechanisms of exocytosis and endocytosis, as well as the significance of extracellular storage of hormones. Those animals that show seasonal changes in parathyroid structure may be useful in the determination of the effects of environmental stimuli (e.g., light, temperature) on parathyroid function. Additionally, the discovery of influences of the carotid body on the function of the avian parathyroid (or vice versa) would be likely to reveal similar actions in mammals which are not obvious, because of the lack of anatomical proximity.

Clearly there are many ways in which basic information about the parathyroid glands of nonmammalian vertebrates can augment our understanding of mammalian parathyroid glands.

What is clear from this study is that the morphology and structure of the parathyroid glands are remarkably similar in all classes of terrestrial vertebrates. The chief cells become activated in response to decreased plasma calcium values and undergo ultrastructural changes indicative of increased protein synthesis. In all groups, parathyroid hormone appears to be produced and secreted rapidly upon demand in order to be a prime regulator of calcium homeostasis.

ACKNOWLEDGMENTS

This work was supported in part by Grant PCM 84-00665 from the National Science Foundation.

REFERENCES

Adams, W. E. (1939). The cervical region of the Lacertilia. *J. Anat.* **74,** 57–71.

Adams, W. E. (1955). The carotid sinus complex, "Parathyroid" III and thymo-parathyroid bodies, with special reference to the Australian opossum, *Trichosurus vulpecula. Am. J. Anat.* **97,** 1–58.

Altenähr, E., and Kampf, E. (1976). Parathyroid function in rats treated with growth hormone. *Virchows Arch. A: Pathol. Anat. Histol.* **371,** 363–371.

Anderson, M. P., and Capen, C. C. (1976). Ultrastructural evaluation of parathyroid and ultimobranchial glands in iguanas with experimental nutritional osteodystrophy. *Gen. Comp. Endocrinol.* **30,** 209–222.

Atwal, O. S. (1981). Myelinated nerve fibers in the parathyroid gland of the dog: a light and electron-microscopic study. *Acta Anat.* **109,** 3–12.

Bargmann, W. (1939). Die Epithelkörperchen. *In* "Handbuch der mikroskopische Anatomie des Menschen" (W. Von Mollendorff, ed.), Vol. 6, Part 2, pp. 137–196. Springer-Verlag, Berlin and New York.

Bennett, T. (1971). The neuronal and extra-neuronal localisations of biogenic amines in the cervical region of the domestic fowl (*Gallus gallus domesticus* L.). *Z. Zellforsch. Mikrosk. Anat.* **112,** 443–464.

Bergdahl, L., and Boquist, L. (1973). Parathyroid morphology in normal dogs. *Pathol. Eur.* **8,** 95–103.

Black, H. E., Capen, C. C., and Arnaud, C. D. (1973). Ultrastructure of parathyroid glands and plasma immunoreactive parthyroid hormone in pregnant cows fed normal and high calcium diets. *Lab. Invest.* **29,** 173–185.

Boquist, L. (1975a). Parathyroid morphology in gerbils after calcium and parathormone administration. *Horm. Metab. Res.* **7,** 261–266.

Boquist, L. (1975b). Occurrence of oxyphil cells in suppressed parathyroid glands. *Cell Tissue Res.* **163,** 465–470.

Boquist, L. (1977). Ultrastructural changes in the parathyroids of Mongolian gerbils induced experimentally *in vitro. Acta Pathol. Microbiol. Scand.* **85,** 203–218.

Boquist, L., and Fahraeus, B. (1975). The parathyroid glands of the Mongolian gerbil (*Meriones unquiculatus*). *Pathol. Eur.* **10,** 135–143.

Boquist, L., and Lundgren, E. (1975). Effects of variations in calcium concentration on parathyroid morphology *in vitro. Lab. Invest.* **33,** 638–647.

Boquist, L., Larsson, S. E., and Lorentzon, R. (1979). Structural changes in parathyroid glands exposed to vitamin D metabolites *in vitro. Endokrinologie* **74,** 177–183.

Boschwitz, D. (1961). The parathyroid glands of *Bufo viridis* Laurenti. *Herpetologica* **17,** 192–199.

Boschwitz, D. (1965). Histological changes in the parathyroids of *Bufo viridis* Laurenti. *Isr. J. Zool.* **14,** 11–23.

Butler, P. J., and Osborne, M. P. (1975). The effect of cervical vagotomy (decentralization) on the ultrastructure of the carotid body of the duck, *Anas platyrhynchos. Cell Tissue Res.* **163,** 491–502.

Capen, C. C. (1971). Fine structural alterations of parathyroid glands in response to experimental and spontaneous changes of calcium in extracellular fluids. *Am. J. Med.* **50,** 598–611.

Capen, C. C., and Roth, S. I. (1973). Ultrastructural and functional relationships of normal and pathological parathyroid cells. *In* "Pathobiology Annual" (H. L. Ioachim, ed.), pp. 129–175. Appleton, New York.

Capen, C. C., and Rowland, G. N. (1968a). The ultrastructure of the parathyroid glands of young cats. *Anat. Rec.* **162,** 327–339.

Capen, C. C., and Rowland, G. N. (1968b). Ultrastructural evaluation of the parathyroid glands of young cats with experimental hyperparathyroidism. *Z. Zellforsch. Mikrosk. Anat.* **90,** 495–506.

Capen, C. C., and Young, D. M. (1967). The ultrastructure of the parathyroid glands and thyroid parafollicular cells of cows with parturient paresis and hypocalcemia. *Lab. Invest.* **17,** 717–737.

Capen, C. C., Koestner, A., and Cole, C. R. (1965a). The ultrastructure and histochemistry of normal parathyroid glands of pregnant and nonpregnant cows. *Lab. Invest.* **14,** 1673–1690.

Capen, C. C., Keostner, A., and Cole, C. R. (1965b). The ultrastructure, histopathology and histochemistry of the parathyroid glands of pregnant and nonpregnant cows fed a high level of vitamin D. *Lab. Invest.* **14,** 1809–1825.

Cave, A. J., and Aumonier, F. J. (1966). Parathyroid histology in the rhinocerotidae. *J. Microsc. Soc.* **86,** 51–57.

Chan, A. S. (1976). Effect of low calcium diet on the ultrastructure of the parathyroid gland of chicks. *Cell Tissue Res.* **173,** 71–76.

Chan, A. S. (1977). Ultrastructure of the parathyroid glands in the chicks. *Acta Anat.* **97,** 205–212.

Chertow, B. S., Buschmann, R. J., and Henderson, W. J. (1975). Subcellular mechanisms of parathyroid hormone secretion: Ultrastructural changes in response to calcium, vitamin A, vinblastine, and cytochalasin B. *Lab. Invest.* **29,** 173–185.

Chin, M. (1974). An electron microscopic study of the turtle parathyroid gland, with special reference to the seasonal changes. *J. Kyoto Prefect. Univ. Med.* **83,** 452–464 (In Japanese with English summary).

Clark, N. B. (1965). Experimental and histological studies of the parathyroid glands of freshwater turtles. *Gen. Comp. Endocrinol.* **5,** 297–312.

Clark, N. B. (1970). The parathyroid. *In* "Biology of the Reptilia" (C. Gans and T. S. Parsons, eds.), pp. 235–262. Academic Press, New York.

Clark, N. B., and Dunn, B. E. (1986). Calcium regulation in the embryonic chick. II. Ultrastructure of the parathyroid glands in shell-less and in ovo embryos. *J. Morphol.* (in press).

Clark, N. B., and Khairallah, L. H. (1972). Ultrastructure of the parathyroid gland of freshwater turtles. *J. Morphol.* **138,** 131–140.

Coleman, R. (1969). Ultrastructural observations on the parathyroid glands of *Xenopus laevis* Daudin. *Z. Zellforsch. Mikrosk. Anat.* **100,** 201–214.

Coleman, R., and Phillips, A. D. (1972). Crystalline bodies in parathyroid gland cells of *Rana temporaria* L. *Z. Zellforsch. Mikrosk. Anat.* **127,** 1–8.

Coleman, R., and Silbermann, M. (1978). Ultrastructure of parathyroid glands in triamcinolone-treated mice. *J. Anat.* **126,** 181–192.

Coleman, R., Silbermann, M., and Bernheim, J. (1980). Fine structure of the parathyroid glands in baboons, *Papio hamadryas* in response to experimental hypercorticoidism. *Acta Anat.* **106,** 424–433.

Cordier, A. C., and Haumont, S. (1980). Development of thymus, parathyroids, and ultimobranchial bodies in NMRI and nude mice. *Am. J. Anat.* **157,** 227–263.

Cortelyou, J. R., and McWhinnie, D. J. (1967). Parathyroid glands of amphibians. I. Parathyroid structure and function in the amphibian, with emphasis on regulation of mineral ions in body fluids. *Am. Zool.* **7,** 843–855.

Cortelyou, J. R., Hibner-Owerko, A., and Mulroy, J. (1960). Blood and urine calcium changes in totally parathyroidectomized *Rana pipiens*. *Endocrinology (Baltimore)* **66,** 441–450.

Forsyth, D. (1908). The comparative anatomy, gross and minute, of the thyroid and parathyroid glands in mammals and birds. *J. Anat.* **42,** 141–169, 302–319.

Frith, C. H., and Fetters, J. (1983). Ectopic parathyroid, mouse. *In* "The Endocrine System" (T. C. Jones, U. Mohr, and R. D. Hunt, eds.), pp. 263–243. Springer-Verlag, Berlin and New York.

Frink, R., Krupp, P. P., and Young, R. A. (1978). The parathyroid gland of the woodchuck

(*Marmota monax*): A study of seasonal variations in the chief cells. *Tissue Cell* **10**, 259–267.

Fujimoto, Y., Matsukawa, K., Inubushi, H., Nakamatsu, M., Satoh, H., and Yamigiwa, S. (1967). Electron microscopic observations of the equine parathyroid glands with particular reference to those with equine osteodystrophia fibrosa. *Jpn. J. Vet. Res.* **15**, 37–69.

Fujita, T., and Kamiya, T. (1963). Zur macroskopischen und mikroskopischen Anatomie de Schilddrüse and Nebenschilddrüse des indischen Elefanten. *Anat. Anz.* **111**, 72–78.

Gorbman, A., and Bern, H. A. (1962). "A Textbook of Comparative Endocrinology." Wiley, New York.

Gould, R. P., and Hodges, R. D. (1971). Studies on the fine structure of the avian parathyroid glands and ultimobranchial bodies. *In* "Subcellular Organization and Function in Endocrine Tissues" (H. Heller and K. Lederis, eds.), pp. 567–603. Cambridge Univ. Press, London and New York.

Habener, J. F., Amherdt, M., and Orci, L. (1977). Subcellular organelles involved in the conversion of biosynthetic precursors of parathyroid hormone. *Trans. Assoc. Am. Physicians* **90**, 366–379.

Hara, J., and Nagatsu-Ishibashi, I. (1964). Electron microscopic study of the parathyroid gland of the mouse. *Nagoya J. Med. Sci.* **26**, 119–125.

Hara, J., and Yamada, K. (1965). Chemocytological observations on the parathyroid gland of the toad (*Bufo vulgaris japonicus*) in specimens taken throughout the year. *Z. Zellforsch. Mikrosk. Anat.* **65**, 814–828.

Hara, J., Isono, H., and Fujii, H. (1959). Electron microscopic observation of the parathyroid gland of the toad, *Bufo vulgaris japonicus*. *Acta Sch. Med. Gifu* **7**, 1548–1556.

Isono, H. (1960). Histological study of the parathyroid gland in the toad (*Bufo vulgaris japonicus*). *Acta Sch. Med. Gifu* **8**, 277–293. (In Japanese with English abstract).

Isono, H., and Shoumura, S. (1973). Fine structure of the newt parathyroid gland after EDTA administration: Mechanism of formation and extrusion of secretory granules. *J. Electron Microsc.* **22**, 191–204.

Isono, H., and Shoumura, S. (1980). Effects of vagotomy on the ultrastructure of the parathyroid gland of the rabbit. *Acta Anat.* **108**, 273–280.

Isono, H., and Shoumura-Sakurai, S. (1973). Electron microscopic study on the parathyroid gland of the parathormone-injected newt, *Triturus pyrrhogaster*(Boie). *Okajimas Folia Anat. Jpn.* **50**, 19–26.

Isono, H., Sakurai, S., Fujii, H., and Aoki, S. (1971). Ultrastructural changes in the parathyroid gland of the phosphate treated newt, *Triturus pyrrhogaster* (Boie). *Arch. Histol. Jpn.* **33**, 357–370.

Isono, H., Mikake, K., and Shoumura, S. (1977). Electron microscopic study on the postnatal development of the mouse parathyroid gland. *Arch. Histol. Jpn.* **40**, 367–380.

Isono, H., Shoumura, S., Ishizaki, N., Hayashi, K., Yamahira, T., and Yamada, S. (1979). Effects of cadmium on the ultrastructure of the mouse parathyroid gland. *Acta Anat.* **105**, 50–55.

Isono, H., Shoumura, S., Hayashi, K., Ishizaki, N., and Emura, S. (1980). Electron microscopic study of the parathyroid gland of the acetazolamine-treated mouse. *Acta Anat.* **107**, 8–17.

Jacobwitz, D. M., and Brown, E. M. (1980). Bovine parathyroid catecholamines: A chemical and histochemical study. *Experientia* **36**, 115–116.

Kapur, S. P. (1977). Fine-structure study of the parathyroid in the gerbil *Meriones unguiculatus*. *Acta Anat.* **97**, 200–204.

Kayser, C., Patrovic, A., and Porte, A. (1961). Variations ultrastructurales de la parathyroïde du hamster ordinaire (*Cricetus cricetus*) au cours du cycle saisonnier. *C. Rd. Seances Soc. Biol. Ses Fil.* **155**, 2178–2184.

Koyama, T., Makita, T., and Enomoto, M. (1984). Parathyroid morphology in rats after administration of active vitamin D_3. *Acta Pathol. Jpn.* **34**, 313–324.

Kraus, W. J., and Cutts, J. H. (1983). Morphological observations on the parathyroid of the opossum (*Didelphis virginiana*). *Gen. Comp. Endocrinol.* **50**, 261–269.

Krook, L., and Lowe, J. E. (1964). Nutritional secondary hyperparathyroidism in the horse, with a description of the normal equine parathyroid gland. *Pathol. Vet.* **1**, 1–98.

Lange, R., and Brehm, H. v. (1965). On the fine structure of the parathyroid gland in the toad and the frog. In "The Parathyroid Glands: Ultrastructure, Secretion and Function" (P. J. Gaillard, R. V. Talmage, and A. M. Budy, eds.), pp. 19–26, 313–334. Univ. of Chicago Press, Chicago, Illinois.

Larsson, H. O., Lorentzon, R., and Boquist, L. (1984). Structure of the parathyroid glands, as revealed by different methods of fixation. *Cell Tissue Res.* **235**, 51–58.

Lopez, E., Tisserand-Jochem, E., Eyquem, A., Milet, C., Hillyard, C., Lallier, F., Vidal, B., and MacIntyre, I. (1984). Immunochemical detection in eel corpuscles of Stannius of a mammalian parathyroid-like hormone. *Gen. Comp. Endocrinol.* **53**, 28–36.

Meuten, D. J., Capen, C. C., Thompson, K. G., and Segre, G. V. (1984). Syncytial cells in canine parathyroid glands. *Vet. Pathol.* **21**, 463–468.

Montskó, T., Tigyi, A., Benedeczky, I., and Lissak, K. (1963). Electron microscopy of parathyroid secretion in *Rana esculenta*. *Acta Biol. Acad. Sci. Hung.* **14**, 81–94.

Mueller, G. L., Anast, C. S., and Breitenbach, R. P. (1970). Dietary calcium and ultimobranchial body and parathyroid gland in the chicken. *Am. J. Physiol.* **218**, 1718–1722.

Munger, B. L., and Roth, S. I. (1963). The cytology of the normal parathyroid glands of man and Virginia deer. A light and electron microscopic study with morphologic evidence of secretory activity. *J. Cell Biol.* **16**, 379–400.

Nakagami, K. (1965). Comparative electron microscopic studies of the parathyroid gland. I. Fine structure of monkey and dog parathyroid glands. *Arch. Histol. Jpn.* **25**, 435–466.

Nakagami, K. (1967). Comparative electron microscopic studies of the parathyroid glands. II. Fine structure of the parathyroid gland of the normal and CaCl$_2$-treated mouse. *Arch. Histol. Jpn.* **28**, 185–205.

Nakagami, K., Warshawsky, H., and LeBlond, C. P. (1971). The elaboration of protein and carbohydrate by rat parathyroid cells as revealed by electron microscope autoradiography. *J. Cell Biol.* **51**, 596–610.

Narbaitz, R. (1972). Submicroscopical aspects of chick embryo parathyroid glands. *Gen. Comp. Endocrinol.* **19**, 253–258.

Nevalainen, T. (1969). Fine structure of the parathyroid gland of the laying hen (*Gallus domesticus*). *Gen. Comp. Endocrinol.* **12**, 561–567.

Norris, E. H. (1937). The parathyroid glands and the lateral thyroid in man: Their morphogenesis, histogenesis, topographic anatomy and prenatal growth. *Contrib. Embryol. Carnegie Inst.* **159**, 249–294.

Nuñez, E. A., Whalen, J. P., and Krook, L. (1972). An ultrastructural study of the natural secretory cycle of the parathyroid gland of the bat. *Am. J. Anat.* **134**, 459–480.

Nuñez, E. A., Hedhammar, A., Wu, F. M., Whalen, J. P., and Krook, L. (1974). Ultrastructure of the parafollicular (C) cells and the parathyroid cell in growing dogs on a high calcium diet. *Lab. Invest.* **31**, 96–108.

Nuñez, E. A., Krook, L., and Whalen, J. P. (1976). Effect of calcium depletion and subsequent repletion on parathyroids, parafollicular (C) cells and bone in the growing pig. *Cell Tissue Res.* **168**, 373–384.

Oguro, C. (1969). Parathyroid glands of the newt, *Cynops pyrrhogaster,* with reference to parathyroidectomy. *Annot. Zool. Jpn.* **42**, 21–29.

Oguro, C. (1973). Parathyroid gland and serum calcium concentration in the giant salamander, *Megalobatrachus davidianus*. *Gen. Comp. Endocrinol.* **21**, 565–568.

Oguro, C., and Sasayama, Y. (1976). Morphology and function of the parathyroid gland of the caiman, *Caiman crocodilus*. *Gen Comp. Endocrinol.* **29**, 161–169.

Oksanen, A. (1980). The ultrastructure of the multi-nucleated cells in canine parathyroid glands. *J. Comp. Pathol.* **90**, 293–301.

Oldham, S. B., Fischer, J. A., Capen, C. C., Sizemore, G. W., and Arnaud, C. D. (1971). Dynamics of parathyroid hormone secretion *in vitro*. *Am. J. Med.* **50,** 650–657.

Pang, P. K. T., Schreibman, M. P., Balbontin, F., and Pang, R. K. (1978). Prolactin and pituitary control of calcium regulation in the killfish, *Fundulus heteroclitus*. *Gen. Comp. Endocrinol.* **36,** 303–316.

Parsons, J. A., Gary, D., Rafferty, B., and Zanelli, J. (1978). Evidence for a hypercalcemic factor in the fish pituitary related to mammalian parathyroid hormone. *In* "Endocrinology of Calcium Metabolism," Proceedings of the 6th Parathyroid Conference, Vancouver, 1977 (D. H. Copp and R. V. Talmage, eds.), pp. 111–114. Excerpta Medica, Amsterdam.

Peters, H. (1941). Morphologische und experimentelle Untersuchungen über die Epithelkörper bei Eidechsen. *Z. Mikrosk.-Anat. Forsch.* **49,** 1–40.

Pischinger, A. (1937). Kiemenanlagen und ihre Schicksale bei Amnioten-Schilddrüse und epitheliale Organe der Pharynxwand bei Tetrapoden. *In* "Handbuch der vergleichenden Anatomie der Wirbeltiere" (L. Bolk, C. Guppert, E. Kallius, and W. Lubosch, eds.), Vol. III, pp. 279–347. Urban & Schwarzenberg, Berlin.

Pour, P. M. (1983). Anatomy, histology, parathyroid, hamster. *In* "The Endocrine System" (T. C. Jones, U. Mohr, and R. D. Hunt, eds.), pp. 248–252. Springer-Verlag, Berlin and New York.

Pour, P. M., Qureshi, S. R., and Salmasi, S. (1983a). Anatomy, histology, ultrastructure, parathyroid, mouse. *In* "The Endocrine System" (T. C. Jones, U. Mohr, and R. D. Hunt, eds.), pp. 252–257. Springer-Verlag, Berlin and New York.

Pour, P. M., Wilson, J. T., and Salmasi, S. (1983b). Anatomy, histology, ultrastructure, parathyroid, rat. *In* "The Endocrine System" (T. C. Jones, U. Mohr, and R. D. Hunt, eds.), pp. 257–262. Springer-Verlag, Berlin and New York.

Ravazzola, M., and Orci, L. (1977). Intercellular junctions in the rat parathyroid gland: A freeze-fracture study. *J. Microsc. Biol. Cell.* **28,** 137–144.

Raybuck, H. E. (1952). The innervation of the parathyroid glands. *Anat. Rec.* **112,** 117–123.

Reaven, E., and Reaven, G. M. (1975). A quantitative ultrastructural study of microtubule content and secretory granules accumulation in parathyroid glands of phosphate- and colchicine-treated rats. *J. Clin. Invest.* **56,** 49–55.

Robertson, D. R. (1977). The annual pattern of plasma calcium in the frog and the seasonal effect of ultimobranchialectomy and parathyroidectomy. *Gen. Comp. Endocrinol.* **33,** 336–343.

Rogers, D. C. (1963). A cytological and cytochemical study of the "epithelial body" on the carotid artery of the lizards, *Trachysaurus rugosus* and *Tiliqua occipitalis*. *J. Microsc. Sci.* **104,** 197–205.

Rogers, D. C. (1965). An electron microscope study of the parathyroid gland of the frog (*Rana clamitans*). *J. Ultrastruct. Res.* **13,** 478–499.

Rosenquist, J., and Boquist, L. (1973). Effects of supply and withdrawal of fluoride: Experimental studies on growing and adult rabbits. 2. Parathyroid morphology and function. *Acta Pathol. Microbiol. Scand.* **81,** 637–644.

Rosof, J. A. (1934). Experimental study of histology and cytology of parathyroid glands in albino rat. *J. Exp. Zool.* **68,** 121–165.

Roth, S. I., and Munger, B. L. (1963). Recent advances in the histology and ultrastructure of the parathyroid glands. *In* "Evaluation of Thyroid and Parathyroid Function" (F. W. Sunderman and F. W. Sunderman, Jr., eds.), pp. 143–148. Lippincott, Philadelphia, Pennsylvania.

Roth, S. I., and Raisz, L. G. (1964). Effect of calcium concentration on the ultrastructure of rat parathyroid in organ culture. *Lab. Invest.* **13,** 331–345.

Roth, S. I., and Raisz, L. G. (1966). The course and reversibility of calcium effect on the ultrastructure of the rat parathyroid gland in organ culture. *Lab. Invest.* **15,** 1187–1211.

Roth, S. I., and Schiller, A. L. (1976). Comparative anatomy of the parathyroid glands. *In* "Endocrinology" (G. D. Aurbach, ed.), Vol. 7, pp. 281–311. Am. Physiol. Soc., Washington, D. C.

Sasayama, Y., and Oguro, C. (1974). Notes on the topography and morphology of the parathyroid glands in some Japanese anurans. *Annot. Zool. Jpn.* **47**, 232–238.

Scholz, J. (1933). Morphologische Untersuchungen über die Epithelkörper der Urodelen. *Z. Mikrosk.-Anat. Forsch.* **34**, 159–200.

Setoguti, T. (1977). Electron microscopic studies of the parathyroid gland of senile dogs. *Am. J. Anat.* **148**, 65–84.

Setoguti, T., and Inoue, Y. (1983). Freeze-fracture study of the rat parathyroid gland under hypo- and hypercalcemic conditions, with special reference to secretory granules. *Cell Tissue Res.* **228**, 219–230.

Setoguti, T., Isono, H., and Sakurai, S. (1970a). Electron microscopic study on the parathyroid gland of the newt *Triturus pyrrhogaster* (Boie) in natural hibernation. *J. Ultrastruct. Res.* **31**, 46–60.

Setoguti, T., Isono, H., Sakurai, S., Yonemoto, Y., and Hagihara, A. (1970b). Ultrastructure of the parathyroid gland of the newt, *Triturus pyrrhogaster* (Boie) in the spring season. *Okajimas Folia Anat. Jpn.* **47**, 1–17.

Setoguti, T., Takagi, M., and Kato, K. (1980). Ultrastructural localization of phosphatases in the rat parathyroid gland. *Arch. Histol. Jpn.* **42**, 45–46.

Setoguti, T., Inoue, Y., and Kato, K. (1981). Electron-microscopic studies on the relationship between the frequency of parathyroid storage granules and serum calcium levels in the rat. *Cell Tissue Res.* **219**, 457–467.

Setoguti, T., Inoue, Y., Shin, M., and Matsumura, H. (1984). Effects of epinephrine treatment on the rat parathyroid gland, with special reference to the frequency of storage granules. *Acta Anat.* **118**, 54–59.

Shannon, W. A., Jr., and Roth, S. I. (1971). Acid phosphatase activity in mammalian parathyroid glands. *Proc.—Annu. Meet., Electron Microsc. Soc. Am.* **29**, 516–517.

Shoumura, S., Iwasaki, Y., Ishizaki, N., Emura, S., Hayashi, K., Yamahira, T., Shoumura, K., and Isono, H. (1983). Origin of autonomic nerve fibers innervating the parathyroid gland in the rabbit. *Acta Anat.* **115**, 289–295.

Sidky, Y. A. (1965). Histological studies on the parathyroid glands of lizards. *Z. Zellforsch.* **65**, 760–769.

Singh, R., and Kar, I. (1983a). Parathyroid and ultimobranchial glands of the sand boa, *Eryx johnii* Daudin. *Gen. Comp. Endocrinol.* **51**, 66–70.

Singh, R., and Kar, I. (1983b). Parathyroid gland of the freshwater snake *Natrix piscator* Schneider. *Gen. Comp. Endocrinol.* **51**, 71–76.

Soji, T., Fujita, T., and Yoshimura, F. (1974). Crystals in hyper-functioning rat parathyroid cells. *Endocrinol. Jpn.* **21**, 551–553.

Srivastav, A. J., and Swarup, K. (1979). Influence of parathormone on C cells, parathyroid glands, serum calcium and serum phosphate levels in the house shrew, *Suncus murinus*. *Arch. Anat. Microsc. Morphol. Exp.* **68**, 227–235.

Stoeckel, M. E., and Porte, A. (1966). Observations ultrastructurales sur la parathyroïde de souris. I. Etude chez la souris normale. *Z. Zellforsch. Mikrosk. Anat.* **73**, 488–502.

Stoeckel, M. E., and Porte, A. (1969). Observations ultrastructurales sur la parathyroïde et les cellules parafolliculaires de la thyroïde de quelques mammiferes. *C. R. Assoc. Anat.* **145**, 362–370.

Stoeckel, M. E., and Porte, A. (1973). Observations ultrastructurales sur la parathyroïde de mammifere et d'oiseau dans les conditions normales et expérimentales. *Arch. Anat. Microsc. Morphol. Exp.* **62**, 55–88.

Stumpf, W. E., Sar, M., Reid, F. A., Huang, S., Narbaitz, R., and DeLuca, H. F. (1981). Autoradiographic studies with ^{3}H 1,25 (OH)$_2$ vitamin D$_3$ and ^{3}H 25 (OH) vitamin D$_3$ in rat parathyroid glands. *Cell Tissue Res.* **221**, 333–338.

Swarup, K., and Tewari, N. P. (1978). Studies of calcitonin cells and parathyroid glands of the Indian palm squirrel, *Funambulus pennanti* in response to experimental hypercalcaemia. *Arch. Anat. Microsc. Morphol. Exp.* **67**, 157–165.

Swarup, K., and Tewari, N. P. (1979). Seasonal changes in the activity of calcitonin cells, parathyroid glands and serum calcium level of the Indian palm squirrel, *Funambulus pennanti* in relation to reproductive cycle. *Arch. Anat. Microsc. Morphol. Exp.* **68**, 159–167.

Swarup, K., Das, S., and Das, V. K. (1979). Thyroid calcitonin cells and parathyroid gland of the Indian rhesus monkey *Macaca mulatta* in response to experimental hypercalcemia. *Ann. Endocrinol.* **40**, 403–414.

Tewari, N. P., and Swarup, K. (1979). Studies of calcitonin cells and parathyroid glands of the Indian jackal *Canis aureus* (Linn.-lex.) in response to experimental hypercalcemia. *Indian J. Exp. Biol.* **17**, 748–751.

Trier, J. S. (1958). The fine structure of the parathyroid gland. *J. Biophys. Biochem. Cytol.* **4**, 13–22.

Tsuchiya, T., and Tamate, H. (1979). Ultrastructural localization of adenyl cyclase activity in sheep parathyroid gland. *Acta Histochem. Cytochem.* **12**, 356–360.

Tsuchiya, T., and Tamate, H. (1981). Cytochemical studies on the follicles in sheep parathyroid gland. *Tohoku J. Agric. Res.* **31**, 198–206.

von Brehm, H. (1963). Morphologische Untersuchungen an Epithelkörperchen (Glandulae parathyreoideae) von Anuren. *Z. Zellforsch. Mikrosk. Anat.* **61**, 376–400.

von Brehm, H. (1964). Experimentelle Studie zur Frage der jahreszyklischen Veränderungen. *Z. Zellforsch. Mikrosk. Anat.* **61**, 725–741.

Waggener, R. A. (1929). A histological study of the parathyroids in the anura. *J. Morphol.* **48**, 1–43.

Weisbrode, S. E., and Capen, C. C. (1974). Effect of uremia and vitamin D on bone and the ultrastructure of thyroid parafollicular cells and parathyroid chief cells in the rat. *Virchows Arch. B* **16**, 231–241.

Welsch, U., and Schubert, C. (1975). Observations on the fine structure, enzyme histochemistry, and innervation of parathyroid gland and ultimobranchial body of *Chthonerpeton indistinctum* (Gymnophiona, Amphibia). *Cell Tissue Res.* **164**, 105–119.

Weymouth, R. J., and Baker, B. L. (1954). Presence of argyrophilic granules in parenchymal cells of parathyroid glands. *Anat. Rec.* **119**, 519–527.

Whittle, L. W., and Dent, J. N. (1979). Effects of parathyroidectomy and of parathyroid extract on levels of calcium and phosphate in the blood and urine of the red-spotted newt. *Gen. Comp. Endocrinol.* **37**, 428–439.

Wideman, R. F., Jr. (1980). Innervation of the parathyroid in the European starling (*Sturnus vulgaris*). *J. Morphol.* **166**, 65–80.

Wild, P. (1980). Correlative light- and electron-microscopic study of parathyroid glands in dogs of different age groups. *Acta Anat.* **108**, 340–349.

Wild, P., and Becker, M. (1980). Response of dog parathyroid glands to short-term alterations of serum calcium. *Acta Anat.* **108**, 361–369.

Wild, P., and Manser, E. (1980). Morphometric analysis of parathyroid glands in neonatal and growing dogs. *Acta Anat.* **108**, 350–360.

Wild, P., Bitterli, D., and Becker, M. (1982). Quantitative changes of membranes in rat parathyroid cells related to variations of serum calcium. *Lab. Invest.* **47**, 370–374.

Wild, P., Schraner, E. M., and Eggenberger, E. (1984). Quantitative aspects of membrane shifts in rat parathyroid cells initiated by decrease in serum calcium. *Biol. Cell.* **50**, 263–272.

Yeghiayan, E., Rojo-Ortega, J. M., and Genest, J. (1972). Parathyroid vessel innervation: an ultrastructural study. *J. Anat.* **112**, 137–142.

Yoshida, R., and Talmage, R. V. (1962). Removal of calcium from frog bone by peritoneal lavage, a study of parathyroid function in amphibians. *Gen. Comp. Endocrinol.* **2**, 551–557.

Young, D. M., Olson, H. M., Prieur, D. J., Cooney, D. A., and Reagan, R. L. (1973). Clinicopathologic and ultrastructural studies of *L*-asparaginase-induced hypocalcemia in rabbits. *Lab. Invest.* **29**, 374–386.

Youshak, M. S., and Capen, C. C. (1970). Fine structural alterations in parathyroid glands of chickens with osteopetrosis. *Am. J. Pathol.* **61**, 257–274.

8

The Ultimobranchial Body

DOUGLAS R. ROBERTSON

Department of Anatomy and Cell Biology
State University of New York
Health Science Center
Syracuse, New York 13210

I. INTRODUCTION

The hypocalcemic hormone calcitonin (CT) is secreted from the ultimo-branchial body and has been isolated from the glands in chickens and sharks (*Squalus suckleyi*) (Copp *et al.*, 1967). The phylogeny of this endocrine system is of particular interest, since immunoreactive CT is found in isolated cells in the pharynx of protochordates. Its morphological history is complex, since the ultimobranchial is a discrete endocrine gland in fish but becomes incorporated as scattered endocrine cells into the mammalian thyroid (Godwin, 1937). A feature of the ultimobranchial body, in addition to its granulated CT-containing cells (C cells), is the arrangement of these C cells in the formation of a follicular or vesicular structure. These follicles also may consist of cells that lack CT but contribute to a holocrine or apocrine secretion (U cells) and have structural characteristics similar to cells of the pharyneal epithelium. The phylogeny and development of the ultimobranchial body throughout the vertebrates illustrates a functional separation of these two components, the granulated endocrine "C cell" and the "U cell," involved in a holocrine secretion.

II. ONTOGENY AND PHYLOGENY

A. Embryological Development

1. Normal Patterns of Development

The ultimobranchial body, as a discrete organ, is not present in the agnatha, but is only found in the jawed vertebrates. In sharks (Camp, 1917),

235

VERTEBRATE ENDOCRINOLOGY:
FUNDAMENTALS AND BIOMEDICAL IMPLICATIONS
Volume 1

various urodele amphibians (Wilder, 1929), lizards, and turtles (Johnson, 1922), the ultimobranchial gland exhibits an asymmetry, it being larger on the left than the right. As its name implies it is derived directly from the last pharyngeal pouch or from the floor of the pharynx at the level of the last pouch, regardless of the number of pouches. It commonly appears as a thickening of the pharyngeal endodermal epithelium, proceeds to expand as a cluster of cells, separates from the epithelium, and forms a central lumen. Subsequently, cell debris associated with a coagulum or colloidlike material will appear in the lumen at a later period. Typically, glandular formation begins with a single follicle, with subsequent budding and the formation of newer, smaller follicles. For example, in the larva of the anuran *Pseudacris,* after the formation of a single follicle the epithelium becomes pseudostratified concomitant with the appearance of cellular debris and a periodic acid Schiff (PAS-)positive coagulum within the central lumen (Robertson and Swartz, 1964b). Membrane-bound granules, in the vicinity of the Golgi apparatus, are first detected in 8-day posthatch larvae. Granules increase in number throughout development and are a dominant feature after metamorphic climax and in the young adult of *Rana temporaria* (Coleman and Phillips, 1974; Coleman, 1975). Smaller secondary follicles may form during larval development but can be induced by maintaining the larvae in a high-calcium environment (Robertson, 1971).

Repitilian development illustrates the beginning stages of glandular incorporation into adjacent pharyngeal tissues. In the freshwater, painted turtles, *Chrysemys marginata* and *C. picta,* after separation from the fourth pouch, the left ultimobranchial migrates caudally to form an independent complex with thymic and parathyroid tissue from the third pouch (Johnson, 1922).

The avian pattern of ultimobranchial development displays three new morphological features. In addition to the incorporation of ultimobranchial tissue into the thymus and parathyroids, as in reptiles, there is a clear separation between the typical granulated C cell and the nongranulated U cell. There also is evidence of two distinct granulated cell types (Stoeckel and Porte, 1969a,b). Finally, the studies of Le Douarin and Lievre (1970) and Le Douarin *et al.* 1974) with chimeric chick–quail grafts provide evidence that the avian ultimobranchial body, as part of the pharyngeal complex, receives contributions from head mesenchyme, which in turn, is of rhombencephalic neural crest origin.

The ultimobranchial gland undergoes considerable transformation in mammals. Some cells are scattered throughout the thyroid and are described as argyophilic cells, which Nonidez (1932) termed "parafollicular" cells, whereas other cells become included into parathyroids (Kameda, 1981) or the thymus (Kameda, 1981; McMillan *et al.,* 1982). Other portions of the gland remain clustered to retain the basic follicular character seen in lower vertebrates.

2. Neural Crest Contribution

Direct evidence that neural crest contributes a cell type to the ultimobranchial gland is confined to ablation and transplant studies in birds. Removal of the rhombencephalic neural crest in the chick at the 7- to 15-somite stage results in a completely involuted ultimobranchial gland or an abnormally small gland if surgery is performed prior to the 11- to 12-somite stage. If performed later the development of the gland is only slightly disturbed, although the number of granulated cells is reduced (Le Douarin and Le Lievre, 1970).

Chimeric chick–quail embryos have been used to demonstrate the migration and differentiation of neural crest cells to cell populations in the ultimobranchial gland (Le Douarin and Le Lievre, 1970). In this procedure, crest cells of the quail donor replace crest cells in the chick host. The prominent nucleolus of the quail nucleus serves as a marker to identify the migration of specific donor neural crest cells. With this technique it was observed that at 4–5 days of incubation, quail crest cells migrate ventrally and anteroposteriorly into the pharyngeal region. The ultimobranchial vesicle is surrounded by quail mesenchymal cells at 5 days, whereas at 8–10 days there is a progressive disruption of the basement membrane and an invasion of mesenchyme. At the end of days 9–10, the ultimobranchial is enlarged and composed of both quail and chick cellular elements. Most and sometimes all of the cells which contain the small, membrane-bound granules belong to the quail species. The original endodermal ultimobranchial follicle consists of cells containing only chick nuclei (Le Douarin and Le Lievre, 1970) that do not takeup L-dopa (Le Douarin et al., 1974) and are nonimmunoreactive for CT (Polak et al., 1974). Thus, in birds it is possible that neural-crest-derived cells assume the major stem cell line for CT-containing cells, whereas the original endodermal anlagen serves as a focus for neural crest migration but retains the cell line for differentiation into small follicles and eventual cyst formation.

Evidence that the neural crest contributes to the ultimobranchial gland in the mouse is based upon the induced fluorescent characteristics of a subpopulation of migrating neural crest cells, rather than neural crest extirpation or transplantation. Neural-crest-derived mesenchyme invades the pharyngeal endoderm prior to the evagination and separation of the ultimobranchial vesicle from the pharyngeal pouch. At the earliest stages available (7- to 8-day embryo), fluorescent cells are found lateral to the neural tube, whereas at 9–10 days these cells are located within the anterior (ventral) portion of the fourth pharyngeal pouch, as well as being scattered in the wall of the definitive pharynx. At 11–12 days the dorsal portion of the fourth pouch, which is destined to become the parathyroid IV, is recognized; however, the fluorescent cells are confined to the ventral, ultimobranchial region (Pearse and Polak, 1971).

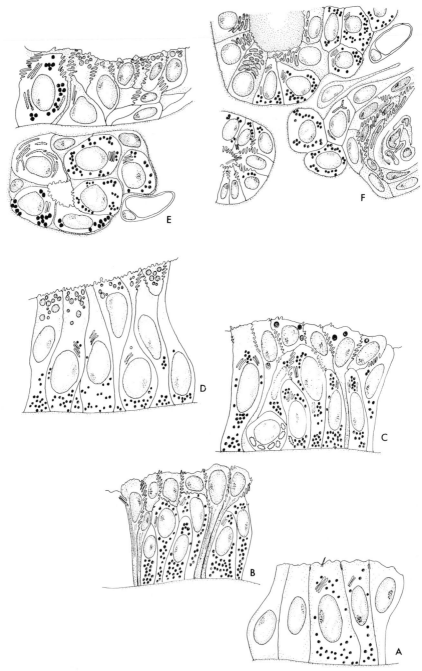

Fig. 1. Diagrammatic representation of cell types containing calcitonin in the protochordate (A) and ultimobranchial gland of vertebrates [trout (B), frog (C), turtle (D), chick (E), and dog (F)]. All cells rest upon a basal lamina; apical surface in (A) borders the pharynx, whereas apical

B. Tissue and Cellular Organization

1. Protochordates

Immunoreactive CT has been observed in the pharyngeal cells of protochordates. In *Styela clava,* these granulated, argyrophilic cells are located in the endostylar region adjacent to iodine-binding thyroidal cells. They rest upon a basal lamina, extend to the pharyngeal lumen, and generally occur in small groups of two or three containing electron-dense, membrane-granules in the basal cytoplasm (Fig. 1A) (Thorndyke and Probert, 1979).

2. Fish

In the shark *Squalus acanthias,* the parenchyma consists of both granulated and nongranulated cell types (Kitoh, 1970). In bony fish like the trout *Salmo,* the paired glands are ventral to the esophagus and lie in the transverse septum of the adult. Nongranulated and granulated cells are seen in *Salmo gairdneri.* A tall columnar type, with light-staining cytoplasm, a supranuclear Golgi apparatus, and a well-developed infranuclear rough endoplasmic reticulum (ER) may contain dense osmophilic inclusions in the apical cytoplasm that are similar to other cell debris in the lumen. In granulated cells, the ER is less conspicuous; however, the Golgi apparatus has numerous dense vesicles, and the infranuclear cytoplasm is filled with dense-cored membrane-bound granules (Table I) (Robertson, 1969). A clearer distinction of these cells shows the nongranulated cell, with an abundance of microfilaments, as a supporting cell which extends to the apical surface with extensive interdigitations between like cells and situated between granulated cells (Fig. 1B). The apical surface has long microvilli encircling the edge of the cell and commonly possesses a single cilium (Hooker *et al.,* 1979). Only the granulated cell is immunoreactive for antisera to CT, whereas the supporting cell is nonimmunoreactive (McMillan *et al.,* 1976). Both cell types are seen in the goldfish (Takagi and Yamada, 1977) and in the zebra fish (Yamane, 1978).

3. Amphibians

Ultimobranchial glands in amphibians are usually located beneath the pharyngeal epithelium and lateral to the glottis. In urodeles the gland has a tubular structure, usually on the left, similar to that seen in elasmobranchs. The pseudostratified epithelium consists of granulated cells, some of which may also contain mucoidlike materials, arranged at the periphery of the

---

surface of cells of ultimobranchial follicles border a closed lumen. For all forms, granules accumulate in basal cytoplasm of C cells, while mucoidlike secretory products accumulate in the apical cytoplasm. Nongranulated U cells exist as supporting cells in (B), as degenerating cells in (C), as "limiting cells" or epithelial types in (E), and as epithelial types in (F). (See text for details.)

TABLE I

Summary of Secretory Granule Size in Granulated Cell Types
in Ultimobranchial Glands of Selected Vertebrates

Group	Cell Type[a] Primary[b]	Secondary[c]	References
Protochordates			
Styela	100–250		Thorndyke and Probert, 1979
Fish			
trout	150–200		Robertson, 1969
	150–250		Hooker et al., 1979
zebrafish	150–250		Yamane, 1978
Amphibia			
frogs	100	300–1000	Robertson and Bell, 1965
	100–200		Coleman, 1972
toads	100	750	Coleman, 1975
	150–200		Treilhou-Lahille et al., 1984a
urodeles	100–200	<2000	Coleman and Phillips, 1972
apodans	350		Welsch and Schubert, 1975
Reptiles			
turtles	150–250		Khairallah and Clark, 1971
	800–1000		
Birds			
chickens (hens)	150–200	250–300	Stoeckel and Porte, 1969a
(hens)	100–1000		Gould and Hodges, 1971
(cockerels)	149 ± 49	407 ± 100	Monsour and Kruger, 1984
quail	150–300	500–800	Treilhou-Lahille et al., 1984b
pigeon	200	150–200	Stoeckel and Porte, 1969a
parakeet	200–320	120–200[d]	Takagi et al., 1984
turtledove	250	<400	Stoeckel and Porte, 1969b
Mammals			
bat	110 ± 20	60 ± 10	Nuñez and Gershon, 1980
dolphin	<150		Young and Harrison, 1969
human (fetal)	90–250		Chan and Conen, 1971
rat	150–200		Calvert, 1972a
	115 ± 25	200 ± 50[e]	Alumets et al., 1980

[a] Diameters given in nanometers (optional as mean ± S.E.).
[b] Membrane-bound, calcitonin-containing cells.
[c] Membrane or non-membrane-bound granule type.
[d] Tyrosine-hydroxylase-containing cells.
[e] Calcitonin- and somatostatin-containing cells.

follicle, while other cells, lacking granules, are found bordering the central lumen. Cells lacking granules but containing mucoidlike materials are found in the terminal follicular body (Coleman and Phillips, 1972).

In the wormlike apodans the glands attain maximal size in the late larval stage. In *Chthonerpeton indistinctum*, the glands are larger than the parathyroids and are composed of small follicles and clusters of cells which are separated by strands of connective tissue. A single large cell type occurs in clusters or in a basal position in follicles characterized by numerous, electron-dense, polymorphic secretory granules, well-developed rough ER and Golgi membranes, microtubules, and small bundles of microfilaments. In larger follicles, relatively flat cells almost devoid of secretory granules are found at the luminal border (Welsch and Schubert, 1975).

In *Rana pipiens* two cell types are recognized at the light microscopic level, the tall elongate cell, with secretory material, and the apical isometric cell, with lipid inclusions (Robertson and Swartz, 1964b). At the ultrastructural level (Robertson, 1965), an additional basal cell is identified that comprises about 2% of the population, rests upon the basal lamina, has unmodified lateral plasma membranes, and has no apical surface at the lumen (Figs. 1C and 2). This basal cell occurs singly or in pairs and is commonly located inferior to degenerating cells. The basal cell is characterized by a large, round or ovoid nucleus, and extensive dilated rough ER, and a Golgi apparatus with associated dense bodies. The dominant secretory cell (75%–85% of the parenchyma) possesses numerous dense-cored, membrane-bound granules in the apical and basal cytoplasm (Table I). These cells, with elongate nuclei, have scant rough ER, rest upon the basal lamina, and have an apical membrane bordering the luminal surface which is modified with pleomorphic microvilli and occasional cilia (Fig. 2). The apical isometric cell, equivalent to a degenerating cell, comprises about 10–15% of the parenchyma and has dark or pycnotic nuclei and an attenuated attachment to the basal lamina, with the nucleus placed towards the upper half of the epithelium. These cells have numerous interdigitations with similar cells adjacent to them and few dense-cored granules, but they contain large, heterogeneous granules, lysosomes, lipid bodies, and associated myelin figures in the apical cytoplasm (Robertson and Bell, 1965). With gluteraldehyde fixation Coleman (1972) also demonstrated large arrays of tonofilaments associated with the lipidlike bodies in *R. temporaria*.

The two cell types described at the light microscopic level for the frog are identical in position to those of the toad *Bufo*. The first type contains the more typical, dense-cored granules, and the second cell type that borders the lumen contains large, pleomorphic, very electron-dense, membrane-bound granules (Coleman, 1972). Treilhou-Lahille *et al.* (1984a) found that whereas both cell types were immunoreactive-positive to antisera for salmon CT, the apical cells were less immunoreactive, and the luminal colloid was negative.

Amphibian ultimobranchial cells are "multipotent" and, based on tritiated thymidine studies (Robertson, 1967a), appear to go through a defined maturation cycle from the basal or "stem" cell in about 7 days. A finite number of CT-containing secretory granules are produced, whereupon the cell enters a senescent phase: the nucleus moves to an apical position, with the addition of new cell products. The apparent migration of maturing cells is reflected in

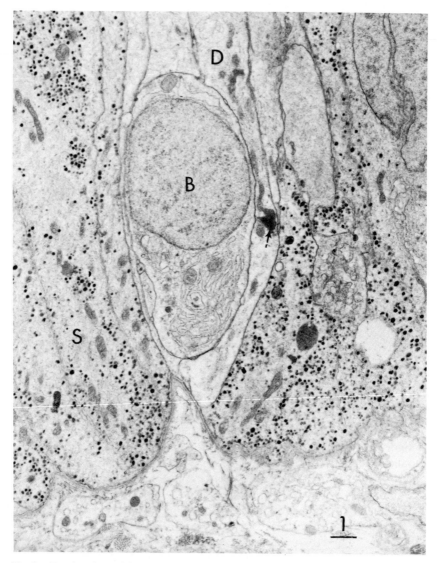

Fig. 2. Basal regions of the secretory parenchyma of ultimobranchial gland in *Rana pipiens*, characterized by cells comprising a pseudostratified epithelium. Three basic cell types are apparent: the basal stem cell (B), commonly found adjacent to the degenerating cell (D). The latter lacks granules and possesses lipid inclusions (arrow). The primary secretory cell (S) possesses numerous membrane-bound granules accumulated in the basal cytoplasm. Bar = 1 μm.

the unmodified plasma membranes of basal and early-maturing cell types, whereas apical cells are characterized by extensive interdigitations of their lateral membranes (Robertson, 1965), suggesting a nonmigratory cell layer. The cellular interdigitations are also seen in trout ultimobranchial parenchyma and are apparent in nongranulated U cells in higher forms.

4. Reptiles

In young turtles (*Pseudemys*) the left ultimobranchial may reach lengths of 800 μm and is composed of numerous small follicles, lined with pseudostratified epithelium or low columnar cells. The gland is dispersed within the connective tissue space adjoining the subclavian and carotid arteries near the bifurcation of the trachea (Sehe, 1965). In *P. scripta* and *Chrysemys picta,* the dual polarity of the ultimobranchial parenchymal cells is clearly represented (Fig. 1D). Cells forming the individual follicles contain basal secretory granules typical of C cells (Table I). In the apical region of these same cells there are larger membrane-bound granules whose density and texture are comparable to PAS-positive material found within the central lumen (Khairallah and Clark, 1971).

Calcitonin has been isolated from the ultimobranchial gland in lizards (*Lacerta muralis*; Galan Galan *et al.,* 1981), which is unpaired on the left and lies closely opposed to the trachea or esophagus. The ultimobranchial commonly consists of one or two large, colloid-filled follicles whose epithelium is usually pseudostratified, low-columnar, or of the squamous variety in older animals. In snakes the glands are paired and lie in the vicinity of parathyroid IV and thymus IV in the caudal region of the thyroid (Sehe, 1965).

5. Birds

The uniform appearance of the ultimobranchial in the adult is the result of the invasion of mesenchyme into the gland and the formation of irregularly anastomosing cords and clusters of cells. In the pigeon and turtledove, granulated cells are small and polygonal and are associated with "limiting cells," whereas in the peripheral region of the thyroid, ultimobranchial cords are situated in both an interfollicular and infracapsular position and may be located in the parathyroids or thymus (Stoeckel and Porte, 1969a). However, Kameda (1984b) noted that in the chick, C cells immunoreactive to anti-CT are restricted to the ultimobranchial gland. Limiting cells surround the granulated cell with thin, flattened cytoplasmic processes. Only small areas of the surface membrane are exposed to the extracellular space (Gould and Hodges, 1971)(Fig. 1E). In the adult domestic fowl, 80% of the cell cords consist of two secretory cell types (Table I) (Stoeckel and Porte, 1969b; Monsour and Kruger, 1984); both types are immunoreactive for CT (Treilhou-Lahille *et al.,* 1984b).

The follicular component of the ultimobranchial complex in chickens is best seen in adults and may be associated with parathyroid nodules and ultimobranchial cord cells (Dudley, 1942; Hodges, 1979; Kameda, 1984b). Transitional cells that are found along cords may be continuous with those lining the lumina that contain occasional granules located in the apical cytoplasm (Hodges, 1979). Cells lining the larger vesicles form either a single layer or multilayered epithelium containing large apical granules which are different from dense-cored, membrane-bound granules associated with C cells (Hodges, 1979; Treilhou-Lahille et al., 1984b) (Fig. 1E). Chan (1978) observed that these follicular cells may vary from squamous to cuboidal to columnar. In some cysts, the entire squamous epithelium is immunopositive to CT, with positive immunoreactive material in the central colloid (Kameda, 1984b). However, in an electron microscopic study, Treilhou-Lahille et al., (1984b) did not find granulated cells bordering the lumen. Chan (1978) confirms the presence of mucous cells, previously described by others (Dudley, 1942; Gould and Hodges, 1971), that contain secretory material within the apical cytoplasm composed of partially fused droplets. Finally, Chan (1978) describes a "basal" cell that is located close to the basal lamina, whose nucleus is round or irregular, with a small Golgi apparatus and scattered rough ER. Whether these basal cells can differentiate into granulated cells, as in the frog (Robertson, 1965, 1967a, 1968), is unknown.

6. Mammals

The literature of the cytophysiology of the parafollicular cells in mammals and the association of these cells with the thyroid is extensive (see review of Nuñez and Gershon, 1978). The distinction of two granulated cell types in the mammalian ultimobranchial, in contrast to birds, is not clear. C cells in adult rats have two distinct granule types: (1) the large granule, with a heterogeneous core and closely applied limiting membrane, and (2) the smaller granule, which is a more dense-cored granule with a lucent space separating the outer membrane (Table I). However, both granule types are found in the same cell, and only occasionally do cells have a predominance of either the large or smaller granule (DeLellis et al., 1979).

To date no studies utilizing tritiated thymidine have shown the "stem" cell origin of granulated C cells, although there are two addition agranular cells that are recognized. These include a "glycogen-rich" and a second "undifferentiated" cell type (Stoeckel and Porte, 1970; Calvert, 1972a). The glycogen-rich cells apparently transform into a thyroid follicular cell but not into C cells, since follicular cells derived from the medial thyroid anlage are structurally similar to glycogen-rich cells (Calvert, 1972b). The undifferentiated cell may initiate the formation of small follicles that then form larger follicles and eventually colloid-filled cysts (Godwin, 1937; Van Dyke, 1945). The development of cysts from small follicles is well documented in the rat. Small follicles are present in 2-day neonates that become more developed and more numerous and are associated with mature cysts in progressively older animals. Follicles consist primarily of an outer layer of small basal cells and the more apically located progeny of these cells; the cells and their nuclei eventually desquamate into the lumen and ultimately disappear (Fig. 3). The debris also includes a PAS-positive colloid that is associated with the "mixed follicle" composed of stratified and columnar cells (Wollman and Neve, 1971a). The cells lining the mature follicles have scant rough ER, abundant free ribosomes, and an irregular cell surface.

Thus, it appears that in the avian and mammalian ultimobranchial complex and the two secretory components of amphibian and reptilian parenchymal cells (e.g., basally secreted, CT-containing granules and apically secreted, mucoid polysaccharide complexes) are clearly represented by two separate cell types, the C cell and U cell (Fig. 1F).

III. IMMUNOHISTOCHEMICAL AND HISTOCHEMICAL CHARACTERISTICS

A. Immunoreactive Proteins

Immunohistochemical localized CT (iCT) has been identified within the ultimobranchial parenchyma of elasmobranchs (Sasayama *et al.,* 1984),

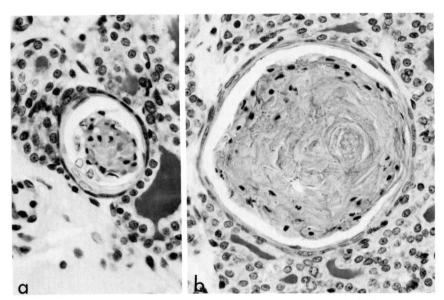

Fig. 3. (a) Ultimobranchial follicle containing weakly PAS-positive debris and some pyknotic nuclei in a 60-day-old rat. (b) Follicle in a 60-day-old rat. Debris in lumen is PAS-positive. Central part of lumen has debris free of pyknotic nuclei, while the outer part has debris with pyknotic nuclei. Both (a) and (b) ×400. [Wollman and Neve, 1971a; Courtesy Dr. S. Wollman.]

bony fish (McMillan *et al.*, 1976; Sasayama *et al.*, 1984), anurans (Van Noorden and Pearse, 1971; Treilhou-Lahille *et al.*, 1984a; Sasayama *et al.*, 1984), birds (Treilhou-Lahille *et al.*, 1984b), and numerous mammals (Blahser, 1978; Kameda, 1981; Kameda *et al.*, 1982). Sasayama *et al.* (1984) were unable to demonstrate iCT in reptiles using either anti-salmon or anti-human calcilonin antisera. Immunoreactivity is affected by fixation and confined to membrane-bound granules (Treilhou-Lahille *et al.*, 1981), and with the use of a protein-A/gold technique, iCT has been localized within newly formed prosecretory granules in the rat (Zabel, 1983). However, not all granules are equally reactive; while granules of the two different cell types in the chick are immunoreactive for CT, only about half of the granules of the larger form show immunoreactivity (Treilhou-Lahille *et al.*, 1984b). Immunoreactive CT in frog is generally confined to mature secretory cells with accumulations in the basal and supranuclear cytoplasm, while cells that border the lumen are iCT-negative (Fig. 4; Robertson, personal observations). The ultimobranchial colloid in amphibians is generally nonreactive to antisera to CT (Van Noorden and Pearse, 1971; Treilhou-Lahille *et al.*, 1984a) but may show positive immunoreactivity in colloid of chick (Kameda, 1984a) and dog (Kameda, 1982).

Somatostatin (somatotrophin-release-inhibiting factor) immunoreactivity is found in scattered D cells of the rat (Alumets *et al.*, 1980) and rabbit thyroid (Buffa *et al.*, 1979). It also may occur simultaneously with cells that are immunoreactive to CT in the rat (Van Noorden *et al.*, 1977), guinea pig, and rabbit (Kameda *et al.*, 1982). Kameda *et al.* (1982) noted in a variety of mammalian species that only a few C cells were immunoreactive for both somatostatin and CT. While CT-containing cells are relatively numerous in the thyroid gland of the rat, somatostatin cells are very scarce except at the time of birth and for a week thereafter, when they are most numerous (Alumets *et al.*, 1980). Somatostatin/calcitonin cells contain granules which are larger than those cells containing the CT granule (Table I). Although the cell relationship is unclear, the granule types may be related to the two types described by DeLellis *et al.* 1979). Chick ultimobranchial is immunoreactively negative for somatostatin (Kameda, 1984a,b).

The premise that mammalian ultimobranchial cells can differentiate into functional thyroid parenchyma has been confusing, since small ultimobranchial follicles superficially resemble thyroid follicles. In addition to the main component of thyroglobulin (M.W. 670,000) with a sedimentation coefficient of 19 S, there is a larger glycoprotein (M.W. 2,600,000) associated with 19 S thyroglobulin that is found in C cells. An antibody developed to this C-cell thyroglobulin (C-Tg) (Kameda and Ikeda, 1979) specifically reacts to secretory granules of C cells and to colloid of ultimobranchial follicles and thyroid (Fig. 5). Ultimobranchial and thyroid colloid are PAS- and immunoreactive-positive for anti-C-Tg, but only thyroid colloid is immunoreactive to antisera of 19 S thyroglobulin (Kameda *et al.*, 1980, 1982) and can incorporate radio-

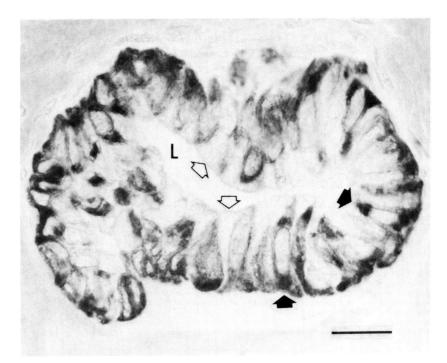

Fig. 4. Immunoreactive calcitonin (iCT) in the adult male frog ultimobranchial gland using antisera to synthetic salmon calcitonin. The primary secretory cell type (filled arrows) contains iCT-positive reaction product that is generally accumulated in the basal cytoplasm with occassional dense accumulations in a supranuclear position. Those iCT-negative cells (open arrows) are usually found to border the central lumen (L) and have a narrow cytoplasmic extension attached to the basal lamina. (Bar = 20 μm).

active iodine (Kameda *et al.*, 1981; Kameda, 1982). In fetal dogs, undifferentiated ultimobranchial cells can give rise to functional thyroid follicles that are immunoreactive to anti-19-S-thyroglobulin and that can incorporate radioactive iodine (Kameda and Ikeda, 1980; Kameda *et al.*, 1981). However, while ultimobranchial colloid resembles thyroid colloid, radioactive iodine is not incorporated into ultimobranchial glands of adult fish and amphibians (Sehe, 1960); reptiles and birds (Sehe, 1966); or mammals (Kameda, 1982; Kameda *et al.*, 1981).

B. Histochemical Characteristics

In 1968 Pearse presented a unifying classification of certain peptide-secreting endocrine cells which shared common histochemical characteristics. These cells, termed APUD for *amine* and *amine precursor uptake* and *decarboxylation,* included not only the mammalian C cells, but pituitary corti-

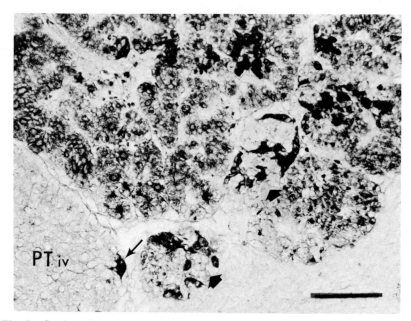

Fig. 5. Section of the thyroid of a dog fetus (41 days of gestation). Immunoperoxidase staining using anti-C-Tg antiserum. Two C-cell complexes (broad arrows) are observed around parathyroid iv. The C cells are packed with numerous dense reaction products and follicular cells with a small amount of granular reaction products. The colloid stored in small follicular lumina shows strong immunoreactivity. Note the C cells (long arrow) in parathyroid iv. PTiv parathyroid iv. Bar = 100 μm. [Kameda *et al.*, 1980; courtesy Dr. Y. Kameda.]

cotrophs and melanotrophs, and β pancreatic islet cells, argyophilic and enterochromaffin cells of the stomach and intestine, the carotid body, adrenal medulla A and NA cells, and Feyrter cells of the lungs. Both the biochemical and ultrastructural features of members of the APUD series allowed Pearse to speculate that such cells may also share a common embryological origin, namely the neural crest cells (Pearse, 1969). However, the usefulness of the APUD classification lies in its explaining the relationship of these cell types to pathological processes. Although the APUD series of cells are characterized as polypeptide-containing endocrine cells (Pearse, 1966), not all cells of the APUD series are derived from neural crest cells. Examples are the enterochromaffin cells and endocrine cells of the gut epithelium (Fontaine *et al.*, 1977), parasympathetic ganglia, and pancreatic endocrine cells (Le Douarin, 1980).

Both mammalian C cells (Larson *et al.*, 1966) and avian ultimobranchial cells have the capacity to incorporate exogenous L-dopa (Melander *et al.*, 1971). Additionally, parafollicular cells of adult sheep, goats, and bats synthesize and store 5-hydroxytryptamine (5-HT)(Falck *et al.*, 1964; Gershon and Nuñez, 1973); in the dog the highest concentrations of 5-HT are found in fetal animals (Gershon *et al.*, 1971). 5-hydroxytryptophan (5-HTP), a precursor of 5-HT, is incorporated into parafollicular cells (Pearse, 1966) and becomes localized in the periphery of mature secretory granules (Gershon and Nuñez, 1973).

Use of antisera to tyrosine hydroxylase (TH), to demonstrate L-dopa formation, and CT in the grass parakeet (*Melopsittacus*) revealed CT-positive cells, some of which are positive to TH (Yamada *et al.*, 1983). At the electron microscopic level three populations of granules based on immunoreactivity and size were recognized: CT-, TH-, and CT/TH-immunoreactive; however, whether such immunoreactive-positive granules exist in the same cell is not resolved (Takagi *et al.*, 1984). In the quail, coincident CT-immunoreactive and TH-immunoreactive cells are not observed, indicating that L-dopa and CT may not occur within the same cell (Yamada *et al.*, 1983).

A consistent feature throughout the vertebrates is the PAS-positive coagulum in the lumen of the ultimobranchial follicle. PAS-positive granules are found within the basal cytoplasm, in the apical cytoplasm bordering the follicular lumen (Deville and Lopez, 1970), and in the follicular colloid of fish (Zaccone, 1977), amphibians (Robertson and Swartz, 1964a,b), reptiles (Sehe, 1965; Khairallah and Clark, 1971), birds (Sehe, 1965; Hodges, 1971; Kameda, 1984a), and mammals (Sehe, 1966; Kameda, 1982). As shown by Kameda (1982) in the dog, the PAS-positive colloid component may be associated with the large-molecular-weight glycoprotein, C-Tg, described in Section III,A. Future studies to determine if this molecule is present in the U cells and follicular colloid of lower vertebrates may show a biochemical association of CT-positive cells that are found adjacent to the endostyle in protochordates.

IV. INNERVATION

Of considerable interest to the comparative endocrinologist is the role of the central nervous system on the modulation of secretion from endocrine systems, specifically on the peripheral circulation of CT. Morphological evidence that the ultimobranchial has direct nerve terminations on secretory cells has only been demonstrated in amphibians and birds. Myelinated and unmyelinated nerve bundles are commonly seen in the underlying connective tissue of fish (Robertson, 1969; Zaccone, 1977) and reptilian (Khairallah and Clark, 1971) ultimobranchials glands, but in neither instance are direct synaptic contacts observed. The only mammal to have direct nerve terminals on C cells is the dolphin (Young and Harrison, 1969).

Although the ultimobranchial glands in the wormlike apodans are innervated by both purinergic and cholinergic fibers (Welsch and Schubert, 1975),
anurans, such as *Rana pipiens,* receive a dual autonomic innervation exclusively from the ipsilateral vagal motor nucleus and the first sympathetic
ganglion (Robertson, 1985, 1986). Retrograde axonal transport of horseradish peroxidase implanted into the ultimobranchial glands also indicates a
large population of sensory neurons in the vicinity of the ultimobranchial
whose cell bodies are located in the vagal jugular ganglion. Both the motor
and sensory fibers follow the laryngeal branch of the vago-sympathetic trunk
and course to the gland (Robertson, 1985, 1986). Some of these fibers terminate directly on secretory cells of the frog ultimobranchial gland (Fig. 6)
(Robertson, 1967b). The terminals may consist of numerous axons that contain mitochondria and large accumulations of granular and agranular vesicles. Actual nerve contact with parenchymal cells is found on relatively few
cells, with occasional nerve fibers passing between them. Other nerve fibers
have membrane specializations associated with an accumulation of the granular and agranular vesicles which do not synapse on parenchymal cells but
apparently release their transmitter substance directly into the pericapillary
space. This has prompted the suggestion that there are two different mechanisms of neural control: one which affects a localized response, whereas the
other is viewed as a "diffuse secretory mechanism" with transmitter substances released in the pericapillary space and affecting a large population of
secretory cells (Robertson, 1967b).

The ultimobranchial body of birds is supplied by nerve fibers from the
vagus nerve and the sympathetic trunk, and with fibers from the ganglion
nodosum and ganglion cells distributed near and within the gland (Dudley,
1942). In the chick embryo nerve fibers first appear at day 13 of development
in contact with secretory cells, and by day 18 the neural elements are similar
to those seen in the adult chicken and pigeon (Stoeckel and Porte, 1969a,b).
According to Gould and Hodges (1971) the fibers end on about 10% of the
granular epithelial cells. Such terminals usually contain mostly agranular
synaptic vesicles 40–50 nm in diameter, but no typical synaptic modifications of their plasma membranes are observed.

V. CELLULAR MECHANISMS OF SECRETION

For fish, amphibians, and reptiles, the polarity of granulated ultimobranchial cells is directed toward the vascular bed, as opposed to the apical
luminal border. In amphibians, granules first appear in Golgi vesicles in the
larvae (Coleman and Phillips, 1974) and accumulate within the vicinity of the
supranuclear Golgi as "deltoid bodies" (Robertson and Swartz, 1964b; Ro-

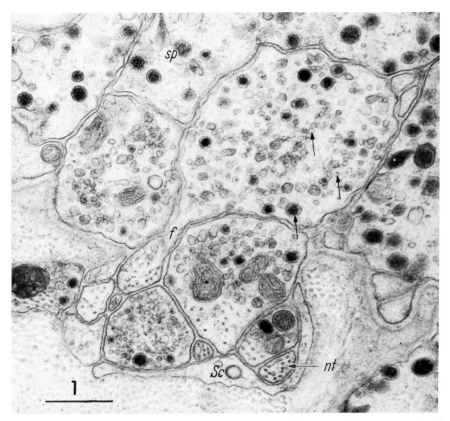

Fig. 6. Cross section through a cluster of nerve fibers which are in contact with the basal region of the secretory parenchyma (sp). Present within the narrow profiles are neurotubules (nt), whereas the bulbous portions contain types 1, 2, and 3 vesicles (arrows). Note the attenuated Schwann cell process (Sc) and a portion of a fiber which is transitional between a narrow axon with neurotubules and the bulbous portion which contains synaptic vesicles (f). Bar = 1 μm. [From Robertson, 1967b.]

bertson and Bell, 1965). Granules migrate to the infranuclear cytoplasm where they accumulate and are eventually released by a process of exocytosis along the basal plasmalemma (Robertson, 1965). There is no immunocytochemical evidence of secretion of CT towards the apical surface into the central lumen (Treilhou-Lahille *et al.,* 1984a). For avian and mammalian C cells, the polarity is not as well defined as in lower vertebrates, since C cells occur in cords and nests or as a parafollicular cell. In the latter case they typically rest upon a basal lamina but do not have an apical surface bordering a follicular lumen. Although scattered throughout the cell, secretory granules tend to be concentrated at the basal lamina toward the vascular bed (DeLellis *et al.,* 1979).

Secretion from follicular ultimobranchial cells also occurs at the apical surface. The prominent lumina of the ultimobranchial glands commonly contain a colloid or heterogeneous material in association with cellular debris (see Section III,B). Ultrastructural features of the luminal contents, in addition to cell debris, include myelin figures and a granular component of similar density to that of the material found within the apical region of degenerating cells of amphibians (Robertson and Bell, 1965; Coleman, 1972, 1975), reptiles (Khairallah and Clark, 1971), birds (Hodges, 1979), and mammals (Neve and Wollman, 1971; Wollman and Neve, 1971b). In cells that border the follicular lumen, for all the species examined, the apical membrane is modified with pleomorphic microvilli and an occasional cilium and generally has few apical vesicles. This suggests that resorption by endocytosis of luminal materials is probably minimal. However, the presence of an apical secretory mechanism and the accumulation of a colloid material throughout the vertebrates indicates the retention of a fundamental function, which to date has not been clarified.

An understanding of the types of secretory cells found in avian and mammalian ultimobranchial follicles and cysts is more readily appreciated when viewed in the context of the ontogenetic history of the gland. During development, CT-containing cells appear during the time when the epithelium becomes columnar. The first appearance of cell debris and colloid occurs after the formation of a pseudostratified or tall columnar epithelium, indicative of a maturation process similar to that seen in the amphibians (Robertson, 1967a). In the amphibian, the period of CT production is finite, whereupon mucoidlike materials are produced and the holocrine secretion is the end product of a senescent cell. The cell renewal cycle is similar to that seen in gut endothelium, except that these mucoid materials are confined to a closed follicular space. In a system in which cells are replaced faster then they are degraded, new, smaller follicles will form while cells of older follicles mature, become squamous, and become associated with degenerative cysts. This process of multiple follicle formation and glandular maturation is readily apparent in the newborn rat (Wollman and Neve, 1971a,b). Such a mechanism partially explains the wide variety of cell types described for the avian and mammalian ultimobranchial complex.

VI. PATTERNS OF SECRETION

What may be considered "normal" morphology of ultimobranchial glands should be viewed in the context of the developmental and physiological state of the animal. In both developmental studies and situations in which animals have been exposed to their natural environmental cycles, wide variations in ultimobranchial morphology can be demonstrated. Among mammals, secretory granule synthesis and storage occur prior to birth (Chan and Conen,

1971; Treilhou-LaHille and Beaumont, 1975), suggesting that secretion occurs during intrauterine life, as in the bat (Nuñez and Gershon, 1980), or just after birth, as in the dog (Nuñez and Gershon, 1976).

In the salmon *Salmo salar* (Deville and Lopez, 1970) and the zebra fish *Brachydanio rerio* (Yamane, 1978), the ultimobranchial glands are relatively inactive in immature fish but become active and degranulated during the spawning phase. In amphibians (Robertson and Swartz, 1964b; Robertson, 1971; Coleman and Phillips, 1972, 1974), secretory activity is markedly enhanced during metamorphosis prior to the adult form. Prominent seasonal cycles in C-cell morphology are observed in a variety of hibernating mammals (see review of Nuñez and Gershon, 1978), in which degranulation occurs during hibernation, with increased synthesis and storage after arousal.

VII. CONCLUSIONS AND BIOMEDICAL IMPLICATIONS

The phylogeny of the ultimobranchial gland(s) reveals an organ which has conserved the fundamental structural attributes of a pharyngeal epithelium while retaining a renewal cell population and production of mucopolysaccharides, certain components of which may be related to thyroglobulins. Its phylogeny is similar to that of the thyroid with respect to the formation of a follicular tissue organization but appears to differ with the production of basally secreted, membrane-bound granules in addition to the apically secreted mucoproteins. Many of these basic features appear to have been retained despite the morphological incorporation of the ultimobranchial into the thyroid in mammals. Assessment of the relationships between the multiple cell types described have, of necessity, relied primarily on morphological criteria. Further studies are necessary to resolve the origin of new C cells, their half-life, cell lineage, and relationship to the ultimobranchial follicles and cysts which are found in the avian and mammalian ultimobranchial glands. Tritiated thymidine labeling studies, combined with immunohistochemical techniques in both mammals and avian embryonic systems, may clarify the fate of "basal" cells seen in birds (Chan, 1978), "undifferentiated" cells in mammals (Godwin, 1937; Van Dyke, 1945; Chan and Conen, 1971), and the ontogeny of multiple cell types.

The neural crest contribution to head and neck tissue development is not limited to the ultimobranchial alone. In birds, the neural crest from the mesencephalon to rhombencephalon regions gives rise to the entire visceral skeleton; dermis; connective tissue of the muscles of the lower jaw, tongue, thymus, thyroid, parathyroids, and ventral part of the neck and walls of the large arteries from the aortic arches; enteric parasympathetic ganglia; carotid body; and neurons of sensory root ganglia (see review of Le Douarin, 1980). While ablation of the neural crest supresses the development of the ultimobranchial, such procedures also cause disruption in the development of the thymus. This has lead to a proposal that the connective tissue elements derived from the neural crest are essential components to the subsequent differentiation of cranial organs, e.g., the heart and portions of the great vessels, which can lead to a variety of clinical syndromes, such as DiGeorge syndrome (Kirby and Bockman, 1984). Thus, the normal development of the ultimobranchial body, represented in birds and mammals, is not unique from other cranial structures in its dependence upon the neural crest.

There is no direct evidence to establish if the neural crest is involved in the embryological development of the ultimobranchial in lower vertebrates. For birds and mammals the neural crest contribution adds a second cell line which is induced and thereby acquires the capacity to synthesize calcitonin (CT) but also incorporates amines. This may partially explain the presence of two cell types in birds, each immunoreactive to CT. Developmental studies in lower vertebrates are required to establish if there is a neural crest contribution. If combined with immunocytochemical techniques, this would establish some continuity to the observations made in birds and mammals of the presence of amines and CT within the same cell.

Endogenous mechanisms of glandular homeostasis may be modulated by the nervous system to influence the systhesis and release of CT. Failure to identify direct neural contact on mammalian C cells, other than that in the dolphin, suggests that if neural contacts exist, they are relatively few in number and/or are functionally insignificant. The presence of somatostatin, a local neuroregulator, in various mammalian species may preclude the requirement of an external neural modulator. Innervation of ultimobranchial parenchyma in birds, which lack somatostatin (Kameda, 1984b), and in amphibians may suggest that exogenous environmental stimuli are more important than local environmental control. Evidence that sensory terminals are in the vicinity of the frog ultimobranchial, with cell bodies in the vagal jugular ganglion (Robertson, 1985), is analogous to the situation seen with the pancreatic islet cells (Sharkey and Williams, 1983). Those afferents found in the frog may represent a specific sensory type related to the function and secretory control of CT. Nonmammalian systems are ideal for pursuing studies on the morphological and physiological relationship of the central and peripheral nervous system to CT secretory mechanisms.

Many questions need resolution. If the ultimobranchial glands are lacking in the Agnatha, are there CT-containing cells in the pharyngeal epithelium? Certainly the presence of a CT-like molecule in the nervous system of the cyclostome *Myxine* (Girgis *et al.*, 1980) is intriguing. What is the explanation for the asymmetry of the ultimobranchial in the lower vertebrates? Why is a follicular structure retained throughout the vertebrates, and what is the functional significance of the apically secreted colloid?

Physiological studies on the biological significance of endogenous CT released from mammalian and avian C cells and the ultimobranchial complex are partially hindered by the anatomical association with the thyroid, parathyroid, or thymic tissues. Evidence that the ultimobranchial is biologically more active in later stages of mammalian development is difficult to attain experimentally. The use of avian and amphibian development models is more readily accessible for physiological studies. In reptiles and amphibians the ultimobranchial glands are ideally suited for direct surgical approach, especially in amphibians, since they are totally isolated from any other endocrine or lymphatic tissue. The application of immunohistochemistry and immunocytochemistry to the cytophysiology of the ultimobranchial gland in the lower vertebrates will surely be rewarded.

ACKNOWLEDGMENTS

Original research by the author was supported by a National Science Foundation Grant, No. PCM 8202699.

REFERENCES

Alumets, J., Hakanson, R., Lundqvist, G., Sundler, F., and Thorell, J. (1980). Ontogeny and ultrastructure of somatostatin and calcitonin cells in the thyroid gland of the rat. *Cell Tissue Res.* **206**, 193–201.

Blahser, S. (1978). Immunocytochemical demonstration of calcitonin-containing C-cells in the thyroid glands of different mammals. *Cell Tissue Res.* **186**, 551–558.

Buffa, R., Chayvialle, J. A., Fontana, P., Usellini, L., Capella, C., and Solcia, E. (1979). Parafollicular cells of rabbit thyroid store both calcitonin and somatostatin and resemble gut D cells ultrastructurally. *Histochemistry* **62**, 281–288.

Calvert. R. (1972a). Electron microscopic observations on the contribution of the ultimobranchial bodies to thyroid histogenesis in the rat. *Am. J. Anat.* **133**, 262–290.

Calvert, R. (1972b). Transitional cells in the postnatal thyroid gland of the rat. *Anat. Rec.* **174**, 341–360.

Camp, W. E. (1917). The development of the suprapericardial (postbranchial, ultimobranchial) body in *Squalus acanthias*. *J. Morphol.* **28**, 369–411.

Chan, A. S. (1978). Ultrastructure of the ultimobranchial follicles of the laying chicken. *Cell Tissue Res.* **195**, 309–316.

Chan, A. S., and Conen, P. E. (1971). Ultrastructural observations on cytodifferentiation of parafollicular cells in the human fetal thyroid. *Lab. Invest.* **25**, 249–259.

Coleman, R. (1972). A comparative ultrastructural study on ultimobranchial glands of some Israeli anurans (*Bufo viridis, Rana ridibunda* and *Hyla arborea*). *Z. Zellforsch. Mikrosk. Anat.* **129**, 40–50.

Coleman, R. (1975). The development and fine structure of ultimobranchial glands in larval anurans. II. *Bufo viridis, Hyla arborea*, and *Rana ribibunda*. *Cell Tissue Res.* **164**, 215–232.

Coleman, R., and Phillips, A. D. (1972). Ultimobranchial gland ultrastructure of larval axolotls, *Ambystoma mexicanum* Shaw, with some observations on the newt, *Pleurodeles waltlii* Micahelles. *Z. Zellforsch. Mikrosk. Anat.* **134**, 183–192.

Coleman, R., and Phillips, A. D. (1974). The development and fine structure of the ultimobranchial glands in larval *Rana temporaria* L. *Cell Tissue Res.* **148**, 69–82.

Copp, D. H., Cockcroft, D. W., and Kueh, Y. (1967). Calcitonin from ultimobranchial glands of dogfish and chickens. *Science* **158**, 924–926.

DeLellis, R. A., Nunnemacher, G., Bitman, W. R., Gagel, R. F., Tashjian, A. H., Jr., Blount, M., and Wolfe, H. J. (1979). C cell hyperplasia and medullary thyroid carcinoma in the rat. *Lab. Invest.* **40**, 140–154.

Deville, J., and Lopez, E. (1970). Le corps ultimobranchial du saumon *Salmo salar* L. Étude histophysiologique a diverses étapes de son cycle vital en eau douce. *Arch. Anat. Microsc. Morphol. Exp.* **59**, 393–402.

Dudley, J. (1942). The development of the ultimobranchial body of the fowl, *Gallus domesticus*. *Am. J. Anat.* **71**, 65–97.

Falck, B., Larson, B., von Mecklenburg, C., Rosengren, E., and Svenaeus, K. (1964). On the presence of a second specific cell system in mammalian thyroid gland. *Acta Physiol. Scand.* **62**, 491–492.

Fontaine, J., Le Lievre, C., and Le Douarin, N. M. (1977). What is the developmental fate of the neural crest cells which migrate into the pancreas in the avian embryo?. *Gen. Comp. Endocrinol.* **33**, 394–404.

Galan Galan, F., Rogers, R. M., Girgis, S. I., Arnett, T. R., Ravazzola, M., Orci, L., and MacIntyre, I. (1981). Immunochemical characterization and distribution of calcitonin in the lizard. *Acta Endocrinol. (Copenhagen)* **97**, 427–432.

Gershon, M. D., and Nuñez, E. A. (1973). Subcellular storage organelles for 5-hydroxytryptamine in parafollicular cells of the thyroid gland. *J. Cell Biol.* **56**, 676–689.

Gershon, M. D., Belshaw, B. E., and Nuñez, E. A. (1971). Biochemical, histochemical and ultrastructural studies of thyroid serotonin, parafollicular and follicular cells during development in the dog. *Am. J. Anat.* **132**, 5–20.

Girgis, S. I., Galan Galan, F., Arnett, T. R., Rogers, R. M., Bone, Q., Ravazzola, M., and MacIntyre, I. (1980). Immunoreactive human calcitonin-like molecule in the nervous systems of protochordates and a cyctostome, *Myxine*. *J. Endocrinol.* **87**, 375–382.

Godwin, M. C. (1937). Complex IV in the dog with special emphasis on the relation of the ultimobranchial body to interfollicular cells in the postnatal thyroid gland. *Am. J. Anat.* **60**, 299–339.

Gould, R. P., and Hodges, R. D. (1971). Studies on the fine structure of the avian parathyroid glands and ultimobranchial bodies. *Mem. Soc. Endocrinol.* **19**, 564–604.

Hodges, R. D. (1971). The histochemistry of the avian parathyroid and ultimobranchial glands. I. Carbohydrates and proteins. *Histochem. J.* **3**, 339–356.

Hodges, R. D. (1979). The fine structure of the vesicular component of the ultimobranchial gland of the domestic fowl. *Cell Tissue Res.* **197**, 113–135.

Hooker, W. M., McMillan, P. J., and Thaete, L. G. (1979). Ultimobranchial gland of the trout (*Salmo gairdneri*). II. Fine structure. *Gen. Comp. Endocrinol.* **38**, 275–284.

Johnson, C. E. (1922). Branchial derivatives in turtles. *J. Morphol.* **36**, 299–329.

Kameda, Y. (1981). Distribution of C-cells in parathyroid gland IV and thymus IV of different mammals studied by immunoperoxidase method using anti-calcitonin and anti-C-thyroglobulin antisera. *Kawasaki Med. J.* **7**, 97–111.

Kameda, Y. (1982). The cysts in C cell complexes of dog thyroids studied by immunoperoxidase staining and autoradiography. *Arch. Histol. Jpn.* **45**, 437–448.

Kameda, Y. (1984a). Immunohistochemical study of cyst structures in chick ultimobranchial glands. *Arch. Histol. Jpn.* **47**, 411–419.

Kameda, Y. (1984b). Ontogeny of chicken ultimobranchial glands studied by an immunoperoxidase method using calcitonin, somatostatin and 19S-thyroglobulin antisera. *Anat. Embryol.* **170**, 139–144.

Kameda, Y., and Ikeda, A. (1979). C cell (parafollicular cell)-immunoreactive thyroglobulin: Purification, identification and immunological characterization. *Histochemistry* **60**, 155–168.

Kameda, Y., and Ikeda, A. (1980). Immunohistochemical study of the C-cell complex of dog thyroid glands with reference to the reactions of calcitonin, C-thyroglobulin and 19S thyroglobulin. *Cell Tissue Res.* **208,** 405–415.

Kameda, Y., Shigemoto, H., and Ikeda, A. (1980). Development and cytodifferentiation of C cell complexes in dog fetal thyroids. *Cell Tissue Res.* **206,** 403–415.

Kameda, Y., Ikeda, K., and Ikeda, A. (1981). Uptake of radioiodine in follicles of dog C-cell complexes studied by autoradiograph and immunoperoxidase staining. *Anat. Rec.* **200,** 461–470.

Kameda, Y., Oyama, H., Endoh, M., and Masaharu, H. (1982). Somatostatin immunoreactive C cells in thyroid glands from various mammalian species. *Anat. Rec.* **204,** 161–170.

Khairallah, L. H., and Clark, N. B. (1971). Ultrastructure and histochemistry of the ultimobranchial body of fresh-water turtles. *Z. Zellforsch. Mikrosk. Anat.* **113,** 311–321.

Kirby, M. L., and Bockman, D. E. (1984). Neural crest and normal development: A new perspective. *Anat. Rec.* **209,** 1–6.

Kitoh, J. (1970). Electron microscopic studies of the ultimobranchial body of the elasmobranchs. *Arch. Histol. Jpn.* **31,** 269–281.

Larson, B., Owman, C., and Sundler, F. (1966). Monoaminergic mechanisms in parafollicular cells of the mouse thyroid gland. *Endocrinology (Baltimore)* **78,** 1109–1114.

Le Douarin, N. (1980). Migration and differentiation of neural crest cells. *Curr. Top. Dev. Biol.* **16,** 31–85.

Le Douarin, N., and Le Lievre, C. (1970). Demonstration de l'origin neural des cellules a calcitonine du corps ultimobranchial chez l'embryon de poulet. *C.R. Hebd. Seances Acad. Sci., Ser.* **270,** 2857–2860.

Le Douarin, N., Fontaine, J., and Le Lievre, C. (1974). New studies on the neural crest origin of the avian ultimobranchial glandular cells- Interspecific combinations and cytochemical characterization of C cells based on the uptake of biogenic amine precursors. *Histochemistry* **38,** 297–305.

McMillan, P. J., Hooker, W. M., Roos, B. A., and Deftos, L. J. (1976). Ultimobranchial gland of the trout (*Salmo gairdneri*). I. Immunological and radioimmunoassay of calcitonin. *Gen. Comp. Endocrinol.* **28,** 313–319.

McMillan, P. J., Heidbuchel, U., and Vollrath, L. (1982). Anamolous occurrence of immunoreactive calcitonin cells in the thymus of the rat. *Cell Tissue Res.* **222,** 629–634.

Melander, A., Owman, C., and Sundler, F. (1971). Concomitant depletion of dopamine and secretory granules from cells in the ultimobranchial gland of Vitamin D2-treated chicken. *Histochemie,* **25,** 21–31.

Monsour, P. A. J., and Kruger, B. J. (1984). Morphology of the ultimobranchial body in cockerels. *Aust. J. Biol. Sci.* **37,** 331–339.

Neve, P., and Wollman, S. H. (1971) Fine structure of ultimobranchial follicles in the thyroid gland of the rat. *Anat. Rec.* **171,** 259–272.

Nonidez, J. F. (1932). The origin of the ´parafollicular´ cell, a second epithelial component of the thyroid gland of the dog. *Am. J. Anat.* **49,** 479–505.

Nuñez, E. A., and Gershon, M. D. (1976). Secretion by parafollicular cells beginning at birth: Ultrastructural evidence from developing canine thyroid. *Am. J. Anat.* **147,** 375–392.

Nuñez, E. A., and Gershon, M. D. (1978). Cytophysiology of thyroid parafollicular cells. *Int. Rev. Cytol.* **52,** 1–80.

Nuñez, E. A., and Gershon, M. D. (1980). Structural remodelling of bat thyroid parafollicular (C) cells during development. *Am. J. Anat.* **157,** 191–204.

Pearse, A. G. E. (1966). 5-Hydroxytryptophan uptake by dog thyroid "C" cells, and its possible significance in polypeptide hormone production. *Nature (London)* **211,** 598–600.

Pearse, A. G. E. (1968). Common cytochemical and ultrastructural characteristics of cells producing polypeptide hormones (the APUD series) and their relevance to thyroid ultimobranchial C cells and calcitonin. *Proc. R. Soc. London, Ser. B* **170,** 71–80.

Pearse, A. G. E. (1969). The cytochemistry and ultrastructure of polypeptide hormone-producing cells of the APUD series and the embryological, physiologic and pathologic implications of the concept. *J. Histochem. Cytochem.* **17**, 303–313.

Pearse. A. G. E., and Polak, J. M. (1971). Cytochemical evidence for the neural crest origin of mammalian ultimobranchial C cells *Histochemie* **27**, 96–102.

Polak, J. M., Pearse, A. G. E., Le Lievre, C., Fontaine, J., and Le Douarin, N. M. (1974). Immunocytochemical confirmation of the neural crest origin of avian calcitonin-producing cells. *Histochemistry* **40**, 209–214.

Robertson, D. R. (1965). The ultimobranchial body in *Rana pipiens*. II. The various cell types and the fate of secretory granules in the parenchyma of the young adult. *Z. Zellforsch. Mikrosk. Anat.* **67**, 584–599.

Robertson, D. R. (1967a). The ultimobranchial body of *Rana pipiens* V. The cell cycle studied with tritiated thymidine. *Trans. Am. Microsc. Soc.* **86**, 195–203.

Robertson, D. R. (1967b). The ultimobranchial body of *Rana pipiens*. III. Sympathetic innervation of the secretory parenchyma. *Z. Zellforsch. Mikrosk. Anat.* **78**, 328–340.

Robertson, D. R. (1968). The ultimobranchial body in *Rana pipiens*. IV. Hypercalcemia and glandular hypertrophy. *Z. Zellforsch. Mikrosk. Anat.* **85**, 441–452.

Robertson, D. R. (1969). Some morphological observations of the ultimobranchial gland in the rainbow trout, *Salmo gairdneri*. *J. Anat.* **105**, 115–127.

Robertson, D. R. (1971). Cytological and physiological activity of the ultimobranchial glands in the premetamorphic anuran, *Rana catesbeiana*. *Gen. Comp. Endocrinol.* **16**, 329–341.

Robertson, D. R. (1985). Retrograde labeling of neurons in the vagal motor nucleus and sympathetic ganglia from the ultimobranchial gland in the frog. *Anat. Rec.* **211**, 161A–162A.

Robertson, D. R. (1986). Autonomic and sensory innervation of anuran ultimobranchial glands—A horseradish peroxidase study. *Neurosci. Lett.* **67**, 181–185.

Robertson, D. R., and Bell, A. L. (1965). The ultimobranchial body in *Rana pipiens*. I. The fine structure. *Z. Zellforsch. Mikrosk. Anat.* **66**, 118–129.

Robertson, D. R., and Swartz, G. E. (1964a). The development of the ultimobranchial body in the frog *Pseudacris nigrita triseriata*. *Trans. Am. Microsc. Soc.* **83**, 330–337.

Robertson, D. R., and Swartz, G. E. (1964b). Observations on the ultimobranchial body in *Rana pipiens*. *Anat. Rec.* **148**, 219–230.

Sasayama, Y., Oguro, C., Yui, R., and Kambegawa, A. (1984). Immunohistochemical demonstration of calcitonin in ultimobranchial glands of some lower vertebrates. *Zool. Sci.* **1**, 755–758.

Sehe, C. T. (1960). Radioautographic studies on the ultimobranchial body and thyroid gland in vertebrates: Fishes and amphibians. *Endocrinology (Baltimore)* **67**, 674–684.

Sehe, C. T. (1965). Comparative studies on the ultimobranchial body in reptiles and birds. *Gen. Comp. Endocrinol.* **5**, 45–59.

Sehe, C. T. (1966). Observations of the ultimobranchial gland in small wild and laboratory mammals, with special reference to the histochemical localization of polysaccharides. *J. Morphol.* **120**, 425–441.

Sharkey, K. A., and Williams, R. G. (1983). Extrinsic innervation of the rat pancreas: Demonstration of vagal sensory neurons in the rat by retrograde tracing. *Neurosci. Lett.* **42**, 131–135.

Stoeckel, M. E., and Porte, A. (1969a). Étude ultrastructurale des corps ultimobranchiaux du poulet. I. Aspect normal et developpement embryonnaire. *Z. Zellforsch. Mikrosk. Anat.* **94**, 495–512.

Stoeckel, M. E., and Porte, A. (1969b). Localisation ultimobranchiale et thyroïdienne des cellules C (cellules a calcitonine) chez deux Columbidae: Le pigeon et le tourtereau. Etude au microscope électronique. *Z. Zellforsch. Mikrosk. Anat.* **102**, 376–386.

Stoeckel, M. E., and Porte, A. (1970). Origine embryonnaire et differenciation secretoire des cellules a calcitonine (cellules C) dans la thyroïde foetale du rat. Étude au microscope électronique. *Z. Zellforsch. Mikrosk. Anat.* **106**, 251–268.

Takagi, I., and Yamada, K. (1977). An electron microscope study of the ultimobranchial body of the crucian carp *(Carassius carassius)*. *Okajimas Folia Anat. Jpn.* **54**, 205–228.

Takagi, I., Yamada, K., Karasawa, N., and Nagatsu, I. (1984). Immunohistocytochemical localization of tyrosine hydroxylase and calcitonin in the ultimobranchial body of the grass parakeet. *Acta Histochem. Cytochem.* **17**, 359–370.

Thorndyke, M. C., and Probert, L. (1979). Calcitonin-like cells in the pharynx of the ascidian *Styela clava*. *Cell Tissue Res.* **203**, 301–309.

Treilhou-Lahille, F., and Beaumont, A. (1975). Étude ultrastructurale du corps ultimobranchial et de l'epithelium pharyngien du foetus de Souris a partir du 11ème jour de view intra-uterine. *J. Ultrastruc. Res.* **52**, 387–403.

Treilhou-Lahille, F., Cressent, M., Taboulet, J., Moukhtar, M. S., and Milhaud, G. (1981). Influences of fixatives on the immuno-detection of calcitonin in mouse "C" cells during pre- and post-development. *J. Histochem. Cytochem.* **29**, 1157–1163.

Treilhou-Lahille, F., Jullienne, A., Aziz, M., Beaumont, A., and Moukhtar, M. S. (1984a). Ultrastructural localization of immunoreactive calcitonin in the two cell types of the ultimobranchial gland of the common toad *(Bufo bufo L.)*. *Gen. Comp. Endocrinol.* **53**, 241–251.

Treilhou-Lahille, F., Lasmoles, F., Taboulet, J., Barlet, J. P., Milhaud, G., and Moukhtar, M. S. (1984b). Ultimobranchial gland of the domestic fowl. *Cell Tissue Res.* **235**, 439–448.

Van Dyke, J. H. (1945) Behavior of ultimobranchial tissue in the postnatal thyroid gland: epithelial cysts, their relation to thyroid parenchyma and to "new-growths" in the thyroid gland of young sheep. *Am. J. Anat.* **76**, 201–251.

Van Noorden, S., and Pearse. A. G. E. (1971). Immunofluorescent localization of calcitonin in the ultimobranchial gland of *Rana temporaria* and *Rana pipiens*. *Histochemie* **26**, 95–97.

Van Noorden, S., Polak, J. M., and Pearse, A. G. E. (1977). Single cellular origin of somatostatin and calcitonin in the rat thyroid gland. *Histochemistry* **53**, 243–247.

Welsch, U., and Schubert, C. (1975). Observations on the fine structure, enzyme histochemistry, and innervation of parathyroid gland and ultimobranchial body of *Chthonerpeton indistinctum* (Gymnophiona, Amphibia). *Cell Tissue Res.* **164**, 105–119.

Wilder, M. C. (1929). The significance of the ultimobranchial body: A comparative study of its occurrence in urodeles. *J. Morphol.* **47**, 383–333.

Wollman, S. H., and Neve, P. (1971a). Postnatal development and properties of ultimobranchial follicles in the rat thyroid. *Anat. Rec.* **171**, 247–258.

Wollman, S. H., and Neve, P. (1971b). Ultimobranchial follicles in the thyroid glands of rats and mice. *Recent Prog. Horm. Res.* **27**, 213–234.

Yamada, K., Takagi, I., Kondo, Y., Karasawa, N., and Nagatsu, I. (1983). Immunofluorescence studies on catecholamine-synthesizing enzymes and calcitonin in ultimobranchial bodies of grass parakeets and quails. *Biomed. Res.* **4**, 1–8.

Yamane, S. (1978). Histology and fine structure of the ultimobranchial gland in the zebrafish. *Brachydanio rerio. Bull. Fac. Fish.* **29**, 213–222.

Young, B. A., and Harrison, R. J. (1969). Ultrastructure of Light cells in the dolphin thyroid. *Z. Zellforsch. Mikrosk. Anat.* **96**, 222–228.

Zabel, M. (1983). Ultrastructural localization of calcitonin in control and stimulated thyroid C cells of the rat using Protein A-gold immunocytochemical technique. *Histochemistry* **77**, 269–273.

Zaccone, G. (1977). Histology, innervation and histochemistry of the UB gland in the Mexican cave fish *Anoptichthys jordani* Hubbs et Innes (Teleostei: Characidae). *Acta Histochem.* **58**, 31–38.

9

Gastrointestinal Tract

STEVEN R. VIGNA

UCLA School of Medicine
Center for Ulcer Research and Education
Veterans Administration Wadsworth Medical Center
Los Angeles, California 90073

I. INTRODUCTION

The gastrointestinal endocrine system is unlike most other vertebrate endocrine systems, because it is diffuse. There are no discrete glands which can be removed surgically to study the physiological perturbations which may follow hormone deprivation. Without the availability of this classical tool of the experimental endocrinologist, basic research into the fundamental properties and physiological significance of the gastrointestinal endocrine system has until recently lagged behind other endocrine systems. Instead of being a glandular endocrine system, the gut endocrine system consists of individual cells distributed widely in the simple epithelium lining the lumen of the gut, forming just a very minor fraction of the cell types comprising this epithelium. Many endocrinologists seem to consider the gut as something less than an authentic endocrine gland, possibly because of their inability to perform classical glandular ablation/hormone replacement therapy types of studies. In the gut, extirpation of the endocrine cells requires removing a segment of the whole gut wall, a radical surgery that causes so many other changes, in addition to the effects of endocrine cell removal, that the results are difficult to interpret. There remains some resistance to the concept that the gut is an endocrine organ in addition to its role as an organ of digestion and nutrition, even in the face of estimates that the endocrine elements in the stomach alone have a cumulative mass greater than that of the pituitary gland (Pearse, 1973). Probably for these historical as well as other reasons, it is still rare for investigators of the gut hormones to be trained in endocrinology or even to publish their research papers in endocrine journals. This is unfortunate, because there is no doubt that the gastrointestinal tract is an important endocrine gland that produces many hormones regulating physiological events within and outside of the digestive tract.

VERTEBRATE ENDOCRINOLOGY:
FUNDAMENTALS AND BIOMEDICAL IMPLICATIONS
Volume 1

An area that deserves much more attention than it has received to date is the exploitation of the variability that nature has provided in the form of the richly diverse gut endocrine systems found in the animal kingdom. As will be discussed in a later section, the vertebrate digestive apparatus is highly plastic, presenting many anatomical variations on the basic theme of a hollow tube involved in extracellular digestion and absorption. Many of these features appear to be potentially advantageous model systems for study of important basic problems that are intractable in standard mammalian experimental systems. Although neglected, there is a wealth of untapped basic information available in the wider animal kingdom that can be harvested for practical applications in the gastrointestinal hormone field as well as in other areas.

A. A Brief History of the Gut Endocrine System

The first published description of gastrointestinal endocrine cells seems to have been by Heidenhain (1870). Following this first description, however, a full century passed before the modern concepts of a diffuse system of true endocrine cells of many distinct types became widely accepted. Feyrter (1938, 1953) is generally credited with recognizing that basal-granulated cells in the gut epithelium shared many cytological features and comprised what he called the *helle-Zellen,* or clear-cell system. Feyrter also seems to have coined the term ''paracrine'' to denote regulatory systems acting by local diffusion of chemicals from the cells which secrete them to nearby target cells upon which they act. It is now known that the clear-cell system (gut endocrine cells appear paler then the majority of gut epithelial cells in routine light microscopical preparations) Feyrter saw consists mainly of endocrine cell types, but there is no doubt that a strong paracrine component exists as well.

Until about 1970, the predominant view was that basal-granulated cells in the gut consisted of a single cell type, the enterochromaffin cell, so-called because it was an enteric (gut), chromium-staining cell. The staining variability seen among gut endocrine cells was interpreted as corresponding to developmental or functional stages in the life cycle of enterochromaffin cells. This monistic view of gut endocrine cells began to break down in the 1960s with increasing electron microscopical evidence for heterogeneity in granular morphology among gut endocrine cells. The final death blow to this interpretation was heralded by the first immunocytochemical identification of the cellular source of a gastrointestinal hormone, gastrin, by McGuigan (1968). This was followed in rapid succession by immunocytochemical localization of all of the gut peptide hormones for which suitable antisera were available.

B. General Features of the Gut Endocrine System

The gastrointestinal endocrine system has been termed a diffuse neuroendocrine system (DNES) by some, because it consists of individual endocrine cells embedded in an overwhelmingly nonendocrine environment and is distributed broadly throughout the epithelial lining of the gut (Pearse and Takor Takor, 1979) (Fig. 1). Pearse first proposed and then championed the concept that gastrointestinal endocrine cells derive embryologically not from endoderm, as do their fellow gut epithelial cells, but instead from neuroectoderm or neurally programed cells of epiblastic origin and thus are truly "neuro"-endocrine (Pearse, 1981). He claimed that the ability of these and other cells to internalize amines from their environment and then decarboxylate them (giving rise to the acronym APUD—amine precursor uptake and decarboxylation) reflects their common ontogenetic derivation. These concepts, however, have been subject to serious experimental criticisms (reviewed in Pearse, 1981) and are by no means established. However, the provocative nature of Pearse's ideas has stimulated much work on gastrointestinal endocrine cells and can be credited in part with rejuvenating a long-neglected research area.

Another acronym often applied to gut endocrine cells refers to the concept that they belong to the so-called GEP (gastroenteropancreatic) endocrine system. This reflects the fact that of the four well-established pancreatic endocrine cells, three of them (glucagon, pancreatic polypeptide, and somatostatin) are also found in the gut epithelium. Also, the fourth pancreatic endocrine cell type, insulin, is present in the gut in invertebrates and protochordates, representing a stage in evolution before the first appearance of the pancreas as a discrete organ (Falkmer et al., 1981).

A distinctive feature of the gastrointestinal endocrine system is its diffuseness. Why should this be so? Consider for a moment the physiological problems that all metazoans, such as vertebrates, face in acquiring nutrition from the environment. Food is not available for nutrition until it has been absorbed across the wall of the gut and enters the circulatory system. In a topographical sense, food in the intestinal lumen is still outside of the body and thus is not available to the organism for metabolic uses. The role of the gut is to regulate the movement and the extracellular breakdown of ingested food so as to render it absorbable in the proper chemical form. Because much of the extracellular digestion used to accomplish this is in the form of potent digestive juices, such as hydrochloric acid and various kinds of enzymes, secreted into the lumen of the gut, the gut constantly faces the physiological peril of digesting itself.

One mechanism used by the gut to orchestrate these important processes is the detection of the presence and quality of foodstuffs in various segments of the gut by combination sensory/effector cells—the gastrointestinal endocrine cells. These cells are rare among endocrine cells in their ability to directly sense features of the environment external to the organism (the gut lumen) and to transduce this information into a hormonal signal secreted into the body. In part because this ability is reminiscent of certain neuronal sensory cells in the body, Fujita (1981) includes gut endocrine cells in his "paraneuron" concept. In any case, it seems clear that the diffuse distribu-

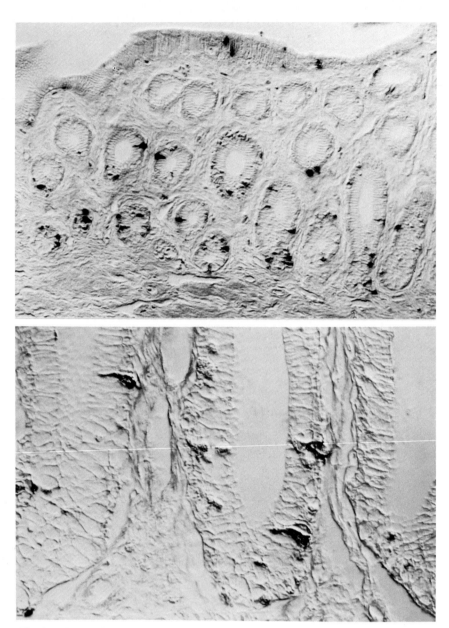

Fig. 1. Nomarski interference photomicrographs of sections of the gastric antrum of the holostean fish, the bowfin (*Amia calva*), immunocytochemically stained for bombesin. At low power (top), the diffuse distribution of the darkly stained, bombesin-containing endocrine cells in the epithelium lining the gastric lumen can be seen. At higher power (bottom), the cells extend from the basal lamina underlying the epithelium to the lumen and thus are classified as "open" endocrine cells. Cells which do not appear to extend to the lumen are probably not sectioned in a plane allowing visualization of the apical cytoplasmic process. [Photomicrographs courtesy of Dr. J. W. Crim and I. Rajjo, University of Georgia.]

tion of endocrine cells in the gut epithelium insures that the signal for hormone release represents an integrated sampling of luminal contents rather than the concentration of releasers at any one point (Fig. 2). This forms a sort of titration system; by linking hormone secretion to intraluminal nutrient load, just enough digestive juice is secreted in response to the hormonal signals to digest completely the food present, and excess digestive juice secretion is avoided. For the other endocrine glands of the body (the pituitary, adrenals, thyroid, etc.), there is no need for such dispersion of endocrine cells, because the signals for regulation of hormone release come from the composition of the blood or from neural stimulation.

A. DIFFUSE CELL DISTRIBUTION

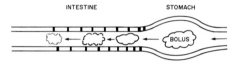

B. CLUSTERED CELL DISTRIBUTION

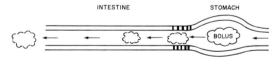

Fig. 2. Diagrammatic representation of the physiological advantages of diffuse versus clustered distribution of gastrointestinal endocrine cells in the epithelium lining the gut lumen. (A) Diffuse endocrine cell distribution allows continuous sampling of the quantity and chemical properties of the food bolus remaining in the gut lumen in any given segment of the intestine. This is a sort of titration system such that the appropriate quantity of hormone is secreted at each level to just effect complete digestion of the food bolus and no more. (B) Problems could arise if the gastrointestinal endocrine system consisted of concentrated clusters of endocrine cells in particular gut segments, such as the anterior small intestine, as shown. If this distribution existed, the endocrine cells would have only a restricted opportunity to sample the quantity and quality of luminal contents as the food bolus passed by on its one-way journey through the digestive tube. Such a system could not operate as a continuously adjustable titration system. The physiological consequences to the organism may include incomplete digestion if insufficient hormone is secreted to stimulate adequate amounts of digestive secretions or excessive digestion if too much endocrine stimulation occurs. Both conditions are maladaptive, one leading to malnutrition and the other possibly to self-digestive disorders such as ulcers.

II. GUT ENDOCRINE CELL TYPES

A. Criteria and Classification

Over the past 15 years or so, application of sophisticated immunostaining and transmission electron microscopical techniques has resulted in the identification of the cell types of origin of most of the biochemically characterized gastrointestinal hormones in mammals. As these techniques developed and came to be applied more and more widely to the gut endocrine system, specialists in the field of morphology of gut endocrine cells met periodically at conferences in various places around the world to discuss their findings and to attempt to reach a consensus acceptable to all of a scheme for classifying the gut endocrine cell types. Thus, there exist the Wiesbaden 1969, Bologna 1973, Lausanne 1977, and Santa Monica 1980 classifications (Solcia *et al.*, 1981). That any such classification scheme can only be considered as provisional is underscored by the observations that all of the formal classifications just listed differ from each other, and that sometimes previously well-accepted cell types are assigned new hormones based on the application of more sophisticated technical analyses. This is not surprising in view of some of the problems faced by gut endocrine morphologists. For example, it has become clear that some cells store more than one secretion product. Also, there are currently more cell types morphologically well-characterized than there are gut hormones to correspond to them. Furthermore, it is clear that there are appreciable differences among species with regard to ultrastructural appearance and distribution of several GEP cells, making it impossible to establish an acceptable classification embracing all species. Notwithstanding all of these limitations,. it probably is of value to attempt such classifications in an effort to bring some order to what would otherwise be chaos to all but specialists in the field. A modified version of the Santa Monica 1980 classification of human GEP cells is presented in Table I; (Solcia *et al.*, 1981). The only things that can be said with some certainty about this scheme are that it is almost undoubtedly wrong in some aspects and that it is also clearly incomplete. Nothing even remotely as complete exists for any nonmammalian species. However, most of the hormones listed in Table I have been identified in nonmammalian vertebrates by immunostaining or assay of tissue extracts. The usual caveat in comparative studies applies here again—very few species have been examined, and none have been surveyed for the presence of all known mammalian gastrointestinal hormones.

B. Functional Morphology

Most gut endocrine cells, notably excepting those present in the fundic mucosa of the stomach, extend from the basal lamina underlying the epithelium to the gut lumen (Fig. 3) and are termed "open" cells (Fujita and

TABLE I.

Modified Santa Monica 1980 Classification of Human GEP Cells

Cell type	Secreted product	Stomach		Small intestine		Large intestine	Pancreas
		Oxyntic	Pyloric	Upper	Lower		
A	Glucagon	+[a]	—	—	—	—	+
B	Insulin	—	—	—	—	—	+
D	Somatostatin	+	+	+	+	+	+
D$_1$	Unknown	+	+	+	+	+	+
EC	Serotonin + various peptides	+	+	+	+	+	+
ECL	Unknown (histamine?)	+	—	—	—	—	—
G	Gastrin	—	+	+	—	—	+[a]
CCK(I)	Cholecystokinin	—	—	+	+	—	—
GIP(K)	Gastric inhibitory peptide	—	—	+	+	—	—
L	Glucagon-like peptides + Peptide YY	—	—	+	+	+	—
N	Neurotensin	—	—	—	+	—	—
PP	Pancreatic polypeptide	—	—	—	—	—	+
S	Secretin	—	—	+	+	—	—

[a] In other species, not usually in man; when present, usually in fetal or neonatal humans, rare in adults.

Kobayashi, 1974) (Fig. 1). Secretory granules are concentrated in the basal portion of the cell, whereas the Golgi complex is supranuclear. The granules are variable in structure and electron density, depending on the cell type of origin. However, all of them are bounded by a membrane derived from the Golgi membrane. Granule size varies from about 50 to 500 nm in diameter and is usually characteristic of the cell type. There seems to be general agreement that secretory granules containing gut hormones are released at the cell base or near its lateral margin by exocytosis. However, obtaining images of exocytosis in progress using the electron microscope is difficult unless the cells have been stimulated by food or by pharmacological manipulations. Probably because of their diffuse distribution in the gut epithelium, GEP endocrine cells do not form close and specialized contacts with blood capillaries or nerves.

GUT LUMEN

SPECIFIC CHEMICAL
STIMULUS

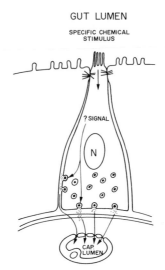

Fig. 3. A diagrammatic representation of a single gut endocrine cell of the open type. The cell extends from basal lamina to gut lumen and extends a tuft of microvilli into the lumen to sense specific chemical stimuli that may be present. The endocrine product of the cell is packaged into membrane-bound secretory granules that collect at the basal pole of the cell and is eventually secreted by the process of exocytosis. Gut endocrine cells combine receptor/ effector functions, because they transduce, by unknown intracellular mechanisms, environmental properties (specific chemicals in the lumen) into hormonal secretions which diffuse into nearby capillaries for distribution within the body by bulk flow in the bloodstream. N, nucleus.

The apical surface of open endocrine cells projects into the gut lumen via a well-developed tuft of microvilli. The apical half of the cells often contains centrioles, and sometimes one of these extends into the gut lumen as a single cilium. In addition, the apical cytoplasm is rich in pinocytotic invaginations and both coated and uncoated vesicles. Near the luminal surface, gut endocrine cells are connected to their (usually nonendocrine) neighboring cells by junctional complexes. The nucleus is usually oval and vesicular in appearance. Microtubules and microfilaments are present, the microfilaments often being associated with lateral desmosomes and the nucleus (Grube and Forssmann, 1979).

These morphological features of all gut endocrine cells of the open type, along with certain physiological and pharmacological observations, have suggested that open-type gut endocrine cells share many functional properties with each other. The luminal contact via microvilli suggests that this surface senses the chemical features of the luminal contents (food plus gut secretions) and transduces this information into an intracellular signal affecting hormone release from secretory granules at the basal surface. A well-known example is the influence of luminal pH on the secretion of gastrin from open cells in the stomach; acid inhibits, and neutral or basic pH solutions favor, gastrin release. The characteristic open type of gut endocrine cell has been observed in nonmammalian species as well, including representatives of the Agnatha (Östberg *et al.*, 1976). This indicates that this regulatory cell type has changed little in evolution, probably because its design is ideally suited to responding directly to intraluminal stimuli and to then using this information to orchestrate the complex digestive machinery.

The finding that some gut hormones, such as gastrin, appear in the gut lumen (Üvnas-Wallensten, 1977) suggests that normal release can also occur at the apical surface of GEP cells. However, such secretion would probably have to occur by a nonexocytotic mechanism, because secretory granules are not seen in the apical region of open cells. An alternative explanation is that luminal gut hormones represent release of cell contents subsequent to cell death or normal wear-and-tear sloughing of the gut epithelium. Further studies will be required to choose between these possible explanations for the intraluminal presence of gut hormones.

The nature of the intracellular signal generated in response to contact with luminal contents is not known for any gut endocrine cell; thus, this concept remains hypothetical. The basally released secretory product enters the connective tissue space beneath the epithelial basal lamina and can diffuse in all directions. This space contains a variety of extracellular connective tissue fiber types, connective tissue cells, blood vessels including capillaries, nerves, and smooth muscle cells, and, of course, the connective tissue space is also contiguous with other cells, both endocrine and nonendocrine, of the gut epithelium. Thus, all of these elements are possible targets (or conduits in the case of the circulatory system) of GEP cell secretions. Diffusion into capillaries certainly explains the mode of delivery of those gut factors serving as hormones. Is there any evidence that other elements in contact with the connective tissue space are targets of GEP cell secretions? Feyrter's paracrine hypothesis was mentioned earlier; recent years have seen accumulating evidence supporting his concept that the secretory products of some GEP cells diffuse to near neighbor cells to regulate their activity (Larsson *et al.*, 1979). Bulk flow transport in the bloodstream or lymphatic vessels is not required for delivery of these paracrine agents to their target cells (Fig. 4). A particularly good example of probable paracrine regulation in the gut endocrine system is the inhibitory effect of somatostatin on gastrin release in the stomach. Somatostatin and gastrin are synthesized and released by separate cell types. Gastrin diffuses into the bloodstream to exert its effects at remote target cells; somatostatin apparently diffuses to immediately adjacent gastrin cells and inhibits secretion of gastrin by this purely local mechanism. These somatostatin cells send out short cytoplasmic processes that end near gastrin cells (Larsson *et al.*, 1979). This adaptation presumably reduces diffusion distance, and therefore diffusion time, and insures delivery of high concentrations of secreted somatostatin. Final proof of a paracrine mode of delivery of GEP cell secretions is lacking, because of technical difficulty in

GUT LUMEN

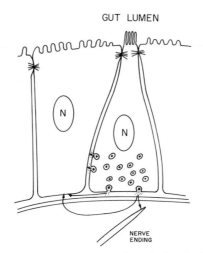

Fig. 4. Diagrammatic representation of a gastrointestinal endocrine cell of the open type, illustrating the concept that some cells of this type exert their regulatory influence on target cells by a paracrine rather than endocrine mode of delivery. Instead of diffusing into capillaries, secreted products of gut paracrine cells diffuse locally to neighboring epithelial cells, which may be either endocrine or nonendocrine, or possibly to nerve endings in the adjacent connective tissue compartment of the gut mucosa. N, nucleus.

measuring locally released peptides in interstitial fluid and proving that the peptides do not enter the bloodstream in sufficiently high concentrations to act as hormones.

The possibility that GEP cell secretions act locally on neural elements of the gut mucosa is attractive but also difficult to prove. Many of the same technical problems that plague studies of gut paracrine secretion make it difficult to assess the importance of this regulatory pathway. For example, many gut nerves contain the same peptides found in endocrine cells in the epithelium, making it difficult to estimate the relative contributions of each in regulatory phenomena. However, this mode of action seems likely to be important in view of recent observations that some actions of GEP hormones can be blocked by, for example, vagotomy (Smith *et al.,* 1981), thereby abolishing the sensory traffic from abdomen to the central nervous system. Further evidence in favor of this mechanism comes from the localization of receptors for the GEP hormone cholecystokinin (CCK) in the vagus nerve (Zarbin *et al.,* 1981).

In contrast to the open type of gastrointestinal endocrine cell, the closed type does not extend an apical process to the gut lumen. Thus, these cell types presumably are not chemosensory like the open cells. One morphological feature that these cell types do seem to share with each other is the presence of basal cell extensions, often two in number, which project in opposite directions from the cell body to come close to the surfaces of the basal margins of other gut epithelial cell types. These anatomical relationships between presumed regulatory cells and their target cells suggest a paracrine mode of delivery of the GEP cell secretory product (Fig. 5).

Closed GEP cells in mammals are found primarily in the fundic epithelium of the stomach (Fujita and Kobayashi, 1977), in which individual cell types have been shown to contain somatostatin (Larsson *et al.*, 1979) and glucagon. Another rich source of closed GEP cells is the avian proventriculus (Iwanaga and Yamada, 1981), where at least a large proportion of the cells contain bombesin (Timson *et al.*, 1979), a peptide discovered in frog skin and later shown to be an important neuropeptide in the mammalian gut. Do closed GEP cells share any functional properties with each other? This question is difficult to answer, because the physiological actions of the peptides secreted by these cells are unknown. Perhaps the best case at the present time can be made for somatostatin-containing cells in the mammalian fundic epithelium. These cells are closed and send basal processes out to contact acid-secreting parietal cells. Exogenously administered somatostatin inhibits acid secretion from parietal cells. It thus seems likely that somatostatin released from these cells at their processes adjoining parietal cells is involved in inhibiting gastric acid secretion. What physiological stim-

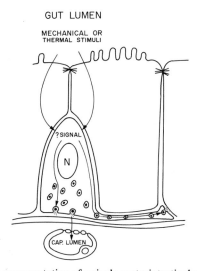

Fig. 5. Diagrammatic representation of a single gastrointestinal endocrine cell of the closed type. These cell types do not extend to the gut lumen from their bases on the basal lamina underlying the epithelium. Because they presumably cannot directly sense the chemical features of the luminal contents, they are not considered to be chemosensory. Instead, some workers have proposed that they are mechanoreceptors or thermoreceptors. If so, this information is transduced intracellularly into unknown signals leading to secretion of granule contents basally. These secretions may diffuse into the bloodstream to act as peripheral hormones. Many closed GEP cells, especially in the stomach, send cytoplasmic processes containing secretory granules out from their bases to contact nearby epithelial cells. These closed cells thus deliver their secreted products by the paracrine mode. N, nucleus.

uli might control these somatostatin cells? The lack of a chemosensory apical projection and the presence of uni- or bipolar basal cellular processes containing secretory granules have suggested that closed gut endocrine or paracrine cells may subserve a mechanosensory or even thermosensory role (Fujita and Kobayashi, 1974; Fujita, 1981). The adaptive value to the organism may be that as food enters the stomach, it necessarily stretches or physically deforms the wall of the stomach and probably also changes its temperature. These stimuli, just as well as the chemical nature of ingested food, may provide accurate signals switching on or off various digestive activities of the organ (Fig. 5). This interpretation is attractive also because it may help explain why closed GEP cell types are seemingly restricted to the stomach—after partial digestion in the stomach, the physical and thermal properties of ingested foodstuffs are changed and then closely resemble the intraluminal environment of the empty gut between meals. Thus, mechanoreceptive or thermoreceptive GEP cells located more distally in the digestive tract would not be presented with stimuli containing appropriate information about the digestive needs of the organism. In support of this scenario, it has been demonstrated that mechanical stimulation of the canine stomach is correlated with elevated concentrations of somatostatin in blood exiting the stomach (Arimura *et al.*, 1978). Nevertheless, in spite of these kinds of circumstantial evidence, the precise role and physiological significance of closed GEP cells cannot yet be considered as established.

III. COMPARATIVE STUDIES

A. Plasticity of the Vertebrate Gut

Gastrointestinal endocrine cells present the same general cytological features in all vertebrate species. Studies of the differences among vertebrates in the kinds of endocrine cells present in the gut are in their infancy. No catalog as complete as Table I can yet be compiled for any nonmammalian species. At the gross anatomical level, all vertebrates also have a tubular gut associated with various glandular organs supplying the lumen with digestive secretions. However, other elements of the gut and its associated organs are highly variable in morphology among the many vertebrate groups and species. This plasticity presents excellent opportunities for developing new model systems for studying fundamental problems in gastrointestinal hormone research.

Examples of gastrointestinal plasticity among vertebrates abound, and there are undoubtedly more waiting to be discovered. Many species lack a stomach altogether. There appear to be at least two reasons for the existence of stomachless vertebrates. In the Agnatha (the cyclostome hagfishes and lampreys), it seems clear that the absence of a stomach is a primitive condition, because there is no evidence that the ancestors of agnathans possessed a stomach (Barrington, 1942), although it is difficult to discount completely the possibility that this is a degenerate condition in living agnathans. On the other hand, the lack of a stomach in such vertebrates as the holocephalans, lungfishes, some teleosts such as the Cyprinidae, and even the mammalian echidna and vampire bat represents a secondary loss (Barrington, 1942).

One digestive organ that is present in all vertebrates is the intestine. This is probably accounted for by the fact that the intestine is the organ providing the surface for the transfer of nutrients from the outside environment into the body of the organism, that is, absorption, and thus is indispensable. Nevertheless, even for the intestine much anatomical, and presumably physiological, variability can be observed. For example, the intestine is highly variable in its length, both in absolute terms and in relation to body size. Indeed, it has recently been shown that the main basis for faster nutrient absorption in mammals versus reptiles is the much greater intestinal surface area in mammals (Karasov and Diamond, 1985). This difference has been proposed as the digestive basis enabling mammals to sustain metabolic rates an order of magnitude higher than those of reptiles and thus fuel the cost of endothermy (Karasov and Diamond, 1985). It seems likely that differences in the regulation of digestion by the gut endocrine system may also play a role in these important adaptations. Other notable anatomical variations seen in the vertebrate intestine include the presence of a spiral valve intestine in some fish, pyloric ceca in others, and the lack of crypts of Lieberkühn in all fish intestines.

Some vertebrates (adult, but not larval, lamprey; pigeon; and rat) lack a gallbladder. Cyclostomes have no discrete pancreas. In the lungfish, the pancreas is embedded in the wall of the intestine instead of being suspended in the peritoneal cavity. Only mammals have Brunner's glands in the small intestine. Chondrichthyeans and coelacanths have a discrete salt-secreting organ, the rectal gland, appended to the distal intestine. Frogs have a more or less clear separation of pepsinogen-secreting (esophagus) and acid-secreting (stomach) regions of the gut. In mammals, hydrochloric acid is secreted by parietal cells in the stomach, whereas pepsinogen (the pepsin precursor) is secreted from a distinct cell type, the chief cell. In nonmammalian vertebrates, acid and pepsinogen secretion are performed by the same cell type, the so-called oxyntic cell. Frogs also have oxyntic cells in the stomach, but most of the pepsinogen is secreted by the specialized esophageal cells which do not secrete acid.

It is likely that among the vertebrates, there is more interesting anatomical variability in the gut than that just described, because only a small fraction of the extant species have been examined carefully from this perspective. The bony fish in particular are poorly studied. In the next section, the opportunities provided by study of lower vertebrates for gaining insight into fundamental problems of biomedical significance in the gastrointestinal endocrine system will be explored.

B. Potential Model Systems of Biomedical Significance

Many fundamental problems concerning the gastrointestinal endocrine system are difficult to study in the commonly used laboratory mammals. For example, the role of gut hormones in controlling gallbladder contraction cannot even be examined in rats, because they lack a gallbladder! While this example is trivial (other common laboratory species such as the guinea pig and rabbit do have a gallbladder), it is clear that some species present certain features of great advantage as potential model systems for use in solving basic problems that are intractable in standard preparations.

The presence of a salt-secreting rectal gland in cartilaginous fishes has provided gut endocrinologists with an important new tool for investigating the control of intestinal secretion. For example, it has recently been shown that dogfish shark intestines contain a peptide which stimulates salt secretion from the dogfish rectal gland (Shuttleworth and Thorndyke, 1984). This peptide is distinct from any known peptide affecting intestinal secretion and has been named "rectin." It seems quite likely that rectin occurs also in other vertebrates, including mammals such as humans, and plays a physiological role in intestinal function and possibly a pathological role in dysfunction. It is possible that understanding of important intestinal secretion disorders such as diarrhea and cholera would benefit from discovery of rectin. The point is that the dogfish shark rectal gland proved to be an ideal model system for discovery of this potentially important new material.

Stomachless species may provide useful model systems for sorting out the complex interactions involved in stimulating the secretion of most intestinal hormones. In stomachless species, one can examine a relatively simple system without the presence of acid and pepsin. This may help sort out the relative contributions of these secretions versus other stimuli for intestinal hormone secretion.

Study of the control of pepsinogen secretion by gut hormones has been aided by the model system found in the frog esophagus (Simpson *et al.,* 1980). In other systems, it is difficult to determine the direct effects of secretagogues on peptic cells, because acid secretion by cells in the same tissue may play a role in pepsinogen release. Thus, the physical separation of peptic cells in the frog esophagus from acid-secreting cells in the stomach makes it possible to dissect the peptic tissue separately and to study it *in vitro*. The basic information learned from such studies may be relevant to ulcer research.

Amphibians also provide useful model systems for fundamental studies of gastric acid secretion. The control of gastric acid secretion is multifactorial, and thus it is a great advantage to have simple systems in which the potential variables are reduced or more easily controlled. Thus, *in vitro* models of acid secretion present clear advantages over *in vivo* models. However, the mammalian gastric-acid-secreting machinery survives poorly *in vitro* and has not proven to be a useful model system. On the other hand, the frog gastric mucosa is easy to separate from the underlying tissue and can be maintained *in vitro* in responsive condition (Hogben, 1965). This system has been used to show, for example, that the GEP cell product bombesin can directly stimulate acid secretion at the oxyntic cell level as well as by an indirect effect on gastrin secretion (Ayalon *et al.,* 1981). Again, these findings may well prove relevant to understanding such stomach-related diseases as ulcers.

Unlike other vertebrates, bony fish seem to lack both gastrin and CCK GEP cells in the antral mucosae of their stomachs (Holmgren *et al.,* 1982). We have already alluded to the fact that endocrine control over gastric acid secretion in vertebrates is extremely complex. Perhaps the bony fish stomach, lacking two of the primary regulators of acid secretion in other species, provides a good model system to evaluate precisely the roles of other GEP cell hormones in controlling gastric acid secretion.

Comparative studies can also illuminate fundamental questions such as the importance and function of closed GEP cells. Falkmer *et al.* (1984) have presented evidence that closed GEP endocrine cells evolved from open-type cells and that this evolutionary transition is associated with the aggregation of the closed cells into a discrete endocrine gland. These workers studied the evolution of insulin in invertebrates and vertebrates. In the invertebrates, insulin cells are found as open GEP endocrine cells in the intestinal epithelium. In the agnathan hagfish, however, all insulin cells have left the gut and are found in the bile duct epithelium where some remain as closed cells. Most of the cells, however, during their development have penetrated the basal lamina and have formed cell clusters comprising a primitive endocrine pancreas. Based on this analogy, we can speculate that the closed GEP cells found in the gastric epithelium of higher vertebrates including mammals may be undergoing a similar phylogenetic change. What remains a challenge for the future is to understand the physiological correlates and adaptive value of these events to the organism. Perhaps, continuing comparative study of a wider range of species will provide information useful in evaluating these concepts.

IV. SUMMARY AND CONCLUSIONS

More than 14 types of endocrine-paracrine gastroenteropancreatic (GEP) cells are present in the mammalian gastrointestinal epithelium. At present, more cell types exist than do peptide hormones that can be assigned to them. The nonmammalian gastrointestinal endocrine system is not known in this much detail. The exact physiological roles of many GEP cells and hormones are not known for any vertebrate, including mammals. Major questions remaining include the number and physiological properties of gut hormones, the cellular mechanisms of chemoreception and secretion in open GEP cells, the role and importance of closed GEP cells, the physiology of paracrine GEP cells, and the roles of the gut endocrine system in disease. Cell biological studies of gut endocrine cells are in their infancy. It seems clear that comparative studies in this field can reveal important model systems for studies designed to unravel the fundamental properties of the gastrointestinal endocrine system.

At present, few if any human diseases have been demonstrated to have a gastrointestinal hormone component. This is undoubtedly because the fundamental properties of the gut endocrine system are poorly understood. Gastrointestinal endocrine cells are very difficult to study, because of their diffuse distribution in the gut epithelium. Evolution and adaptive radiation among the living vertebrates have provided a multitude of anatomical and physiological variations upon the common theme of the need to digest food efficiently while simultaneously avoiding self-digestion. Clever investigators will make use of this rich natural variability to exploit certain preparations ideally suited to the study of fundamentally important phenomena intractable to examination in common mammalian laboratory species. For example, the presence of a rectal gland in sharks stimulated the discovery of a new gastrointestinal peptide, rectin, that may play an important role in mammalian physiology and pathophysiology. The physical separation of the primary acid- and pepsinogen-secreting cells in the frog stomach and esophagus may allow physiological dissection of the mechanisms separately controlling these potent digestive secretions; such dissection is not possible in mammalian stomach preparations, because the two cell types are interdispersed in the gastric epithelium. It seems likely that many other advantageous model systems will be discovered in the future when the gastrointestinal endocrine systems of more species have been studied. Only a clear understanding of fundamental mechanisms can lead to appreciation of the biomedical implications of this large endocrine system found in the epithelium lining the digestive tube.

REFERENCES

Arimura, A., Itoh, Z., Aizawa, I., and Rothman, J. (1978). Radioimmunoassay for plasma somatostatin and its application in studying somatostatin release from the dog stomach. *Endocrinology (Baltimore)* **102**, 152.

Ayalon, A., Yazigi, R., Devitt, P. G., Rayford, P. L., and Thompson, J. C. (1981). Direct effect of bombesin on isolated gastric mucosa. *Biochem. Biophys. Res. Commun.* **99**, 1390–1397.

Barrington, E. J. W. (1942). Gastric digestion in the lower vertebrates. *Biol. Rev. Cambridge Philos. Soc.* **17**, 1–27.

Falkmer, S., Carraway, R. E., El-Salhy, M., Emdin, S. O., Grimelius, L., Rehfeld, J. F., Reinecke, M., and Schwartz, T. W. (1981). Phylogeny of the gastroenteropancreatic neuroendocrine system: A review. *In* "Cellular Basis of Chemical Messengers in the Digestive

System" (M. I. Grossman, M. A. B. Brazier, and J. Lechago, eds.), pp. 13–42. Academic Press, New York.

Falkmer, S., El-Salhy, M., and Titlbach, M. (1984). Evolution of the neuroendocrine system in vertebrates. A review with particular reference to the phylogeny and postnatal maturation of the islet parenchyma. *In* "Evolution and Tumour Pathology of the Neuroendocrine System" (S. Falkmer, R. Håkanson, and F. Sundler, eds.), pp. 59–87. Elsevier, Amsterdam.

Feyrter, F. (1938). "Über diffuse endokrine epitheliale Organe." Barth, Leipzig.

Feyrter, F. (1953). "Über die peripheren endokrinen (parakrinen) Drüsen des Menschen." Verlag W. Maudrich, Wien-Dusseldorf.

Fujita, T. (1981). Paraneuron, its current implications. *In* "Paraneurons, Their Features and Functions" (T. Kanno, ed.), pp. 3–9. Excerpta Medica, Amsterdam.

Fujita, T., and Kobayashi, S. (1974). The cells and hormones of the GEP endocrine system. *In* "Gastro-Entero-Pancreatic Endocrine System. A Cell Biological Approach" (T. Fujita, ed.), pp. 1–16. Williams & Wilkins, Baltimore, Maryland.

Fujita, T., and Kobayashi, S. (1977). Structure and function of gut endocrine cells. *Int. Rev. Cytol., Suppl.* **6,** 187–233.

Grube, D., and Forssmann, W. G. (1979). Morphology and function of the entero-endocrine cells. *Horm. Metab. Res.* **11,** 589–606.

Heidenhain, R. (1870). Untersuchungen über den Bau der Labdrüsen. *Arch. Mikrosk. Anat.* **6,** 368–406.

Hogben, C. A. M. (1965). The natural history of the isolated bullfrog gastric mucosa. *Fed. Proc., Fed. Am. Soc. Exp. Biol.* **24,** 1353–1359.

Holmgren, S., Vaillant, C., and Dimaline, R. (1982). VIP-, substance P-, gastrin/CCK-, bombesin-, somatostatin- and glucagon-like immunoreactivities in the gut of the rainbow trout, *Salmo gairdneri. Cell Tissue Res.* **223,** 141–153.

Iwanaga, T., and Yamada, J. (1981). Endocrine-like cells of peculiar shape in the proventribulus of the chicken—possible mechanoreceptors? *In* "Paraneurons, Their Features and Functions" (T. Kanno, ed.), pp. 28–32. Excerpta Medica, Amsterdam.

Karasov, W. H., and Diamond, J. M. (1985). Digestive adaptations for fueling the cost of endothermy. *Science* **228,** 202–204.

Larsson, L.-I., Goltermann, N., De Magistris, L., Rehfeld, J. F., and Schwartz, T. W. (1979). Somatostatin cell processes as pathways for paracine secretion. *Science* **205,** 1393–1395.

McGuigan, J. E. (1968). Gastric mucosal intracellular localization of gastrin by immunofluorescence. *Gastroenterology* **55,** 315–327.

Östberg, Y., Van Noorden, S., Pearse, A. G. E., and Thomas, N. W. (1976). Cytochemical, immunofluorescence, and ultrastructural investigations on polypeptide hormone containing cells in the intestinal mucosa of a cyclostome, *Myxine glutinosa. Gen. Comp. Endocrinol.* **28,** 213–227.

Pearse, A. G. E. (1973). Cell migration and the alimentary system: Endocrine contributions of the neural crest to the gut and its derivatives. *Digestion* **8,** 372–385.

Pearse, A. G. E. (1981). The diffuse neuroendocrine system: Falsification and verification of a concept. *In* "Cellular Basis of Chemical Messengers in the Digestive System" (M. I. Grossman, M. A. B. Brazier, and J. Lechago, eds.), pp. 13–19. Academic Press, New York.

Pearse, A. G. E., and Takor Takor, T. (1979). Embryology of the diffuse neuroendocrine system and its relationship to the common peptides. *Fed. Proc., Fed. Am. Soc. Exp. Biol.* **38,** 2288–2294.

Shuttleworth, T. J., and Thorndyke, M. C. (1984). An endogenous peptide stimulates secretory activity in the elasmobranch rectal gland. *Science* **225,** 319–321.

Simpson, L., Goldenberg, D., and Hirschowitz, B. I. (1980). Pepsinogen secretion by the frog esophagus in vitro. *Am. J. Physiol.* **238,** G79–G84.

Smith, G. P., Jerome, C., Cushin, B. J., Eterno, R., and Simansky, K. J. (1981). Abdominal vagotomy blocks the satiety effect of cholecystokinin in the rat. *Science* **213,** 1036–1037.

Solcia, E., Creutzfeldt, W., Falkmer, S., Fujita, T., Greider, M. H., Grossman, M. I., Grube, D., Håkanson, R., Larsson, L.-I., Lechago, J., Lewin, K., Polak, J. M., and Rubin, W. (1981). Human gastroenteropancreatic endocrine-paracrine cells: Santa Monica 1980 classification. *In* "Cellular Basis of Chemical Messengers in the Digestive System" (M. I. Grossman, M. A. B. Brazier, and J. Lechago, eds.), pp. 159–165. Academic Press, New York.

Timson, C. M., Polak, J. M., Wharton, J., Ghatei, M. A., Bloom, S. R., Usellini, L., Capella, C., Solcia, E., Brown, M. R., and Pearse, A. G. E. (1979). Bombesin-like immunoreactivity in the avian gut and its localisation to a distinct cell type. *Histochemistry* **61,** 312–221.

Üvnas-Wallensten, K. (1977). Occurrence of gastrin in gastric juice, in antral secretion, and in antral perfusates of cats. *Gastroenterology* **73,** 487–491.

Zarbin, M. A., Wamsley, J. K., Innis, R. B., and Kuhar, M. J. (1981). Cholecystokinin receptors: Presence and axonal flow in the rat vagus nerve. *Life Sci.* **29,** 697–705.

10

Pancreatic Islets

A. EPPLE
Daniel Baugh Institute of Anatomy
Jefferson Medical College of
Thomas Jefferson University
Philadelphia, Pennsylvania 19107

J. E. BRINN
Department of Anatomy
East Carolina University
School of Medicine
Greenville, North Carolina 27834

I. THE EVOLVING ISLET ORGAN: A NEW GLAND USES ANCIENT HORMONES

After 100 years of intensive study, it is now generally accepted that the pancreatic islets of the higher vertebrates (gnathostomes) contain four major cell types, which produce glucagon (A cells), insulin (B cells), somatostatin (D cells) and pancreatic polypeptide (F cells). Probably, these four cells occur in all gnathostome species, though evidence for some of the major groups (e.g., lungfishes and apodan amphibians) is incomplete. On the other hand, additional, obviously very important cell types occur in isolated taxa (holocephalans, garpikes). The current situation is summarized in Fig. 1.

Thus, while the basic cellular composition of the islet organ appears reasonably clear, its early embryonic origin remains a matter of debate (cf. Pearse, 1982, 1984; Andrew, 1982), and we are faced with two options: (1) the endocrine tissue which is associated with the exocrine pancreas (variously referred to as pancreatic islets, islet organ, islets of Langerhans, or endocrine pancrease) is of the same embryonic origin as the latter; (2) the islets are a tissue *sui generis,* topographically associated with the exocrine pancreas. The issue is difficult and at the same time of fundamental importance. If, indeed, the islet organ is formed from an embryonic substrate different from the exocrine pancreas, then there should be important consequences for both tumor and diabetes research (cf. Falkmer *et al.,* 1984a). A

279

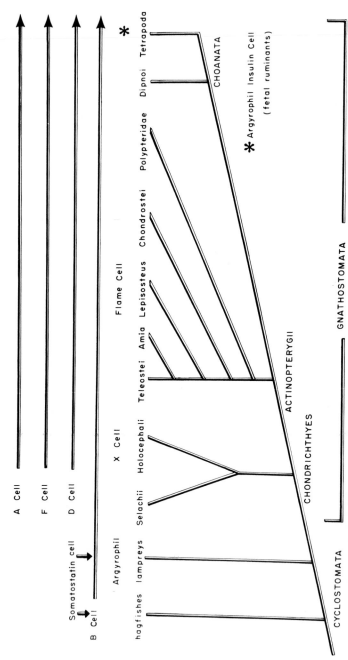

Fig. 1. The phylogenetic distribution of the major types of islet cells. *Latimeria* has been omitted, because of the limited information on its islet cytology (see Section IV,B,2).

TABLE I

Messenger Peptides of the Islet Organ Other Than the Four Major Hormones[a]

Substance	Cell type	Species	Citations
Gastrin	D cell	Human	Erlandsen et al., 1976
	Independent?	Lizard	El-Salhy and Grimelius, 1981
GIP	A cell	Rat	Smith et al., 1977
	A cell	Human, pig, dog, cat, guinea pig	Alumets et al., 1978
	Independent?	Lizard	El-Salhy and Grimelius, 1981
Cholecystokinin	A cell	Human, rat	Grube et al., 1978a
Luteinizing hormone releasing hormone	B cell?	Rat	Seppälä et al., 1979
Thyrotropin-releasing factor	A, B cells	Rat	Kawano et al., 1983
Corticotropin-releasing factor	A cell	Catfish, toad, lizard, chicken, rat, mouse, cat, monkey, human	Petrusz et al., 1983
Endorphin	A cell	Rat	Grube et al., 1978b
	D cell	Rat, guinea pig, human	Watkins et al., 1980
	F cell	Lizard	El-Salhy and Grimelius, 1981
	?	Human	Bruni et al., 1979
Adrenocortico-tropin	?	Human, dog, rat	Larsson, 1977
	A cell	Rat	Graf, 1981
Prolactin	B cell	Rat	Meuris et al., 1983
Neurotensin	?	Mouse	Berelowitz and Frohman, 1983
Pancreatic polypeptide	A cell	Frog	Kaung and Elde, 1980
			El-Salhy et al., 1982
Pancreatic polypeptide	A cell	Fetal human and rat	Ali-Rachedi et al., 1984

[a] This compilation is based on immunoreactivites. Hence, the precise nature of some of these peptides remains to be confirmed.

neural crest origin of the islet tissue, originally suggested by Pearse and Polak (1971), appears to be no longer accepted by the majority of investigators, who now favor an endodermal origin of the islet organ (cf. Rutter, 1980; Ayer-LeLievre and Fontaine-Perus, 1982; Andrew, 1984). In particular, the elegant work of Andrew and co-workers (see Andrew, 1984; Rawdon, 1984; Rawdon *et al.*, 1984) seems to have settled the issue for both islet cells and gastrointestinal endocrines. These investigators used the greatly different nucleoli of quail and chicken cells as biological markers. In several sets of transplantation experiments with early embryonic material, they showed that in appropriately combined quail–chicken chimeras, only endoderm gives rise to islet cells. The endodermal origin of the pancreatic islets and related gut endocrines is interesting, since they share their hormones with endocrines and neurons of definitely ectodermal origin. Thus, insulin, glucagon, somatostatin, and pancreatic polypeptide (PP) (or closely related substances) have been found in the nervous systems of various vertebrates and invertebrates (cf. Falkmer *et al.*, 1984a; Epple and Brinn, 1987), and conversely, peptides (immunoreactivities) originally identified in the nervous system also occur in the islets as cosecretions (Table I). The seemingly unorthodox distribution of these messenger substances in derivatives of two different germ layers becomes understandable if we recall that these "regulatory peptides" phylogenetically antedate the appearance of multicellular organisms. Like steroids and biogenic amines, several peptide hormones (immunoreactivities), including insulin and somatostatin, have been found in protozoa (Le Roith and Roth, 1984; Rosenzweig *et al.*, 1985). From these findings, it appears that we must look at the vertebrate endocrines as organs which evolved when evolutionary factors called for *increased or special use of already existing messenger substances.* The availability of an array of ancient messenger substances may also explain the surprising ease with which, in phylogenetic terms, the APUD (amine precursor uptake and decarboxylation) system (cf. Pearse, 1984) creates, modifies, or disposes of its members (Epple and Brinn, 1980)—a phenomenon particularly obvious in the islet phylogeny.

II. EARLY PHYLOGENETIC RELATIONSHIPS BETWEEN ISLET ORGAN, EXOCRINE PANCREAS, AND LIVER

While a common endodermal origin of both islet organ and exocrine pancreas appears almost certain, this does not necessarily mean that both tissues share an immediate, common precursor cell. Rather, the possibility remains that they evolved independently from each other as much as both seem to have evolved independently from the liver. This unexpected problem, outlined in Fig. 2, arose from recent studies on the islet organ of the most primitive vertebrates, the Southern hemisphere lampreys *Geotria australis* and *Mordacia mordax* (Hilliard *et al.*, 1985; and unpublished data).

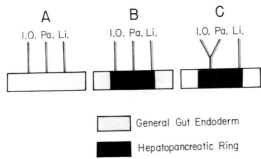

Fig. 2. Proposed ontogenetic–phylogenetic interrelationships between the islet organ, exocrine pancreas, and liver. The independent origin of the precursors of all three organs from "general" gut endoderm of the ancestral chordates is likely, though no data for the liver are available (A). However, widely scattered endocrine cells in the gut of *Branchiostoma* and diffusely distributed zymogen cells in the gut of the Myxinidae are compatible with this interpretation. At the cyclostome level, liver and islet organ develop from a restricted region (equivalent of the hepatopancreatic ring), and in the Petromyzontidea this also holds for the exocrine pancreas equivalents (B). As a consequence of the common, restricted embryonic substrate, the three organs form a variety of associations. For example, in the Myxinidae, the islet organ develops from bile duct material; in larval *Geotria,* the bile duct opens into an intestinal diverticulum (exocrine pancreas equivalent); in *Petromyzon,* islet material develops in close contact with the ceca (probably rudiments of the diverticula), as well as with hepatic material. At the gnathostome level (C), formation of both liver and pancreas tissues is restricted to the hepatopancreatic ring material. However, their immediate precursors are now programed for separate lines: pancreas tissues and liver system. The potential of the gut endoderm to produce elements closely related to islet cells persists probably in all vertebrates, though intestinal insulin cells may be generally absent above the selachian level (see Section IV,B,1). See Siwe (1926), Bargmann (1939), Ferner (1952), Barrington (1972), Rutter (1980), Andrew (1984), Falkmer *et al.* (1984a), Rawdon (1984), Hilliard *et al.* (1985), and Epple and Brinn (1986). Abbreviations: I.O., islet organ and mucosal precursor cells; Pa., exocrine pancreas and phylogenetic precursors (diffuse zymogen cells, secretory diverticuli, protopancreas, intestinal ceca of lampreys); Li., liver system and precursors.

Apart from its potential relevance to oncogenic and other aberrations of the endocrine pancreas (see Pour *et al.,* 1982; Bosman, 1984; Bosman *et al.,* 1985), an understanding of the early steps in islet evolution may help to explain the phylogenetically persistent, intimate vascular connection between the islet organ and the liver.

Based on the data of Yamada (1951), Maclean (1965), and Hilliard *et al.* (1985), it appears that the evolution of the interrelations between the islet organ, exocrine pancreas, and liver can be reconstructed as follows (Fig. 3):

1. *Branchiostoma (Amphioxus)* stage. At this level, diffuse (open) endocrine gut cells containing insulin, glucagon, somatostatin, and PP-like immunoreactivity are present (cf. Van Noorden, 1984) and so is an intestinal diverticulum. The diverticulum is vascularized by the venous drainage of the gut and therefore must receive gut hormones and nutrients in the highest concentration to reach any organ.

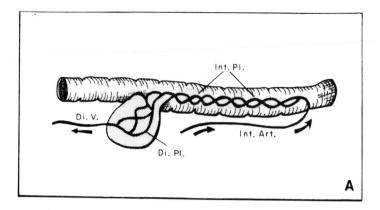

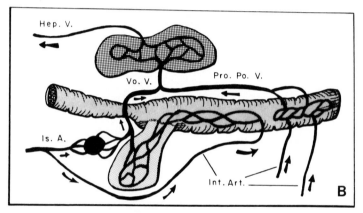

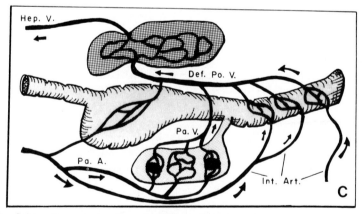

Fig. 3. Schematic reconstruction of phylogenetic stages of the vascular connections between the islet organ, pancreas, and liver. (A) Branchiostoma stage. A secretory diverticulum receives blood from the intestinal plexus, which is continuous with the diverticular plexus. The ancestors of the islet cells reside in the intestinal mucosa (not shown). (B) Larval mordacia stage. The newly evolved islet organ drains via a vorticose vein into a provisional hepatic portal vein, which carries blood from the lower intestinal region. The protopancreas is still supplied by

Fig. 4. Hypothetical stages of the evolution of "closed" intestinal endocrines from "open" cells. The "open" cells (left) monitor the intestinal contents with receptor sites (asterisks), located on large microvilli, and release their messenger substance at the basal pole into capillaries (fine granules). "Closed" cells (right) monitor messages coming from the blood; they are separated from the intestinal lumen by tight junctions. In lampreys, such cells give rise to islets which leave the intestinal mucosa. An intermediate stage (center) may have responded to metabolites released from intestinal absorbing cells (small microvilli) "below" tight junctions. Exocrine intestinal cells contain large secretory granules. [From Epple *et al.,* 1980.]

2. Ammocete (larval lamprey) stage, as seen in the larva of a Southern hemisphere lamprey (*Mordacia mordax*). Here, some endocrine gut cells (B cells) have become organized as an islet organ; the diverticulum has become an (exocrine) protopancreas (Strahan and Maclean, 1969), and a liver is present. The islet organ receives arterial blood, and its venous blood drains *downstream*(!) from the protopancreas into the hepatic portal circulation. Hence, there is no direct functional link between the islet organ and the early equivalent of the pancreas.

3. Gnathostome stage. The islet organ ("endocrine pancreas") and the exocrine pancreas form a combined organ, which retains a close vascular link with the liver.

In the preceding sequence, the most important functional changes occurred during the transition from the *Branchiostoma* to the ammocete stage and include (1) the creation of an islet organ, whose cells switched from the monitoring of intestinal, luminal signals to the monitoring of blood-borne signals (Fig. 4) and (2) the supply of the islet organ with arterial blood (cf.

blood from the upper intestinal plexus. (C) Gnathostome stage. Both islets and exocrine pancreas receive arterial blood, and drain via pancreatic vein(s) into the definitive hepatic portal vein (of the adult lampreys and gnathostomes). The latter has evolved from the intestinal plexus and replaces the provisional portal vein, which has disappeared. No vascular connection between intestine and pancreas. For clarity, hepatic artery and biliary drainage were omitted. For additional literature see Youson (1981a). Abbreviations: Int.Art., intestinal artery; Int.Pl., intestinal plexus; Di.Pl., diverticular plexus; Di.V., diverticular vein; Is.A., insular artery; Vo.V., vorticose vein; Pro.Po.V., provisional hepatic portal vein; Hep.V., hepatic vein; Pa.A., pancreatic artery; Pa.V., pancreatic vein; Def.Po.V., definitive hepatic portal vein.

Epple and Brinn, 1980; Epple *et al.*, 1980). These two steps created a highly efficient dual control system of digestive and metabolic processes, in which the former can be integrated with the state of the blood metabolites via the combined signals from open intestinal endocrines and islet cells. It is difficult to imagine how larger vertebrates with their discontinuous, often irregular feeding habits could have evolved without such a coordinating system. From this phylogenetic perspective, it becomes clear why the same hormones may be released in response to gastrointestinal, as well as blood-borne, signals. Examples are somatostatin and PP, which are secreted by both open endocrine gut cells and islets. And it becomes also understandable why, on the other hand, two different hormones may affect a target tissue in the same way as, e.g., gastrin (from the open G cells of the pyloric region) and PP (probably from both "closed" islet cells and "open" gut cells) in their stimulatory effect on gastric secretions (cf. Epple and Brinn, 1987).

III. THE ASSOCIATION OF ISLET ORGAN AND EXOCRINE PANCREAS: RAISON D'ÊTRE?

From the preceding it is obvious that the islet organ evolved independently from the exocrine pancreas. Inevitably, this raises the question as to the possible functional significance of its close association with the exocrine pancreas in the gnathostomes, as well as to the great morphological variations of this association. Two (mutually not exclusive) explanations have been proposed: (1) Since the growing exocrine pancreas uses the vascular branches of the portal vein system as a trellis, the associated islet organ is assured fast and concentrated delivery of its hormones to the liver. The latter is both target and checkpoint of the islet hormones, whose final concentration in the systemic blood is determined by the liver (Epple and Brinn, 1980, 1986; Epple *et al.*, 1980); in addition, the exocrine pancreas provides a support for the islet tissue which, in this extramural location, can greatly increase its volume. (2) The dissemination of the islet tissue serves to control the surrounding exocrine acini by islet hormones which reach the latter via "insuloacinar portal systems" (cf. Henderson *et al.*, 1981), i.e., it follows a principle similar to that of the intratesticular distribution of the Leydig cells (Ferner, 1957). The following analysis, based on embryonic and comparative data, will address these questions.

In all vertebrates "above" the Chondrichthyes, there seem to be three embryonic pancreatic anlagen. Of these, a single dorsal one usually develops first, and in many species it already contains endocrine-determined cells as it begins to leave the gut (Siwe, 1926; Bargmann, 1939; Andrew, 1984). On the other hand, the development of the ventral anlagen is inconsistent and may be totally absent. The latter is the case in the guinea pig (Siwe, 1926), which nevertheless has a typical mammalian pancreas with scattered islets (Baskin *et al.*, 1984). In humans, only the right ventral anlage differentiates and forms the pancreas structures proximal to the duodenum (Ferner, 1952). The ventral anlagen are poor in islet tissue, and they often appear late (cf. Bargmann, 1939). However, tissue derived from the ventral anlagen always seems to contain a high percentage of scattered F cells (cf. Orci, 1983; Andrew, 1984). At least in some amphibian larvae, typical islets are totally absent in the derivatives of the ventral anlagen (cf. Frye, 1962). An immuno-cytochemical study of adult *Ambystoma* confirms the absence of islets in the ventral anlage; however, there are scattered A and F cells (Brinn *et al.,* 1986). Andrew (1984) suggests that F cells are of ventral anlage origin only; this view contrasts with the presence of F cells in the guinea pig and shark pancreas (Baskin *et al.,* 1984; El-Salhy, 1984), which seem to develop from dorsal anlage material only (Siwe, 1926). There is apparently no explanation for the common time gap in the islet development between the dorsal and ventral anlagen, and the early precursor cells of the islets in the ventral anlagen are unidentified, though an origin from dorsal material need not necessarily be invoked (cf. Bargmann, 1939; Andrew, 1984). This problem may be related to or identical with that of the small pancreatic ductules which give rise to both exocrine and endocrine cells in pre- and postnatal gnathostomes (see Bjorenson, 1985; Bosman *et al.,* 1985).

The close ontogenetic relationship between the endocrine cells of the dorsal anlage and the developing exocrine pancreas explains most of the phylogenetic variations in the islet size and distribution of the gnathostomes "above" the Chondrichthyes (cf. Siwe, 1926). Islets, which are presumably of ventral anlagen origin, are usually small and relatively few in number (cf. Bargmann, 1939). In many species, islet tissue first appears as a compact mass in the very early dorsal pancreas anlage. If this endocrine material is not split up by invading exocrine tissue, it becomes a large principal islet, as in the swordtail fish (*Xiphophorus helleri*: see Epple and Brinn, 1975). In other cases, the degree of mingling of both tissues determines the final morphology of the islet organ, which then varies from widespread, relatively small islets (as in, e.g., sturgeons, the bowfin, and many mammals) to concentrations of larger islets close to or inside the spleen (as in some teleosts, squamate reptiles, and birds). These options are illustrated in Fig. 5, and it must be noted that in adult gnathostomes the islet tissue is also virtually always associated with exocrine pancreas. Rare exceptions have been reported for a teleost with (aberrant?) intraovarial islet tissue (Lepori, 1959), another teleost with intramucosal islets (Punetha and Singh, 1983), and the intrasplenic islets of some other teleosts (cf. Bargmann, 1939), lizards, and snakes (see Section IV,B). In these cases, it is likely that the islet tissue has been "cut off" from the tip of the growing dorsal anlage. However, there is also a second route of islet formation, which begins with proliferation of small islet buds from many regions of the developing pancreatic duct system (cf. Bargmann, 1939).

The comparative data presented here do not favor the idea that an intra-

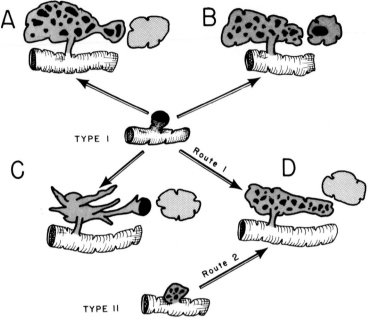

Fig. 5. Development of dorsal-anlage-derived pancrease tissues in different vertebrates. The type I anlage (with compact endocrine material) gives rise to the third and splenic lobe of the galliform pancreas (A), large pancreas regions, the intrasplenic islet of certain Squamata (B), large pancreas regions, usually (?) the Brockmann bodies of the Actinopterygii (C), and the total pancreas or pancreas tail regions (Muridae) of certain mammals (D). However, the configuration shown in D may also develop from the type II anlage with early diffuse islet tissue formation (e.g., guinea pig, human, roe deer, cow). Shaded region, exocrine pancrease; black region, islet tissues; stippled region, spleen. [Compiled and simplified from Siwe (1926), Bargmann (1939), and Ferner (1952).]

pancreatic islet dissemination is a basic functional principle, by which the islet tissue controls the exocrine pancreas (cf. Bonner-Weir and Weir, 1979; Henderson *et al.*, 1981; Kawai *et al.*, 1982a; Trimble *et al.*, 1985). If this were so, then evolutionary pressures should have created an uniform intrapancreatic islet distribution in "higher" forms (Epple *et al.*, 1980). Instead, no phylogenetic trend is recognizable, and small, disseminated and large, locally restricted islets occur, even in closely related taxa of actinopterygians and reptiles (see Section IV,B). There is no question that large regions of the exocrine pancreas of the Polypteridae, *Lepisosteus,* many teleosts, and probably also the turtles (see Section IV,B) have no vascular relations with the islet tissue except, perhaps, for scattered D and/or PP cells. Even in mammals, only a small percentage (20% or less) of the total pancreatic blood (Lifson *et al.*, 1980, 1985; Jansson and Hellerström, 1983) passes through the islets, and both in mammals and in the chicken the acinar regions supplied by islet-derived capillaries are very small (cf. Bonner-Weir and Orci, 1982;

Syed Ali, 1982; Iwanaga *et al.*, 1983). If there are functionally significant insuloacinar portal systems at all, they must be restricted to a few groups of vertebrates in which they developed secondarily. Clearly, the mammalian pancreas with its disseminated islets would be a favorable organ for such a development. Nevertheless, there is good evidence that the islet organ affects the surrounding exocrine tissue in many species of fishes and tetrapods, as shown by the presence of so-called halos or zymogen mantles, i.e., periinsular acini whose enzyme content is different from those further away (see Leclerq-Meyer *et al.*, 1985; Epple and Brinn, 1986). Thus, one could argue that in some species or taxa the endocrine pancreas secretions create two functional compartments of the exocrine pancreas. However, there is no proof yet that the "halos" are anything else but by-products of the topographic association of both pancreas tissues, and at least in the chicken, they must be due largely or exclusively to paracrine secretions(s) (see Section IV,B,7), a possibility ignored in mammalian studies. Furthermore, the islet substances responsible for the halo phenomenon have not yet been identified, and it is thus unknown if the halos are due to hormones, co-released islet substances, or even neurosecretions, released into the islet circulation. The last have been found by Fujita and co-workers in the islets of the dog, the mink, and a snake (*Elaphe quadrivirgata*), as well as in the poison gland of another snake. Hence, they may represent a *peripheral neurosecretory system* (Fijii *et al.*, 1980; Fujita *et al.*, 1982) that so far has been overlooked.

In summary, there is no convincing evidence that the association between the islet organ and exocrine pancreas is anything but a topographical one. It appears that this association (1) serves in the supply of highly concentrated iselt hormones to the liver (for further discussion, see Epple and Brinn, 1987 and (2) provides a structural support for the islet tissue. Figure 6 shows the basic types of associations between the islets and pancreas (Epple, 1969).

IV. COMPARATIVE ISLET ANATOMY

A. Agnathan Islet Organ

The two groups of extant cyclostomes are the only survivors of the agnathan level of vertebrate evolution (cf. Hardisty, 1979, 1983).

In both, the lampreys (Petromyzontiformes) and the hagfishes (Myxinidae), the islet organ is separated from the equivalents of the exocrine pancreas. In the hagfishes and Northern hemisphere lampreys (Petromyzontidae), the latter are represented by zymogen cells in the intestinal mucosa (Barrington, 1972; Youson, 1981a); however, the studies on the protopancreas of larval *Mordacia* (see Section II) suggest that, at least in the Petro-

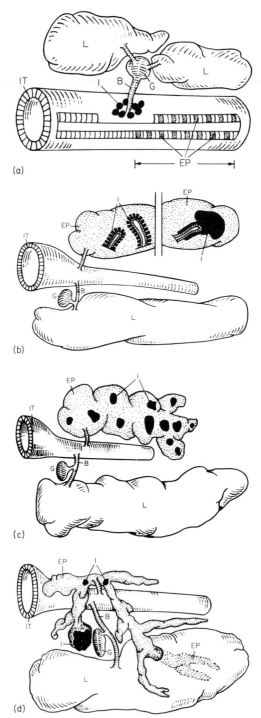

Fig. 6. The pancreas types of vertebrates. (a) Cyclostome type, *Myxine*. Ringlike arrangement of islet cell follicles near the distal end of the bile duct. The exocrine pancreas is represented by specialized cells in the midgut epithelium. (b) Primitive gnathostome type. Many

myzontidae, this is a secondary situation. The islet organ of lampreys often develops directly from cells of the intestine (see, e.g., Brinn and Epple, 1976; Hilliard *et al.*, 1985), a process which recapitulates in a textbook fashion the probable phylogenetic history of the early islet organ. In the adults of the Petromyzontidae, the islet tissue ultimately forms three accumulations, one of which occurs in the liver (Barrington, 1972). This happens during the metamorphosis at the time when, due to disappearance of the bile duct, the liver loses its connection with the intestine (cf. Youson, 1981b). Small, intrahepatic groups of islet cells are sometimes completely separated from the main hepatic islet (Epple and Brinn, 1975; Brinn and Epple, 1976), which raises the question of whether they are derived by transformation of biliary duct material, similar to the situation in the Myxinidae (see below). The cranial islet of prespawning river lampreys has a tendency to form cycts and hamartomas, whose incidence increases towards the end of the migratory period (Hardisty, 1976). This observation is particularly noteworthy because of similar neoplasias in hagfish (see below).

It is possible to destroy the islet tissue of the Petromyzontidae by electrocautery or surgery, and this procedure results in hyperglycemia (cf. Hardisty, 1979). Nevertheless, it is doubtful whether all islet tissue can be removed in such experiments. However, the islet tissue of one, perhaps both, of the Southern hemisphere lampreys (*Geotria* and *Mordacia*) accumulates in a single mass. In the adult of *Geotria australis*, the islet tissue lies on top of the cartilagenous pericardium and can be removed easily, a situation most favorable for the collection of "pure" islet tissue (Hilliard *et al.*, 1985). The prespawning *Geotria* spend many months in a fasting state in fresh water, and the males (about the weight of a laboratory rat) remain very sturdy during this time. The females, on the other hand, use most of their body mass for the developing ovary and finally appear like living egg cases (Potter *et al.*, 1983). We have recently developed a simple and rapid technique for the cannulation of the tail vessels in male *Geotria*, which allows serial blood sampling (Macey *et al.*, 1984). Together with the simplicity of islectectomy, this should make the male *Geotria* an ideal model for islet-related studies at the level of the most primitive extant vertebrates.

The islet organ of the larval lampreys seems to consist of B cells and some undifferentiated agranular elements. However, during metamorphosis, so-

---

sharks have a compact pancreas with islet cells occurring as an outer layer of small ducts (left). The endocrine pancreas of the holocephalians (right) exemplifies an intermediate stage to the tetrapod type. (c) Tetrapod type. A compact pancreas with scattered endocrine islets, similar to that of the tetrapods, is also found in a few teleosts, e.g., *Anguilla*. The pancreas of the lungfishes is basically of the tetrapod type, but surrounded by folds of the stomach and intestine. (d) Actinopterygian type (teleost). In most actinopterygians the exocrine pancreas tends to split into small strands of tissue which sometimes invade other organs. Abbreviations: L, liver; EP, exocrine pancreas; I, islet tissue; G, gallbladder; B, bile duct; and IT, intestinal tract. [From Epple, 1969.]

matostatin cells appear in large numbers (cf. Hardisty and Baker, 1983); in addition, a smaller number of argyrophil (Grimelius-positive) cells do also form, at least in *Geotria* and *Petromyzon* (Hilliard *et al.*, 1985; S. Van Noorden *et al.*, unpublished data). Since the somatostatin cells of *Petromyzon marinus* stain with tryptophan reactions (often considered specific for A cells) but not with the Hellman-Hellerström silver technique, their identity with the D- cells of the gnathostomes is uncertain (Brinn and Epple, 1976). The nature of the newly discovered argyrophil cells is totally unknown. There is no unequivocal evidence for glucagon in lamprey islets (cf. Hardisty and Baker, 1983; Falkmer *et al.*, 1984b), and one wonders if the close association of the argyrophil cells with the B cells indicates a relationship of these elements with the argyrophil insulin cells of fetal sheep or insulinomas (Falkmer *et al.*, 1984b). In prespawning *P. marinus*, additional cell types appear in the islet organ (Brinn and Epple, 1976); their presumptive hormones have not yet been identified. The appearance of somatostatin cells coincides with the change of the liver into an "endocrine" organ (cf. Youson, 1981b). At the same time, however, the lamprey switches from filter feeding to the parasitic phase, which may involve irregular, heavy food uptake. Possibly, this switch is related to the appearance of islet somatostatin cells, which might serve to provide a more efficient dual control system of digestive and metabolic functions (see Section I). However, the persistence of insular somatostatin cells in prespawning, fasting lampreys, as well as in adults of species which do not feed after the metamorphosis, remains to be reconciled with this hypothesis.

The hagfishes are bottom dwellers of the oceans, and most forms live in the lower shelf regions (Hardisty, 1979). Unlike the lampreys, they do not have a larval stage. Throughout their life cycle, the islet organ contains both insulin and somatostatin cells, but the percentage of the latter is exceedingly small in *Myxine* (Falkmer and Van Noorden, 1983; Van Noorden, 1984). The islet tissue forms a single, compact mass which surrounds the lower bile duct, from which its precursor cells are recruited (cf. Falkmer *et al.*, 1984b). The islets of *Myxine glutinosa* seem to be rather susceptible to carcinogens, as suggested by a high frequency of islet cysts, harmartomas, adenomas, and adenocarcinomas in animals from polluted waters (Falkmer *et al.*, 1976).

Total isletectomy is possible but requires ligation of the bile duct. Contrary to the findings in lampreys, the operation is not followed by a distinct hyperglycemia (cf. Hardisty, 1979; Epple and Brinn, 1986). This discrepancy is so far unexplained, though one could invoke a role of intestinal endocrine cells, which produce varying ratios of hyper- and/or hypoglycemic hormones.

While the islet organs of both groups of cyclostomes can easily be collected for the study of their hormones, only hagfish insulin has been investigated in detail (Emdin, 1981). In general, it appears that the cyclostomes are a largely unexplored gold mine for basic studies on the islet biology.

B. Gnathostome Pancreas

1. Chondrichthyes

The Chondrichthyes have many extant species which can be assigned to two major subdivisions: the elasmobranchs (sharks, rays, and skates) and the holocephalans (chimaeras). Both elasmobranchs and holocephalans have a distinct extramural pancreas whose surgical removal is easy. However, the structure of the islet organ varies greatly between both groups (Fig. 5b). In the elasmobranchs, the pancreas develops from the dorsal anlage only (Siwe, 1926). Its islet cells often form a layer surrounding small, agranular ductules whose possible relationship with the ductules of the mammalian pancreas has been mentioned in Section II. In some forms, there are also smaller islets which seem to bud off from the exocrine duct system. The islets are innervated (Epple and Brinn, 1975), but no detailed study of the innervation has been carried out. In general, the islet organ of the elasmobranchs shows a surprising similarity with certain stages of the embryonic mammalian pancreas (cf. Bargmann, 1939). Furthermore, the number of islet cell types is considerably higher than in other gnathostomes, possibly amounting to eight (Epple, 1967; Kobayashi and Syed Ali, 1981). Immunocytochemically, A, B, D, F, and GIP (gastric inhibitory peptide) cells have been identified in the islet organs of elasmobranchs (Sekine and Yui, 1981; El-Salhy, 1984). The digestive tract contains a large number of endocrine cells, including glucagon, somatostatin, PP, and insulin cells (El-Salhy, 1984). It is remarkable that the latter are of the "open" type, though gastrointestinal mucosal insulin cells are usually absent in the "higher" vertebrates. The presence of a large number of islet cells in a very ancient group of vertebrates may appear paradoxical unless one considers this as a primitive situation in which, similar to the embryonic pancreas, the elasmobranch pancreas carries a large "sampling" of gastrointestinal endocrines (or their precursors). While such cells persist in the elasmobranchs, they are usually reduced to four in higher vertebrates. Pancreatectomy in elasmobranchs is followed in some cases by hyperglycemia, but in others by normoglycemia or even hypoglycemia (cf. Patent, 1973; Epple and Lewis, 1973). It is not clear whether these erratic results are due to the physiological state of the animals, experimental factors (stress?), or perhaps to the impact of intestinal insulin secretion. Vascular cannulation is possible and has been used in studies on the effects of exogenous islet hormones (Patent, 1970, 1973).

The holocephalans form a sister group of the elasmobranchs which, similar to the hagfishes, usually lives in the lower sea-shelf regions. Their islet organ differs from that of the elasmobranchs in the presence of many obvious islets, whose prevailing cell type is the X cell (Fujita, 1962; Patent and Epple, 1967). Studies have shown that this peculiar element produces a substances related to glucagon and whose precise nature requires further study (Stefan et al., 1981). A, B, D, and F cells are also present, though F cells are scarce. In addition, smaller numbers of enkephalin, GIP, and serotonin cells have been found. The intestine contains somatostatin, glucagon, and PP cells, as well as other endocrines, but apparently no insulin cells (Falkmer et al., 1984b). In contrast to the elasmobranchs, the islet cells do not seem to be innervated (Patent, 1976). The peculiarities of the islet organ of the holocephalans are probably specializations related to the life-style of this group. However, as a sister group of the elasmobranchs (Fig. 1), the holocephalans cannot be considered a priori representatives of an islet organization which phylogenetically precedes that of the former, as recently suggested (Falkmer and Van Noorden, 1983; Falkmer et al., 1984b). On the contrary, the morphology of the holocephalan islet organ is intermediate between that of the elasmobranchs and higher fishes (Fujita, 1962). One species (Hydrolagus colliei) enters the upper shelf regions of the U.S. northwest coast, particularly the Puget Sound, where it can be collected in large numbers. While its frailty in captivity requires careful design of physiological studies (Patent, 1970; and personal communication), its pancreas may provide useful material for the study of evolution, biosynthesis, and interrelationships of molecules of the glucagon family.

2. *Latimeria chalumnae*

The phylogenetic status of this living fossil, *Latimeria chalumnae*, is a matter of heated debate. Based on traditional morphological and paleontological criteria, it has been considered a close relative of the tetrapods, but this view has been challenged, since its soft anatomy, its reproduction, and some biochemical parameters show great similarities with the elasmobranchs (cf. Lagios, 1979; Forey, 1980). Its scarcity and poor survival after capture (Locket, 1980) preclude the use of *Latimeria* in experimental endocrine work. However, its compact pancreas and islet organ show surprising similarities with those of the elasmobranchs. The precise number of islet cells is not yet known (cf. Epple and Brinn, 1975).

3. *Actinopterygians*

This group of ray-finned fishes comprises the vast majority of living vertebrates, and they form a distinct side branch from the phylogenetic main line leading to the tetrapods. It is a common mistake to interpret more advanced actinopterygians (e.g., the teleosts) as closer to the lungfishes or tetrapods. However, a look at Fig. 1 shows that the teleosts are farthest away from the common ancestors of both tetrapods and actinopterygians. The actinopterygians are a morphologically and physiologically plastic group, and it is not surprising, therefore, that the topographic relations between their islet organ, exocrine pancreas, and liver show a variety unsurpassed by other vertebrates. Nevertheless, two typical trends are obvious: (1) to concentrate the islet tissue in one or a few large accumulations and (2) to mingle the exocrine pancreas and liver tissue (Epple, 1969). These two trends are very pronounced in the Polypteridae, which are considered the most primitive actinopterygians (or a closely related sister group of the latter: Patterson, 1982) as well as in many of the most advanced teleosts (Epple and Brinn, 1975). The large accumulations of islet tissue are referred to as "principal islets," and since these are almost always in contact with some exocrine tissue (see Section III) principal islets together with the closely associated exocrine pancreas are termed "Brockmann bodies."

a. Polypteridae (Bichirs and Ropefish). These African freshwater fishes are bottom dwellers: *Polypterus,* whose six species have well-developed, strong pectoral and pelvic fins, and *Calamoichthys calabaricus,* which is more eellike and lacks pelvic fins. Unfortunately, the rather small size of the specimens usually available makes them unsuitable for islet research which requires vascular cannulation. Their exocrine pancreas intermingles with the hepatic tissue to a degree which is only surpassed by the garpikes (see Section IV,B,3,c). A large principal islet with a complete rim of exocrine pancreas is buried deep inside the liver. A few smaller islets occur in nearby exocrine pancreas regions. All four major types of islet cells appear to be present, and the islets are well innervated (Epple and Brinn, 1975; Mazzi, 1976). Detailed studies with immunocytological techniques have yet to be conducted. Perhaps the Polypteridae will prove interesting for the study of the interactions between islet organ, pancreas, liver, and their major blood vessels during the embryonic development.

b. Chondrostei (Sturgeons and Paddlefishes). These fishes are often of considerable size, and some species undergo extensive spawning migrations from their oceanic feeding grounds to the smaller tributaries of Eurasian and North American rivers. The chondrostean pancreas is rather diffuse and is partly included in the liver; the islet organ seems to consist of scattered, inconspicuous islets of unknown cytology (Epple and Brinn, 1975).

c. Lepisosteus (Garpikes). The seven extant species of garpikes all belong to one genus, and they are restricted to North and Central America. Their exocrine pancreas mingles with the liver to an even greater degree than in the Polypteridae and extends along the bile duct. Here, it surrounds large accumulations of islet cells, forming an aggregation of Brockmann bodies. Complete surgical removal of the islet tissue appears possible but must inevitably involve ligation of the biliary and pancreatic drainage. Since the blood vessels of larger specimens should be easily cannulated, this may allow short-term studies of isletectomized garpikes. The islet cytology of the garpikes is puzzling, since in addition to the four main cell types, there is a very frequent, fifth, flame-shaped cell of unknown function (Epple and Brinn, 1975; Van Noorden and Patent, 1978; J. E. Brinn and A. Epple, unpublished data). One has difficulties in accepting the isolated *de novo* appearance of an apparently very important islet cell in a rather advanced actinopterygian, but speculations as to its significance may be rather meaningless until its hormone has been identified. The islets of *Lepisosteus* are highly innervated by fibers resembling the classical cholinergic type (J. E. Brinn, unpublished data).

d. Amia calva (Bowfin). The bowfin (*Amia calva*) is the single extant species of the Halecomorphi, a group closely related to the teleosts (Patterson, 1982), whose geographic distribution is limited to North America. Larger specimens grow to a size sufficient for repeated blood sampling. However, the pancreas is rather diffuse and partly invades the liver; the islet tissue is scattered and does not form a distinct accumulation, which makes an effective isletectomy impossible. *Amia* has the four major types of islet cells, and as in *Lepisosteus,* there is a well-developed innervation which ultrastructurally appears to be "cholinergic" (Epple and Brinn, 1975).

e. Teleostei. This group comprises the most advanced actinopterygians, and its many thousand extant members can be assigned to some 12 major branches (Rosen, 1982). Teleostean species which have principal islets with little associated exocrine tissue (see Fig. 5C) in particular have contributed important data in islet research. Very large and almost purely endocrine Brockmann bodies are bound in the genera *Lophius* (anglerfish) and *Cottus* (sculpin); these, and the Brockmann bodies of *Ictalurus* (catfish) species, are probably the most widely studied teleostean islets. A comprehensive description of the islet organs of North American fishes has been given by McCormick (1925). Rennie and Fraser tried to use Brockmann body extracts as early as 1907 in the treatment of diabetics; in 1922, Macleod succeeded by the extraction of Brockmann bodies to prove the islet origin of insulin, and in 1925 McCormick and Mcleod confirmed the relationship between islet tissue deficiency and diabetes mellitus [better labeled, perhaps, hyperglycemia (see Epple, 1969)] by isletectomy in *Myoxocephalus* (=*Cottus*). Brockmann bodies have been used extensively in studies of insulin biosynthesis, and in recent years much of the progress in our understanding of biosynthesis and molecular heterogeneity of somatostatin was achieved by use of this tissue source (see Albert, 1982; Sorokin *et al.,* 1982; Fletcher *et al.,* 1983; Shen and Rutter, 1984). On the other hand, the extremely strong peptidergic [probably vasoactive intestinal polypeptidergic (VIP-ergic)] innervation of the Brockmann bodies lends itself to investigations on the role of the islet innervation, as well as to retrograde tracing studies on the origin of the neurons involved (Brinn, 1973; Epple and Brinn, 1975, 1986; Van Noorden and Patent, 1980). This latter possibility has hardly been exploited (Patent *et al.,* 1978), though it should be interesting to see if, and to what extent, direct ("preganglionic") fibers are present. This question has been raised by the discovery of such direct islet nerves in the rat, where they originate from the thoracic spinal cord (Luiten *et al.*, 1984). The "reason" for the extremely strong islet innervation in the teleosts is unknown, but it may play an important role in the integration of osmoregulatory and metabolic signals (cf. Epple *et al.*, 1980; Epple, 1985). A peculiar syndrome similar to diabetes has been discovered in cultured Japanese carps (so-called Sekoke disease), but its precise etiology and relevance to the human pathological conditions remain yet to be established (cf. Yokote, 1970; Nakamura *et al.*, 1971). On the other hand, it is uncertain if a specific insulin deficiency (type I) diabetes can be produced surgically in teleosts. As far as we are aware, there is no unequivocal proof of a selective destruction of teleost B cells with cytotoxins (alloxan or streptozotocin), and another problem arises with the existence of small islets in pancreas regions which are virtually inaccessible to surgical removal. A preponderance of A cells in these smaller islets may lead to a glucagon-induced hyperglycemia (Epple and Lewis, 1977). Furthermore, surgical stress, water temperature, and seasonal fluctuations in nutrition and fuel storage may all contribute to the degree of postoperative hyperglycemia. Thus, in contrast to their enormous value for biochemical work, the usefulness of teleosts for experimental islet research, particularly studies requiring isletectomy, is limited. Eels of the genus *Anguilla* are perhaps the best teleost models available; they have a compact pancreas, whose surgical removal allows in most cases complete isletectomy without damage to the biliary system (Lewis *et al.*, 1977). Using pancreatectomized American eels, we have shown, *inter alia,* that in these fishes the impact of the islet hormones on most metabolic parameters is balanced so that the complete removal of the islet tissue is not followed by diabetes mellitus; instead, the operation results in an undampened equilibrium which can easily be affected by stress (Epple and Lewis, 1977; Lewis and Epple, 1984). Furthermore, we have shown by pancreatectomy that one or more islet hormone(s) are necessary for the survival of the yellow American eel in seawater (Epple and Miller, 1981; Epple, 1985). Recent findings (cf. Davis and Shuttleworth, 1985) suggest that glucagon is the hormone in question. In other teleosts in which total isletectomy appears possible, either by pancreatectomy (pike, genus *Esox*) or removal of the total, concentrated islet tissue (*Xiphophorus*), the operation would probably require ligation of the bile duct. In addition, *Xiphophorus* would be too small for repeated blood sampling. The shape of the teleost pancreas varies from compact with scattered islets (e.g., eel, pike) to extremely diffuse with Brockmann bodies (cf. Epple, 1969; Epple and Brinn, 1975; Punetha and Singh, 1983). In the salmonids, there is a Brockmann body near the gallbladder, and considerable quantities of exocrine and endocrine pancreas occur between the numerous pyloric ceca. During the summer, the intercecal pancreas of *Salmo trutta* is widely dispersed by adipose tissue; however, during the spawning time, this adipose tissue disappears completely except for small aggregations of cells which give the superficial impression of lymphatic tissue. Consequently, the intercecal pancreas now consists of a collection of compact lobules (Epple and Schneider, 1974). Such seasonal variations of adipose tissue are not uncommon in teleosts and *Amia* and may strongly interfere with the localization of pancreas tissue (A. Epple and J. E. Brinn, personal observation). As in many tetrapods, the teleost islets tend to have a core of B and D cells and a rim of A and F cells. However, the percentage of A and F cells in smaller islets varies considerably, and in *Cottus,* Stefan and Falkmer (1980) report that only one (juxtapyloric) of the two Brockmann bodies contains all four major types of islet cells, while the other (juxtasplenic) lacks F cells. According to Lange (1984) there is no cellular arrangement conducive to an organized intrainsular blood flow, as postulated for the rat (see Section IV,B,7).

4. Dipnoi (Lungfishes)

The extant lungfishes are living fossils which belong to two rather different branches of this once varied group (South American and African lungfishes, Lepidosirenidae, and the Australian lungfish, Ceratodontidae). They are very similar in the peculiar gross anatomy of their digestive systems. The latter involves an almost complete "wrapping" of the pancreas and adjacent spleen regions by parts of the stomach and the intestine (Rafn and Wingstrand, 1981), which makes pancreatectomy impossible. Interestingly, the islet organ of *Protopterus* consists of a few well-encapsulated "principal islets," while in *Neoceratodus* the islets are scattered in the mammalian fashion. So far, only A, B, and D cells have been described (Brinn, 1973; Epple and Brinn, 1975), but further studies are needed to identify the full number of islet cells and the precise nature of their innervation (Brinn, 1973). The lungfishes received no attention in physiological or biochemical islet studies, which is surprising considering the key phylogenetic position of this group (Rosen *et al.,* 1981), and the habit of *Protopterus* to estivate (i.e., retire to a self-built cocoon for summer sleep during the dry season).

5. Amphibia

There are three extant orders of amphibians. The Gymnophiona (Apoda) are snakelike, limbless animals which mostly burrow in the soil of the tropical regions of South America, Africa, and Southeast Asia. Little is known about their islet organ except that there are at least A, B, and D cells (Gabe, 1968; Welsch and Storch, 1972). The Urodela (salamanders, newts) have a basically compact pancreas which varies greatly with the species. There is also a considerable variation in the structure of their islets (Epple, 1966b) whose relationship to the taxonomy or biology of these animals is unknown. Ultrastructurally, the islets of *Necturus maculosus* contain four cell types (Epple and Brinn, 1976) which may correspond to the A, B, D, and F cells. Amphiphil islet cells were discovered simultaneously in certain urodeles and

anurans in which the islet cells have particularly favorable histological staining properties (Epple, 1966a,b). Probably, these elements contain more than
one peptide hormone (see Table I). The urodele islets appear poorly innervated (Trandaburu, 1976), though we recently found presumptive peptidergic terminals in the islets of *N. maculosus* (J. E. Brinn, unpublished). The
only attempt at pancreatectomy in urodeles was made in a newt, *Taricha
torosa,* and involved removal of part of the duodenum and the bile duct
(Miller, 1961). Due to the short experimental period (24 hr) it is not possible
to decide to what extent the ensuing hyperglycemia was stress related. The
usefulness of the urodeles for islet research remains to be assessed, but
considering their wide range of ecological adaptations, from permanent
neotenous life in subtropical waters (e.g., *Amphiuma, Siredon*) to a completely terrestrial life without a larval stage (*Salamandra atra*), some members of this group should be interesting. It is sufficient to point out that the
islets of *Amphiuma* are well suited for long-term organ culture (Gater and
Balls, 1977) and that *N. maculosus* may not show hypoglycemic symptoms,
even when the blood sugar becomes nondetectable (Copeland and DeRoos,
1971).

The Anura (frogs, toads) have been widely used in early islet research,
particularly in studies involving pancreatectomy (Penhos and Ramey, 1973);
it is noteworthy that some of the basic data on pituitary–islet interactions
were obtained by Houssay and collaborators during work with toads (cf.
Houssay, 1959; Miller, 1961; Penhos and Ramey, 1973). Like the urodelan
pancreas (see Section III) the anuran pancreas has also been used in studies
of the embryonic origin of the islet tissue (cf. Bargmann, 1939; Frye, 1964).
According to Frye (1964), larval anurans, like larval urodeles, can regulate
their carbohydrate metabolism in the absence of islets. Complete pancreatectomy in adult anurans is very difficult, and even in cases where it was
believed to have been achieved, islet regeneration was observed (cf. Penhos
and Ramey, 1973).

Recent interest in the anuran pancreas has centered around the islet cytology. A, B, D, and F cells have been identified (Tomita and Pollock, 1981), as
have cells which react to both anti-glucagon and anti-PP (Kaung and Elde,
1980; El-Salhy *et al.,* 1982). The identity of the latter elements with am-
phiphils (Epple, 1966a) appears possible but is unproven. Of particular interest is a report by Fujita *et al.* (1981), which describes the presence of a
conspicuous number of secretin-immunoreactive cells in the pancreas of
Rana catesbeiana. In contrast to the urodele islets, anuran islets are heavily
innervated, mainly with fibers of the classical "cholinergic" type (Trandaburu, 1976). The role of the pancreatic islets during the metamorphosis is
uncertain, though some decrease in the B-cell population seems to occur at
this time (Frye, 1964; Kaung, 1983).

In all, it appears that the islet organ of the amphibians has never received
the attention it may deserve.

6. Reptilia

Despite their important phylogenetic position the Reptilia have been used only to a limited extent in pancreatic islet research. The four extant orders, i.e., the Chelonia (turtles, tortoises), Crocodilia (crocodiles and their relations), Squamata (lizards and snakes), and Rhynchocephalia (one species: *Sphenodon punctatus*) all have a basically compact pancreas (Penhos and Ramey, 1973) but differ considerably in their islet structure and responses to pancreatectomy (cf. Epple *et al.,* 1980).

The Chelonia are the oldest surviving group, and despite their large number of species and easy availability, little is known about their islet organs. Total pancreatectomy results in hyperglycemia, just as in mammals, but survival is so poor that a clear picture of the specific effects of the operation cannot be drawn (cf. Epple and Lewis, 1973; Epple *et al.,* 1980). The islets seem to be rather small and scattered in some, but not all, pancreas regions, and they are largest near the spleen. However, individual islet cells (especially D and F cells) are very widespread in *Pseudemys scripta* (Gapp and Polak, 1983; Agulleiro *et al.,* 1985), which makes a paracrine interaction between D and other islet cells unlikely (see Section IV,B,8). At least four cell types (A, B, D, and F cells) occur, but innervation may be absent (cf. Epple and Brinn, 1975; Epple *et al.,* 1980; Gapp and Polak, 1983; Buchan *et al.,* 1985).

The Crocodilia are the survivors of the Archosauria, a group which gave rise also to the birds. Pancreatectomy of *Alligator mississippiensis* resulted in a progressive hyperglycemia and ketonemia, which ultimately reached 589 and 134 mg% (Penhos *et al.,* 1967), respectively. These ketone levels may be the highest ever seen in any vertebrate (Penhos and Ramey, 1973), and the pancreatectomized alligator may thus be of considerable interest for further study. The islets of the alligator are scattered throughout the pancreas, without regional differences in the cytology, and contain A, B, D, and F cells (Buchan *et al.,* 1982). The F cells show an immunoreaction with anti-avian-PP but not with anti-mammalian-PP antibodies (Buchan *et al.,* 1982). This possibly can be explained by the phylogenetic proximity between both crocodilians and birds, which is reflected in the similarity of the molecular structures of chicken and alligator PP (Lance *et al.,* 1984). The crocodilian islets receive VIP-ergic innervation (Buchan *et al.,* 1982).

Though the Rhynchocephalia are closely related to the Squamata, the islets of *Sphenodon punctatus* seem to be as scattered as in the Chelonia, Crocodilia, and mammals (Gabe, 1970). A, B, and D cells are present, but studies with modern techniques are desirable. Nothing seems to be known about the islet physiology of this living fossil.

Among the large number of Squamata, some lizards and very few snakes have been used in islet research, though several peculiarities make this group noteworthy. In most species, there is a trend to form large islet accumulations in the juxtasplenic pancreas regions, and in some snakes and at least one monitor lizard (*Varanus exanthematicus*) there is even a large mass of islet tissue inside the spleen (Hellerström and Asplund, 1966; Buchan, 1984; Dupé-Godet and Adjovi, 1983) (see also Section III). In all species studied so far with specific methods, A, B, D, and F cells have been found (Rhoten and Smith, 1978; Rhoten and Hall, 1981), though the immunocytochemical demonstration of F cells may present difficulties (compare Buchan, 1984, and Rhoten, 1984). The degree of islet innervation of the Squamata varies with the species, and VIP has been identified in islet nerves. Some species may provide particularly good models for the study of "peripheral neurosecretion" (cf. Fujii *et al.*, 1980; Fujita *et al.*, 1982; Buchan, 1984; Buchan *et al.*, 1985). It seems that in almost all Squamata, islet cells other than B cells prevail, and the high blood-sugar level (which is similar to that of birds; see Section IV,B,6) has been related to the large number of A cells (cf. Epple and Lewis, 1973; Epple *et al.*, 1980). Exceptions to this rule seem to be the amphisbaenid lizards (Gabe, 1970; Rhoten, 1970) and the water snake, *Natrix piscator* (Rangneker and Padgaonkar, 1972). The latter species also seems to be exceptional in its response to pancreatectomy, which results in a progressive hyperglycemia. In the other Squamata, the operation was followed by a transient hypoglycemia and subsequent hyperglycemia (cf. Epple *et al.*, 1980). In some species, it is difficult to separate A and D cells by histological staining techniques (Gabe, 1970; Theret *et al.*, 1975). One wonders if this is related to the presence of additional, histologically unidentified islet cells, since El-Salhy and Grimelius (1981) found in the hibernating lizard, *Uromastix aegyptia*, a transient population of GIP and gastrin cells. This perplexing observation, if confirmed, should stimulate further investigations, since it may provide insights into both ontogenetic and oncogenic mechanisms of APUD-type endocrine cells. Another interesting observation has been reported in *Varanus exanthematicus*. Here, the D cells(?) of a large, intrasplenic islet appear to produce only a large form of somatostatin (possibly S-28), while the blood from the pancreas proper contains a smaller form (possibly S-14). Since the monitor reaches a considerable size, separate cannulations of a number of blood vessels (including the pancreas and splenic vein) are possible; this species may be a model for the *in vivo* study of the release and functions of different somatostatins (Dupé-Godet, 1984). On the other hand, it is obvious that the juxtasplenic and intrasplenic islet accumulations of many Squamata should be useful for hormone extractions and a variety of *in vitro* investigations. However, it seems that this has been done in a larger scale only by Rhoten (1978, 1984), who used the lizard *Anolis carolinensis*.

In summary, the islet organ of the reptiles varies greatly between the groups, and both islet structure and physiology of the birds resemble more those of the distantly related Squamata than those of the crocodilians. On the other hand, the molecular structures of crocodilian and avian PP closely resemble each other—as one would expect from the phylogenetic relationship of the two groups.

7. Aves (Birds)

Since the avian pancreas is rather compact and easily recognized by its characteristic location in a loop of the duodenum, it has been removed in numerous studies to observe the metabolic sequelae of pancreatectomy. However, many data have been contradictory, which may be due to the failure of several early studies to remove the inconspicuous splenic lobe (see Mialhe, 1958; Sirek, 1969; Langslow and Hales, 1971). The splenic lobe corresponds to the juxta- and intrasplenic pancreatic regions of the Squamata, as well as to the Brockmann bodies of the Actinopterygii, and it contains a large amount of islet tissue. Complete pancreatectomy in the duck and goose and removal of the major A-cell-containing regions of the chicken pancreas (third and splenic lobe) are followed by a fatal hypoglycemia (Mialhe, 1958; Sitbon, 1967; Mikami and Ono, 1962). However, Mialhe (1958) and Karmann and Mialhe (1976) have shown that in the duck and goose, surgically induced selective deficiency of B cells can lead to hyperglycemia, and Mihail et al. (1963) report hyperglycemia in totally pancreatectomized pigeons. The findings in the duck, goose, and chicken can be explained by the observation that, on an overall scale, the A cells far outnumber the B cells of the avian pancreas, whereas in the derivatives of the ventral anlagen, B cells prevail. Thus, removal of only the B cells will cause diabetes (Sitbon et al., 1980). However, the observations in the pigeon require further studies. Both A and B cells are largely, though not exclusively, restricted to two different types of islets which are unique to birds: "dark" A islets, which are large and contain, besides A cells, a considerable number of D cells; and "light" B islets, which contain mostly B cells and a varying number of D cells. PP cells are widely scattered over the exocrine pancreas and have a tendency to be more frequent in pancreas regions derived from the ventral anlagen (Iwanaga et al., 1983; Tomita et al., 1985). In the Phasianidae (chicken, quail, and pheasant family), the A islets are virtually restricted to the third and splenic lobes, which are largely derived from the dorsal anlage (Bargmann, 1939; Andrew, 1984). However, it must be noted that in many birds the third lobe is absent (Guha and Ghosh, 1978). The relative numbers of the islet cell types are reflected in the plasma titers of their hormones. Thus, compared with mammals, the plasma titers of glucagon, somatostatin, and PP are rather high, while the opposite is true for insulin. As in the Squamata, the high titer of plasma glucagon in correlated with that of the blood sugar (Langslow and Hales, 1971; Sitbon et al., 1980; Hazelwood, 1981, 1984). The innervation of the islet organ, if any, lacks synaptic specializations, and its extent is still controversial (Epple et al., 1980; Watanabe, 1983). Because the glucagon content parallels the high number of A cells, the avian pancreas became important in vehement debates in the early 1950s concerning the hormonal status of glucagon (Vuyl-steke and de Duve, 1953). In the early 1960s, it was shown in cytophysiological studies on the avian pancreas that the D cell must be the source of a third islet hormone (Epple, 1963); in 1969, Hellman and Lernmark observed *in vitro* that pigeon islets released a substance which inhibits insulin secretion (i.e., a somatostatin effect), but it took another 7 years until somatostatin became fully recognized as the hormone of the D cell (Dubois, 1975; Polak *et al.*, 1975). Perhaps one of the most remarkable events in recent islet research was the accidental discovery of the fourth islet hormone, PP, in extracts of the chicken pancreas (Kimmel *et al.*, 1968). The chicken pancreas should also make an excellent model for the study of the "halo" phenomenon, which can be induced in the pancreas of pullets by an unbalanced diet. Here, halos spread from the periphery of both dark and light islets continuously, until wide regions of the exocrine pancreas are filled with pseudoisocyanin-positive granules (Epple, 1968). Since these regions have certainly no vascular connections with the islets, it is obvious that this phenomenon must be mediated by a nonvascular route, either paracrine secretion or (less likely) transmission via gap junctions. It is remarkable that in more than 15 years of speculations on insulo–acinar interactions (see Section III) these basic observations have been ignored. The restriction of A and B cells largely to islets of their own should also make the avian pancreas an interesting object for studies on the interactions between the islet hormones. As pointed out in Section IV,B,6, the model of islet hormone interactions in the rat (see Section IV,B,8) cannot be applied to the turtle pancreas, and the same holds true for the birds. On the other hand, one wonders if the D cells of the A and B islets, respectively, secrete different somatostatins (or ratios of somatostatins) as shown for the different islets of *Varanus* (see Section IV,B,6). In this connection, it must be recalled that differential effects for S-14 and S-28 have been established for both mammals and birds (cf. Conlon, 1983; Marco *et al.*, 1983; Strosser *et al.*, 1984).

8. *Mammalia*

In contrast to the ample literature on the anatomy of the islet organ of many "placental" (eutherian) mammals, surprisingly little work has been done on the pancreatic islets of the marsupials (metatheria), and possibly none at all on those of egg-laying monotremes. Among the marsupials, the opossum (*Didelphys virginiana*) is the best-studied species. Its islets contain all four major cell types, plus a peculiar E cell of uncertain function. F cells seem to occur predominantly in the islets of the duodenal lobe and also as scattered elements in the splenic portion (Munger *et al.*, 1965; Larsson *et al.*, 1976). The islets of the kangaroos were found to contain only 8–15% B cells, while the percentage was much higher (53%) in the brush-tailed opossum (White and Harrop, 1975). The pancreas of all eutherian mammals seems to contain A, B, D, and F cells (cf. Ferner and Kern, 1969; Falkmer and Östberg, 1977; Munger, 1981). Usually the F cells are more common in the derivatives of the ventral anlagen (Baetens et al., 1979; Malaisse-Lagae et al., 1979; Fiocca et al., 1983; Orci, 1983). In the human, about 15% of the pancreatic PP cells strongly differ ultrastructurally from the F cells (Solcia et al., 1984). Perhaps, the cosecretions in the granules of these "D cells" are responsible for this phenomenon. In fetal ruminants, there is a peculiar population of argyrophil insulin cells (see also Section IV,A) which disappears after birth. These elements amount to about 90% of the islet cells of fetal sheep, and they may be identical with cells found in certain islet tumors in the human (cf. Falkmer et al., 1984b). There are considerable variations in the islet architecture; for example, the well-known islets of the rat contain a core of B cells, and an outer rim of A and D cells, while the islets of the horse (Fujita, 1973) have a core of A cells, and a rim of B and D cells. The elucidation of the functional significance, if any, of these arrangements is very difficult, since it requires a tedious analysis of the vascularization and in particular of the direction of blood flow (cf. Samols et al., 1983). According to Bonner-Weir and Orci (1982), blood first reaches the B cells of the rat and then carries insulin in high concentrations to A and D cells, while the B cells are exposed to low titers of systemic glucagon and insulin. The proposed functional principle of this system, outlined in Fig. 7, provides for a negative feedback mechanism (Kawai et al., 1982b; Unger, 1983; Maruyama et al., 1984). However, it is unknown to what extent this applies to mammals other than the Muridae and the dog. In the horse, the situation is apparently opposite of that in the rat, i.e., blood carries high titers of glucagon from the central regions of the islet to peripheral B and D cells (Fujita, 1973). An interesting question arises concerning the role of the presumptive vascular sphincter mechanisms at the periphery of the cat islets (Syed Ali, 1984). Do they, perhaps, intermittently redirect the blood flow?

Though there has been a growing interest in the role of the innervation of the mammalian islet organ (cf. R. E. Miller, 1981; Smith and Madson, 1981; Palmer and Porte, 1983; Smith and Davis, 1983) several fundamental aspects are yet to be clarified. These include (1) the relationship between the islet innervation and the enteric nervous system (Gershon et al., 1983), (2) the presence of hypothalamic factors with insulinotropic and glucagonotropic factors (Knip et al., 1983; Moltz and Fawcett, 1983), (3) the existence of neurosecretory fibers which release their secretions into the islet circulation (Fujii et al., 1980; Fujita et al., 1982), and (4) species variations in the type of neurotransmitter(s). So far, VIP and the carboxy-terminal peptide of cholecystokinin have been identified immunohistologically in islet nerves (Larsson, 1980; Rehfeld et al., 1980). Possibly, the spiny mouse (Acomys cahirinus) may be a useful model for pertinent work. This species has been studied mainly because of some diabetes-related metabolic aberrations (cf. Nesher et al., 1985). However, the absence of an islet innervation in Acomys

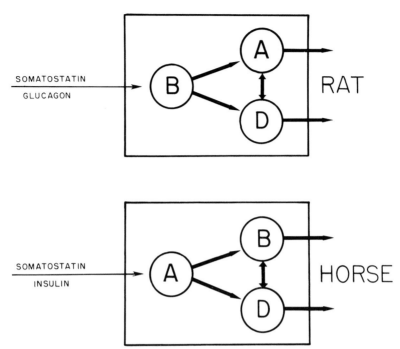

Fig. 7.   The hormonal concentrations arriving at the islet cells of rat and horse. Unger's (1983) proposed model of islet hormone interactions is based on the situation in the rat. In it, the B cell responds to low, systemic titers of somatostatin and glucagon and controls A- and D-cell activities with high, intrainsular, ''portal'' titers of insulin. Note that the situation in the horse islet requires a different interpretation. Also note that the D cells in both cases may be exposed to high titers of insulin, while high titers of somatostatin should reach the A cells of the rat and the B cells of the horse.

(Hahn von Dorsche *et al.,* 1976) may turn out to be of special investigative value.

In the preceding, we have provided some glimpses of the usefulness of certain mammals for islet research. We have omitted many aspects relating to the availability of special laboratory models for diabetes research, since their discussion was beyond the scope of this chapter. However, it must be emphasized that one of the most important features of many mammals for diabetes research is the unique susceptibility of their B cells to alloxan and streptozotocin (Dulin and Soret, 1977; Cooperstein and Watkins, 1981). This enables investigators to induce insulin-deficiency diabetes in forms which are rather closely related to the human. On the other hand, the literature appears to contain no report on cytotoxin-induced diabetes in nonmammalian vertebrates, in which complete, selective B-cell destruction was documented. As a rule, in reports of such work with nonmammals there was incomplete destruction of B cells and/or damage to other organs (kidney, liver). It follows that in mammals as well, induction of chemical diabetes requires full histological documentation of selective B-cell destruction whenever a new species or strain is used. Otherwise, general, toxic drug effects may be confused with diabetic hyperglycemia.

V. CONCLUDING REMARKS

It is difficult to overrate the importance of comparative studies for our understanding of the islet organ, and in retrospect, it is clear that progress in the field would have been much faster if findings in nonmammalian vertebrates would have received the attention they deserved. The following compilation should make the point.

Brockmann bodies, containing the islet organ of teleosts, were discovered in 1846 by both Brockmann and Oppel (cf. Epple, 1969), long before the concept of an endocrine pancreas was formulated (cf. Bargmann, 1939). A and D cells were first observed in teleosts (Masssari, 1898; Bowie, 1925). The pancreatic islet origin of insulin was established by the extraction of Brockmann bodies (Macleod, 1922), and studies on the avian pancreas were instrumental for the recognition of the hormonal status of glucagon (see Section IV,B,7; and Epple and Brinn, 1986). The functional autonomy of the D cell was clearly shown by comparative histophysiological and histochemical studies (Epple, 1963, 1965), years before the discovery of somatostatin. More recently, *in vitro* work with Brockmann bodies led to the identification of S-22, a new form of somatostatin (cf. Fletcher *et al.*, 1983). The wide distribution and high frequency of the F cell in the chicken pancreas certainly contributed to the accidental discovery of PP during the extraction of avian insulin (Kimmel *et al.*, 1968). Several unique types of islet cells in adult lampreys, holocephalans, and *Lepisosteus* (see Sections IV,A, IV,B,1 and IV,B,3,c) may foreshadow the isolation of new messenger peptides. The description of "amphiphil" islet cells in a selachian and in amphibians (cf. Epple, 1967) seems to have been the first hint of the presence of multiple secretions within individual types of islet cells—a phenomenon now almost taken for granted (see Table I). Large, unique, synaptic granules were first described in the islets of a snake (Trandaburu and Calugareanu, 1966) and later were also recognized in teleosts, leading to the suggestion that there is a third, "unconventional" type of islet innervation (cf. Epple and Brinn, 1975, 1986). Today, it is clear that in many—though not all—species, these (and even far less conspicuous islet nerve terminals) contain VIP (cf. Epple and Brinn, 1986). The knowledge of the structural peculiarities of the eel pancreas ultimately led to the discovery of the osmoregulatory role of the islet organ (see Section IV,B,3,e); now, it may not be too farfetched to raise the question whether islet hormones also play an osmoregulatory role in prenatal amniotes, including the human fetus, which live in an "aquatic" environment.

Studies on the digestive tract of *Branchiostoma* and of larval lampreys have been basic for our understanding of the evolution of the endocrine pancreas and its interrelations with gastrointestinal endocrines, the exocrine pancreas, and the liver (cf. Van Noorden, 1984) (see Sections II and III). The development of Pearse's (1968) APUD and Fujita's (1977) paraneuron concept was strongly influenced by comparative islet research; new perspectives upon the role of islet secretions in both intrainsular and insuloacinar interactions, as outlined in several of the previous sections, arose from studies in "lower" vertebrates. Last but not least, work on the larval lamprey has provided us with a basic understanding of the vascular links between islets, the exocrine pancreas, and the liver (see Sections II and III).

In summary, comparative pancreatic islet studies have yielded many discoveries of practical value or clinical relevance, and they have been of fundamental significance in the development of a conceptual framework for the understanding of both messenger peptides and the cells of their origin (cf. Federlin and Scholtholt, 1983; Steiner *et al.*, 1984; Epple and Brinn, 1987). One hopes that future studies will pay attention to the advantages offered by the islet organ of many nonmammalian vertebrates.

ACKNOWLEDGMENTS

The authors' studies referred to in this chapter have been supported by NIA grant 5R01-AG01148 and NSF grants GB 38146, PCM 76-01453, and PCM 8209263 to AE. We are indebted to Dr. J. W. Atz for his advice on taxonomic questions. The skillful typing of Mrs. S. Parsons is gratefully acknowledged.

REFERENCES

Agulleiro, A., Garcia Ayala, A., and Abad, M. A. (1985). An immunocytochemical and ultrastructural study of the endocrine pancreas of *Pseudemys scripta elegans* (Chelonia). *Gen. Comp. Endocrinol.* **60**, 95–103.

Ali-Rachedi, A., Varndell, I. M., Adrian, T. E., Gapp, D., Van Noorden, S., Bloom, S. R., and Polak, J. M. (1984). Peptide YY (PYY) immunoreactivity is co-stored with glucagon-related immunoreactants in endocrine cells of the gut and pancreas. *Histochemistry* **80**, 487–491.

Albert, S. G. (1982). Insulin biosynthesis in isolated channel catfish (*Ictalurus punctatus*) islets: The effect of glucose. *Gen. Comp. Endocrinol.* **48**, 167–173.

Alumets, J., Håkanson, R., and Sundler, F. (1978). Distribution, ontogeny and ultrastructure of pancreatic polypeptide (PP) cells in the pancreas and gut of the chicken. *Cell Tissue Res.* **194**, 377–386.

Andrew, A. (1982). The APUD concept: Where has it led us? *Br. Med. Bull.* **38**, 221–225.

Andrew, A. (1984). The development of the gastro-entero-pancreatic neuroendocrine system in birds. *In* "Evolution and Tumour Pathology of the Neuroendocrine System" (S. Falkmer, R. Håkanson, and F. Sundler, eds.), pp. 91–109. Elsevier, Amsterdam.

Ayer-LeLievre, C., and Fontaine-Perus, J. (1982). The neural crest: Its relation with APUD and paraneuron concepts. *Arch. Histol. Jpn.* **45**, 409–427.

Baetens, D., Malaisse-Lagae, F., Perrelet, A., and Orci, L. (1979). Endocrine pancreas: Three-dimensional reconstruction shows two types of islets of Langerhans. *Science* **206**, 1323–1325.

Bargmann, W. (1939). Die Langerhansschen Inseln des Pankreas. *In* "Handbuch der mikrokospischen Anatomie des Menschen" (W. von Möllendorf, ed.), Vol. VI. Part 2, pp. 197–288. Springer-Verlag, Berlin and New York.

Barrington, E. J. W. (1972). The pancreas and intestine. *In* "The Biology of Lampreys" (M. W. Hardisty and I. C. Potter, eds.), Vol. 2, pp. 135–169. Academic Press, London.

Baskin, D. G., Gorray, K. C., and Fujimoto, W. Y. (1984). Immunocytochemical identification of cells containing insulin, glucagon, somatostatin, and pancreatic polypeptide in the islets of Langerhans of the guinea pig pancreas with light and electron microscopy. *Anat. Rec.* **208**, 567–578.

Berelowitz, M., and Frohman, L. (1983). Immunoreactive neurotensin in the pancreas of genetically obese and diabetic mice. *Diabetes* **32**, 51–54.

Bjorenson, J. E. (1985). Effect of initial developmental stage on future morphology of transplanted embryonic chick pancreas. *Cell Tissue Res.* **240**, 367–373.

Bonner-Weir, S., and Orci, L. (1982). New perspectives on the microvasculature of the islets of Langerhans in the rat. *Diabetes* **31**, 883–889.

Bonner-Weir, S., and Weir, G. C. (1979). The organization of the endocrine pancreas: a hypothetical unifying view of the phylogenetic differences. *Gen. Comp. Endocrinol.* **38**, 28–37.

Bosman, F. T. (1984). Neuroendocrine cells in non-neuroendocrine tumours. *In* Evolution and Tumour Pathology of the Neuroendocrine System" (S. Falkmer, R. Håkanson and F. Sundler, eds.), pp. 545–580. Elsevier, Amsterdam.

Bosman, F. T., Blankenstein, M., Daxenbichler, G., Falkmer, S., Heitz, P. U., and Kracht, J. (1985). What's new in endocrine factors of tumor growth? *Pathol. Res. Pract.* **180**, 81–92.

Bowie, D. J. (1925). Cytological studies of the islets of Langerhans in a teleost, *Neomaenis griseus*. *Anat. Rec.* **29**, 57–73.

Brinn, J. E., Jr. (1973). The pancreatic islets of bony fishes. *Am. Zool.* **13**, 653–665.

Brinn, J. E., Jr., and Epple, A. (1976). New types of islet cells in a cyclostome, *Petromyzon marinus* L. *Cell Tissue Res.* **171**, 317–329.

Brinn, J. E., Jr. *et al.* (1986). In preparation.

Bruni, J. F., Watkins, W. B., and Yen, S. S. C. (1979). β-endorphin in the human pancreas. *J. Clin. Endocrinol. Metab.* **49**, 649–651.

Buchan, A. M. J. (1984). An immunocytochemical study of endocrine pancreas of snakes. *Cell Tissue Res.* **235**, 657–661.

Buchan, A. M. J., Lance, V., and Polak, J. M. (1982). The endocrine pancreas of *Alligator mississippiensis*. *Cell Tissue Res.* **224**, 117–128.

Buchan, A. M. J., Lance, V., and Polak, J. M. (1985). The reptilian endocrine pancreas: An immunocytochemical study. *In* "Current Trends in Comparative Endocrinology" (B. Lofts and W. N. Holmes, eds.), pp. 1065–1069. Hong Kong Univ. Press, Hong Kong.

Conlon, J. M. (1983). Biosynthesis and molecular forms of somatostatin. *In* "Diabetes 1982" (E. N. Mngola, ed.), pp. 243–249. Excerpta Medica, Amsterdam.

Cooperstein, S. J., and Watkins, D. (1981). Action of toxic drugs on islet cells. *In* "The Islets of Langerhans. Biochemistry, Physiology and Pathology" (S. J. Cooperstein and D. Watkins, eds.), pp. 387–425. Academic Press, New York.

Copeland, D. L., and DeRoos, R. (1971). Effect of mammalian insulin on plasma glucose in the mudpuppy (*Necturus maculosus*). *J. Exp. Zool.* **178**, 35–44.

Davis, M. S., and Shuttleworth, T. J. (1985). Peptidergic and adrenergic regulation of electrogenic ion transport in isolated gills of the flounder (*Platichthys flesus L.*). *J. Comp. Physiol. B* **155**, 471–478.

Dubois, M. P. (1975). Presence of immunoreactive somatostatin in discrete cells of the endocrine pancreas. *Proc. Natl. Acad. Sci. U.S.A.* **72**, 1340–1343.

Dulin, W. E., and Soret, M. G. (1977). Chemically and hormonally induced diabetes. *In* "The Diabetic Pancreas" (B. W. Volk and K. F. Wellman, eds.), pp. 425–465. Plenum, New York.

Dupé-Godet, M. (1984). Characterization and measurement of plasma somatostatin-like immunoreactivity in a Sahelian lizard (*Varanus exanthematicus*) during starvation. *Comp. Biochem. Physiol. A.* **78A**, 53–58.

Dupé-Godet, M., and Adjovi, Y. (1983). Somatostatin-like immunoreactivity in pancreatic extracts of a Sahelian lizard, (*Varanus exanthematicus*) during starvation. *Comp. Biochem. Physiol. A.* **75A**, 347–352.

El-Salhy, M. (1984). Immunocytochemical investigation of the gastro-entero-pancreatic (GEP) neurohormonal peptides in the pancreas and gastrointestinal tract of the dogfish *Squalus acanthus*. *Histochemistry* **80**, 193–205.

El-Salhy, M., and Grimelius, L. (1981). Immunohistochemical localization of gastrin C-terminus, gastric inhibitory peptide (GIP) and endorphin in the pancreas of lizards with special reference to the hibernation period. *Regul. Pep.* **2**, 97–111.

El-Salhy, M., Wilander, E., and Abu-Sinna, G. (1982). The endocrine pancreas of anuran amphibians: A histological and immunocytochemical study. *Biomed. Res.* **3**, 579–589.

Emdin, S. O. (1981). Myxine insulin. Amino-acid sequence, three dimensional structure, biosynthesis, release, physiological role, receptor binding affinity, and biological activity. M. D. Thesis, Umea University, Sweden.

Epple, A. (1963). Zur vergleichenden Zytologie des Inselorgans. *Zool. Anz., Suppl.* **27**, 461–470.

Epple, A. (1965). Weitere Untersuchungen über ein drittes Pankreashormon. *Zool. Anz., Suppl.* **29**, 459–470.

Epple, A. (1966a). Cytology of pancreatic islet tissue in the toad, *Bufo bufo* (L). *Gen. Comp. Endocrinol.* **7**, 191–196.

Epple, A. (1966b). Islet cytology in urodele amphibians. *Gen. Comp. Endocrinol.* **7**, 207–214.

Epple, A. (1967). Further observations on amphiphil cells in the pancreatic islets. *Gen. Comp. Endocrinol.* **9**, 137–142.

Epple, A. (1968). Körpergewicht and Pankreas von Küken bei einseitiger Diät. *Zool. Anz.* **181**, 190–195.

Epple, A. (1969). The endocrine pancreas. *In* "Fish Physiology" (W. Hoar and D. J. Randall, eds.), Vol. 2, pp. 271–319. Academic Press, New York.

Epple, A. (1985). Osmoregulatory effects of pancreatic islet hormones. *In* "Current Trends in Comparative Endocrinology" (B. Lofts and W. N. Holmes, eds.), pp. 915–916. Hong Kong Univ. Press, Hong Kong.

Epple, A., and Brinn, J. E. (1975). Islet histophysiology: Evolutionary correlations. *Gen. Comp. Endocrinol.* **27**, 320–349.

Epple, A., and Brinn, J. E. (1976). New perspectives in comparative islet research. *In* "The Evolution of Pancreatic Islets" (T. A. I. Grillo, L. Leibson, and A. Epple, eds.), pp. 83–95. Pergamon, Oxford.

Epple, A., and Brinn, J. E. (1980). Morphology of the islet organ. *In* "Hormones, Adaptation, and Evolution" (S. Ishii, T. Hirano, and M. Wada, eds.), pp. 213–220. Jpn. Sci. Soc. Press, Tokyo.

Epple, A., and Brinn, J. E. (1987). "The Comparative Physiology of the Pancreatic Islets" (D. S. Farner, coord. ed.), Zoophysiol. Ser. Springer-Verlag, Berlin and New York (in preparation).

Epple, A., and Lewis, T. L. (1973). Comparative histophysiology of the pancreatic islets. *Am. Zool.* **13**, 567–590.

Epple, A., and Lewis, T. L. (1977). Metabolic effects of pancreatectomy and hypophysectomy in the yellow American eel, *Anguilla rostrata* Le Sueur. *Gen. Comp. Endocrinol.* **32,** 294–315.

Epple, A., and Miller, S. B. (1981). Pancreatectomy in the eel: Osmoregulatory effects. *Gen. Comp. Endocrinol.* **45,** 453–457.

Epple, A., and Schneider, U. (1974). Annual cycle of intrapancreatic adipose tissue in the European brown trout (*Salmo trutta*). *Anat. Rec.* **178,** 350.

Epple, A., Brinn, J. E., and Young, J. B. (1980). Evolution of pancreatic islet functions. *In* "Evolution of Vertebrate Endocrine Systems" (P. K. T. Pang and A. Epple, eds.), pp. 269–321. Texas Tech Univ. Press, Lubbock.

Erlandsen, S. L., Hegre, O. D., Parsons, J. A., McEvoy, R. C., and Elde, R. P. (1976). Pancreatic islet cell hormones. Distribution of cell types in the islet and evidence for the presence of somatostatin and gastrin within the D cell. *J. Histochem. Cytochem.* **24,** 883–897.

Falkmer, S., and Östberg, Y. (1977). Comparative morphology of pancreatic islets in animals. *In* "The Diabetic Pancreas" (B. W. Volk and K. F. Wellman, eds.), pp. 15–58. Plenum, New York.

Falkmer, S., and van Noorden, S. (1983). Ontogeny and phylogeny of the glucagon cell. *In* "Glucagon I" (P. J. Lefebvre, ed.), pp. 81–119. Springer-Verlag, Berlin and New York.

Falkmer, S., Emdin, S. O., Östberg, Y., Mattisson, A., Johansson-Sjöbeck, M.-L., and Fänge, R. (1976). Tumor pathology of the hagfish, *Myxine glutinosa,* and the river lamprey, *Lampetra fluviatilis. Prog. Exp. Tumor Res.* **20,** 217–250.

Falkmer, S., Håkanson, R., and Sundler, F., eds. (1984a). "Evolution and Tumour Pathology of the Neuroendocrine System." Elsevier, Amsterdam.

Falkmer, S., El-Salhy, M., and Titlbach, M. (1984b). Evolution of the neuroendocrine system in vertebrates. *In* "Evolution and Tumour Pathology of the Neuroendocrine System" (S. Falkmer, R. Håkanson, and F. Sundler, eds.), pp. 59–87. Elsevier, Amsterdam.

Federlin, K. F., and Scholtholt, J., eds. (1983). "The Importance of Islets of Langerhans for Modern Endocrinology." Raven Press, New York.

Ferner, H. (1952). "Das Inselsystem des Pankreas." Thieme, Stuttgart.

Ferner, H. (1957). Die Dissemination der Hodenzwischenzellen und Langerhansschen Inseln als funktionelles Prinzip für die Samenkanälchen und das exokrine Pankreas. *Z. Mikrosk.-Anat. Forsch.* **63,** 35.

Ferner, H., and Kern, H. (1969). Die vergleichende Morphologie der Langerhansschen Inseln. Fische bis Vögel. *In* "Handbuch des Diabetes mellitus. Pathophysiologie und Klinik" (E. F. Pfeiffer, ed.), Vol. I, pp. 11–38. Lehmann, München.

Fiocca, R., Sessa, F., Tenti, P., *et al.* (1983). Pancreatic polypeptide (PP) cells in the PP-rich lobe of the human pancreas are identified ultrastructurally and immunocytochemically as F cells. *Histochemistry* **77,** 511–523.

Fletcher, D. J., Trent, D. F., and Weir, G. C. (1983). Catfish somatostatin is unique to piscine tissues. *Regul. Pept.* **5,** 181–187.

Forey, P. L. (1980). Latimeria: A paradoxical fish. *Proc. R. Soc. London, Ser. B* **208,** 369–384.

Frye, B. E. (1962). Extirpation and transplantation of the pancreatic rudiments of the salamanders, *Amblystoma punctatum* and *Eurycea bislineata. Anat. Rec.* **144,** 97–107.

Frye, B. E. (1964). Metamorphic changes in the blood sugar and the pancreatic islets of the frog. *Rana clamitans. J. Exp. Zool.* **155,** 215–224.

Fujii, S., Kobayashi, S., Fujita, T., and Yanaihara, N. (1980). VIP-immunoreactive nerves in the pancreas of the snake, *Elaphe quadrivirgata* (Boie): Another model for insular neurosecretion. *Biomed. Res.* **1,** 180–184.

Fujita, T. (1962). Über das Inselsystem des Pankreas von *Chimaera monstrosa. Z. Zellforsch. Mikrosk. Anat.* **57,** 487–494.

Fujita, T. (1973). Insulo-acinar portal system in the horse pancreas. *Arch. Histol. Jpn.* **35,** 161–171.

Fujita, T. (1977). Concept of paraneurons. *Arch. Histol. Jpn.* **40**, 1–12.

Fujita, T., Yui, R., Iwanaga, T., Nishiitsutsuji-Uwo, J., Endo, Y., and Yanaihara, N. (1981). Evolutionary aspects of "Brain-Gut Peptides": An immunohistochemical study. *Peptides* **2**, 123–131.

Fujita, T., Iwanga, T., Kusumoto, Y., and Yoshie, S. (1982). Paraneurons and neurosecretion. *In* "Neurosecretion: Molecules, Cells, Systems" (D. S. Farner and K. Lederis, eds.), pp. 2–13. Plenum, New York.

Gabe, M. (1968). Données histologiques sur le pancréas endocrine *d'Ichthyophis glutinosis* (L.), *(Batracien gymnophione)*. *Arch. Anat. Histol. Embryol.* **51**, 232–246.

Gabe, M. (1970). Données histologiques sur le pancréas endocrine des Lepido-sauriens (Reptiles). *Ergeb. Anat. Entwichlungsgesch.* **42**, 7–62.

Gapp, D. A., and Polak, J. M. (1983). The endocrine pancreas of the turtle, *Chrysemys picta*. *Am. Zool.* **23**, 910.

Gater, S., and Balls, M. (1977). Amphibian pancreas function in long-term organ culture. Control of insulin release. *Gen. Comp. Endocrinol.* **31**, 249–256.

Gershon, M. D., Payette, R. F., and Rothman, T. P. (1983). Development of the enteric nervous system. *Fed. Proc., Fed. Am. Soc. Exp. Biol.* **42**, 1620–1625.

Graf, R. (1981). Immunocytochemical detection of anti-ACTH reactivity in pancreatic islet cells of normal and steroid diabetic rats. *Histochemistry* **73**, 233–238.

Grube, D., Maier, V., Raptis, S., and Schlegel, W. (1978a). Immunoreactivity of the endocrine pancreas. Evidence for the presence of cholecystokinin-pancreozymin with the A-cell. *Histochemistry* **56**, 13–35.

Grube, D., Voigt, K. H., and Weber, E. (1978b). Pancreatic glucagon cells contain endorphin-like immunoreactivity. *Histochemistry* **59**, 75–79.

Guha, B., and Ghosh, A. (1978). A cytomorphological study of the endocrine pancreas of some Indian birds. *Gen. Comp. Endocrinol.* **34**, 38–44.

Hahn von Dorsche, H., Krause, R., Fehrmann, P., and Sulzmann, R. (1976). The verification of neurons in the pancreas of the spiny mice (*Acomys cahirinus*). *Endokrinologie* **67**, 115–118.

Hardisty, M. W. (1976). Cysts and neoplastic lesions in the endocrine pancreas of the lamprey. *J. Zool.* **178**, 305–317.

Hardisty, M. W. (1979). "Biology of the Cyclostomes." Chapman & Hall, London.

Hardisty, M. W. (1983). Lampreys and hagfishes: Analysis of cyclostome relationships. *In* "The Biology of Lampreys" (M. W. Hardisty and I. C. Potter, eds.), Vol. 4B, pp. 166–259. Academic Press, New York.

Hardisty, M. W., and Baker, B. I. (1983). Endocrinology of Lampreys. *In* "The Biology of Lampreys" (M. W. Hardisty and I. C. Potter, eds.), Vol. 4B, pp. 1–115. Academic Press, New York.

Hazelwood, R. L. (1981). Synthesis, storage, secretion, and significance of pancreatic polypeptide in vertebrates. *In* "The islets of Langerhans" (S. J. Cooperstein and D. Watkins, eds.), pp. 275–318. Academic Press, New York.

Hazelwood, R. L. (1984). Pancreatic hormones, insulin/glucagon molar ratios, and somatostatin as determinants of avian carbohydrate metabolism. *J. Exp. Zool.* **232**, 647–652.

Hellerström, C. L., and Asplund, K. (1966). The two types of A-cells in the pancreatic islets of snakes. *Z. Zellforsch. Mikrosk. Anat.* **70**, 68–80.

Hellman, B., and Lernmark, A. (1969). Inhibition of the *in vitro* secretion of insulin by an extract of pancreatic α_1 cells. *Endocrinology* **84**, 1484–1487.

Henderson, J. R., Daniel, P. M., and Fraser, P. A. (1981). The pancreas as a single organ: The influence of the endocrine upon the exocrine part of the gland. *Gut* **22**, 158–167.

Hilliard, R. W., Epple, A., and Potter, I. C. (1985). The morphology and histology of the endocrine pancreas of the Southern Hemisphere lamprey, *Geotria australis* Gray. *J. Morphol.* **184**, 253–261.

Houssay, B. A. (1959). Comparative physiology of the endocrine pancreas. *In* "Comparative Endocrinology" (A. Gorbman, ed.), pp. 639–667. Wiley, New York.

Iwanaga, T., Yui, R., and Fujita, T. (1983). The pancreatic islets of the chicken. *In* "Avian Endocrinology" (S. Mikami, K. Homma, and M. Wada, eds.), pp. 81–94. Springer-Verlag, Berlin and New York.

Jansson, L., and Hellerström, D. (1983). Stimulation by glucose of the blood flow to the pancreatic islets of the rat. *Diabetolologia* **25**, 45–50.

Karmann, H., and Mialhe, P. (1976). Glucose, insulin, and glucagon in the diabetic goose. *Horm. Metab. Res.* **8**, 419–426.

Kaung, H.-L. C. (1983). Changes of pancreatic beta cell population during larval development of *Rana pipiens*. *Gen. Comp. Endocrinol.* **49**, 50–56.

Kaung, H.-L. C., and Elde, R. (1980). Distribution and morphometric quantitation of pancreatic endocrine cell types in the frog, *Rana pipiens*. *Anat. Rec.* **196**, 173–181.

Kawai, K., Orci, L., and Unger, R. H. (1982a). High somatostatin uptake by the isolated perfused dog pancreas consistent with an "insulo-acinar" axis. *Endocrinology (Baltimore)* **110**, 660–662.

Kawai, K., Ipp, E., Orci, L., Perrelet, A., and Unger, R. H. (1982b). Circulating somatostatin acts on the islets of Langerhans by way of a somatostatin-poor compartment. *Science* **218**, 477–478.

Kawano, H., Daikohu, S., Saito, S. (1983). Location of thyrotropin releasing hormone-like immunoreactivity in rat pancreas. *Endocrinology (Baltimore)* **112**, 951–955.

Kimmel, J. R., Pollock, H. G., and Hazelwood, R. L. (1968). Isolation and characterization of chicken insulin. *Endocrinology (Baltimore)* **83**, 1323–1330.

Knip, M., Lautala, P., Åkerblom, H. K., Kouvalainen, K., and Martin, J. M. (1983). Partial purification of an insulin-releasing activity in human serum. *Life Sci.* **33**, 2311–2319.

Kobayashi, K., and Syed Ali, S. (1981). Cell types of the endocrine pancreas in the shark *Scyliorhinus stellaris* as revealed by correlative light and electron microscopy. *Cell Tissue Res.* **215**, 475–490.

Lagios, M. D. (1979). The coelacanth and the chondrichthyes as sister groups: A review of shared apomorph characters and a cladistic analysis and reinterpretation. *Occas. Pap. Calif. Acad. Sci.* **134**, 25–44.

Lance, V., Hamilton, J. W., Rouse, J. B., Kimmel, J. R., and Pollock, H. G. (1984). Isolation and characterization of reptilian insulin, glucagon, and pancreatic polypeptide: Complete amino acid sequence of alligator (*Alligator mississippiensis*) insulin and pancreatic polypeptide. *Gen. Comp. Endocrinol.* **55**, 112–124.

Lange, R. H. (1984). The vascular system of principal islets: Semithin-section studies in teleosts fixed by perfusion. *Gen. Comp. Endocrinol.* **54**, 270–276.

Langslow, D. R., and Hales, C. N. (1971). The role of the endocrine pancreas and catecholamines in the control of carbohydrate and lipid metabolism. *In* "Physiology and Biochemistry of the Domestic Fowl" (D. J. Bell and B. M. Freeman, eds.), Vol. I, pp. 521–548. Academic Press, London.

Larsson, L.-I. (1977). Corticotropin-like peptides in central nerves and in endocrine cells of gut and pancreas. *Lancet*, 1321–1323.

Larsson, L.-I. (1980). New aspects on the neural, paracrine and endocrine regulation of islet function. *Front. Horm. Res.* **7**, 14–29.

Larsson, L.-I., Sundler, F., and Håkanson, R. (1976). Pancreatic polypeptide—a postulated new hormone: Identification of its cellular storage site by light and electron microscopic immunocytochemistry. *Diabetologia* **12**, 211–226.

Leclercq-Meyer, V., Marchand, J., and Malaisse, W. J. (1985). Insulin and glucagon release from the ventral and dorsal parts of the perfused pancreas of the rat. *Horm. Res.* **21**, 19–32.

Lepori, N. G. (1959). Sulla presenza di tessuto pancreatico negli ovari di alcuni *Centracanthidae* (Pisces, Perciformes). *Studi Sassar., Sez. 1* **37**, 244–248.

Le Roith, D., and Roth, J. (1984). Evolutionary origins of messenger peptides: Materials in microbes that resemble vertebrate hormones. *In* "Evolution and Tumour Pathology of the

Neuroendocrine System" (S. Falkmer, R. Håkanson, and F. Sundler, eds.), pp. 147–164. Elsevier, Amsterdam.

Lewis, T. L., and Epple, A. (1984). Effects of fasting, pancreatectomy and hypophysectomy in the yellow eel, *Auguilla rostrata*. *Gen. Comp. Endocrinol.* **55**, 182–194.

Lewis, T. L., Parke, W. W., and Epple, A. (1977). Pancreatectomy in a teleost fish, *Anguilla rostrata*. *Lab. Anim. Sci.* **27**, 102–109.

Lifson, N., Kramlinger, K. G., Mayrand, R. R., and Lender, E. J. (1980). Blood flow to the rabbit pancreas with special reference to the islets of Langerhans. *Gastroenterology* **79**, 466–473.

Lifson, N., Lassa, C. V., and Dixit, P. K. (1985). Relation between blood flow and morphology in islet organ of rat pancreas. *Am. J. Physiol.* **249**, E43–E48.

Locket, N. A. (1980). Some advances in coelacanth biology. *Proc. R. Soc. London, Ser. B* **208**, 265–307.

Luiten, P. G. M., ter Horst, G. J., Koopmans, S. J., Rietberg, M., and Steffens, A. B. (1984). Preganglionic innervation of the pancreas islet cells in the rat. *J. Auton. Nerv. Syst.* **10**, 27–42.

Macey, D. J., Epple, A., Potter, J. C., and Hilliard, R. W. (1984). The effect of catecholamines on branchial and cardiac electrical recordings in adult lampreys (*Geotria australis*). *Comp. Biochem. Physiol. C.* **73C**, 295–300.

McCormick, N. A. (1925). The distribution and structure of the islands of Langerhans in certain fresh-water and marine fishes. *Trans. R. Can. Inst.* **15**, 57–81.

McCormick, N. A., and Macleod, J. J. R. (1925). The effect on the blood sugar of fish of various conditions, including removal of the principal islets (isletectomy). *Proc. R. Soc. London, Ser. B* **98**, 1–29.

Maclean, J. L. (1965). Studies on the intestinal diverticulum of the East Australian lamprey, *Mordacia mordax*. Bachelor of Science Thesis, University of New South Wales, Australia.

Macleod, J. J. R. (1922). The source of insulin. A study of the effect produced on blood sugar by extracts of the pancreas and the principal islets of fishes. *J. Metab. Res.* **2**, 149–172.

Malaisse-Lagae, F., Stefan, Y., Cox, J., Perrelet, A., and Orci, L. (1979). Identification of a lobe in the adult human pancreas rich in pancreatic polypeptide. *Diabetologia* **17**, 361–365.

Marco, J., Correas, I., Zylueta, M. A., Vincent, E., Coy, D. A. Comarri-Schally, A. M., Schally, A. V., Rodriguez-Arnao, M. D., and Gomez-Pan, A. (1983). Inhibitory effect of somatostatin-28 on pancreatic polypeptide, glucagon and insulin secretion in normal man. *Horm. Metab. Res.* **15**, 363–366.

Maruyama, H., Hisatomi, A., Orci, L., Grodsky, G. M., and Unger, R. H. (1984). Insulin within islets is a physiologic glucagon release inhibitor. *J. Clin. Invest.* **74**, 2296–2299.

Massari, G. (1898). Sul pancreas di pesci. *Atti Acad. Naz. Lincei, Cl. Sci. Fis., Mat. Nat., Rend.* [5] **7**, 134–137.

Mazzi, V. (1976). Note sul pancreas endocrino del polipteriforme *Calamoichthys calabaricus*. *Atti. Acad. Sci. Torino., Cl. Sci. Fis., Mat. Nat.* **110**, 387–392.

Meuris, S., Verloes, A., and Robyn, C. (1983). Immunocytochemical localization of prolactin-like immunoreactivity in rat pancreatic islets. *Endocrinology (Baltimore)* **112**, 2221–2223.

Mialhe, P. (1958). Glucagon, insuline et régulation endocrine chez le canard. *Acta Endocrinol. (Copenhagen)* **28**, Suppl. 36, 9–134.

Mihail, N., Ionescu, D., and Dusa, L. (1963). Morphologische Auswirkungen der Pankreatektomie bei der Taube. *Anat. Anz.* **112**, 97–100.

Mikami, S., and Ono, K. (1962). Glucagon deficiency induced by extirpation of alpha islets of the fowl pancreas. *Endocrinology (Baltimore)* **71**, 464–474.

Miller, M. R. (1961). Carbohydrate metabolism in amphibians and reptiles. *In* "Comparative Physiology of Carbohydrate Metabolism of Heterothermic Animals" (A. W. Martin, ed.), pp. 125–144. Univ. of Washington Press, Seattle.

Miller, R. E. (1981). Pancreatic neuroendocrinology: Peripheral neural mechanisms in the regulation of the islets of Langerhans. *Endocr. Rev.* **2,** 471–494.

Moltz, J. H., and Fawcett, C. P. (1983). Purification of a glucagon releasing factor from the rat hypothalamus. *Life Sci.* **32,** 1271–1278.

Munger, B. L. (1981). Morphological characterization of islet cell diversity. *In* "The Islets of Langerhans. Biochemistry, Physiology and Pathology" (S. J. Cooperstein and D. Watkins, eds.), pp. 3–49. Academic Press, New York.

Munger, B. L., Caramia, F., and Lacy, P. E. (1965). The ultrastructural basis for the identification of cell types in the pancreatic islets II. Rabbit, dog and opossum. *Z. Zellforsch. Mikrosk. Anat.* **67,** 776–798.

Nakamura, M., Yamada, K., and Yokote, J. (1971). Ultrastructural aspects of the pancreatic islets in carps of spontaneous Diabetes mellitus. *Experientia* **27,** 75–76.

Nesher, R., Abramovitch, E., and Cerasi, E. (1985). Correction of diabetic pattern of insulin release from islets of the spiny mouse (*Acomys cahirinus*) by glucose priming in vitro. *Diabetologia* **28,** 233–236.

Orci, L. (1983). Cellular relationships in the islet of Langerhans: A regulatory perspective. *In* "The Importance of Islets of Langerhans for Modern Endocrinology" (K. Federlin and J. Scholtholt, eds.), pp. 11–26. Raven Press, New York.

Palmer, J. P., and Porte, D. (1983). Neural control of glucagon secretion. *In* "Glucagon II" (P. J. Lefebvre, ed.), pp. 115–132. Springer-Verlag, Berlin and New York.

Patent, G. J. (1970). Comparison of some hormonal effects on carbohydrate metabolism in an elasmobranch (*Squalus acanthus*) and a holocephalan (*Hydrolagus colliei*). *Gen. Comp. Endocrinol.* **14,** 215–242.

Patent, G. J. (1973). The chondrichthyean endocrine pancreas: What are its functions? *Am. Zool.* **13,** 639–651.

Patent, G. J. (1976). The ultrastructure and innervation of the pancreatic islets of the holocephalan ratfish, *Hydrolagus colliei*. *In* "The Evolution of Pancreatic Islets" (T. A. I. Grillo, L. Leibson, and A. Epple, eds.), pp. 131–140. Pergamon, Oxford.

Patent, G. J., and Epple, A. (1967). On the occurrence of two types of argyrophil cells in the pancreatic islets of the holocephalan fish, *Hydrolagus colliei*. *Gen. Comp. Endocrinol.* **9,** 325–333.

Patent, G. J., Kechele, P. O., and Carrano, V. T. (1978). Nonconventional innervation of the pancreatic islets of the teleost fish, *Gillichthys mirabilis*. *Cell Tissue Res.* **191,** 305–315.

Patterson, C. (1982). Morphology and interrelationships of primitive actino-pterygian fishes. *Am. Zool.* **22,** 241–259.

Pearse, A. G. E. (1968). Common cytochemical and ultrastructural characteristics of cells producing polypeptide hormones (the APUD series) and their relevance to thyroid and ultimobranchial C cells and calcitonin. *Proc. R. Soc. London* **170,** 71–80.

Pearse, A. G. E. (1982). Islet cell precursors are neurons. *Nature (London)* **295,** 96–97.

Pearse, A. G. E. (1984). The diffuse neuroendocrine system: Historical review. *Front. Horm. Res.* **12,** 1–7.

Pearse, A. G. E., and Polak, J. M. (1971). Neural crest origin of the endocrine polypeptide (APUD) cells of the gastrointestinal tract and pancreas. *Gut* **12,** 783–788.

Penhos, J. C., and Ramey, E. (1973). Studies on the endocrine pancreas of amphibians and reptiles. *Am. Zool.* **13,** 667–698.

Penhos, J. C., Wee, C. H., Reitman, M., Sodero, E., White, R., and Levine, R. (1967). Effects of several hormones after total pancreatectomy in alligators. *Gen. Comp. Endocrinol.* **8,** 32–43.

Petrusz, P., Merchenthaler, I., Maderdrut, J. L., Vigh, S., and Schally, A. V. (1983). Corticotropin-releasing factor (CRF)-like immunoreactivity in the vertebrate endocrine pancreas. *Proc. Natl. Acad. Sci. U.S.A.* **80,** 1721–1725.

Polak, J. M., Pearse, A. G. E., Grimelius, L., Bloom, S. R., and Arimura, A. (1975). Growth hormone releasing inhibiting hormone (GHRH) in gastrointestinal and pancreatic D-cells. *Lancet* **1**, 1220–1222.

Potter, I. C., Hilliard, R. W., Bird, D. J., and Macey, D. J. (1983). Quantitative data on morphology and organ weight during the protracted spawning-run period of the Southern Hemisphere lamprey *Geotria australis J. Zool.* **200**, 1–20.

Pour, P. M., Sayed, S., and Sayed, G. (1982). Hyperplastic, preneoplastic and neoplastic lesions found in 83 human pancreases. *Am. J. Clin. Pathol.* **77**, 137–152.

Punetha, J. C., and Singh, T. (1983). Cytomorphology of the principal islets in two hillstream cyprinids. *Zool. Pol.* **30**, 149–154.

Rafn, S., and Wingstrand, K. G. (1981). Structure of intestine, pancreas, and spleen of the Australian lungfish, *Neoceratodus forsteri* (Krefft). *Zool. Scr.* **10**, 223–239.

Rangneker, P. V., and Padgaonkar, A. S. (1972). Effect of total hypophysectomy on the glycemic and plasma cholesterol levels in the snake, *Natrix piscator* (Russell). *Acta Zool.* **53**, 1–7.

Rawdon, B. B. (1984). Gastrointestinal hormones in birds: Morphological, chemical and developmental aspects. *J. Exp. Zool.* **232**, 659–670.

Rawdon, B. B., Kramer, B., and Andrew, A. (1984). The distribution of endocrine cell progenitors in the gut of chick embryos. *J. Embryol. Exp. Morphol.* **82**, 131–145.

Rehfeld, J. F., Larsson, L.-I., Yolterman, N. R. *et al.* (1980). Neural regulation of pancreatic hormone secretion by the C-terminal tetrapeptide of CCK. *Nature (London)* **184**, 33–38.

Rennie, J., and Fraser, T. (1907). The islets of Langerhans in relation to diabetes. *Biochem. J.* **2**, 7–12.

Rhoten, W. B. (1970). The cell population in pancreatic islets of amphisbaenidae-a light and electron microscopic study. *Anat. Rec.* **167**, 401–423.

Rhoten, W. B. (1978). Effects of glucose on glucagon secretion by the anolian splenic pancreas. *Proc. Soc. Exp. Biol. Med.* **157**, 180–183.

Rhoten, W. B. (1984). Immunocytochemical localization of four hormones in the pancreas of the gartersnake, *Thamnophis sirtalis. Anat. Rec.* **209**, 233–242.

Rhoten, W. B., and Hall, C. E. (1981). Four hormones in the pancreas of the lizard, *Anolis carolinensis. Anat. Rec.* **199**, 89–97.

Rhoten, W. B., and Smith, P. H. (1978). Localization of four polypeptide hormones in saurian pancreas. *Am. J. Anat.* **151**, 595–601.

Rosen, D. E. (1982). Teleostean interrelationships, morphological function and evolutionary inference. *Am. Zool.* **22**, 261–273.

Rosen, D. E., Forey, P. L., Gardiner, B. G., and Patterson, C. (1971). Lungfishes, tetrapods, paleontology, and plesiomorphy. *Bull. Am. Mus. Nat. Hist.* **167**, 159–276.

Rosenzweig, J. L., LeRoith, D., Lesniak, M. A., Yip, C. C. *et al.* (1985). Two distinct insulin-related molecules in the guinea pig: Immunological and biochemical characterization of insulin-like immunoreactivity from extra-pancreatic tissues of the guinea pig. *Diabetologia* **28**, 237–243.

Rutter, W. J. (1980). The development of the endocrine and exocrine pancreas. *In* "The Pancreas" (P. J. Fitzgerald and A. M. Morrison, eds.), pp. 30–38. Williams & Wilkins, Baltimore, Maryland.

Samols, E., Weir, G. C., and Bonner-Weir, S. (1983). Intraislet insulin-glucagon-somatostatin relationships. *In* "Glucagon" (P. J. Lefebvre, ed.), Vol. II, pp. 133–173. Springer-Verlag, Berlin and New York.

Sekine, Y., and Yui, R. (1981). Immunohistochemical study of the pancreatic endocrine cells of the ray, *Dasyatis akajei. Arch. Histol. Jpn.* **44**, 95–101.

Seppälä, M., Wahlström, T., and Leppaluto, J. (1979). Luteinizing hormone-releasing factor (LRF)-like immunoreactivity in rat pancreatic islet cells. *Life Sci.* **25**, 1489–1496.

Shen, L.-P., and Rutter, W. J. (1984). Sequence of the human somatostatin I gene. *Science* **224**, 168–171.

Sirek, A. (1969). Pancreatectomy and diabetes. In "Handbuch des Diabetes mellitus. Patho-physiologie und Klinik" (E. F. Pfeiffer, ed.), Vol. I, pp. 727–743. Lehmann, München.

Sitbon, G. (1967). la Pancréatectomie total chez l'oie. Diabetologia 3, 427–434.

Sitbon, G., Strosser, M.-T., Gross, R., Laurent, F., Foltzer, C., Karmann, H., Cohen, L., Jean-Marie, P., and Mialhe, P. (1980). Endocrine factors in intermediary metabolism, with special reference to pancreatic hormones. In "Avian Endocrinology" (A. Epple and M. H. Stetson, eds.), pp. 251–270. Academic Press, New York.

Siwe, S. A. (1926). Pankreasstudien. Morphol. Jahrb. 57, 84–307.

Smith, P. H., and Davis, B. J. (1983). Morphological and functional aspects of pancreatic islet innervation. J. Auton. Nerv. Syst. 9, 53–66.

Smith, P. H., and Madson, K. L. (1981). Interactions between autonomic nerves and endocrine cells of the gatroenteropancreatic system. Diabetologia 20, 314–324.

Smith, P. H., Merchant, F. W., Johnson, D. G., Fujimoto, W. Y., and Williams, R. H. (1977). Immunocytochemical localization of gastric inhibitory polypeptide-like material within A-cells of the endocrine pancreas. Am. J. Anat. 149, 585–590.

Solcia, E., Capella, C., Fiocca, R., Sessa, F. et al. (1984). Ultrastructural and immunohisto-chemical characterization of F-type and D_1-type PP cells and their distribution in normal, annular, chronically inflammed, heterotropic or tumor pancreas. Front. Horm. Res. 12, 31–40.

Sorokin, A. V., Petrenko, O. I., Kavsan, V. M., Kozlov, Y. I., Debabov, V. G., and Zlo-chevskij, M. L. (1982). Nucleotide sequence analysis of the cloned salmon preproinsulin cDNA. Gene 20, 367–376.

Stefan, Y., and Falkmer, S. (1980). Identification of four endocrine cell types in the pancreas of Cottus scorpius (Teleostei) by immunofluorescence and electron microscopy. Gen. Comp. Endocrinol. 42, 171–178.

Stefan, Y., Ravazzola, M., and Orci, L. (1981). Primitive islets contain two populations of cells with differing glucagon immunoreactivity. Diabetes 30, 192–195.

Steiner, D. F., Chan, S. J., Docherty, K., Emdin, S. O., Dodson, G. G., and Falkmer, S. (1984). Evolution of polypeptide hormones and their precursor processing mechanisms. In "Evolution and Tumour Pathology of the Neuroendocrine System" (S. Falkmer, R. Kåkanson and F. Sundler, eds.), pp. 203–223. Elsevier, Amsterdam.

Strahan, R., and Maclean, J. L. (1969). A pancreas-like organ in the larva of the lamprey, Mordacia mordax. Aust. J. Sci. 32, 54–55.

Strosser, M. T., Harvey, S., Foltzer, C., and Mialhe, P. (1984). Comparative effects of soma-tostatin-28 and somatostatin-14 on basal growth hormone release and pancreatic function in immature ducks (Anas platyrhynchos). Gen. Comp. Endocrinol. 56, 265–270.

Syed Ali, S. (1982). Vascular pattern in the pancreas of the cat. Cell Tissue Res. 223, 221–234.

Syed Ali, S. (1984). Angioarchitecture of the pancreas of the cat. Light-, scanning- and trans-mission electron microscopy. Cell Tissue Res. 235, 675–682.

Theret, C., Alliet, J., Comlan, G., and Gourdier, D. (1975). Endocrine pancreas of Varanus niloticus–ultrastructural and cytochemical study with electron microscopy and x-ray mi-crodiffraction. C. R. Hebd. Seances Acad. Sci. 280, 2125–2128.

Tomita, T., and Pollock, H. G. (1981). Four pancreatic endocrine cells in the bullfrog (Rana catesbeiana). Gen. Comp. Endocrinol. 45, 355–363.

Tomita, T., Doull, V., Pollock, H. G., and Kimmel, J. R. (1985). Regional distribution of pancreatic polypeptide and other hormones in chicken pancreas: reciprocal relationship between pancreatic polypeptide and glucagon. Gen. Comp. Endocrinol. 58, 303–310.

Trandaburu, T. (1976). Intrinsic innervation, monoamines and acetylcholinesterase activity in the pancreatic islets of some poikilothermic vertebrates and birds. In "The Evolution of Pancreatic Islets" (T. A. I. Grillo, L. Leibson, and A. Epple, eds.), pp. 121–130. Perga-mon, Oxford.

Trandaburu, T., and Calugareanu, L. (1966). Light and electron microscopic investigation of the endocrine pancreas of the grass-snake (*Natrix n. natrix* [L.]). *Z. Zellforsch. Mikrosk. Anat.* **97,** 212–225.

Trimble, E. R., Bruzzone, R., Gjinovci, A., and Renold, S. E. (1985). Activity of the insuloacinar axis in the isolated perfused rat pancreas. *Endocrinology (Baltimore)* **117,** 1246–1252.

Unger, R. H. (1983). Insulin-glucagon relationships in the defense against hypoglycemia. *Diabetes* **32,** 575–583.

Van Noorden, S. (1984). The neuroendocrine system in protostomian and deuterostomian invertebrates and lower vertebrates. *In* "Evolution and Tumour Pathology of the Neuroendocrine System" (S. Falkmer, R. Håkanson, and F. Sundler, eds.), pp. 7–38. Elsevier, Amsterdam.

Van Noorden, S., and Patent, G. J. (1978). Localization of pancreatic polypeptide (PP)-like immunoreactivity in the pancreatic islets of some teleost fishes. *Cell Tissue Res.* **188,** 521–525.

Van Noorden, S., and Patent, G. J. (1980). Vasoactive intestinal polypeptide-like immunoreactivity in nerves of the pancreatic islet of the teleost fish, *Gillichthys mirabilis. Cell Tissue Res.* **212,** 139–146.

Vuylsteke, C. A., and de Duve, C. (1953). Le contenu en glucagon du pancreas aviaire. *Arch. Int. Physiol. Biochim.* **41,** 273–274.

Watanabe, T. (1983). Ultrastructure of the chicken pancreatic islets with special reference to neural control. *In* "Avian Endocrinology" (S. Mikami, K. Komma, and M. Wada, eds.), pp. 95–104. Springer-Verlag, Berlin and New York.

Watkins, W. B., Bruni, J. F., and Yen, S. S. C. (1980). β-endorphin and somatostatin in the pancreatic C-cell colocalization by immmunocytochemistry. *J. Histochem. Cytochem.* **28,** 1170–1174.

Welsch, U. N., and Storch, V. N. (1972). The fine structure of the endocrine pancreatic cells of *Ichthyophis kohtaoensis* (Gymnophiona, Amphibia). *Arch. Histol. Jpn.* **34,** 73–85.

White, A. W., and Harrop, C. J. F. (1975). The islets of Langerhans of the pancreas of macropodid marsupials: a comparison with eutherian species. *Aust. J. Zool.* **23,** 309–319.

Yamada, H. (1951). The postbranchial gut of *Lampetra planeri,* Block, with special reference to its vascular system. *Okayama Igakkai Zasshi* **63,** 1–52.

Yokote, M. (1970). Sekoke disease, spontaneous diabetes in carp, *Cyprinus carpio,* found in fish farms. I. Pathological study. *Bull. Freshwater Fish. Res. Lab.* **20,** 39–72.

Youson, J. H. (1981a). The alimentary canal. *In* "The Biology of Lampreys" (M. W. Hardisty and I. C. Potter, eds.), Vol. 3, pp. 95–189. Academic Press, New York.

Youson, J. H. (1981b). The liver. *In* "The Biology of Lampreys" (M. W. Hardisty and I. C. Potter, eds.), Vol. 3, pp. 263–332. Academic Press, New York.

11

The Adrenal and
Interrenal Glands

I. CHESTER JONES*,† AND J. G. PHILLIPS†

Department of Zoology
The University of Sheffield
Sheffield S10 2TN, England

† The Wolfson Institute
Hull University
Hull HU6 7RK, England

I. TERMINOLOGY

The adrenal gland is an organ near but separate from the kidney on each side of the body. The term suprarenal gland and, rarely, suprarenal capsule, is also used. The adrenal gland is characteristic of the Amniota of the Gnathostomata (jawed vertebrates with fetal membranes), comprising the Mammalia, Aves, and Reptilia. The interrenal gland intermingles, to varying degrees, with the kidney, as seen in the Amamniota (without fetal membranes), which are represented by the Amphibia and Pisces (a convenient but loose generalization for fishes). The Agnatha (without jaws) possess living members of the order Cyclostomata (round mouths) and may have groups of cells representing the interrenal gland.

II. THE ADRENAL GLAND

A. Mammalia

1. Eutheria

The Eutheria comprise the dominant group of mammals, some 4500 genera, including humans and laboratory and agricultural animals, from which the greater part of our endocrinological knowledge derives.

The adrenal glands are paired structures lying in the region of the anterior pole of each kidney (Chester Jones, 1957; Idelman, 1978). There is variation,

319

though conceptually unimportant, in the closeness of the adrenal glands' renal juxtaposition and vascularization arising from the main abdominal vessels. Figure 1 gives the general eutherian plan. The glands receive branches from the main abdominal arteries which pass near them, and drainage is into the local, chiefly renal, veins. Numerous adrenal arteries run to the surface of the gland to lie in the loose external connective tissue from which their twigs, in an ill-defined plexus, vascularize the gland. The arteries arise from the abdominal aorta and the phrenic and renal arteries. For the dog, Flint (1900) describes the anterior adrenal dorsal and ventral surfaces receiving three to five branches of the phrenic artery and the posterior two to four branches of the abdominal aorta. In addition, the posterior dorsal surface has two to six branches of the lumbar artery and, more caudally, four to six branches from the renal artery. In humans, the vein emerges from the hilus of each gland; the right gland opens into the inferior vena cava, the left into the left renal vein. One drainage vein is characteristic of the eutherian adrenal; the dog is atypical in having up to four adrenal veins reaching the renal and phrenic vessels (Flint, 1900).

The adrenal gland is made up of two separate secretory tissues; their interplay is ill-defined (Chester Jones, 1957; Brudieux *et al.,* 1966). The center of the gland, the medulla, comprises chromaffin cells grouped together [chromaffin cells are so called because their coloration by potassium

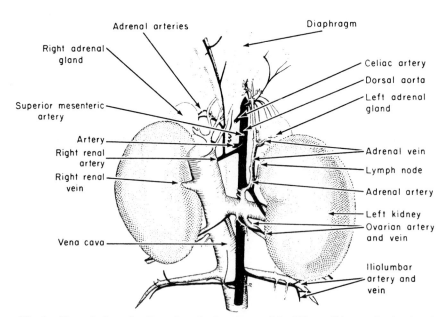

Fig. 1. Ventral view of a dissection of a female rat of the Wistar albino strain showing the vascular system associated with the adrenal glands after latex injection. [Prepared and drawn by I. Carthy and J. G. Phillips.]

bichromate, the "chromaffin reaction" of Henle (1865)] (Fig. 2a,b). They secrete epinephrine and norepinephrine, whereas the surrounding adrenal cortex produces steroid hormones (Idelman, 1978). Initially, chromaffin cells arise from neighboring paraganglion cells of the neural crest complex and then migrate to lie adjacent to the adrenocortical cell groups. These latter are mesodermal in origin (Chester Jones, 1976).

In the fully formed gland, the cortex is divided, on histological criteria, into three zones. The terms used have been retained from Arnold (1866): zona glomerulosa, zona fasciculata, and zona reticularis (Fig. 2b). Nomenclature for subdivisions is also found (see Chester Jones, 1957).

The cells of the zona glomerulosa consist of ovoid groups or "balls" immediately inside the outer connective tissue capsule. This region may be particularly obvious, as in sheep, or difficult to discern, as in small rodents (e.g., mouse, gerbil). The nuclei vary in shape from a sausage to an oval appearance, with one or two nucleoli. The cytoplasm contains a varying

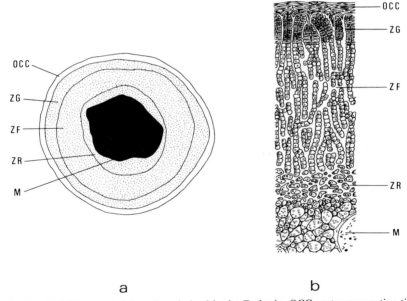

a   b

Fig. 2. (a) Midsection of the adrenal gland in the Eutheria. OCC, outer connective tissue capsule; ZG, zona glomerulosa; ZF, zona fasciculata; ZR, zona reticularis; M, centrally coalesced chromaffin tissue, medulla. These are the main divisions. Subdivisions are often named, e.g., zona intermedia between the ZG and ZF. The medulla is often surrounded by a connective tissue capsule of varying thickness. (b) A portion of (a), showing OCC, the outer connective tissue layer which encapsulates the adrenal gland. The ZG is shown as the looped ends of the cords of the ZF, obviously not so clearly seen in all species and dependent on the angle of sectioning. There is usually more transition from the inner ZF to the ZR, which may be in some phases a large, active (on histological bases) zone. [Drawings prepared by P. Groves.]

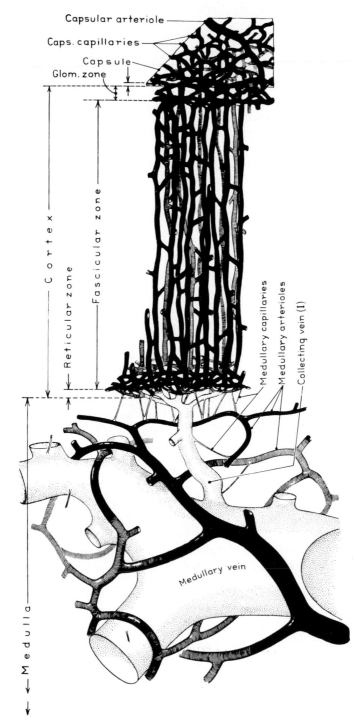

Fig. 3. Vasculature displayed in part of the cortex and adjacent medulla of a normal adult mouse. × 200. [From Gersh and Grollman, 1941.]

number of lipid droplets (liposomes), depending on the species and its physiological status; the liposomes lie between the nucleus and that side of the cells which abuts on its local capillary (Idelman, 1978).

The zona fasciculata appears histologically in radial lines when the sections are cut in appropriate orientations. The cells of the zone are polyhedral, with round nuclei and one or two nucleoli. They are packed with cytoplasmic liposomes in some species (e.g., the rat, as bred for laboratory use, which provides so great a basis for the study of endocrinology) and not in others (e.g., the hamster; Chester Jones, 1957).

The zona reticularis presents a variety of aspects. Generally, it comprises anastomosing networks of reticular cell cords surrounding the large blood sinuses. The cells are smaller than those of the rest of the cortex. At different ages and in a range of species, the reticular cells appear degenerative. Sometimes a zona reticularis is not apparent, and at other times a histologically "healthy" zone is to be seen (Long, 1975).

The vasculature of the adrenal cortex is all-important in the determination of structure and activity (Fig. 3) (Gersh and Grollman, 1941). In general, in the Eutheria, arterioles (1) form a network to embrace the balls of cells of the zona glomerulosa, (2) proceed radially concomitant with the columns of the zona fasciculata and enter the zona reticularis to splay out into sinuses, (3) pass through the adrenal cortex to vascularize the medulla in addition to the blood seeping in from reticularis sinuses, and (4) occasionally loop back to the outer connective tissue capsule (e.g., in humans) (Idelman, 1978). Blood drains to the major medullary vein which carries away formed steroid hormones, their metabolites, and the catecholamines (epinephrine, norepinephrine) from the chromaffin tissue (Idelman, 1978). The possibility of interplay between the products of the different areas can readily be envisaged but not precisely delineated. See Section III for a general discussion of morphological interrelationships. The ultrastructure of only a few species has been studied and much work has been concentrated on the adrenal of the rat (Figs. 4–8).

The zona glomerulosa cells (Fig. 4) are columnar with a spherical nucleus and a well-developed nucleolus. A basement lamina (BL) coats the perivascular surface. There is extensive, smooth-surfaced endoplasmic reticulum (SER) as a network of branching and anastomosing tubules. There are a few short segments of rough-surfaced endoplasmic reticulum and some free cytoplasmic ribosomes occurring in clusters, spirals, or linear chains. Mitochondria (M) are round, oval, or elongated, with cristae in the form of lamellar infoldings of the inner mitochondrial membrane. Lipid droplets (LD) are spherical or irregular in shape. The Golgi apparatus is well developed and adjacent to the nucleus. Other structures, including lysomes, multivesicular bodies, and glycogen grannules are also seen.

The zona fasciculata (Fig. 5) cells are large, polyhedral cells with a central nucleus which contains peripherally condensed chromatin, a nucleolus, and

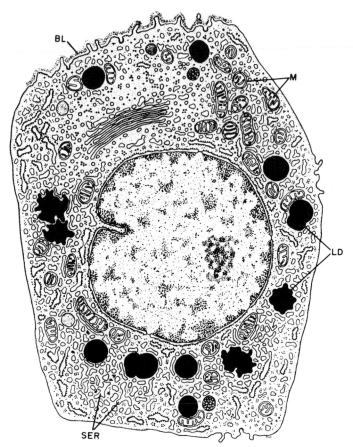

Fig. 4. Diagram of the ultrastructure of a cell of the zona glomerulosa. BL, basement lamina; SER, smooth-surfaced endoplasmic reticulum; M, mitochondria; LD, lipid droplets. Further description is found in the text. [From Lentz, 1971, based on the rat.]

an internal fibrous lamina (FL). Branching and anastomosing tubules of smooth-surfaced endoplasmic reticulum (SER) are extensive. The rough-surfaced endoplasmic reticulum (ER) is more abundant in this zone than in the zona glomerulosa, consisting of lamellar stacks of ribosome-studded cisternae. The mitochondria are large and generally spherical. The cristae occur as vesicular inpocketings of the inner mitochondrial membrane and as vesicles apparently free in the mitochondrial matrix. The vesicular cristae of the zona fasciculata cells are in contrast to the lamellar cristae of the zona glomerulosa. Lipid droplets are frequent and large. The Golgi apparatus is well developed, and there are lysosomes (Ly) and lipofuscin pigment granules (LPG).

The cells of the zona reticularis (Fig. 6, 7) have much in common with

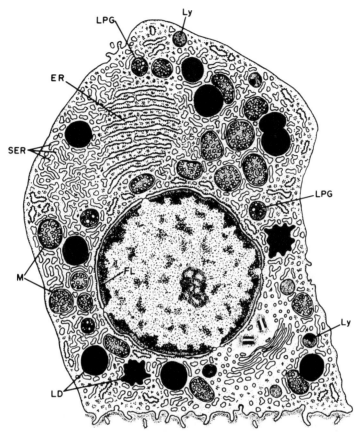

Fig. 5. Diagram of the ultrastructure of a cell of the zona fasciculata. FL, nuclear internal fibrous lamina; SER, smooth-surfaced endoplasmic reticulum; ER, rough-surfaced endoplasmic reticulum; M, mitochondria; LD, lipid droplets; Ly, lysosomes; LPG, lipofuchsin pigment granules. Further description is found in the text. [From Lentz, 1971, based on the rat.]

those of the zona fasciculata, including tubules of smooth-surfaced endoplasmic reticulum, stacks or whorls of rough-surfaced lamellae, cytoplasmic ribosomes, Golgi apparatus, lipid droplets, and lysosomes. The differences lie in the structure of the mitochondria and in the number and size of lipofuscin pigment bodies. The mitochondria are more elongated than in those of other zones. Two types of cristae are seen: vesicular cristae and lamellar invaginations of the inner mitochondrial membrane. The lipofuscin pigment granules are present in large numbers and are often quite big. Their shape is irregular, comprising a granular matrix within which are embedded droplets of varying size, probably lipid with a dense or translucent rim. Smaller vesicles and vacuoles containing material similar to the granular component of the lipofuscin body occur in the cytoplasm. The pigment in the granules

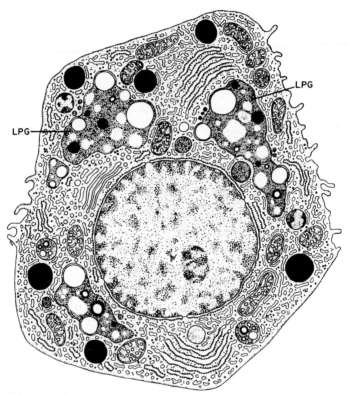

Fig. 6. Diagram of the ultrastructure of one type of cell of the zona reticularis. LPG, lipofuchsin pigment granules, vary in number and frequently are very numerous. In some conditions a reticularis cell has close resemblance to one of the zona fasciculata. Further description is found in the text. [From Lentz, 1971, based on the rat.]

gives the golden brown color of these cells of the zona reticularis, as seen with the light microscope. It is interesting to see catecholamine granules in cells of the zona reticularis (Fig. 7). These granules may have an important, though presently unknown, function.

The medulla (chromaffin cells) consists of pheochromocytes arranged in nests, alveoli, or cords, presenting a meshlike anastomosing network, between which are rich, vascular plexuses of capillaries and sinusoids with walls containing elastic tissue. The cells are ovoid or polygonal; the nucleus is eccentric and may be spherical, ovoid, or elongated in shape. There are one or more prominent nucleoli (Figs. 7, 8). Sympathetic ganglion cells lie singly or in groups, and the cells are innervated by preganglionic fibers which aid the control of secretion. There are two types of chromaffin cells, though they are very similar. Cells which contain norepinephrine have opaque granules after glutaraldehyde and osmium fixation, while those with epinephrine are only moderately dense (Fig. 8). They are membrane-bound

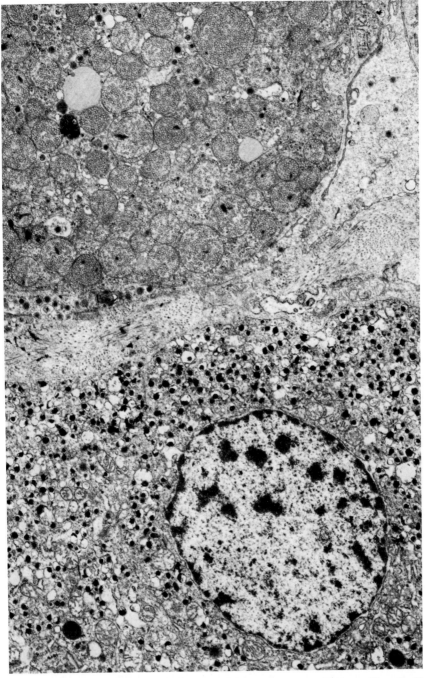

Fig. 7. Microphotograph of ultrastructure in the rat. Bottom: part of a chromaffin cell which demonstrates the rounded nucleus and the numerous opaque granules containing norepinephrine. Top: part of zona reticularis cells. Note that chromaffin cell granules are scattered within these cells. × 8000. [Courtesy of Patricia M. Ingleton.]

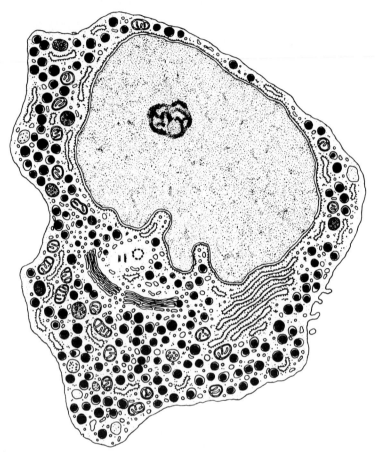

Fig. 8. Diagram of the ultrastructure of a norepinephrine-storing chromaffin cell. The granules, after glutaraldehyde and osmium fixation, are opaque. Further description is found in the text. Chromaffin cells containing epinephrine are similar in general structure with the granules that are only moderately dense. [From Lentz, 1971, based on the rat.]

at 1000–3000 Å in diameter. The characteristics of chromaffin cells are similar throughout the Vertebrata.

Another layer of cells, the X zone, lies in some species between the inner zona fasciculata and the medulla and is exemplified by the mouse adrenal cortex, particularly the laboratory strain Swiss albino (Chester Jones, 1957; Deacon *et al.*, 1986). The zone consists of cells with acidophilic cytoplasm and prominent basophilic nuclei. In the male, the X zone collapses by the direct action of androgens produced at maturity, giving pycnotic nuclei and shrunken cell cytoplasm (Chester Jones, 1949a,b). A similar picture is presented during first pregnancy, induced, perhaps, by ovarian androgens. In the virgin female, the X zone gradually declines over many months by "fatty degeneration,'' that is, accumulation of lipid droplets. The gonadotropin, luteinizing hormone (LH), is tropic for the X zone and not adrenocorticotropin (ACTH), which controls the zona fasciculata (Chester Jones, 1949b; Deacon *et al.*, 1986). Castration allows a persistent X zone, and when done after puberty, a secondary X zone appears under the influence of LH not then opposed by androgens (Chester Jones, 1949a). No function has been assigned to the X zone. The pattern shown by this strain of mice is not a universal one. Changes in genetic makeup produce variations so that, for example, female mice homozygous for the hypothyroid (*hyt*) mutation have poor development of the X zone (Shire and Beamer, 1984). Furthermore, certain wild rodents, especially voles and shrews, which have an X zone histologically similar, display varied reactions. In some species, the X zone does not involute at male puberty, and in the red squirrel (*Sciurus vulgaris*) it is a wide zone, markedly changing concomitant with breeding cycles (Delost and Guérin, 1962). The X zone might be expected to produce steroids, but there is no evidence for this. The mitochondria are not similar to those of known steroid-producing cells and, indeed, form peculiar complexes.

Another transient zone is seen in the adrenal gland of the human embryo. It is referred to as the fetal or transient cortex and occupies some 90% of the developing gland, the outermost part being the "permanent cortex," that is, a zone which will give rise to the adrenal of the postnatal and adult humans. The fetal cortex disappears around the time of birth. It has some relationship to the pituitary (perhaps ACTH?), as it does not fully develop in the anencephalic. No function has been proven to be associated with the zone, and it certainly does not appear to have steroidogenic properties (Jost, 1975).

2. *Metatheria (Marsupials)*

In general, the adrenal gland of the Metatheria (marsupials) has centralized chromaffin tissue, like a medulla with a rim of adrenocortex proportionally smaller than that seen in the Eutheria (Idelman, 1978). Nevertheless, zonae glomerulosa, fasciculata, and reticularis can be seen, though they are not clearly distinctive. In the American opossum (*Didelphys virginia*), cells of the zona glomerulosa have numerous short and rod-shaped mitochondria (Idelman, 1978). Their cristae are in the shape of lamellae. The SER makes up a richly developed network of long tubules with a cross-diameter of about 500 Å. The rough endoplasmic reticulum occurs as tubules scattered in the cytoplasm. Liposomes are rare. In the zona fasciculata, the mitochondria have tubules rather than cristae. Smooth endoplasmic reticulum fills up the cell with an anastomosing system of tubules with a diameter between 500 and 1000 Å. There are also some liposomes.

In one Australian marsupial (*Trichosurus vulpecula,* the brush-tailed possum) there is, in the female, a large additional adrenocortical zone. It lies between the definitive cortex and the medulla and may come to occupy about 45% of the whole gland. The volume of the zone may change in relation to reproductive status and may produce androgens, though neither of these properties are certain (Call and Janssens, 1984). It is, therefore, another transient zone, seen in some Mammalia, without ascribed function.

3. Prototheria (Monotremata)

The females of the Prototheria lay large, yolky, shelled eggs. The family Tachyglossidae (echidna or spiny anteater) has two genera and five species occurring in Australia and New Guinea. Little is known about the adrenal gland. It is an organ separate from the kidney on each side. The cortex appears homogeneous, made up of closely packed, irregularly placed cords, often separated by blood sinuses. The cells in the center of the gland are more closely packed. The chromaffin tissue is confined to the lower bulbous pole of the gland and is separated from the cortex by a marked development of connective tissue. The family Ornithorhynchidae—duckbilled platypus— occurs in eastern Australia and has one genus and one species. The separate adrenal glands of the platypus are encapsulated with connective tissue. The chromaffin tissue occupies the venous pole of the gland. The cortical tissue seems to be composed of three groups of cells in a rather complicated way. As nothing of functional significance is known of these cell types, speculation is not of value (Chester Jones, 1957). Griffiths (1978) brings together observations on the biology of monotremes and gives a discussion on the question of their relationships in the Vertebrata. In this regard, the similarity of the structure of the adrenal of echidna to that of *Sphenodon* is interesting.

B. Aves (Birds)

The paired adrenal glands of birds are found, each by a kidney, as separate encapsulated organs, composed of adrenocortical and chromaffin tissues. These are intermingled, to varying degrees, depending on the species. There is no coalescence of chromaffin tissue to form a medulla (as previously noted to be characteristic of the Eutheria and Metatheria). The glands lie, often wholly or partly covered by the gonads, at the anterior end of the kidneys (metanephric in embryological origin) and just behind the lungs. The location and vascularization of the adrenal of the gull (*Larus argentatus*) is shown in Fig. 9, as characteristic of those genera of birds so far examined (Holmes and Phillips, 1976).

The basic unit of avian adrenocortical tissue is a cord composed of a double row of parenchymal cells, with their long axes in the transverse plane of the strand. The cords radiate from the center of the gland and branch and anastomose frequently. At the periphery, they loop against the inner surface of the connective tissue capsule (Fig. 10a,b). The path of any one individual cord is tortuous and therefore difficult to follow throughout its length without three-dimensional pictorial display. The nucleus of each parenchymal cell is frequently situated towards the basal lamina on the outer margin of the

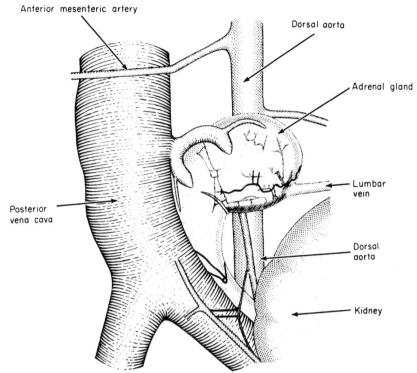

Fig. 9. Ventral view of the left adrenal gland (*in situ*) of the gull (*Larus argentatus*). [Preparation and drawing by I. Carthy and J. G. Phillips.]

cord. The cells at the periphery of the gland may be large with a considerable amount of cytoplasm. The nuclei are round or slightly oval, the chromatin network is thin and diffuse, and there are usually two nucleoli. Towards the middle of the gland, the cells are more elongated and smaller. Adrenal and interrenal lipid is characteristic of many species of all vertebrate groups. Thus in birds, also, we find sudanophilic and osmophilic droplets in the cells, though clearly the amounts in each vary according to genus and physiological status. In ultrastructural studies, adrenal cells are characterized by the presence of numerous liposomes, many mitochondria, and a moderate amount of SER. Frequent interdigitations occur between the plasma membranes of adjacent cells. All cells contain many mitochondria with mainly tubular cristae. In the matrix of a mitochondrion, paracrystalline aggregates of parallel tubules, about 120 Å in diameter, are often seen. Cisternae of SER and lesser amounts of rough endoplasmic reticulum are distributed throughout the cells. Close associations between cisternae and endoplasmic reticulum, mitochondrial surfaces, and lipid droplets are to be observed. An extensive Golgi complex, with three to six cisternae in each body, is situated close

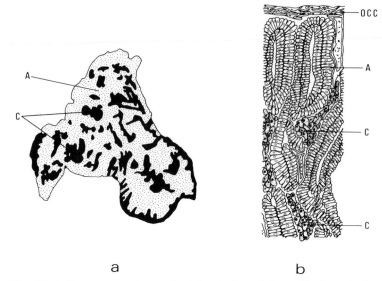

a   b

Fig. 10. (a) A diagram to emphasize the intermingling of chromaffin tissue in the typical avian adrenal gland. C, chromaffin islets; A, adrenocortical cells. In some bird species the areas of chromaffin tissue are large and numerous, in others slight. (b) A part of (a). OCC, outer connective tissue layer, which envelopes the gland; A, the adrenocortical cords with intermingled chromaffin islets (C). [Drawing prepared by P. Groves.]

to, and on the apical side of, the nucleus. Some smooth and some coated vesicles, about 1000 Å in diameter, are found in the apical region of the cell. Microtubules, 250 Å in diameter, are found in the peripheral regions of the cell. Close to the nucleus are microfilaments, 100 Å in diameter (Holmes and Phillips, 1976).

Groups of chromaffin cells lie between the cords of parenchymal cells and are also ensheathed within the connective tissue reticulum. In some species, the chromaffin tissue lies towards the outer portion of the gland and in others it is scattered as islets. A chromaffin cell has numerous secretory granules and sparse lipid droplets, and mitochondria are not abundant (as for Fig. 7, 8).

The possibility of zonation of the avian adrenal has intrigued workers, with aldosterone secreted from peripheral cells and corticosterone from cells more centrally situated. Following the theory of the basic cord (the following), it is not surprising to find that the outer subcapsular cells, looping in a manner typical of the zona glomerulosa, are designated as predominantly aldosterone secretors. Those cells reaching towards the central part of the gland form corticosterone (Pearce *et al.*, 1978).

C. Reptilia

Reptiles are the third tetrapod group with an amnion (hence, the Amniota). The group characteristically has the "new" kidney, the metanephros in the evolutionary sequence. As a consequence, the mesodermal cells which form the adrenal gland come to lie, as in birds and mammals, near the kidney. The adrenal gland is a discrete encapsulated organ and, basically, it does not differ substantially from the varying degrees of intermingling of adrenocortical and chromaffin tissues observed in birds. In the Squamata, the adrenal glands are generally elongate yellowish-colored bodies, enclosed within the folds of the mesorchium or mesovarium, in close contiguity with the gonads. In lizards and snakes, the right adrenal is caphalad of the left (Fig. 11) (Lofts, 1978). Crocodilia are similar, but in the Chelonia the glands are located on the ventral surface of the kidneys, though they are still sepa-

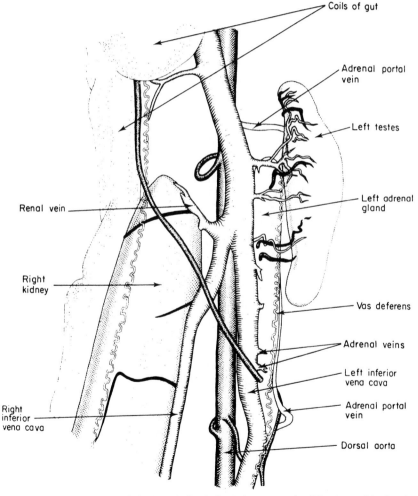

Fig. 11. Dissection of the left adrenal gland of a male grass snake (*Natrix natrix*), showing a ventral view of the vascular system after latex injection, with the gut displaced laterally to the right side. × 3. [Preparation and drawing by I. Carthy and J. G. Phillips.]

rate organs. Vascularization of the reptilian adrenal gland falls into the general pattern seen in the Amniota. The arterial supply is provided by a series of small arteries branching off the dorsal aorta (Fig. 11). Anteriorly an artery branches off from the dorsal aorta and joins the adrenal gland dorsolaterally approximately a third of the way along its length. It then divides into two main branches, one anterior, the other posterior, to the gland. Arterial vessels penetrate the enclosing adrenal capsule and provide branches to supply the gonads, situated nearby. The venous drainage of each gland is by numerous short efferent veins, given off along the length of the gland and entering directly into the vena cava on each side. In addition, there are portal vessels, that is, two veins, one anterior and one posterior, running from the body wall to each adrenal. The morphological schema of arterial supply and venous drainage which applies to mammals, birds, and reptiles has a few exceptions (such as *Alligator mississippiensis* and *Lacerta veridis*) in which the adrenal venous return is by a number of small ports (i.e., outlets) which interconnect the gland intimately to the venae cavae.

Structural examinations show that the reptilian adrenal is completely enclosed in a thin connective tissue capsule, consisting of collagenous, reticular, and elastic fibers surrounding the inner mass of cells (Lofts, 1978). These cells are organized into a loosely connected network of cortical cords which are generally orientated longitudinally along the length of the gland. The avian pattern is seen once more with pronounced islets of chromaffin tissue intermixed with cortical cords. As would be expected, the degree of intermingling varies from species to species. No functional significance has been assigned to avian and reptilian variations in this regard. Once it is accepted that chromaffin cells migrate between the adrenocortical cells as they form the adrenal gland, and this for unknown reasons, then varying degrees of association must be anticipated solely on morphological considerations (Fig. 12a,b). Nevertheless, an expression of aggregation of chromaffin tissue into clumps is seen from a variety of species assigned to Rhynchocephalia and Ophidia. By and large, the pattern of intermix in Crocodilia and Chelonia is similar to that of birds. The whole glandular mass is permeated with numerous small lenticular blood sinuses which ramify between the cortical cords, some passing to the periphery and others coalescing to form one or more of the larger central sinuses which eventually lead into the tributaries of the short adrenal veins entering the posterior vena cava.

Adrenocortical tissue displays a homogeneity of microscopic structure throughout the Reptilia (Lofts, 1978). As we have seen in the Aves, the basic unit is a cord of a double row of cortical cells to form a continuous network (Fig. 12b). These irregularly anastomosing cords run in all three planes, oriented, in general, parallel to the anterior–posterior axis of the gland. The cords are separated by vascular channels outlined by thin collagen fibers which separate the cortical cells from the endothelial lining. Reticular fibers run irregularly to form a lattice encompassing groups of cortical cells.

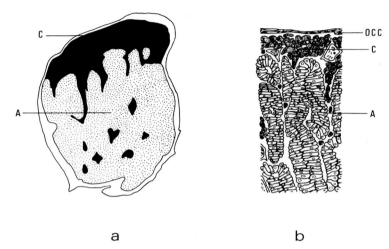

a   b

Fig. 12. (a) The Reptilian gland has, in most species, the same intermingling of cortical and chromaffin tissue as demonstrated in birds. This diagram of *Sphenodon punctatus* illustrates the tendency for chromaffin tissue (C) to have peripheral aggregations in addition to the islets scattered throughout the adrenocortical cells (A). The adrenal of *Echidna* (Prototherian) is similar. (b) A portion of (a) with the encapsulating OCC; the adrenocortical tissue (A) cast into cords with chromaffin tissue (C). [Drawings by P. Groves.]

The usual cortical cell is columnar and has a round or oval basophilic nucleus but shows seasonal variations. It has a scattering of fine chromatin particles and one or two conspicuous nucleoli. The cells have numerous lipid droplets. The properties of the tetrapod cell are displayed in its ascorbic acid content and enzymatic sequences concerned with steroid hormone secretion. As to the possibility of zonation, there has been rigorous investigation of the Reptilia. It is, however, implicit in the general theory of the basic cortical cord that peripheral cells contain the enzyme systems leading to mineralocorticoid formation, and the central ones, to glucocorticoids. Evidence is sparse, though the cobra (*Naja naja*) presents an easily distinguishable subcapsular zone, and it would be an extrapolation to assign mineralocorticoid secretion to this. In ultrastructural terms, the reptilian adrenocortical cell has all the characteristics of typical steroid-secreting cells. It is relevant, therefore, to mention that the peripheral zone of the cobra differs from the appearance of the inner cortical cords of the gland.

D. Amphibia

Amphibians belong to the Tetrapoda, and the major interrenal secretions are aldosterone and corticosterone. The clues to the morphological appearance of the interrenal lie in embryological development (Witschi, 1956). We have already mentioned the importance of the "new" kidney, the metanephros in mammals, birds, and reptiles. The mesonephros for the most part disappears in the adult, though represented principally by efferent tubules of the testes. However, in all vertebrates the interrenal/adrenal, gonads, and mesonephros arise in early development from a central block of mesoderm. The metanephric ducts originate as outpockets of mesonephric ducts, one on each side; they take an upward dorsolateral direction and eventually form the "new" kidney. Thus, the adrenocortical/interrenal tissue is separate spatially from the metanphros and, as we have seen, mammals, birds, and reptiles (Amniota) have a distinct encapsulated adrenal gland. This is not so in the Amphibia, because no metanephros is formed. The mesonephros remains in adult life. Since it is formed from the block of mesoderm which also gives the interrenal, they remain intermingled in adult life. Thus we find the interrenal lying ventrally on the mesonephric kidney and vascularized by common interrenal and renal blood vessels. The basic microstructure of the interrenal displays the general vertebrate pattern; this would be expected, because despite gross anatomical differences, the embryological formation is similar. Thus, it can be seen that the interrenal, or adrenocortical homologue, comprises discrete or scattered tissues set about the ventral surface of the functional opisthonephros (that is, a term to include the mesonephros and its posterior parts) (Figs. 13 and 14a,b). Bilateral branches of the dorsal aorta supply both the gland and kidneys. The blood exits through venules into renal veins to enter the posterior vena cava. Venous blood from the renal portal vein enters numerous small vessels to reach the outer edges of the kidneys and is also carried to the vena cava by the renal veins. In the Anura, islets of interrenal tissue are found on the ventromedial surface of each kidney, while in the Urodela they are set in the midline. Anomalies occur; for example, the genera of the Pipidae (which includes *Xenopus laevis*) of the Anura have an urodele type of arrangement of the interrenal islets.

Histologically, the interrenal is made up of cords of the polygonal cells interspersed with blood sinuses; chromaffin cells, mostly in separate small groups, intermingle. (Stilling cells occur in the Ranidae; they are clearly characterized by prominent cytoplasmic eosinophilic granulation, but their function and possible homologies remain unresolved problems.) Like interrenal and adrenocortical cells, there are numerous lipid droplets, varying concomitantly with activity and particularly with seasonal changes. Throughout the vertebrate interrenal/adrenocortical cell, characteristic enzymes are found: 5-ene-3β-01-steroid dehydrogenase, cholesterol, and glucose-6-phosphate dehydrogenase. The ultrastructure shows tubular mitochondria, SER, and liposomes.

The possibility that the amphibian interrenal has zonation has been examined. In *Rana* species (e.g. *R. temporaria, R. esculenta*), the subcapsular zone and the zone nearest the kidney have active cells with large nuclei and nucleoli, considerable amounts of RNA, and fine lipid granules. Between these zones, the central part of the gland has cells with smaller nuclei and

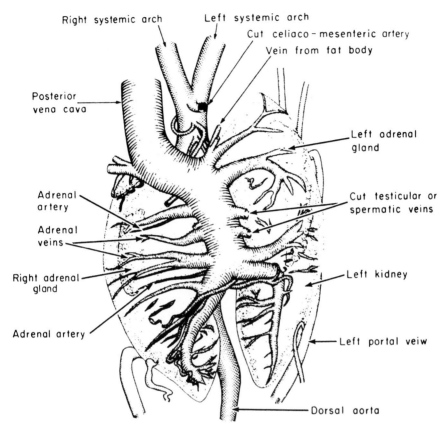

Right systemic arch

Left systemic arch

Cut celiaco - mesenteric artery

Vein from fat body

Posterior vena cava

Left adrenal gland

Adrenal artery

Cut testicular or spermatic veins

Adrenal veins

Right adrenal gland

Left kidney

Adrenal artery

Left portal veiw

Dorsal aorta

Fig. 13. Dissection of the adrenal gland of a male frog (*Rana*) showing arterial and venous systems following latex injection. Ventral view with testes removed. × 3. [Preparation and drawing by I. Carthy and J. G. Phillips.]

larger lipid droplets. The whole interrenal can be given an active appearance, histologically, by seasonal activity and, experimentally, by exogenous ACTH injection. In times of hibernation and seasonal inactivity, zonation is not very obvious, since cells throughout are small, and lipid droplets are large and coarse, indicative, in general, of low secretory capacity.

The life histories of the Amphibia underscore the importance of this group in functional studies. All members of the Amphibia are dependent on water at some stage, some genera to a slight degree and others totally aquatic. Experimental studies have contributed much to the understanding of osmoregulatory mechanisms, the control of water, and electrolyte movements.

E. Pisces (Fish)

Pisces, or fishes, is used as a convenient term for nontetrapod vertebrates, but neither term has real standing in established systematics. The whole

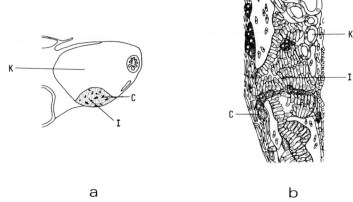

a   b

Fig. 14. Interrenal apparatus of the Amphibia. (a) There is no discrete adrenal gland, separated from the kidney. The interrenal lies on the ventral surface of the mesonephros with chromaffin tissue (C) intermingling with the interrenal tissue (I). K, kidney. (b) The interrenal cords of cells (I) are adposed to the kidney tubules (K); chromaffin tissue (C) intermingles with the adrenocortical homologous cells (I). [Drawings by P. Groves.]

group contains about half the total number of vertebrates, perhaps about 20,000 living species. Comprised within the whole lie the jawed fish (predominantly the Teleostei, with bones) the Elasmobranchii (cartilaginous fish), and the Agnatha (jawless fish). The interrenal glands cannot be considered apart from the excretory systems, the basic involvement of which we have seen in the wide-ranging group of Tetrapoda. Basically, in vertebrate embryology, the renal, gonadal, and interrenal tissues are derived, in varying proportions, from mesodermal blocks, referred to by Witschi (1956) as nephrogenic blastema. The fundamental, similar plan is, of course, differentially expressed. Potentially it may extend from the level of the second postotic somite to some distance behind the anus. Confusion has arisen by the use of pronephros and mesonephros (and in the Amniota, metanephros) supposedly to imply a succession along the length. However, it is only possible to distinguish a pronephros when the young occur as a larval stage, thus requiring an early functional excretory system.

1. Teleostei

These bony fish, an infraclass of the class osteichthyes, are the most abundant and diversified group of all vertebrates. They are placed in approximately 31 orders, 415 families, and 3869 genera, and 18,007 species and represent about 96% of the known living fish.

In general, the teleost interrenal consists of groups, cords, or strands of cells set around the posterior cardinal veins and their branches, vascularized by twigs of the dorsal aorta (Fig. 15a,b) (Chester Jones and Mosley, 1980). The interrenal is usually confined to the anterior part of the kidney, named

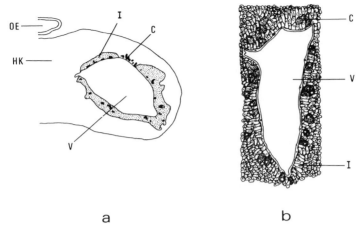

a   b

Fig. 15. Interrenal tissue of the Teleostei (bony fish). (a) The interrenal tissue (I) lies anteriorly in the head kidney (HK), which is the nonfunctional part of the mesonephros, rarely displaying tubules, and mostly lymphocytes. The interrenal = adrenocortical homologue is disposed about branches of the posterior cardinal veins (V), and the diagram shows the fish's left side. The esophagus (E) is central. Chromaffin tissue (C, black dots) is scattered throughout the interrenal tissue in many species, though the majority of the c. 20,000 species has not been examined. (b) A typical set of interrenal cells (I) set about a branch of the posterior cardinal vein (V) with intermingling of chromaffin cells (C). [Drawings by P. Groves.]

the "head kidney," commencing caudad to the last gill slit. Equivalence to the pronephros is not, for the most part, valid. In the majority of fish, the head kidney, representing the cephalad part of the mesonephros, is mostly lymphoid with melanin-type pigment and hemopoietic cells. We can simplify the main patterns into the following types: type I, interrenal tissue surrounds the postcardinal veins or their largest branches; type II, interrenal tissue surrounds small or medium-sized branches of the veins and is widely dispersed throughout the head kidney; type III, associated with the venous sinuses, scattered throughout the lymphoid and hemopoietic tissue; type IV, a solid mass of cells in a localized area. The histology of the interrenal cell conforms to the tetrapod pattern, though with less display of lipid droplets. The presence of cholestrol, ascorbic acid, glucose-6-phosphate dehydrogenase, 5-ene-3β-01-steroid dehydrogenase has been demonstrated in the few species examined. The ultrastructure is so similar to that normally occurring in a typical cell of the eutherian zona fasciculata that it does not demand a separate description. Hypophysectomy results in various degrees of atrophy, depending on species, the length of time after operation, and the environment. There is clear evidence for pituitary control by some type of fish ACTH. Mammalian ACTH is also very effective—an interesting emphasis on evolutionary uniformity in the midst or morphological diversity (Chester Jones and Mosley, 1981). The major hormonal steroid is cortisol; additionally aldosterone debatedly occurs. Indeed there is a view that aldosterone, at least as a dominant mineralocorticoid, arose with the emergence of the Tetrapoda. The formidable osmoregulatory tasks facing teleosts in an enormous range of salinities may well depend, *inter alia,* on a type of fish prolactin plus cortisol.

Apparent stressful reactions may be involved in such natural processes as migration and spawning. The interrenal in teleosts undergoes seasonal variations, with enlargement at the reproductive period. The salmonids are prime examples. The universal postspawning mortality of the Pacific salmon has been associated with hyperplasia of the interrenal. Many tissues show degenerative changes similar to those found in Cushing's syndrome in humans, in which adrenal hyperplasia and hypersecretion are well documented. In the Atlantic salmon, immature parr show increasing interrenal activity, which increases at the smolt stage. Once the fish is in the sea, its interrenal appears to show considerable activity (Chester Jones and Mosley, 1980). In another genus of salmoniforms, most Ayu (*Plecoglossus altivelis*) die after spawning. Some overwinter if the degree of fatigue and debility is not too excessive. Interrenal tissue increases then in the prespawning period and becomes marked in the breeding season. There is, in general, interrenal hypertrophy associated with gonadal maturation (Chester Jones and Mosley, 1980).

2. Dipneusti (Dipnoi) (Lungfish)

There are three extant genera of lungfishes—the Australian *Neoceratodus,* the South American *Lepidosiren,* and the African *Protopterus.* The presumptive interrenal consists of small groups of cells closely associated with the postcardinal veins and their tributaries where the vessels pass through the kidneys. The cells contain lipid, cholesterol, and 5-ene-3β-hydroxysteroid dehydrogenase. The morphological site is more amphibian than teleostean, akin to the interrenal tissue of the urodele *Pleurodeles waltii* and tadpoles of *Rana* (Chester Jones and Mosley, 1980).

3. Holostei, Chondrostei, and Crossopterygii

Other groups in the Pisces—Holostei (*Amia calva*); Chondrostei (sturgeons); and Crossopterygii (coelacanth)—are not well enough known to be included in this chapter.

4. Chondrichthyes

The class Chondrichthyes, or cartilaginous fish, includes the Elasmobranchii (sharks and rays) and Holocephali (e.g., ratfish) and comprises dominant, successful fish of evolutionary antiquity. It is important to note that the interrenal gland is completely separated from the chromaffin cells, which lie in paired groups segmentally arranged. The interrenal itself lies in the region of the posterior kidney as blocks of cells, horseshoe-shaped (rays) or rod-shaped (sharks), or is concentrated in one oval form (*Torpedo*) (Fig. 16a). It

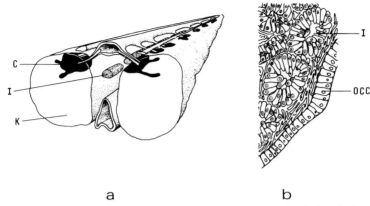

a   b

Fig. 16. Interrenal tissue of the Elasmobranchs. (a) The main point is that the interrenal = adrenocortical homologue (I) is completely separated from chromaffin tissue (C), which is segmentally arranged. In the dogfish type the interrenal is rod-shaped and is found midway between the mesonephric groups. K, kidney. (b) The interrenal = adrenocortical homologue is cast into lobules and cords of cells. It is a discrete gland bounded by a connective tissue capsule (OCC). [Drawings by P. Groves.]

is generally found that interrenal cells are homogenous, all grouped in cords, and are round or polygonal, with a large prominent nucleus containing basophilic chromatin and having one or more nucleoli (Fig. 16b). Cells have lipid droplets, cholesterol, and 5-ene-3β-hydroxysteroid dehydrogenase. Ultrastructurally, two types of cell, dark and light, are indicated in immature *Raia clavata*. Dark cells are packed with liposomes and mitochondria, whereas in light cells the organelles are more widely spaced. Smooth endoplasmic reticulum and Golgi material are evident. Liposomes are of variable size, with or without a single limiting membrane. The numerous mitochondria are round or oval, with cristae mostly of the tubulovesicular type. Polyribosomes are dispersed throughout the cytoplasm. There are suggestions of interrenal cell changes with sexual activity and viviparity, though further investigation is needed (Chester Jones and Mosley, 1980). Zonation within the interrenal mass of cells may well exist, though it is not usually obvious. It is, however, emphasized in nurse sharks (*Ginglymostoma cirratum*) by Taylor *et al.* (1975). The interrenal of elasmobranchs is regarded as sluggish in its response to the usual experimental procedures. In the common species of *Scyliorhinus,* total removal of the pituitary gland may produce some histological changes about a year after operation.

5. Agnatha (Cyclostomata) (Jawless fish)

Living jawless vertebrates (Fig. 17) are divided into the Petromyzonidae (lampreys) and Myxinidae (hagfishes). These two groups may represent in their living forms those types which heralded the transition from protochordate to vertebrate somatic organization, perhaps as far back as the Cambrian

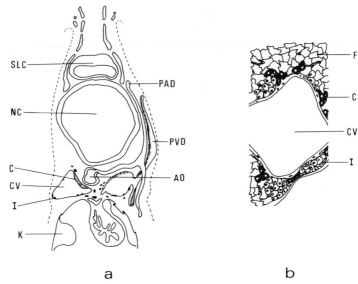

a   b

Fig. 17. A transverse section of *Petromyzon* (Cyclostomes, Petromyzonidae) which shows the presumptive interrenal = adrenocortical homologue cells (I) arranged about the cardinal veins. Chromaffin cells (C) stippled, are nearby. SLC, spinal cord; PAD and PVD, posterior parietal arteries and veins; AO, aorta; K, kidney = mesonephros pronephros; NC, notochord; CV, cardinal vein. (b) Presumptive interrenal cells (I) of *Lampetra* (Cyclostomes) occurring about a cardinal vein (CV) with adjacent chromaffin cells (C) with fatty tissue (F). [Drawings by P. Groves.]

era. However this may be, the Petromyzonidae and Myxinidae present difficulties in identification of interrenal tissue which, if it exists, may or may not secrete known interrenal/adrenocortical steroids.

For the Petromyzonidae, presumptive interrenal tissue may be concentrated within the pronephric region on the dorsal side of the pericardium and along the walls of the cardinal veins. Seiler and his colleagues (1973) found, in *Lampetra planeri,* cells with organelles characteristic of steroid production. The definitive enzyme 5-ene-3β-hydroxysteroid dehydrogenase is not readily demonstrated, but it may occur at certain stages of the lamprey life cycle, for example, during migration. There may be agreed-upon pathways of vertebrate steroidogenesis in certain lamprey cells, but the theory has yet to be proven. Lampreys clearly have to osmoregulate and may well rely on nonsteroidal mechanisms still to be identified, as, for example, in *Amphioxus.*

The Myxinidae may be related to the Petromyzonidae but differ markedly in that the serum has a similar ionic composition to that of the environmental seawater. Again, the presumptive interrenal tissue may lie in the region of the pronephros, but we are here on even weaker ground than in the discussion about the lampreys. There are no certain results.

III. GENERAL OBSERVATIONS ON MORPHOLOGY

The Gnathostomata present a recognizable morphological evolution of the adrenal and interrenal glands, inextricably bound up with the embryological formation of the kidneys, gonads, and neural crest. If we consider the Tetrapoda, then there is a sharp division between those with an amnion (the Amniota—mammals, birds, and reptiles) and those without (the Anamniota—amphibians). The latter have a persistent mesonephros (opisthonephros) in the adult, carved out from the mesonephric blastema, associated with the gonads. The interrenal tissue arises concomitantly from the same source and is thus intimately associated. On the other hand, the Amniota have, in the adult, a separate metanephros, leaving behind, as it were, the adrenocortical tissue to form an individual gland on each side. Gnathostomatous fish comprise a variety of types. The two major groups, teleosts and elasmobranchs, have great differences in the morphological arrangement of the interrenal tissue. These groups are Anamniota, with functional mesonephroi in the adult, and the interrenal tissue of both groups is associated with them and the cardinal veins. In the teleosts the potential segmental contribution is realised anteriorly in the head kidney, while in the elasmobranchs the interrenal tissues are posterior structures. From the point of view of microanatomy, wherever topographically situated, the basic unit of adrenocortical/interrenal tissue is a cord of cells of mesodermal origin (Fig. 18). These tissues are associated in varying degrees with chromaffin cells migrating from the neural crest. The amount of intermingling, with its interesting absence in the elasmobranchs, culminates in the complete centralization of the chromaffin cells in the eutherian mammal as the medulla.

Given that the adrenal cortex and interrenal glands are composed of cords of cells throughout the jawed vertebrates (Gnathostomata)—and this is difficult to refute (Balment *et al.,* 1980)—then it is only a question of morphological arrangement during evolution. The cells of a cord, usually in the form of a double strand, are able to show cell division at some point along their length at various rates throughout the animal's life. The gland, therefore, to a greater or less degree, is zoned. The histological picture of species of Pisces (Chester Jones and Mosley, 1980), Amphibia (Varma, 1977; Hanke, 1978), Reptilia (Lofts, 1978), and Aves (Holmes and Phillips, 1976) gives solid support to this zonation. The zonation is not completely obvious without careful inspection, until one reaches the Eutheria (Idelman, 1978; Balment *et al.,* 1980). The aggregation of chromaffin tissue into a certain medulla inevitably imposes order on the cortical cords. The classical zones, glomerulosa, fasciculata, and reticularis, are then seen.

In this observation lies the basis of a controversy yet to be resolved. Briefly, there is the "zonal" theory and the "escalator" or "cell migration" theory to account for the interrelationships of the classical zones of the eutherian adrenal cortex. The accumulation of literature on the subject since

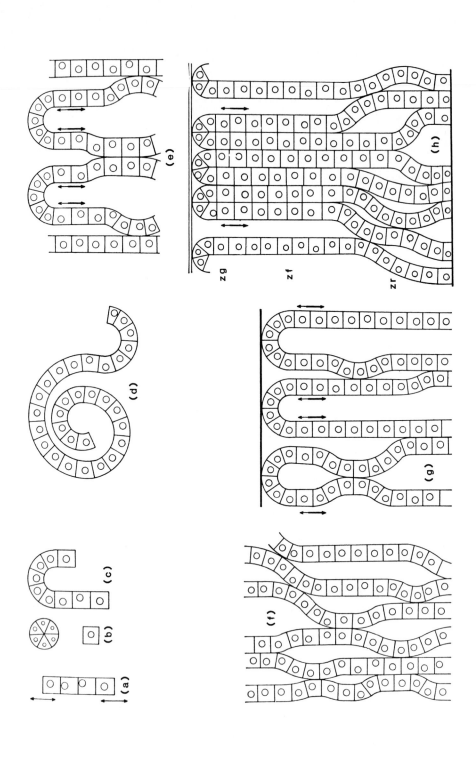

Gottschau (1883) is enormous, and reference may be made to Long (1975) and Idelman (1978). Any theory must embody agreed-upon experimental facts:

1. In the absence of the anterior lobe of the pituitary, the zonae fasciculata and reticularis degenerate; the zona glomerulosa remains and persists actively to produce aldosterone, the major mineralocorticoid. The animal has sodium and potassium levels within the normal range. Water loads are not readily excreted.

2. Appropriate administration of ACTH to the hypophysectomized animal (the rat and mouse are especially documented) leads to the reappearance of the zonae fasciculata and reticularis, and cortisol (or corticosterone) is again secreted. Water loads are readily excreted.

3. Stripping the adrenals *in situ,* called enucleation, leaving only the outer connective tissue capsule and few adherent adrenocortical cells, is followed in due course by the regrowth of the adrenal cortex, cast into zones, though without the medulla (Chester Jones and Spalding, 1954; Brudieux *et al., 1966; Peaker* et al., 1967).

The cell migration theory offers the view that there is cell division around the area of the periphery of the gland. The new supply of cells slowly migrates centripetally, eventually reaching the zona reticularis, where degeneration occurs. It is noted that in some, but not necessarily discernible in all, eutherian species there is a zona intermedia (transitional zone; sudanophobic zone) lying between the zona glomerulosa and zona fasciculata. Once evidence about the number and site of mitotic figures and the site of such cells pulse-labeled and counted some time thereafter is taken, we come upon areas of controversy. The rat and the mouse provide the majority of observations. Care must be taken about the age of the animal. For example, prepubertal growth will demand more cell divisions than that in the more stable adult animal. Cell divisions occur in the zona glomerulosa and outer zona fasciculata. The cells so formed (identified by labeling with tritiated thymidine) do not migrate, according to Walker and Rennels (1961). On the

---

Fig. 18. The organization of the basic cord of adrenocortical cells throughout the vertebrates. (a) The basic unit: a cord of cells capable of growth in each direction. (b) Transverse sections of (a) showing that the cord can be either a single line of cells or grouped to give a radial pattern. (c) Simple looping of the basic cord. (d) More complex looping (as in teleosts). (e) Illustrating the capacity to form loops, all parts of which are not equally capable of growth: it is easier for growth to take place in the direction away from the loop. (f) Organization of cords of cells in longitudinal runs (as in amphibians and reptiles). (g) Looping of cords peripherally against a well-formed outer connective tissue capsule (c) (as in birds) when the consideration given in (e) applies. (h) Looping of cords peripherally against a well-formed capsule (c) when (e) again applies, and also the termination of the cords against a central medulla (with necessary alteration of growth pattern) as in the Metatheria and Eutheria. zg, zona glomerulosa; zf, zona fasciculata; zr, zona reticularis. [From Chester Jones, 1957.]

other hand, Mitchell (1948) took a compromised view by supposing that the new cells in the zona glomerulosa sustained this zone and did not pass through the zona intermedia, whereas those from the outer zona fasciculata did move down to the zona reticularis. Wright (1971) in essence supports this compromised view of the escalator theory. The zona reticularis presents another difficulty in that it is by no means always made up of degenerating cells.

The zonal theory [named by Chester Jones (1948), following the work of Deane and Greep (1946)] claims simply that once the zones are established in the adult animal, in normal life each zone is replenished by cell division within itself. The zona glomerulosa produces aldosterone subject to various controls (renin–angiotensin system, electrolyte levels, serotonin) but not ACTH. The zona fasciculata secretes glucocorticoids (cortisol/corticosterone) and is under the influence of ACTH both for steroidogenesis and for morphology. The zona reticularis still presents difficulties, but by and large it may produce sex steroids (perhaps mostly androgens). However, in some species the zona reticularis appears to be degenerative; in others, active ("healthy"); and in some species the zona is not histologically discernible. Often the zona reticularis has accumulation of lipofuchsin granules, and this is taken as indicative of aging in many other tissues (Cole *et al.*, 1982)

It is shown in this chapter that the adrenal/interrenal of the Gnathostomata consists of cords of cells. This supposes that the cords differ in activity along their length. The Tetrapod gland secretes aldosterone and cortisol/corticosterone. Some zonation is shown. The difficulty then is that we have to envisage the peripheral part of a cord with enzyme systems for aldosterone and, later, for glucocorticoids, and that at this point the cord has become particularly responsive to ACTH. One supposition is that this is a factor of age, so that the cells after division are equipped for aldosterone formation and, with age, change to glucocorticoid production. Such a view would be in keeping with the cell migration theory. The final situation would be the cords of cells arranged in obvious zones in the Eutheria confined between the outer connective tissue capsule and the medulla. In the Cyclostomata, it is possible that the gonads and kidney tissues arising embryologically together also leave potentially interrenal cells around about the cardinal veins. The potential to produce steroids may not, or may rarely, be realized. These fish are the living representatives of ancient forms which donated pathways to gnathostomatous genera wherein steroidogenesis is patently established.

IV. BIOMEDICAL IMPLICATIONS

The natural abnormalities of function in humans, in addition to iatrogenic manifestations, provide a variety of morphological changes. These can often be simulated in experimental animals, though clearly not with rigorous equivalence. Although the adrenal cortex/interrenal is involved in many, if not all, metabolic functions, we may for simplicity mention some examples of clinical morbidity (Cullen *et al.*, 1981): (1) Cushing's disease and syndrome, (2) Addison's disease and other adrenal insufficiencies, (3) adrenal–genital syndromes of various presentations in the child and adult (Kime *et al.*, 1980, and (4) primary and secondary aldosteronism.

Morphological changes must be related, as far as possible, to physiology. Enhanced corticosteroid production and action may be produced by ACTH administration, and consequent morphological changes can be followed (e.g., Pudney *et al.*, 1984). This approach aids, to some degree, in understanding the Waterhouse-Friderischsen syndrome. This adrenal apoplexy may include cortical collapse after excessive ACTH stimulation consequent on, for example, fulminating septicemia.

Regarding lower vertebrates, a naturally occurring phenomenon in the Pacific salmon mimics Cushing's syndrome and experimental hyperadrenocorticism (Table I) The endocrinology of fish can also be put to good use in

TABLE I

Histological Changes in Spawning Pacific Salmon, Cushing's Syndrome, and Experimental Hyperadrenocorticism

	Spawning salmon	Cushing's syndrome	Experimental hyperadrenocorticism
Adrenal	Hyperplasia and degeneration	Hyperplasia and tumors	Hyperplasia
Pituitary	Degeneration	Hyalin change in basophils	Minimal degeneration
Spleen	Depletion of lymphocytes, fibrosis	No reported change	Depletion of lymphocytes
Thymus	Involution, depletion of thymocytes	Involution, occasional tumor	Depletion of thymocytes
Liver	Degeneration	Fatty degeneration	Fatty degeneration
Kidney	Degeneration	Degeneration	Degeneration
Pancreas	Hypertrophy of islets	Hypertrophy of islets variable	Hypertrophy of islets
Stomach	Atrophy and degeneration	Occasional ulcers	Atrophy of epithelium and occasional ulcers
Thyroid	Atrophy and degeneration	Atrophy of follicular epithelium	Atrophy of follicular epithelium
Gonads	Degeneration of testes	Atrophy	Degeneration
Muscle	Degeneration of masseter	Atrophy	Atrophy
Cardiovascular system	Degeneration and arteriosclerosis	Arteriosclerosis	Arteriosclerosis
Skin	Hypertrophy followed by atrophy	Atrophy	Atrophy

other areas, for example, the adrenogenital syndrome. The freshwater rice field eel (*Monopterus*) offers the opportunity to investigate the interrenal, an opportunity yet to be followed. There are small eels, which are female, and large ones, male; medium-sized have intersexual gonads. Elucidation of the interrenal involvement in this species and others showing sex reversal (e.g., serranids and sparids) would be helpful as an experimental model of this phenomenon which extends to humans.

Addison's disease is associated with the gradual loss of adrenocortical secretions which results from such destructive agents of the cortices as tuberculosis, fungi, and idiopathic atrophy. Except for terminal patients, experimental adrenalectomy is too precipitous an approach. However, there are drugs which affect the secretory capacity of the adrenal cortex, and these have been used in all classes of vertebrates, e.g., the DDD type : 2,2-bis(parachlorophenyl)-1,1-dichloroethane and Amphenone-B : 1,2-bis(*p*-aminophenyl)-2-methylpropanone-1, and others. Determining the structural consequences of their administration in any of the vertebrate species would be rewarding.

Aldosteronism in its various types has been described by Edmonds (1978). Of the Tetrapoda, the Amphibia have proved most useful (Bentley and Scott, 1978). Antagonists, such as the spirolactones, give morphological effects, and these should be further examined. Thus, Varma (1977) has drawn attention to differences in ultrastructure between aldosterone- and corticosterone-secreting cells in the bullfrog, and studies of the modification by the use of drugs would be rewarding.

Finally, it is difficult to overlook the importance of sodium, generally as the chloride, in the life of animals. Denton (1982) has drawn together a very full account. To take but a couple of examples: the wild kangaroos in the Snowy Mountains, Australia, show experimentally a specific appetite for sodium, with consequent adrenocortical changes; at the lower end of the vertebrate scale, the eel in fresh water may be transferred to seawater and may adapt rapidly with interrenal involvement.

It is clear that vertebrate species from Pisces to Mammalia offer a richness of knowledge to endow biomedical advances, but the opportunities have been rarely taken.

REFERENCES

Arnold, J. (1886). Ein Beitrag zu feineren Structur und dem Chemismus der Nebennieren. *Arch. Pathol. Anat. Physiol. Klin. Med.* **35,** 64–107.

Balment, R. J., Henderson, I. W., and Chester Jones, I. (1980). The adrenal cortex and its homologues in vertebrates; evolutionary considerations. *In* "General, Comparative and Clinical Endocrinology of the Adrenal Cortex" (I. Chester Jones and I. W. Henderson, eds.), Vol. 3, pp. 525–562. Academic Press, New York.

Bentley, P. J., and Scott, W. N. (1978). The actions of aldosterone. *In* "General, Comparative and Clinical Endocrinology of the Adrenal Cortex" (I. Chester Jones and I. W. Henderson, eds.), Vol. 2, pp. 497–564. Academic Press, New York.

Brudieux, R., Chirvan-Nia, P., and Delost, P. (1966). Sur les relations directes entre la médullo-surrénale et le cortex surrénal. *J. Physiol. (Paris)* **58**, 213–217.

Call, R. N., and Janssens, P. A. (1984). Hypertrophic adrenocortical tissue of the Australian brush-tailed possum (*Trichosurus vulpecula*): Uniformity during reproduction. *J. Endocrinol.* **101**, 263–267.

Chester Jones, I. (1948). Variation in the mouse adrenal cortex with special reference to the zona reticularis and to brown degeneration, together with a discussion of the "cell migration" theory. *J. Microsc. Sci.* **89**, 53–74.

Chester Jones, I. (1949a). The action of testosterone on the adrenal cortex of the hypophysecto-mized, prepuberally castrated male mouse. *Endocrinology (Baltimore)* **44**, 427–438.

Chester Jones, I. (1949b). The relationship of the mouse adrenal cortex to the pituitary. *Endocrinology (Baltimore)* **45**, 514–536.

Chester Jones, I. (1957). "The Adrenal Cortex." Cambridge Univ. Press, London and New York.

Chester Jones, I. (1976). Evolutionary aspects of the adrenal cortex and its homologues. The Dale Lecture. *J. Endocrinol.* **71**, 1P–31P.

Chester Jones, I., and Mosley, W. (1980). The interrenal gland in Pisces. *In* "General, Comparative and Clinical Endocrinology of the Adrenal Cortex" (I. Chester Jones and I. W. Henderson, eds.), Vol. 3, pp. 395–472. Academic Press, New York.

Chester Jones, I., and Spalding, M. H. (1954). Some aspects of zonation and function of the adrenal cortex. II. The rat adrenal after enucleation. *J. Endocrinol.* **10**, 251–261.

Cole, G. M., Segall, P. E., and Timiras, P. S. (1982). Hormones during ageing. *In* "Hormones in Development and Ageing" (A. Vernadakis and P. S. Timiras, eds.), Chapter 14. MTP Press Ltd., Lancaster, England.

Cullen, D. R., Reckless, J. P. D., and McLaren, E. H. (1980). Clinical disorders involving adrenocortical insufficiency and overactivity. *In* "General, Comparative and Clinical Endocrinology of the Adrenal Cortex" (I. Chester Jones and I. W. Henderson, eds.), Vol. 3, pp. 57–116. Academic Press, New York.

Deacon, C. F., Mosley, W., and Chester Jones, I. (1986). The X zone of the mouse adrenal cortex of the Swiss albino strain. *Gen. Comp. Endocrinol.* **61**, 87–99.

Dean, H. W., and Greep, R. O. (1946). A morphological and histochemical study of the rat's adrenal cortex after hypophysectomy, with comments on the liver. *Am. J. Anat.* **79**, 117–145.

Delost, P., and Guerin, M. (1962). Variations ponderales saisonières des glandes surrénales de l'écureuil (*Sciurus vulgaris*) dans le Tarn. *C. R. Séances Soc. Biol.* **156**, 1305–1308.

Denton, D. (1982). "The Hunger for Salt." Springer-Verlag, Berlin and New York.

Edmonds, C. J. (1978). Aldosterone secretion and its clinical disorders. *In* "General, Comparative and Clinical Endocrinology of the Adrenal Cortex" (I. Chester Jones and I. W. Henderson, eds.), Vol. 2, pp. 565–599. Academic Press, New York.

Flint, J. M. (1900). The blood vessels, angiogenesis, organogenesis, reticulum, and histology, of the adrenal. *Johns Hopkins Hosp. Rep.* **9**, 153–229.

Gersh, I., and Grollman, A. (1941). The vascular pattern of the adrenal gland of the mouse and rat and its physiological response to changes in glandular activity. *Carnegie Inst. Washington Publ.* **525**, Contrib. Embryol. 29, No. 183, 112–125.

Gottschau, M. (1883). Structur und Embryonale Entwicklung der Nebennieren bei Säugetieren. *Arch. Anat. Physiol., Anat. Abt. Lpz.* **9**, 412–458.

Griffiths, M. (1978). "The Biology of the Monotremes." Academic Press, New York.

Hanke, W. (1978). The adrenal cortex of Amphibia. *In* "General, Comparative and Clinical Endocrinology of the Adrenal Cortex" (I. Chester Jones and I. W. Henderson, eds.), Vol. 2, pp. 419–495. Academic Press, New York.

Henle, F. G. J. (1865). Über das Gewebe der Nebennieren und der Hypophyse. *Z. Rat. Med.* **24**, 143–165.

Holmes, W. N., and Phillips, J. G. (1976). The adrenal cortex of birds. *In* "General, Comparative and Clinical Endocrinology of the Adrenal Cortex" (I. Chester Jones and I. W. Henderson, eds.), Vol. 1, pp. 293–413. Academic Press, New York.

Idelman, S. (1978). The structure of the mammalian adrenal cortex. *In* "General, Comparative and Clinical Endocrinology of the Adrenal Cortex" (I. Chester Jones and I. W. Henderson, eds.), Vol. 2, pp. 1–199. Academic Press, New York.

Jost, A. (1975). The fetal adrenal cortex. *In* "Handbook of Physiology" (H. Blaschko, G. Sayers, and A. D. Smith, eds.), Sect. 7, Vol. VI, pp. 107–115. Am. Physiol. Soc., Washington, D. C.

Kime, D. E., Vinson, G. P., Major, P. W., and Kilpatrick, K. (1980). Adrenal-gonad relationships. *In* "General, Comparative and Clinical Endocrinology of the Adrenal Cortex" (I. Chester Jones and I. W. Henderson, eds.), Vol. 3, pp. 183–264. Academic Press, New York.

Lentz, T. L. (1971). "Cell Fine Structure." Saunders, Philadelphia, Pennsylvania.

Lofts, B. (1978). The adrenal gland in Reptilia. Part 1. Structure. *In* "General, Comparative and Clinical Endocrinology of the Adrenal Cortes" (I. Chester Jones and I. W. Henderson, eds.), Vol. 2, pp. 292–360. Academic Press, New York.

Long, J. A. (1975). Zonation of the mammalian adrenal cortex. *In* "Handbook of Physiology" (H. Blaschko, G. Sayers, and A. D. Smith, eds.), Sect. 7, Vol. VI, pp. 13–24. Am. Physiol. Soc., Washington D. C.

Mitchell, R. M. (1948). Histological changes and mitotic activity in the rat adrenal during postnatal development. *Anat. Rec.* **101**, 161–185.

Peaker, M., Phillips, J. G., and Peaker, S. J. (1967). A relationship between the medulla and the X zone of the mouse adrenal. *Proc. Asia Oceania Congr. Endocrinol., 3rd, 1967*, pp. 317–321.

Pearce, R. B., Cronshaw, J., and Holmes, W. N. (1978). Evidence for the zonation of interrenal tissue in the adrenal gland of the duck (*Anas platyrynchos*). *Cell Tissue Res.* **192**, 363–379.

Pudney, J., Price, G. M., Whitehouse, B. J., and Vinson, G. P. (1984). Effects of chronic ACTH stimulation on the morphology of the rat adrenal cortex. *Anat. Rec.* **210**, 603–615.

Seiler, K., Seiler, R., and Hoheisel, G. (1973). Zur Cytologie des Interrenal-Systems beim Bachneunauge (*Lampetra planeri* Bloch). *Morphol. Jahrb.* **119**,(6), 823–856.

Shire, J. G. M., and Beamer, W. G. (1984). Adrenal changes in genetically hypothyroid mice. *J. Endocrinol.* **102**, 277–280.

Taylor, J. D., Honn, K. V., and Chavin, W. (1975). Adrenocortial ultrastructure in the squaliform elasmobranch (*Ginglymostomata cirratum* Bonnaterre): Cell death, a postulate for holocrine secretion. *Gen. Comp. Endocrinol.* **27**, 358–370.

Varma, M. M. (1977). Ultrastructural evidence for aldosterone- and corticosterone-secreting cells in the adrenocortical tissue of the American bullfrog (*Rana catesbeiana*). *Gen. Comp. Endocrinol.* **33**, 61–75.

Walker, B. E., and Rennels, E. G. (1961). Adrenal cortical cell replacement in the mouse. *Endocrinology* (*Baltimore*) **68**, 365–374.

Witschi, E. (1956). "Development of Vertebrates." Saunders, Philadelphia, Pennsylvania.

Wright, N. A. (1971). Cell proliferation in the prepubertal male rat adrenal cortex: An autoradiographic study. *J. Endocrinol.* **49**, 599–609.

12

The Ovary

J. M. DODD

School of Animal Biology
University College of North Wales
Bangor, Gwynedd LL57 2UW, Wales

I. INTRODUCTION

The vertebrate ovary commands a vast literature which, like the present account, is full of generalizations, many of which are vulnerable because so few species have been investigated; only ignorance allows us to generalize! It is quite impossible to do justice to the subject in a short chapter, and several important aspects, like gonadogenesis, have had to be omitted. Fortunately, however, all aspects of gonadogenesis, oogenesis, and ovarian morphology, across the vertebrate spectrum, have been exhaustively treated in two recent volumes: "The Ovary" (Zuckerman and Weir, 1977) and "The Vertebrate Ovary: Comparative Biology and Evolution" (Jones, 1978). Furthermore, Mossman and Duke (1973) have provided a comprehensive treatise on "The Comparative Morphology of the Mammalian Ovary," and Brambell (1956) has reviewed and synthesized the older literature. The present account owes a great deal to these publications, and they provide the detail that cannot be accommodated in a chapter of this length.

General Considerations and Terminology

It is generally accepted that the vertebrate ovary is derived from three major sources: primordial germ cells, the "germinal" epithelium, and mesenchyme cells, usually said, at least in nonmammals, to arise from the developing mesonephros (mesonephric blastema). The origin and fate of the primordial germ cells is well established, but there is no unanimous view as to the role of the "germinal" epithelium other than that it does not contribute germ cells. The part played by mesenchyme-derived cells is also in doubt, as is their origin, though the evidence that they are implicated in medulla formation is strong. Steroid-secreting cells are present early in gonadogenesis though again, their origins have not been firmly established.

351

VERTEBRATE ENDOCRINOLOGY:
FUNDAMENTALS AND BIOMEDICAL IMPLICATIONS
Volume 1

Ovarian terminology is confused and confusing. To avoid ambiguity we shall define the terms and their synonyms used in the chapter and give an introductory consideration to the various aspects of ovarian development and structure examined in more detail later. *Germinal epithelium*: Use of this term has caused a great deal of confusion. As the name suggests, it was originally believed to be the source of germ cells both during embryogenesis and in the mature ovary. However, it is now accepted that this is not the case and that it consists solely of nongerminal cells derived from coelomic (peritoneal) epithelium. The term persists in the literature and will be used here in this restricted sense. *Primordial germ cells (gonocytes)*: Primordial germ cells (PGCs) are the sexually bipotential cells that are segregated early in development and are usually first recognizable in extraembryonic endoderm from which they migrate, actively or passively, into the genital ridges (primordial gonads). PGCs have been described in all classes of vertebrates. *Oogenesis*: Once the primordial gonads have developed into recognizable ovaries, the progeny of the PGCs are termed *primary oogonia*. In the early stages they form nests, or germinal beds, and are characterized by the presence of intercellular cytoplasmic bridges. Primary oogonia undergo a number of mitotic divisions, at the end of which they become *secondary oogonia,* which, usually without delay, enter the prophase of the first meiotic division and become *primary oocytes*. The transformation of oogonia into primary oocytes constitutes oogenesis *sensu stricto* (Zuckerman and Baker, 1977; Gondos, 1978). Meiosis in primary oocytes proceeds as far as the diplotene stage of prophase and then enters a protracted period of arrest which, in the human female, may last for 40 or more years. Even in vertebrates with yolky eggs, the first meiotic prophase is protracted, once vitellogenesis is completed, oocytes must either mature or become atretic. The nuclei of oocytes that do not become vitellogenic appear to be protected and remain in early prophase. *Folliculogenesis* starts when small oocytes become invested with a single layer of squamous cells. Mature and maturing follicles consist of an oocyte surrounded by a layer of follicle cells here called the *granulosa (membrana granulosa)* and a *theca,* usually of two layers, *interna* and *externa*. Between the oocyte and granulosa there is an acellular *zona pellucida,* which becomes a *zona radiata* when microvilli from granulosa and oocyte enter it, and between theca and granulosa there is a *basement membrane (membrana propria)* on which lies a rich vasculature and which appears to play an important role in species with yolky eggs. The thecal layers of the follicle have been relatively little studied. They appear to develop from mesenchymatous cells and in the early stages form a single layered investment. The theca externa consists largely of fibroblasts, connective tissue, and collagen. The interna contains cells which from histochemical and fine-structural studies are believed to be steroidogenic. The thecal layers carry the main vascular supply of the follicle, the granulosa being avascular. *Corpora atretica*: Atresia is a universal phenomenon in vertebrate ovaries. Indeed, as Branca (1925) has written: "l'atrésie, phénomène anormal, est la règle; la ponte, processus physiologique, l'exception". It is achieved in a number of different ways and affects oocytes of all sizes, although small and postmature ones are most at risk. Ingram (1962) has defined atresia as the process or processes by which oocytes are lost from the ovary other than by ovulation. Comparative aspects of its manifestation, functional significance, and incidence will be considered later. Comprehensive reviews have been published by Ingram (1962), Weir and Rowlands (1977), and Byskov (1978). *Corpora lutea (postovulatory follicles)*: Except in the mammal section below, the term "postovulatory follicle" is used in preference to "corpus luteum" for the structure formed by the follicle after ovulation. The changes it undergoes, primitively, and still in the vast majority of nonmammalian vertebrates, lead only to the resorption and disappearance of the follicular remains. However, in some cold-blooded vertebrates, especially viviparous elasmobranchs and reptiles, histological and histochemical evidence points to a possible endocrine function for the postovulatory follicle, as in mammals. If direct evidence for such a function is forthcoming, then the term "corpus luteum" will be appropriate for these follicles also. The *ovarian stroma* has been defined by Duke (1978) as "those cellular and fibrous connective tissue elements which occupy the space between the follicles, corpora lutea, interstitial gland cells, blood vessels, lymphatics and nerves." In the nonmammalian groups, there is relatively little stroma, and the large follicles become pedunculate and project into the lymphoidal cavities which occupy the central regions of the ovaries; contractile elements may be present. Stroma appears to be derived from the regressing medulla and mesenchymatous cells; it may also receive a contribution from the germinal epithelium. The literature has been critically reviewed by Duke (1978). *Interstitial tissue (interstitial gland)*: Interstitial tissue is mainly a feature of the mammalian ovary and it varies widely in amount in different species. The interstitial cell has been defined as "any cell of an endocrine gland type that occurs in, or closely associated with, the ovary that is not part of a thecal gland or a luteal gland" (Duke, 1978; also see Mossman and Duke, 1973; Harrison and Weir, 1977). *Ovarian fecundity,* defined as the number of eggs produced in each reproductive cycle, varies widely between the one egg produced every second year by an albatross to the 28 million or so that may be produced each year by a mature ling (*Molva* sp.; see Section II,C,1). However, when comparisons are made between the numbers of oocytes produced by different species, the differences are much less striking.

II. COMPARATIVE SURVEY OF OVARIAN MORPHOLOGY

A. Cyclostomes

Cyclostomes are phylogenetically the most ancient of living vertebrates, and they are of two main kinds: lampreys (Petromyzonidae) and hagfishes (Myxinoidea). The two groups show a number of major differences and they

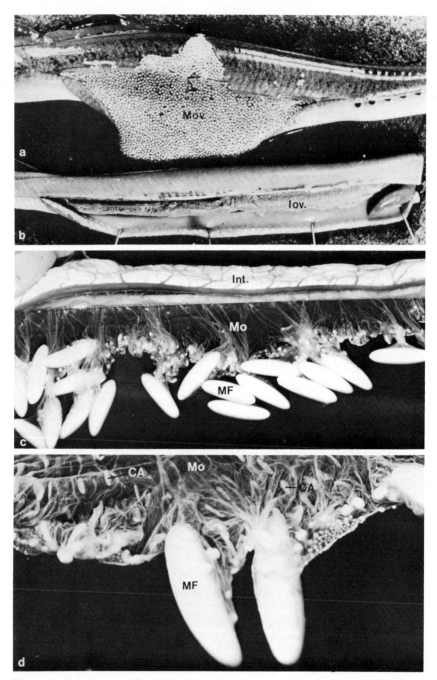

Fig. 1. (a,b) *Petromyzon fluviatilis* (lamprey). (a) Ovary of a normal female in April, at the time of ovulation (Mov). (b) Ovary of a lamprey hypophysectomized in October, immature at the time of operation and showing no subsequent development (Iov). (b,c) *Eptatretus stouti* (hagfish). (c) Part of the ovary of a mature female, suspended from the intestine (Int.) by the mesovarium (Mo), in which there were 20 oocytes in an advanced stage of vitellogenesis. Each

are believed to be diphyletic in origin (Dodd and Dodd, 1986). These differences are nowhere more striking than in the gross structure of their ovaries (Fig. 1), and they appear to be largely due to the different reproductive strategies employed by the two groups. The lampreys (Fig. 1a,b) produce a large number of small yolky eggs and reproduce only once in their lifetime, whereas hagfish ovaries, at any one time, contain only a few large yolky eggs (Figs. 1c,d), and most are believed to reproduce continuously over several seasons, although *Eptatretus burgeri,* an exception, shows an annual cycle of ovarian development (Kobayashi *et al.,* 1972).

Ovarian Morphology

a. The Lamprey Ovary (Fig. 1a,b). Oogenesis and ovarian structure have been described in a number of lampreys (Dodd *et al.,* 1960; Dodd, 1977; Fukayama and Takahashi, 1982, 1983). The lamprey ovary, like that of the hagfish, is unpaired. At maturity, it is elongated and lobate and virtually fills the body cavity since, by this time, most of the other coelomic organs have atrophied. York and McMillan (1980) state that in *Petromyzon marinus* the fully developed ovary comprises 20–25% of total body weight. Most of the growth and development of the lamprey ovary takes place during the anadromous spawning migration. Lanzing (1959) found that the gonadosomatic index of females increases from 3.9 to 13.9 during the anadromous migration of *Lampetra fluviatilis.* The stroma of the mature lamprey ovary is sparse and consists merely of a thin meshwork of fibroblasts and collagen fibers which support the follicles and carry blood vessels and, possibly, nerve fibers to them. The paucity of the stroma is best illustrated after ovulation when virtually nothing of the ovary remains. The oocytes, prior to ovulation, are compressed and irregular in shape; after ovulation they are ellipsoidal. Ultrastructural studies by Busson-Mabillot (1967a,b) on the *Lampetra planeri* ovary have shown that the basic structure of the lamprey follicle is virtually identical to that of other vertebrates. A two-layered theca lies under a covering of peritoneal epithelium and is separated from the granulosa by a basement membrane on which the blood vessels lie. The granulosa is unique among vertebrates in that it invests only one pole of the oocyte, the latter sitting like an egg in an egg cup. The thecal layers are typical of other vertebrates, the externa consisting of well-vascularized connective tissue and collagen, whereas the interna becomes glandular during development

---

oocyte lies in a sac of mesovarium attached by a stalk to the more dorsal region of the ovary. There are many small oocytes lying in the free ventral edge of the mesovarium, none more than about 4 mm long. There are no oocytes intermediate in size between these and the oocytes that are almost mature (22–24 mm.). (d) Enlarged view of the ventral edge of a mature ovary. Note large mature and maturing follicles (MF), small oocytes, and numerous shadowy corpora atretica (CA). [Parts (a) and (b), reproduced by permission from Evennett (1963). Part (c) from Gorbman (1983). Part (d), unpublished, kindly provided by Professor A. Gorbman.]

and is believed to secrete estrogens. Ovarian fecundity in lampreys appears to be related both to size at maturity and to the exigencies of the anadromous migration. It varies between 260,000 eggs in *P. marinus* and 1000–3000 in *L. planeri.*

Corpora atretica and postovulatory follicles. Atresia is said by Hardisty (1965) to occur at all stages in the development of the germ cells of *L. planeri* but to be most frequent during early meiotic prophase. Hardisty (1971) has estimated that as many as 80% of the oocytes that enter meiotic prophase undergo atresia. In *P. marinus,* unlike *L. fluviatilis* and *L. planeri,* atresia affects the oocytes of adult, as well as immature, ovaries.

b. The Hagfish Ovary. Gorbman (1983) has described the ovary of *Eptatretus stouti* in detail and has surveyed the earlier literature. The hagfish ovary (Fig. 1c,d) is an unpaired ribbonlike structure, its upper margin being continuous with the mesovarium by which it is suspended on the right-hand side of the gut. Oogonia (Fig. 1d) occupy the ventral margin of the ovary, dividing by mitosis and migrating dorsally to become oocytes and undergo folliculogenesis. Patzner (1978) has briefly described early oogenesis in *E. burgeri,* the only hagfish so far known that is seasonal in its reproduction, in consequence of which the ovary changes markedly in appearance throughout the year. Immediately after spawning (October), resting oocytes 1–2 mm in diameter are present, together with ruptured follicles about 11 mm long. The oocytes start to grow in November, and by April the granulosa shows thickened areas at both poles of the oocyte. The granulosa then thickens around the entire oocyte to form about four layers of cells. A multilayered theca surrounds the granulosa. Gorbman (1983) has concluded that the gonads of both lampreys and hagfish produce 3β-hydroxysteroid dehydrogenase (3β-HSD) and that low concentrations of sex steroids may be produced, although there is doubt as to whether the hagfish ovary produces the aromatase necessary for estrogen synthesis. In the mature ovary, the large ovoid vitellogenic oocytes of up to 2.5 cm in length and numbering about 20 appear to be the ventral-most elements, since, by this time, they are suspended from the upper regions of the ovary by "stalks" of mesovarium. It should be noted that hagfish eggs are unique in the animal kingdom in having a shell with anchoring hooks at both poles and a precisely positioned micropyle, all produced within the follicle by the granulosa.

Corpora atretica and postovulatory follicles. As in all other vertebrates, follicular atresia is a feature of the hagfish ovary (Fig. 1d); it appears to be the method by which some 100 oocytes are reduced to the approximately 20 that grow and are ovulated. Lyngnes (1936) has described the two main routes by which developing follicles become atretic in *Myxine glutinosa*; this work is discussed in detail in Walvig (1963). Gorbman and Dickhoff (1978) have investigated the question of atresia and found that in *E. stouti,* oocytes other than the group selected for further growth appeared to develop up to a length of about 4.5 mm prior to atresia, there being no follicles present between this size and the 20–30 undergoing active vitellogenesis and comprising the "clutch" destined for ovulation (see also Gorbman, 1983). Nothing is known of the histochemistry of atretic follicles, or whether they secrete or merely resorb.

Following ovulation, the follicle wall contracts and appears to thicken (Walvig, 1963). According to Patzner (1978) the empty follicle does not develop into a corpus luteum; it merely shrinks, the cells being removed, presumably by phagocytic action.

Thus the ovaries of the most ancient group of extant vertebrates are basically similar to those of the rest of the vertebrate series. The follicles of the lamprey ovary, even in fine structure, closely resemble those of other vertebrates. They have a zona radiata, a single-layered, albeit incomplete, granulosa, a two-layered theca, and the biochemical machinery for steroid synthesis and estrogen-stimulated, liver-based, vitellogenesis (Larsen, 1978), including the presence of high-affinity nuclear receptors for estrogen on liver cells (*E. stouti*; Turner *et al.,* 1982). Furthermore, not only do the follicles provide a home for the developing oocytes, they also secrete a shell with a precisely positioned micropyle and a tuft of anchoring filaments at each pole of the shell. Atresia is a feature of, at least, developing ovaries, and the absence of corpora lutea is not suprising, since the animals die immediately after spawning. The hagfish ovary is particularly remarkable, since these animals are usually rated as primitive, and yet their ovaries are arguably the most sophisticated in the animal kingdom. Atresia in hagfish shows a remarkable degree of quantitative refinement; a number of developing oocytes, possibly 100, is reduced to between 20 and 30. This may well be the basic function of early atresia in all vertebrates.

B. Cartilaginous Fishes (Chondrichthyes)

The cartilaginous fishes are the most ancient of the jawed vertebrates. Like the lampreys, they comprise two main groups: Holocephali (Chimaeroids), with 6 extant genera and 28 species, and Elasmobranchii (sharks, skates, and rays), with 128 genera and 600 species (see Dodd, 1983). The former are oviparous, whereas the elasmobranchs, although predominantly ovoviviparous, also embrace species that are oviparous and truly viviparous (placental). However, all so far examined, with the possible exception of the basking shark, *Cetorhinus maximus* (see Section II,B,1,b), produce yolk-laden eggs, usually large, in ovaries that may be either single or paired. Discrete breeding seasons are usual, though not universal; in consequence, the ovary of a cartilaginous fish can vary greatly in gross appearance at different times of the year (Wourms, 1977; Dodd, 1983; Dodd and Sumpter, 1984).

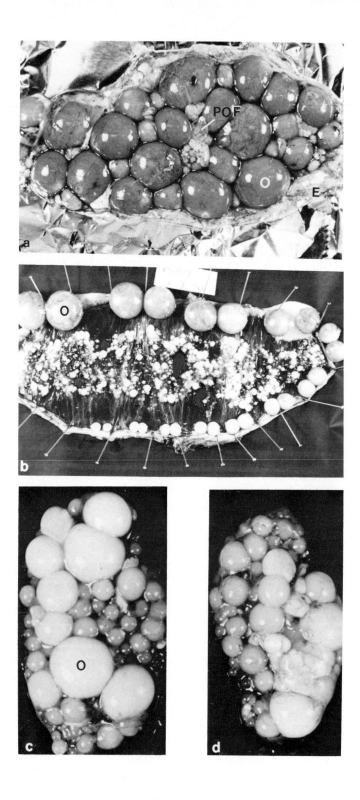

Ovarian Morphology

The mature ovary in most of the elasmobranchs that have been examined, probably fewer than 30 of the 600 extant species, develops from the right-hand-side primordium. However, Dodd (1983) has summarized the relevant literature and has shown that in some species only the left ovary develops, whereas in others, as in the Holocephali, both ovaries develop and become functional. Clearly, this is not an important difference.

Most chondrichthyan ovaries are of low fecundity. In *Squalus acanthias,* which has a 2-year cycle (Hisaw and Albert, 1947), approximately 50 oocytes start vitellogenic growth in each ovary, but by the end of vitellogenesis only 2 or 3 of these persist and are ovulated, each weighing about 50 g. In *Mustelus canis,* 12–24 oocytes reach ovulable size (Te Winkel, 1950), whereas in *Carcharhinus dussumieri* only 2 reach ovulable size (Teshima and Mizue, 1972). Furthermore, there is a striking variability in the size of oocytes at ovulation. In *Scoliodon sorrakowah,* the oocytes have a diameter of only 1.0 mm (Ranzi, 1934), whereas *Chlamydoselachus anguineus,* the most primitive living shark, produces the largest oocytes in the animal kingdom; they reach a diameter of 120 mm, and each female may contain 12 of these (Castro, 1983). The main features of ovarian structure in elasmobranchs may be illustrated by reference to an oviparous species, the lesser spotted dogfish, *Scyliorhinus canicula,* and a vivaparous species, the basking shark, *Cetorhinus maximus.*

a. The Ovary of an Oviparous Dogfish, *Scyliorhinus canicula* (Figs. 2a,b and Fig. 3). The ovary of *S. canicula* is probably typical of the vast majority of elasmobranch ovaries, since even in ovoviviparous (aplacental) and placental species the eggs are usually heavily yolked. The ovary is unpaired, develops from the right primordium, and is suspended from the dorsal body wall by an elongated mesovarium carrying blood vessels and nerves. Development is cyclical, consequently ovarian morphology varies markedly throughout the year. During summer (temperatures between 13°C and 18°C) it presents a picture of atresia of all large follicles, with only the previtellogenic and early vitellogenic follicles appearing normal. But in winter and spring (temperatures below ca. 13°C) it contains a striking array of 40 or so oocytes, arranged in a paired hierarchy and ranging in size from large, yolk-laden oocytes of up to 3 g in weight, ready for ovulation, to early vitellogenic oocytes weighing a few milligrams (Fig. 2a,b). It also contains a variable

Fig. 2. *Scyliorhinus canicula* (the lesser spotted dogfish). (a) Ventral view of a partly dissected mature ovary. (b) Vitellogenic oocytes dissected from a mature ovary arranged to show their paired size hierarchy. *Hydrolagus colliei* (the eastern Pacific ratfish). (c) and (d) Paired ovaries, note size hierarchy of follicles. E, epigonal organ; O, vitellogenic follicle; POF, postovulatory follicle. [Parts (a) and (b) from Dodd (1983). Parts (c) and (d) from Dodd and Dodd (1986).]

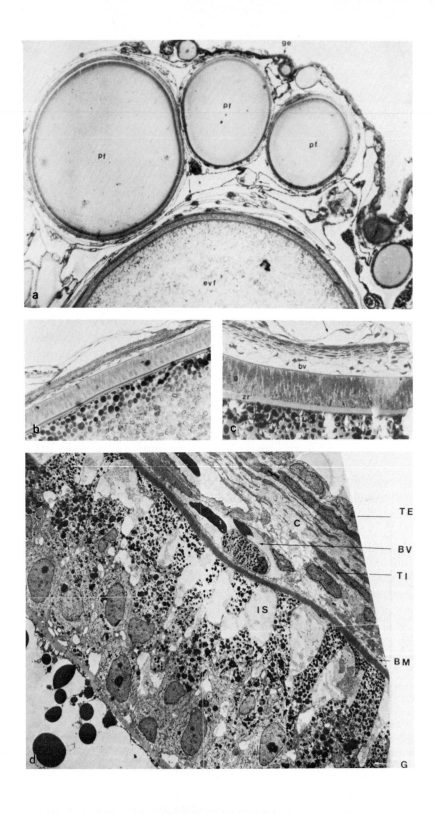

number of previtellogenic oocytes, possibly a few atretic follicles, and a number of postovulatory follicles (corpora lutea) at various stages of resorption. The histology of the wall of vitellogenic follicles is closely similar to that of all vertebrates with yolky eggs (Fig. 3) (see Dodd, 1977). The oocyte communicates with the granulosa by means of interdigitating microvilli to form a zona radiata. In mature follicles, the granulosa is single-layered and has extensive intercellular spaces (Fig. 3d). It is separated from the two-layered theca by a basement membrane. The histology of the thecal layers has been described in a number of elasmobranch species by Guraya (1978); they are overlain by a squamous investment of peritoneal cells. The only ultrastructural observations on the follicles of the elasmobranch ovary are those of Dodd and Dodd (1980) (Fig. 3d).

b. The Ovary of a Viviparous Shark, *Cetorhinus maximus***, the Basking Shark.** This ovary is strikingly different from that of *S. canicula* and all other elasmobranch ovaries described to date. Matthews (1950), who has discussed the morphology and histology of the *Cetorhinus maximus* ovary in considerable detail, was inclined to explain its uniqueness by postulating that the species was viviparous. This is certainly the case (Sund, 1943), however, all known viviparous elasmobranchs, both placental and aplacental, have sizeable yolky eggs. It is possible that Matthews described the ovary of an immature female; the general description and photomicrographs of follicular histology and atresia support this suggestion. However, Matthews also described "post ovulatory follicles" present in large numbers, and these, of course, would not be present in an immature ovary. If the ovary he described was, in fact, that of a mature specimen then we can only suggest that large numbers of follicles ovulate and that the majority of the eggs provide food for the few embryos that develop.

c. The Holocephalan Ovary (Fig 2c,d). General accounts of the development and histological structure of the ovaries of *Chimaera monstrosa* (Vu Tan Tue, 1972) and *Hydrolagus colliei* (Stanley, 1963; Dodd and Dodd, 1986) are available. In both species, the ovaries are paired and they show the

---

Fig. 3. *Scyliorhinus canicula* (the lesser spotted dogfish). Histological and fine structural details of ovarian structure. (a) Thin section of ovary showing follicles at various stages of development. pf, Previtellogenic follicle; evf, early vitellogenic follicle; ge, "germinal epithelium." (b) Thin section of wall of early vitellogenic follicle. (c) Thin section of wall of late vitellogenic follicle in early atresia. Note early dissolution of peripheral yolk immediately below zona radiata. bv, blood vessel lying on basement membrane; g, granulosa; zr, zona radiata. (d) Electron micrograph of a section through the wall of a vitellogenic follicle. Note yolk platelets in lower left corner; basement membrane between blood vessel (BV) and granulosa (BM); collagen in theca interna (C); granulosa (G); space between contiguous granulosa cells (IS); theca externa (TE); theca interna (TI). (Figs. 3a,b and c from Dodd and Sumpter, 1984. Figs. by courtesy of Dr. M. H. I. Dodd.)

hierarchical arrangement of follicles characteristic of elasmobranchs. Interestingly, the size hierarchy is shared between the two ovaries. However, until detailed accounts are available of the histology and fine structure of the holocephalan ovary, it is not possible to make detailed comparisons between the ovaries of the two orders of cartilaginous fishes, a point of some interest since the two have evolved separately for so long.

i. Corpora atretica and postovulatory follicles. Atresia of follicles of all sizes is seen in the dogfish ovary; it is particularly striking in large yolky follicles which, in the later stages of atresia, become bright yellow and hypervascular. They are rare at the height of the egg-laying season, although at its end, yolky follicles that have failed to ovulate become atretic, and corpora atretica are then the most obvious ovarian constituents. Hisaw and Hisaw (1959) have described the histology of corpora atretica and postovulatory follicles in the ovaries of a number of cartilaginous fishes, and Lance and Callard (1969) have described four stages in the development of these structures in *Squalus acanthias*. Dodd (1983) has reviewed the relevant literature in some detail.

ii. Nongerminal constituents of the ovary. Stroma is sparse in the chondrichthyan ovary and there are no reports of interstitial cells being present. There is a rich blood supply from the ovarian artery, and a large venous sinus fills with blood *post mortem,* but neither the vascularization nor the innervation of the chondrichthyan ovary have been described. The ovary is closely associated with a prominent lymphomyeloid epigonal organ (Fig. 2a).

C. Teleosts

The teleosts are not only the largest assemblage of bony fishes but also, by far, the largest vertebrate group, embracing more than 20,000 species. Dodd (1960), Dodd and Sumpter (1984), and Harrington (1974), among others, have emphasized that the teleosts are an unusually diverse group and the number of species examined is small; thus, generalizations are vulnerable. In an encyclopedic volume on the vertebrate ovary (Jones, 1978), reference is made to only about 100 species of teleosts; this may be taken as a good indication of the number examined in any detail in the context of the ovary of the more than 20,000 species available. Reproduction, including ovarian morphology, in teleosts has been reviewed frequently (Barr, 1963, 1969; Hoar, 1969; Dodd, 1977; Jones, 1978; Dodd and Sumpter, 1984; Nagahama, 1983).

Ovarian Morphology

Most teleosts are oviparous, though a unique kind of viviparity commonly occurs in eight families belonging to the two orders Cyprinodontes and Acanthopteri. Many oviparous teleosts are characterized by a remarkably high ovarian fecundity; a single specimen of *Molva* sp., the ling, weighing 24.5 kg is on record as having produced more than 28 million eggs in a single year (Table I). All the basic constituents of the vertebrate ovary are present in teleosts, and they show few special modifications in oviparous species, though in viviparous forms the follicles are uniquely modified to accommodate fertilization and gestation. However, the gross morphology of the telost ovary is different from that of other vertebrates, including other bony fish. The ovary has a central cavity, the ovocoele, into which the oocytes are ovulated. This cavity is part of the coelom and is lined by coelomic epithelium, unlike the ovarian cavities in other nonmammalian vertebrates, which are lined by mesenchyme-derived epithelia. Teleost ovaries are said to be "cystovarian" (Fig. 4b,c) as opposed to the "gymnovarian" (Fig. 4a) ovaries of other bony fish, like sturgeons, and of elasmobranchs in which eggs are ovulated into the coelom and carried to the exterior by oviducts derived from the paramesonephric Müllerian ducts which arise from coelomic epithelium. Furthermore, the ovarian walls of teleosts extend backwards into sterile tubes which form the oviducts, opening to the exterior by the genital pore. In all viviparous, and some oviparous, species, these oviducts persist in their entirety (Fig. 4b). In others, there is a discontinuity between the posterior margin of the ovary and its duct (Fig. 4c); only a short posterior segment of the latter persists, and this opens into the coelom via a funnel. In viviparous species, oocytes, or embryos, pass from the follicles into the

TABLE I

Fecundity in Teleosts[a]

Species	Weight in kg (lb)	No. of eggs	Notes
Ling (*Molva*)	24.5 (54)	28,361,000	155 cm (61 in)
Turbot (*Scophthalmus*)	7.5 (17)	9,000,000	
Cod (*Gadus*)	9.75 (21.5)	6,652,000	
Flounder (*Platichthys*)		1,000,000	
Sole (*Solea*)		570,000	
Salmon (*Salmo*)		1,300–1,500	No. of eggs per kg body weight
Herring (*Clupea*)		21,000–47,000	Eggs laid in clumps on sea floor
Carp (*Cyprinus*)	1.75 (4)	400,000	
	7.5 (16.5)	2,000,000	

[a] From Dodd and Sumpter (1984).

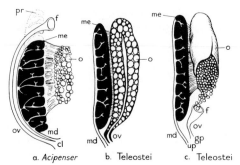

a. *Acipenser*   b. Teleostei   c. Teleostei

Fig. 4. Ovaries and reproductive ducts of *Acipenser* (the sturgeon) and various teleosts. Diagrammatic representations of gymnovarian (a) and cystovarian (b and c) ovaries in bony fishes. cl, Cloaca; f, ostium of the oviduct; gp, genital papilla; md, mesonephric duct; me, mesonephros; o, ovary; ov, oviduct; pr, pronephros up, urinary pore. [Reproduced by permission from Hoar (1957).]

ovarian cavity and are conveyed to the outside directly by the ducts. Thus, like cyclostomes, the teleosts do not use Müllerian ducts as oviducts, and, in consequence, the ovary has become uniquely modified to provide the oviduct in oviparous forms and to allow for gestation, intrafollicular or within the ovocoele, in viviparous species. These appear to be major divergences from the usual vertebrate pattern, yet the basic structure of the ovary remains unchanged. Hermaphroditism and gonadal sex change are also features encountered in some teleosti (Chan, 1970; Remacle, 1970; Dodd, 1977).

a. The Ovary in Oviparous Species (Fig. 5). Ovarian morphology in oviparous teleosts has received a good deal of attention since the time of Brock (1878), who classified the ovaries of 57 species into eight different morphological types. Many other species have been studied in the meantime, though only a small number of the 20,000 available. Polder (1961) has described eight stages in the development of the herring, *Clupea harengus* (see also Bowers and Holliday, 1961). Other species in which cyclical changes in the ovary have been described in detail are whiting and Norway pout, *Gadus merlangus* and *G. esmarkii* (Gokhale, 1957); stickleback, *Eucalia inconstans* (Braekvelt and McMillan, 1967); perch, *Perca fluviatilis* (Treasurer and Holliday, 1981); and gizzard shad, *Nematalosa vlaminghi* (Chubb and Potter, 1984). These studies have been supplemented by investigations in which the structure of oocyte and follicle wall have been described in a number of species at both histological and fine-ultrastructural levels (Yamamoto and Onozato, 1965; Yamamoto and Oota, 1967; Anderson, 1967, 1968, 1974; Gupta and Yamamoto, 1972; Beams and Kessel, 1973; Tan, 1976; Schulz and Blüm, 1983; Kessel *et al.,* 1984).

The ovary is surrounded by a squamous layer of peritoneal epithelium, as is the ovocoele. The walls of the latter are thrown into conspicuous "ovigerous" folds carrying large numbers of follicles at various stages of development, lying in a loose connective tissue stroma (Fig. 5a). Since most oviparous teleosts are cyclical breeders, the gross morphology of the ovary varies widely throughout the year. This variability is even more striking, because of the extreme fecundity of many teleosts already mentioned (Table I). Oogonia are usually aggregated into small groups or "nests" with which are associated prefollicular cells derived from coelomic epithelium (Fig. 5b), although, as in other vertebrates, additional sources of these cells are not precluded. Investment of the germ cells by a single layer of prefollicular cells coincides with their transition to primary oocytes in the diplotene stage of the first meiotic prophase (Beach, 1959; Wiebe, 1968). Contacts between the follicular cells and the oocyte in the primordial follicles are increased by the formation of intermeshing villi (Azevedo, 1974). As the follicles grow, their walls become differentiated into an outer theca, a basal lamina, single-layered granulosa, and zona radiata.

In the primordial follicles, the oocyte and the plasma membrane of the granulosa cells are virtually in direct contact (Flügel, 1967; Busson-Mabillot, 1973), but as the follicles grow, a zona pellucida, consisting of carbohydrates and proteins, is secreted between them (Guaraya, 1965a); it is soon penetrated by microvilli from both oocyte and granulosa and becomes the zona radiata (Fig. 5c). The genesis and structure of the pellucida and radiata have received a good deal of attention; the relevant literature has been reviewed by Guraya (1978). The importance of the zona radiata extends beyond the intrafollicular life of the oocyte. It is implicated in the formation of the chorion, or primary envelope, of the ovulated egg, and usually this contains a specialized region, the micropyle, the product of a single specialized cell in the granulosa (Ohta and Takano, 1982; Ohta and Teranishi, 1982), which appears to be designed to facilitate fertilization. Kuchnow and Scott (1977) have described the ultrastructure of the chorion and micropyle in *Fundulus heteroclitus*.

Guraya (1978) has reviewed the literature on follicular structure (Fig. 5c) and has shown that it is uniform throughout the group. The granulosa cells become cuboidal and then columnar as the follicles grow, and as pressure of the vitellogenic oocyte increases, the cells become squamous. Ultrastructural studies show that their internal structure increases in complexity; there is abundant rough endoplasmic reticulum, Golgi apparatus, mitochondria, lysosomes, lipid droplets, and other assorted cytological signs of high secretory activity. Histochemical tests for steroidogenic enzymes in the granulosa have yielded variable results, though it is generally accepted that the granulosa cells secrete proteins rather than steroids (see Nagahama, 1983). Intercellular spaces into which microvilli from the oocyte penetrate seem to be a common feature of the teleost granulosa (Nicholls and Maple, 1972).

The theca of the teleost follicle, which is separated from the granulosa by a

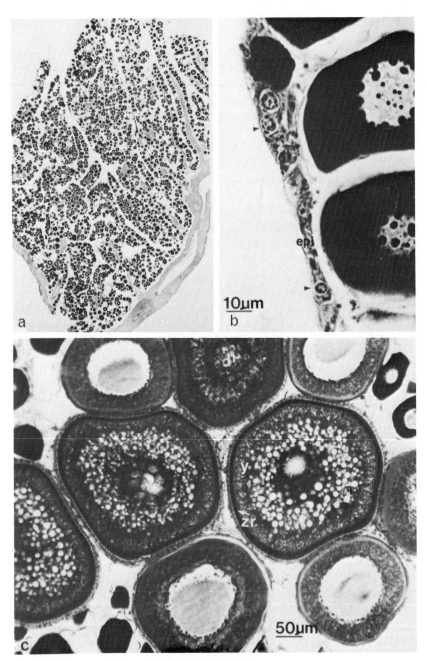

Fig. 5. *Dicentrarchus labrax* (the sea bass). Details of ovarian structure. (a) Thin section of stage I ovary of a fish caught in August. Note ovigerous lamellae carrying only previtellogenic oocytes. (b) Thin section of oogonia (arrowed) lying within the surface layers of an ovigerous lamella; they may occur as nests or as solitary cells. (c) Thin section of vitellogenic oocytes with a distinct zona radiata (Zr), an investment of follicle cells, and numerous yolk granules (y) within the ooplasm. The white vesicles are the remains of either lipid droplets or yolk granules that have been affected by the histological techniques. Note also previtellogenic oocytes in early stages of folliculogenesis. [Reproduced by permission from Tan (1978).]

basement membrane, is usually differentiated into two layers, theca externa and theca interna. The cytoplasm of the thecal cells is less differentiated than that of the granulosa cells, though rough endoplasmic reticulum, mitochondria, and Golgi apparatus are present (Yamamoto and Onozato, 1968; Szöllösi and Jalabert, 1974; Pendergrass and Schroeder, 1976). The theca may be heterogeneous; Nicholls and Maple (1972) and Hoar and Nagahama (1978) have described "special" thecal cells in two species of cichlid fishes and the amago salmon (*Oncorhynchus rhodurus*), respectively, which have abundant smooth endoplasmic reticulum and mitochondria with tubular cristae. Nagahama (1983) has reviewed this and other evidence derived from *in vitro* studies which supports the view that the theca in teleosts is steroidogenic. Other studies on the putative steroidogenic sites in the teleost ovary have been contributed by Saidapur and Nadkarni (1976), Nagahama *et al.* (1978), and van den Hurk and Peute (1979).

Corpora atretica and postovulatory follicles. Some degree of atresia is univeral in the teleost ovary, although estimates of its incidence vary widely. For example, in the salmonid *Salvelinus fontinalis,* Vladykov (1956) estimated that 40% of the largest oocytes become atretic, while Henderson (1963) reported an incidence of only 3–5% in the same species. This discrepancy is almost certainly due to differences in the levels of adverse environmental factors (stress, nutrition, temperature, etc.) operating on the two populations sampled. Yaron (1971), for example, has reported that in *Acanthobrama terrae-sanctae,* corpora atretica are rarely encountered in lake fish, but frequently in fish kept in the laboratory. There is an extensive literature on atresia in teleosts (Bretschneider and Duyvené de Wit, 1947; Polder, 1964; Hoar, 1965; Rai, 1966; Rajalakshmi, 1966; Saidapur, 1978). Postovulatory follicles may contain small amounts of yolk and debris. This is removed by the phagocytic granulosa which becomes folded and pseudostratified when the pressure exerted by the distended oocyte is removed at ovulation. This release of tension produces a spurious appearance of hypertrophy and hyperplasia in the granulosa cells, though there have been no reports of mitotic activity. Also, the presence of steroidogenic enzymes in the postovulatory follicles of some species has been held to indicate that they secrete steroids, though the enzymes are usually restricted to the day or two following ovulation, and there is no reason to suppose that they represent anything other than a carryover from the preovulatory state. Since postovulatory follicles are usually transient, there are few studies on their histology (see Nagahama, 1983).

b. The Ovary in Viviparous Species (Fig. 6). Viviparous teleosts contribute a new dimension to the realms of ovarian structure and function; in all other vertebrates ovaries have only germinal and steroidogenic functions, but in viviparous teleosts they also have a role in gestation, supplying the

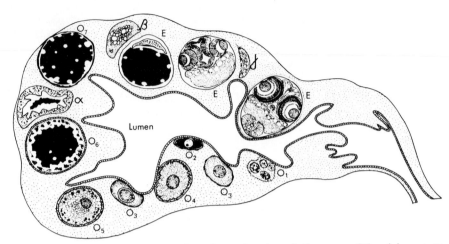

Fig. 6. Diagrammatic representation of a section through the ovary of the viviparous teleost, *Poecilia reticulata*. All stages of follicular growth and atresia are shown, though they are
not normally present simultaneously. O_1–O_7, oocytes in various stages of development; E,
embryos; α, β, γ stages in follicular atresia. [Reproduced by permission from Lambert (1970a).]

milieu, intraovarian or intrafollicular or both, in which sperm is stored and in
which the young develop, sometimes until they are sexually mature (Turner,
1940, 1947; Mendoza, 1940; Lambert, 1970a,b; also see Wourms, 1981, for a
detailed review). Viviparity is encountered in 13 families of teleosts, comprising 122 genera and 510 species (Wourms, 1981). The nonpregnant viviparous ovary is basically similar to that of oviparous species, being a hollow
sac with an anterior ovigerous region and a shorter, posterior tubular region
which puts the ovarian cavity in communication with the outside via the
genital pore. The degree to which the ovary is modified in viviparous species
depends on a number of variables, such as the amount of yolk present in the
oocytes and the length of time spent by the embryo within the follicle relative to the time spent in the ovarian cavity. In the Goodeiidae, the germinal
tissue is reduced to two dorsolateral folds projecting into the ovocoele, and
oogenesis is continuous throughout the year (Turner, 1933). The follicles are
only 0.5 mm in diameter, and the oocytes have little yolk; if they do not
mature or are not fertilized they become atretic. Fertilization is intrafollicular, and the embryos "hatch" (i.e., escape from the follicle), although they
may already have developed nutritive villi which protrude from the anal
region before they leave the follicle. The follicles show few modifications for
accommodating the early embryos, but the ovocoele changes markedly to
assume a nutritive function. These changes start at the time of fertilization.
The entire ovary becomes more vascular, the epithelium lining the ovocoele
hypertrophies and penetrates the stroma, which changes from being a compact tissue to one with a loose, spongy texture due to the development of
extensive fluid-filled areas. Gardiner (1978) has described cyclical changes in

the fine structure of the epithelium of the ovocoele in *Cymatogaster aggregata*. Vesicles are present in the cells throughout much of the year; these increase in size prior to ovulation and fertilization, and their secretion is released into the ovarian lumen. Also, pockets are present in the epithelium, lined by cells lacking the secretory vesicles; sperm are maintained in these pockets. In the gravid ovary there is almost total atresia of oocytes, although nests of oogonia remain intact.

Corpora atretica and corpora lutea. These are usually present in the mature viviparous ovary. Lambert (1970a), using the eight stages recognized by Takano (1964), has provided a detailed description of the ovary of *Poecilia reticulata,* the guppy (Fig. 6), including follicular growth, enzyme cytochemistry of follicles, corpora lutea, and postovulatory follicles. Several steroidogenic enzymes were located in the granulosa, and it is concluded that steroidogenesis is one of the functions of granulosa cells.

Another unique feature of the ovary of some viviparous teleosts is superfetation (Turner, 1937), in which two or more broods of embryos at different stages of development are present in the ovary at the same time. In *Heterandria formosa* and several species of *Poeciliopsis* and *Aulophallus,* as many as nine broods, each at a different stage of development, may be present simultaneously.

D. Amphibians

Amphibians are of two main kinds, anurans (frogs and toads) and urodeles (newts and salamanders); rather little is known of a third, much smaller group, the Gymnophiona. Amphibian ovaries vary considerably in gross morphology both within and between the groups. The anurans are usually more fecund and produce larger numbers of smaller eggs than do the urodeles, though there appear to be no fundamental differences in ovarian structure. Anuran ovaries are large, paired, and lobulated; when fully grown they occupy most of the coelomic cavity and distend the body wall. They are saclike and have extensive fluid-filled cavities produced by lysis of the medullary regions of the developing gonads. The cavities are therefore mesenchymal in origin unlike the ovocoele in teleosts, which is coelomic. Stromal tissue is sparse and is restricted to the walls of the ovarian sacs; developing follicles protrude into both the internal cavities of the ovary and into the coelomic cavity.

Ovarian Morphology

There have been many studies on various aspects of ovarian morphology, especially in anurans, but also in urodeles (for reviews, see Dodd, 1977; Jones, 1978). Dumont (1972) has summarized and compared the several methods that have been proposed for classifying ovarian development in

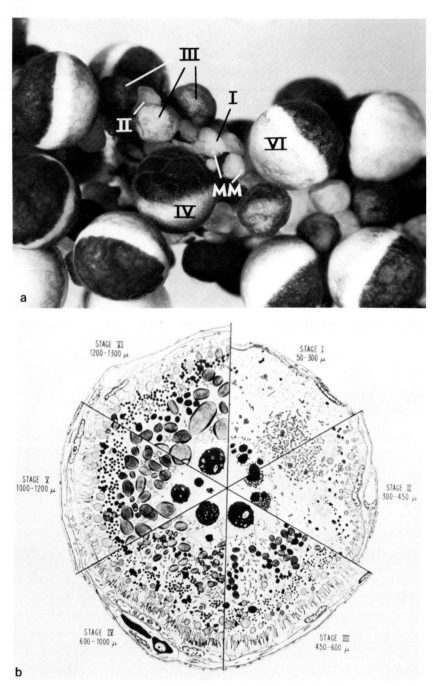

Fig. 7. Ovarian structure in *Xenopus laevis* (the South African clawed toad). (a) Surface view of part of a mature ovary of *X. laevis* showing gross appearance of oocytes at various stages of development. MM, mitochondrial mass. (b) Diagrammatic representation of the six stages of oocyte development in *X. laevis*. Stage I oocytes are characterized by a very thin covering of follicle cells, a large mitochondrial mass, a few lipid droplets (light spheres), and

anurans, based on oogenesis in species of *Rana*, and has provided a detailed, six-stage analysis of the process in *Xenopus laevis* (Fig. 7). The ovaries in *Xenopus* are paired, each consisting of some 24 lobes in which all stages in oogenesis are present throughout the year; this is atypical if not unique. Stage I oocytes are 50–300 μm in diameter and are previtellogenic. They have a prominent yolk nucleus (Balbiani body) consisting of dense aggregations of particles, mitochondria, small Golgi complexes, and cisternae of endoplasmic reticulum. They are surrounded by an outer thecal layer consisting of collagen, fibrocytes, small blood vessels, and an inner single layer of granulosa cells with microvilli which intermesh with microvilli from the oocyte. Oocyte growth (vitellogenesis) is divided into four stages during which the diameter of the oocyte increases from 300 to 1200 μm, and cortical granules, yolk platelets, and pigment appear in the ooplasm. The single-layered granulosa becomes cuboidal, then columnar, and finally squamous as the oocyte grows. The theca, also single-layered, is covered by a layer of squamous cells of peritoneal origin. Developmental stage VI is postvitellogenic and the oocyte reaches a diameter of about 1300 μm.

Guraya (1978) has reviewed the extensive ultrastructural studies that have been carried out on anuran and urodele follicles and has shown that they are closely similar, though the structure of the zona pellucida appears to differ in the two groups. Granulosa cells contain organelles associated with active secretion: ribosomes, granular and agranular endoplasmic reticulum, mitochondria with laminar cristae, lipid droplets, glycogen granules, multivesicular bodies, Golgi apparatus, and microtubules (Wartenberg, 1962; Hope *et al.*, 1963; Wischnitzer, 1963, 1964a,b; Kessel and Panje, 1968; Anderson and Yatvin, 1970; Joly and Picheral, 1972; Thornton and Evennett, 1973). According to Guraya (1978) the histochemical and fine-structural characteristics of amphibian granulosa cells indicate that they are actively secreting proteins, lipids, and carbohydrates, including glycogen, and there are several reports on an increase in the number of lipid droplets and in the amount of agranular endoplasmic reticulum in granulosa cells in the preovulatory period. The steroidogenic enzymes 3β-, 3α-, 17β-, and 17α-HSD have been demonstrated in these cells in vitellogenic follicles of *X. laevis* (Redshaw and Nicholls, 1971; see also Saidapur and Nadkarni, 1974). The theca has received relatively little attention. It consists of fibroblasts, bundles of colla-

---

small Golgi complexes. During stage II the follicle cells increase in thickness and arch over the oocyte surface, which now has microvilli. The vitelline envelope forms beneath the arches of the follicle cells. Cortical granules, premelanosomes (small dark spheres), and some yolk appear. Vitellogenesis begins in stage III, when pigment and cortical granules increase and the vitelline envelope forms a continuous layer over the oocyte. Irregularly shaped yolk platelets are present in the peripheral cytoplasm of stage IV oocytes; microvilli are large and numerous. During stage V, the accumulation of yolk gradually ceases, and the microvilli become shorter and less numerous. During stage VI many of the microvilli are lost (retracted?), and the follicle cells decrease in thickness. [Reproduced by permission from Dumont (1972).]

gen fibers, and blood vessels. The fibroblasts are the sole cellular constituents. They are fusiform cells sparse in cytoplasm, containing granular endoplasmic reticulum, ribosomes, few mitochondria, lipid droplets, glycogen granules, and small Golgi apparatus (Wartenberg, 1962; Hope *et al.*, 1963; Wischitzer, 1966; Guraya, 1965a; Joly and Picheral, 1972). Chieffi and Botte (1963) have identified 3β-HSD in the thecal cells of *Triturus cristatus* and *Rana esculenta*. According to Guraya (1976), thecal gland cells have not as yet been demonstrated in the amphibian ovary.

Hope *et al.* (1963, 1964a,b) have carried out ultrastructural studies on follicular development in the urodele *Triturus viridescens*, with special reference to oocyte–granulosa interrelations and yolk and pigment deposition; there are no major differences between these processes in *Triturus* compared with other urodeles and anurans so far described, most of which are oviparous. However, *Salamandra salamandra* is ovoviviparous, and its ovaries have been described by Joly and Picheral (1972). At the time of ovulation each ovary contains between 10 and 20 yolky oocytes, 5 mm in diameter, a number of smaller vitellogenic oocytes, and large numbers of previtellogenic oocytes. Joly and Picheral (1972) have also described the histology, histochemistry, and fine structure of the follicles and have found them to be closely similar to those already described for oviparous species.

a. Corpora Atretica and Postovulatory Follicles. According to Byskov (1978), in the immature female toad, *Bufo bufo,* 30% of follicles smaller than 0.1 mm show signs of atresia, whereas only 8% of follicles between 0.1 and 0.4 mm are affected. Various aspects of atresia in amphibians have been reviewed by a number of authors (Dodd, 1977; Barr, 1969; Lofts and Bern, 1972; Redshaw, 1972; Lofts, 1974; Saidapur, 1978; Byskov, 1978). Genesis of the atretic follicle is virtually identical to that seen in other vertebrates, and its function, if any, is questionable. A number of workers have reported histochemical tests for hydroxysteroid dehydrogenases, especially 3 β-HSD, and these have yielded conflicting results; the literature has been reviewed by Saidapur (1978). Ultrastructural studies on the corpora atretica of *B. bufo* have failed to demonstrate the presence of organelles usually associated with steroidogenesis (Lofts and Bern, 1972). As in many other nonmammalian species, postovulatory follicles in amphibians are usually transitory; mitoses are not seen, and the follicles soon atrophy and disappear. Cunningham and Smart (1934) found that in *Xenopus laevis,* absorption of the follicle wall starts immediately after ovulation, and the process is completed in about 6 days. In the urodele, *Triturus vulgaris,* Hett (1923) found that some yolk continues to be secreted into the lumen of the empty follicle and this is an important point, since it demonstrates that cells of the follicle wall do not necessarily stop functioning immediately after ovulation. If the same applies to the steroidogenic cells, this may explain reports of the presence of 3 β-HSD in them in the early stages of follicle wall resorption, without necessarily indicating a functional role. Hett (1923) found that contraction of the follicle wall at ovulation produced signs of thickening in both granulosa and theca, but that these were due to release of tension and were not accompanied by mitoses; the cells degenerated and disappeared in a few days. Kessel and Panje (1968) have described the fine structure of the postovulatory follicles of *Necturus maculosus* (Urodela) and found more of the organelles usually associated with active secretion than in the cells of the preovulatory follicle wall.

b. Nongerminal Elements. The ovaries of amphibians are closely invested by a layer of squamous epithelium which may be ciliated. Stroma is sparse; it consists of scattered fibroblasts and collagen and is restricted to the septa that support the walls of the ovarian sacs and carry blood vessels and nerves to the follicles. Guraya (1969) has shown that interstitial gland cells are present in the ovaries of *Bufo stomaticus* and *Rana pipiens* and that they are formed by hypertrophy of thecal elements in the walls of large degenerating (atretic) vitellogenic follicles. These interstitial cells occur in isolated groups in the ovarian stroma and are said to resemble closely the interstitial cells in the ovaries of other vertebrates, including mammals. At certain times of the year they give positive histochemical reactions for cholesterol and its esters, triglycerides, and phospholipids.

E. Reptiles

Reptile ovaries, like those of other vertebrates with large yolky oocytes, vary greatly in gross morphology in different species and within species throughout the breeding season, depending on the number and size of the vitellogenic oocytes present (Fig. 8a,b). Licht (1984) has reviewed the range of reproductive cycles in reptiles and has shown that in some tropical species, the ovaries are continuously active, although in the vast majority of species, vitellogenesis and oocyte growth alternate with periods of inactivity. However, in basic structure, the differences are few. In *Natrix rhombifera,* a viviparous snake, the ovaries are asymmetrically placed in the body cavity, the right one being anterior (Betz, 1963). Each is suspended by a mesovarium and is saccular and elongated, the lymph-filled sacs being derived from the rete complex (Franchi *et al.,* 1962). The stroma is sparse, oogonia and primordial follicles being present in it in small scattered groups, and at certain times of the year, corpora atretica and corpora lutea are also present. Lizards produce few oocytes, two to four in *Agama agama,* and only one per ovary in geckos (Boyd, 1940). In *Anolis carolinensis,* a cyclical breeding species, the largest follicles during autumn and winter have a diameter of only 1 mm, whereas at the height of reproductive activity they reach a diameter of 8 mm, and there is a size hierarchy of 9–12 yolky follicles (Neaves, 1971; Jones *et al.,* 1975).

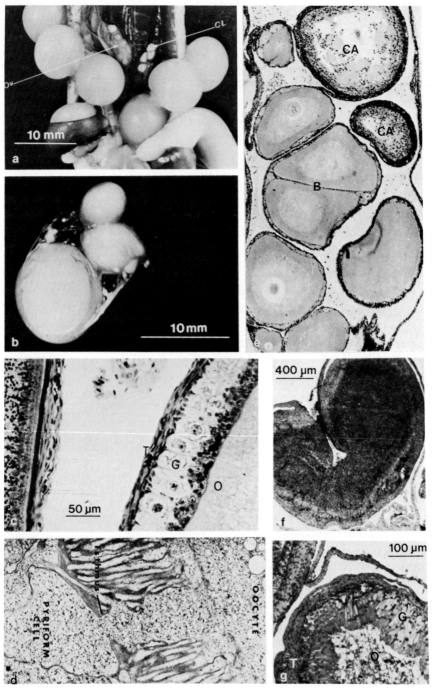

Fig. 8. Lizard ovaries; gross structure and histology. (a) *Chalcides chalcides* dissected to show the two ovaries shortly after ovulation and the ovulated oocytes in the paired oviducts. Ov, Right ovary; CL, corpora lutea. (b) *C. chalcides* ovary dissected from a vitellogenic female.

Ovarian Morphology (Fig. 8)

Primoridal follicles are formed, as in other vertebrates, when oocytes in early prophase are invested by a single layer of granulosa cells. The latter then proliferate and, in lizards and snakes, produce a polymorphic granulosa consisting of three distinct cell types termed "small," "intermediate or transitional," and "large" (Loyez, 1906; Tokarz, 1977). In chelonians the granulosa is said to be single-layered and monomorphic. The large cells are pyriform, and in the early stages of growth of the follicle they are in cytoplasmic continuity with the oocyte (Neaves, 1971; Andreucetti *et al.*, 1978; also Fig. 8d). Filosa *et al.* (1979) have studied the differentiation and proliferation of follicle cells during oocyte growth in *Lacerta sicula* and conclude that the pyriform cells arise from small cells by progressive differentiation. Fine-structural studies on pyriform cells in a number of reptile species have been carried out by Neaves (1971) and Hubert (1973, 1975, 1976). Neaves (1971), in a fine-structural study on *Anolis carolinensis,* found that in follicles of 1 mm diameter, pyriform cells are always present; they traverse the zone pellucida by means of fine protoplasmic fibrils and fuse with the cytoplasm of the oocyte (Fig. 8d). When exogenous vitellogenesis starts in the follicles of *A. carolinensis* the pyriform cells degenerate and have disappeared by the time the follicles reach the diameter of 2.0 mm. As vitellogenesis proceeds, the granulosa becomes single-layered and monomorphic (Fig. 8c,e) (Gerrard *et al.,* 1973; Guraya and Varma, 1976). Ultrastructural studies on the granulosa cells of a number of reptile species have shown them to have all the organelles associated with secretory cells, rough and smooth endoplasmic reticulum, ribosomes, Golgi apparatus, and mitochondria (Hubert, 1971a,b). Furthermore, histochemical investigations have revealed the presence of the steroidogenic enzyme 3β-HSD in reptilian granulosa cells (Lance and Callard, 1978), though it is variable both in location and in stage of follicular development (Guraya, 1978). In most reptiles the theca is two-layered (Guraya, 1965b), though in lizards, the distinction between theca externa and theca interna may be obvious only after ovulation (Weekes,

---

Note size hierarchy of follicles and follicle wall in the ruptured largest follicle (optical section). The many small oocytes are inactive and previtellogenic; they will provide future generations of yolked oocytes. (c) *Agama agama.* Transverse section of a vitellogenic follicle (left) and previtellogenic follicle (right). G, granulosa; O, oocyte; T, theca. (d) *Anolis carolinensis.* Follicle wall at the time of its maximum structural complexity (November, follicle diameter of 1.0 mm) with the pyriform cell extending toward the oocyte. (e) *C. chalcides.* Longitudinal section of a developing ovary. Note size hierarchy of follicles, biovular follicle (B), and corpora atretica (CA). (f) *A. agama.* Transverse section of a corpus luteum. (g) *A. agama.* Transverse section of an atretic vitellogenic follicle. Note granulosa cells (G) invading an oocytel (O). T, theca. [Parts (a), (b), and (e) by kind permission of Dr. I. B. Wilson and M. M. Ibrahim. Parts (c), (f), and (g) reproduced by permission from Eyeson (1971); Part (d) reproduced by permission from Neaves (1971).]

1934). The theca externa has a fibrous appearance and contains elastic fibers and collagen, whereas the interna is cellular and glandular, and its morphology, cytology, and histochemistry undergo changes that are correlated with stages in the reproductive cycle. Laughran *et al.* (1981) have shown that two types of microfilaments (4–6 and 7–10 nm in diameter) are present in the surface epithelial cells and in the fibroblasts of the theca of vitellogenic follicles in the lizard *A. carolinensis.*

a. Corpora Atretica and Postovulatory Follicles (Corpora Lutea). Atresia in the reptile ovary may affect follicles of any size, though it is most frequently seen in follicles with a polymorphic granulosa and in post mature follicles that fail to ovulate (Fig. 8c,g) (Bragdon, 1952; Betz, 1963; Gerrard *et al.,* 1973; Jones *et al.,* 1975; Byskov, 1978). The histogenesis of atresia has been described in a number of species and appears to follow the usual vertebrate pattern. Corpora lutea (Fig. 8a,f) are common features of the reptile ovary; Cyrus *et al.* (1978) state that there may be up to 80 at various stages of development and involution at any one time in the ovary of the snapping turtle *Chelydra serpentina,* and these ovaries have received a good deal of histological and histochemical attention. The literature has been reviewed by Badir (1968), Browning (1973), and Fox (1977). Cyrus *et al.* (1978) have studied the fine structure of the corpus luteum in *C. serpentina.* The granulosa cells are polygonal with an excentric nucleus; their cytoplasm contains organelles typical of steroid-secreting cells: agranular endoplasmic reticulum, frequently whorled and associated with lipid droplets, and mitochondria with tubular or vesicular cristae. Golgi complexes with associated vesicles, bundles of microfilaments, and scattered microtubules are also present. These findings provide circumstantial evidence that the corpora lutea may be steroidogenic. In viviparous species they appear to regress shortly after oviposition (Weekes, 1934; Mayhew, 1963, 1966), whereas in viviparous species they are better developed and more persistent (Betz, 1963; Cunningham and Smart, 1934; Weekes, 1934).

b. Nonfollicular Components. Miller (1948) states that the ovarian stroma in *Xantusia* is sparse and is composed of collagenous fibers and scattered fibroblasts. This is probably true of most reptiles, though the nonfollicular elements usually receive little attention in studies on ovarian morphology. Interstitial gland cells are clearly not a major constituent of the reptile ovary, though Guraya (1965b) claims that patches of cells in the theca interna, which show high 3β-HSD activity during atresia, persist and become interstitial gland cells in the stroma.

The vasculature of the reptile ovary has received rather little attention apart from the studies of Gerrard *et al.* (1973) and Jones *et al.* (1983).

F. Birds

Although there are two gonadal primordia only the left one usually develops into a functional ovary in birds, but there are exceptions in which both become functional. This appears to be the case in a number of birds of prey (see Gilbert, 1979, for a review). Most birds have only one breeding season per year, some have two, and a number of sparrows and the mourning dove may have four or five; the albatross and king penguin breed every second year. Ovarian fecundity is usually low, in some species only a single egg is produced, though a few wild birds approach the fecundity of the domestic hen (Gilbert, 1979).

Ovarian Morphology

The ovary varies greatly in gross morphology during development and at different times of the year, as in other vertebrates with yolky oocytes. The differences depend, *inter alia,* on clutch size and whether the species breeds cyclically or more or less continuously, since these variables affect the number of yolky oocytes present at any one time. Primordial follicles are recognizable in the chick 4 or 5 days after hatching, earlier in the quail. Each consists of an oocyte surrounded by a single-layered investment of prefollicular cells. Stromal cells soon become associated with the primordial follicles and, during development, give rise to a two-layered theca. The morphology and histology of the mature ovary, especially that of the domestic hen (Fig. 9), are well known and have frequently been described (Gilbert, 1971, 1979; Lofts and Murton, 1973; Perry *et al.,* 1978; Perry and Gilbert, 1979; Gilbert *et al.,* 1980).

Oogonia are not present in the mature ovary, though, of course, oocytes, which may be previtellogenic or vitellogenic, are. Pearl and Schoppe (1921) counted 1906 oocytes visible to the naked eye in the chick ovary, and Hutt (1949) estimated that there are several million in total. The vitellogenic oocytes conform to a precise size hierarchy in the domestic hen. The largest oocyte just prior to ovulation reaches a diameter of some 40 mm and is contained in a pedunculate follicle. Mature follicles in the fowl have a prominent stigma, the site of follicular rupture (Fig. 9). It is a rectangular area extending almost halfway around the circumference of the follicle and is poorly vascularized, whereas the rest of the follicle wall is richly supplied with blood. Recent work on the histology and ultrastructure of the follicle wall in birds has been comprehensively reviewed by Guraya (1978) and Gilbert (1979). Gilbert and Wells (1984), Bellairs (1965), and Perry *et al.* (1978) have described the ultrastructure of the granulosa cells at different stages of development. The cells show signs of high activity, presumably in connection with yolk deposition, though their precise function in this context is not known. They contain vesicles of different sizes, vacuoles, fila-

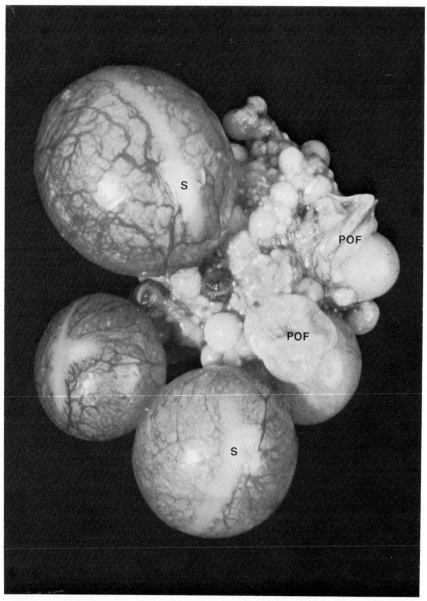

Fig. 9. *Gallus domesticus* (the domestic hen). Ovary of a mature, egg-laying hen. Note size hierarchy of vitellogenic follicles, abundant follicular blood supply, stigma (S), and postovulatory follicles (POF). [Photograph by courtesy of Dr. A. B. Gilbert. Reproduced by permission of the Agricultural and Food Research Council.]

ments, mitochondria, granules, and a "complex mass" consisting of paired centrioles, microtubules, stacked membranes, Golgi apparatus, and vesicles. Bellairs (1964, 1965) has clarified the situation regarding certain specialized membranous structures which lie at the tips of granulosa microvilli and are virtually injected into the oocyte when the villi penetrate the oocyte cytoplasm. These were first described by Das (1931), but their nature was elucidated only after fine-structural studies. They have been given a number of different names but Bellairs' term, "lining bodies," seems to be the most widely used. They are osmophilic structures closely associated with the cell membrane and may lie between follicle cells or between follicle cell and oocyte. When the zona radiata forms, the lining bodies are carried towards the oocyte by the growing microvilli, indenting the oocyte membrane and finally entering the ooplasm. They have been seen in association with oocytes up to 8 mm in diameter, but not larger. Press (1964) believed that the lining bodies remained attached to the villi, but most workers (see Guraya, 1978, for references) say that they become sequestered within the ooplasm in a vesicle derived from the oocyte cell membrane. They are said to be rich in ribosomes and therefore they may act by stimulating enzyme synthesis, but no certain function has as yet been ascribed to them.

The zona pellucida first appears between oocyte and granulosa as the follicle starts to grow; according to Bellairs (1965) it is secreted by the oocyte. It consists at first of a homogeneous layer which later, as villi develop from both oocyte and granulosa, differentiates into the striated zona radiata. The fine structure of the zona radiata in the fowl has been described by Wyburn et al. (1965).

The theca of the follicle in birds is divisible into two zones, although Brambell (1956) states, without giving reasons, that it is doubtful if the two thecal layers are strictly comparable with those of mammals. Separation of the two layers occurs in the domestic fowl when the follicle reaches a diameter of about 2 mm. The theca interna consists of fibroblasts, glandular cells, connective tissue fibers, and collagen and is well vascularized (Gilbert, 1971, 1979).

a. Corpora Atretica and Postovulatory Follicles. These structures have been comprehensively reviewed by Gilbert (1979) and Gilbert and Wells (1984). Stieve (1919) estimated that not more than 0.01% of the oocytes of the jackdaw (*Colaeus monedula*) are ovulated, and Gilbert (1971) states that the adult hen probably utilizes less than 1% of its oocyte potential to produce eggs; the rest become atretic. Brambell (1925) states that most previtellogenic oocytes in the fowl become atretic before they reach a diameter of 400 μm; atresia of yolky follicles is extremely rare.

Postovulatory follicles (Fig. 9) are transient, and knowledge of them is virtually restricted to the domestic fowl. Stieve (1918) found that in this species the follicular capsule contracts immediately after ovulation, no new cells being formed. Wyburn *et al.* (1966) do not rate postovulatory follicles in the hen as corpora lutea; they find that the cells of the granulosa and theca interna become vacuolated and do not persist beyond 72 hr after ovulation. Gilbert (1979), summarizing the situation, states: "In contrast with the mammals, it must be concluded that in *Gallus,* the post-ovulatory follicle is an active structure during only the first 24 hours of its life and that no known function has been found for the older stages."

b. Vasculature and Innervation. Gilbert (1979) has reviewed what is known of the blood supply to the ovary and has shown it to be complex and to vary with the functional state of the tissue. The arterial supply usually arises from the ovario-oviductal branch of the left anterior renal artery, although in some species, it may arise directly from the renal artery or even from the dorsal aorta. Scanes *et al.* (1982) have reported a quantitative study on blood flow in the pre- and postovulatory follicles in the ovary of the domestic fowl and have shown that the five major preovulatory follicles receive about half of the ovarian blood flow and that the supply to each follicle increases with increasing maturity. On the other hand, the supply to postovulatory follicles was low.

More is known of the innervation of the avian ovary than of any other nonmammalian group, though this is based mainly on the domestic fowl. Components arise from a number of sources including adrenal, renal, aortic, and pelvic plexuses (Bradley and Grahame, 1961; Gilbert, 1967). In addition, the follicle walls and stroma are copiously innervated by adrenergic fibers (Bennett and Malmfors, 1979). The fibers are particularly plentiful in the theca interna, and Dahl (1970), in fine-structural studies has shown that at least some of these fibers innervate steroid secretory cells (the thecal gland), smooth muscle cells, and blood vessels. Gilbert (1965, 1969) also has contributed detailed studies of follicular innervation in the domestic fowl and reviewed the field (Gilbert, 1979).

G. Mammals

Mammals are classified into three orders of very different sizes: Monotremata, Marsupialia, and Eutheria. The ovary in monotremes and marsupials has been little studied until 1976 (Lintern-Moore *et al.,* 1976; see Brambell, 1956, for the early work), whereas the eutherian ovary has received more research attention than that of the rest of the vertebrates put together. Even so, as Mossman and Duke (1973) have pointed out, significant knowledge of ovarian morphology is available for only 29 of the 121 families of extant eutherians. The account that follows is perforce a generalized one; for greater detail and taxonomic coverage, the following reviews should be consulted: Brambell (1956), Mossman and Duke (1973), and the appropriate chapters in the volumes on the ovary, edited by Zuckerman and Weir (1977) and Jones (1978).

Ovarian Morphology (Fig. 10)

a. The Eutherian Ovary. The ovary of eutherian and metatherian mammals is small in proportion to body size, the ovary of a cow of 600 kg weighing only some 10–15 g, whereas in a dogfish of 1 kg the mature ovary may weigh 40 g. The ovary is spherical, ovoid, or flattened and is usually smooth-surfaced, though in mature ovaries, corpora lutea and corpora albicantia may project above the general surface (Harrison and Weir, 1977). Unlike many nonmammalian ovaries, which have extensive central fluid-filled cavities and sparse stroma, mammal ovaries are firm in texture. They are invested by peritoneal epithelium which is separated from the thick underlying cortex by a basement membrane. The cortex is always extensive and contains the germinal elements in all stages of growth and change, including primordial follicles, Graafian follicles, atretic follicles, and corpora lutea (Fig. 10a, b). The stroma occupies all the nongerminal areas of the ovary; it is dense and consists of connective tissue, fibers, smooth muscle cells, and usually, glandular interstitial tissue. A central mass of spongy tissue, the medulla, completes the constituents of the mammal ovary. This consists of connective tissue and a few cell cords deriving from the rete and medullary cords of the embryo and contains lymphatic sinuses and large blood vessels. The mature and maturing follicles have a multilayered granulosa and lie deep in the cortex. They have acquired an investment of stromal cells which constitute the theca. The theca later differentiates into a well-vascularized and glandular inner layer, the theca interna, and an outer theca externa of connective tissue which may also contain smooth muscle fibers (O'Shea, 1970, 1971; Burden, 1972). As in all other vetebrates, a membrana propria, or basement membrane, separates the theca interna from the granulosa. Hope (1965) and Hertig and Adams (1967) have described the fine structure of developing follicles. As the follicles grow, small, fluid-filled, intercellular cavities appear in the granulosa; these coalesce to form the antrum, which is bounded by granulosa cells. These cells, on one side of the antrum, contain the oocyte and form the discus proligerus or cumulus which projects into the antrum (Fig. 10a); this, with its contained oocyte becomes detached from the granulosa prior to ovulation and floats free in the follicular fluid. The outer cells disperse, and the inner cells become club-shaped and radially arranged around the oocyte, their nuclei being located peripherally, to form the corona radiata. Studies on the follicular fluid have been reviewed by McNatty (1978).

b. The Monotreme Ovary (Fig. 11). The ovaries of the two extant monotreme mammals, *Tachyglossus* (*Echidna*), the spiny anteater, and *Ornithorhynchus,* the duckbilled platypus, have received little attention since the early studies of Solomons and Gatenby (1924), Hill and Gatenby (1926), Garde (1930), and Flynn and Hill (1939). The ovaries are paired, both being

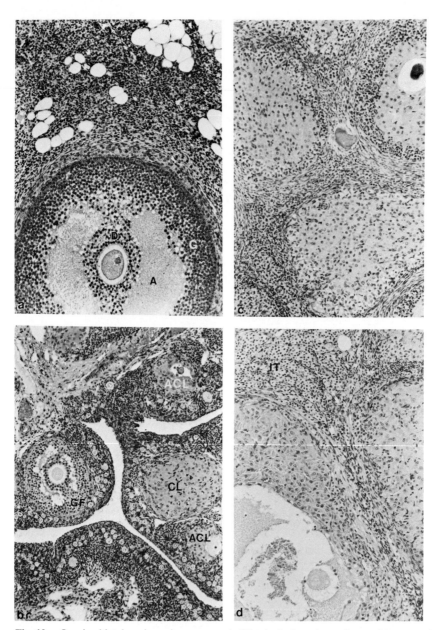

Fig. 10. Ovarian histology in various mammals (histological sections of part of the ovary in each case). (a) *Galea musteloides,* the cuis. The maturing Graafian follicle shows early cumulus formation (D), a distinct membrana propria, antrum (A), granulosa (G), and thecae interna and externa. The large vacuoles in the stroma are the remains of lipid droplets. (b) *Lagostomus maximus,* the plains viscacha. Note the extensive surface area due to infolding of the ovarian surface, numerous oocytes, mature Graafian follicle with small antrum (GF), mature true corpus luteum (CL), and two small accessory corpora lutea (ACL). (c) *Myoprocta pratti,* the acouchi. Note three atretic follicles (accessory corpora lutea) and remains of oocyte in upper right follicle. (d) *Dasyprocta aguti,* the agouti. Note atretic follicle (lower left) and abundant interstitial tissue (IT). [Parts (a) and (b) from Harrison and Weir (1977). Parts (c) and (d) by courtesy of Dr. Barbara Weir.]

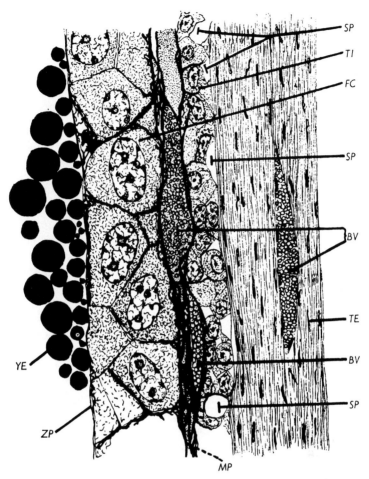

Fig. 11. *Ornithorhynchus* (the duckbilled platypus): follicular structure. Diagrammatic representation of part of the follicle of a full-grown oocyte of *Ornithorhynchus*. BV, capillaries; FC, follicular epithelium; MP, membrana propria; SP, spaces in the theca interna; TE, theca externa; TI, theca interna; YE, yolk spheres in cytoplasm of oocyte; ZP, zona pellucida. [Reproduced from Hill and Gatenby (1926).]

functional in *Tachyglossus*, but only the left in *Ornithorhynchus*. The monotreme ovary is similar to that of reptiles, consisting of a narrow cortical zone in which the oocytes are located and a central spongy medulla consisting mainly of blood vessels and lymphatics. Follicular structure shows no differences from that in other mammals that cannot be explained by the relatively large yolky oocytes (4.0–4.5 mm in diameter) that are produced by monotremes. Both granulosa and the two-layered theca are thicker than in reptiles; this may be due to the smaller oocyte. The thecal layers are separated from the granulosa by a prominent basement membrane on which lies an abundant blood supply as in all vertebrates with yolky oocytes. A layer of fluid lies between oocyte and granulosa in the mature follicles of both mono-treme species (Flynn and Hill, 1939), and it has been suggested that this is the homologue of the follicular fluid of eutherian mammals.

c. The Marsupial Ovary. What is known of the marsupial ovary has been reviewed by Brambell (1956) and Harrison and Weir (1977); corpora lutea have received special attention. The oocytes are yolk-free and, in conse-quence, the follicles show only minor differences from those of the *Eutheria*. A corona radiata does not appear to be formed in *Dasyurus* (Hill, 1910), and naked oocytes are ovulated in some species. In *Ericulus* (Strauss, 1938), fertilization of the oocyte takes place within the follicle, an interesting point of comparison with the viviparous teleosts. The work of Lintern-Moore *et al.* (1976) on follicular growth patterns supports the view that the marsupial ovary is basically similar to that of the Eutheria. In both, in the first growth phase, the oocyte and follicle undergo coincident growth, the follicle becom-ing multilayered; in the second phase, follicular fluid appears and only the follicle grows.

i. Corpora Atretica (Fig. 10c,d). Atresia occurs at all ages, though more commonly at certain times than at others. Follicles that start to grow soon after birth are at risk (Krarup *et al.,* 1969), and atresia before puberty ex-ceeds that during maturity (see Byskov, 1978). Weir and Rowlands (1977) point to the fact that of the 2 million oocytes said to be present in the human ovary at birth, only about 400 could possible be ovulated. The discrepancy is largely accounted for by atresia, which takes a number of different forms and adds appreciably to the morphological variety manifested by the ovary. It has been calculated that in the mouse, 200–300 small follicles undergo atre-sia each day in the first 2 weeks of life (Pedersen, 1970), and in the rat, about 500 follicles are lost per day during the same period. The total loss of follicles through atresia ranges from ca. 77% in mouse and rat to 99.9% in humans and the dog. Table II illustrates the relationship between oocytes that are ovulated and those that are lost through atresia in seven mammalian species (Byskov, 1978). The cellular changes that occur during atresia are different in follicles of different sizes. Ultrastructural changes in the oocyte during atresia have been described in the rat and mouse by Vazques-Nin and Sotelo (1967) and Belterman (1965), respectively. Comprehensive reviews by Weir and Rowlands (1977) and Byskov (1978) should be consulted for a more detailed discussion of atretic structures in the mammal ovary.

ii. Corpora Lutea (Fig. 10b). A range of structures in the mammalian ovary are termed "corpora lutea"; these have been discussed in detail by Weir and Rowlands (1977). True corpora lutea (also called corpora lutea of

TABLE II

**Number of Oocytes Lost by Ovulation and
Atresia during the Fertile Period in Different
Mammalian Species[a]**

Species	Total number of oocytes by the onset of the fertile period	Total number of oocytes ovulated during the fertile period	Oocytes that disappeared by atresia as percentage of total number of oocytes
Human	390,000	360	99.9
Monkey	200,000	280	99.9
Dog	70,000	60	99.9
Guinea pig	15,000	270	99
Pig	100,000	1190	98
Mouse	3,000	630	77
Rat	5,000	1260	75

[a] From Byskov (1978).

pregnancy or corpora lutea graviditatis) are those usually referred to; they are universal in mammals, though they are possibly not uniquely mammalian. Secondary corpora lutea are those developing from follicles which ovulate during pregnancy. Accessory corpora lutea develop by the luteinization of unruptured atretic follicles, and corpora haemorrhagica are corpora lutea with a central blood clot. Corpora lutea found in nonpregnant animals are termed false corpora lutea, corpora lutea of the cycle, or corpora lutea of menstruation. Only true corpora lutea will be considered here. The relevant literature, both early and more recent, has been fully reviewed by Harrison (1948), Brambell (1956), and Weir and Rowlands (1977). The fully developed corpus luteum consists of large polygonal cells, closely packed and containing a yellow pigment; the luteal cells, supported by a framework of connective tissue probably derived from the theca externa; and a rich vasculature. Corpora lutea are usually solid, though they may be hollow and they frequently project from the surface of the ovary. The extent to which the various constituent layers of the follicle contribute to their formation has been the subject of controversy. This has centered largely on the role of the theca interna in the genesis of corpora lutea. Brambell (1956) has comprehensively reviewed and synthesized the literature and has come to the widely accepted conclusion that the evidence for the granulosa cells giving rise to the luteal cells is overwhelming. The role of the theca interna is more difficult to establish, though he concludes that it, also, may be involved. The theca externa seems to provide the connective tissue skeleton of the corpus luteum.

iii. Interstitial Cells (Fig. 10d). Interstitial cells are an important component of the mammal ovary. Their numbers vary with species and vary cyclically, and their function is problematical (Harrison and Weir, 1977). They usually form aggregations which are sometimes referred to as interstitial glands. In some species, like the agouti (Weir, 1971, and Fig. 10d), they are the predominant nongerminal constituent of the ovary, and during pregnancy they become luteinized (Harrison, 1962). Harrison and Weir (1977) have reviewed the literature on the histological and histochemical characteristics of mammalian interstitial cells, which shows them to be typical of steroid-secreting cells in having abundant smooth endoplasmic reticulum, spherical mitochondria with tubular cristae, lipid droplets, and conspicuous Golgi apparatus.

iv. Blood Supply and Innervation. The blood supply to the ovaries of a range of mammals has been reviewed by Harrison and Weir (1977) and Ellinwood *et al.* [1978; see also Scanes *et al.,* 1982, for more recent references]. The ovarian artery arises as a branch of the dorsal aorta posterior to the renal arteries. It forms a network of capillaries, below the connective tissue capsule, from which the ovarian vasculature springs. The follicular blood supply becomes specialized only after antrum formation. Then, probably as in all vertebrates (see Section II,B), an extensive meshwork arises on the thecal side of the basement membrane lying between theca interna and granulosa. A second vascular meshwork lies in the theca externa (Andersen, 1926); the granulosa is avascular. Corpora lutea have a rich blood supply and in some cases have blood in a central lumen. Antral follicles have an extensive lymphatic supply consisting of valved channels, first formed in the theca externa; they frequently surround both layers of arterioles (Andersen, 1926). Ovarian innervation originates in the ovarian ganglion and lumbar sympathetic chain and consists of both sympathetic and parasympathetic fibers which innervate both blood vessels and follicles and appear to be randomly distributed in the stroma; they do not penetrate the granulosa (Owman and Sjöberg, 1966).

III. SUMMARY AND COMMENT

This comparative survey of ovarian morphology supports the initial contention that it is the basic similarities between the ovaries of different vertebrates that are remarkable, not the superficial variations. Yet, a number of differences have emerged, and it is important that their significance should be properly assessed, although it is often difficult to decide whether the putative differences arise from ignorance, faulty interpretation of the data, or are basic phylogenetic divergences. The data stem mainly from the techniques of descriptive morphology, which are acknowledged to provide an inadequate basis for interpreting the dynamic changes occurring during embryogenesis. Furthermore, research interests have moved steadily away from structural studies like anatomy, morphology, and histology, and some of the discrepancies may never be resolved. It is clear that electron microscopy and, especially, microsurgery and transplantation (Byskov *et al.*, 1977) and the experimental techniques recently applied to teleosts (see Nagahama, 1983) could be used with advantage to investigate some of the outstanding problems in ovarian morphology.

The basic similarities between ovarian origins, development, and structure across the vertebrates can be generalized within the constraints discussed in the Introduction, as follows: Primordial germ cells are closely similar in origin, cytology, and fate; oogonia form nests and are in cellular contact with each other through intercellular bridges; oogenesis, *sensu stricto*, is identical, though not in timing; the first meiotic prophase is always protracted; oocyte ultrastructure, both cytoplasmic and nuclear, is closely similar; a recognizable germinal epithelium is always present and it contributes most of the cortical tissue; folliculogenesis gives rise to follicles that are basically similar in structure from cyclostomes to mammals; atresia, and probably the methods by which it is achieved, is universal in the vertebrate ovary. The differences are about equally numerous, although, apart from the presence or absence of yolk, they show little allegiance to phylogenetic relationships, and some differences may be apparent rather than real, since they fall in the gray areas that have been insufficiently researched.

Ovaries are usually paired, but sometimes single. In general, it is species with large yolky eggs that have single ovaries; presumably, constraints of space and streamlining have provided the selection pressure favoring one ovary rather than two, and it is clear that this is a superficial difference. Oogenesis, *sensu stricto*, is restricted to the early stages of gonad development in some vertebrate groups, and in others it occurs cyclically during adult life. This also would appear to be a basic difference, yet the two types appear to be haphazardly distributed among the vertebrates, and there are exceptions even in the mammals to the precocious, once only, oogeneiss that is the rule. One of the most difficult areas in which to assess the significance of differences is folliculogenesis, since there is no consensus as to the origins of thecal and granulosa cells. Most workers appear to believe that the granulosa cells arise from the germinal epithelium. However, it has also been suggested that the cells may arise from mesenchyme, the rete ovarii, or the interrenal blastema. The granulosa is multifunctional, nutritive, enzyme-secretory, steroidogenic, and phagocytic. There is therefore, a *prima facie* case for considering that its cells may be polymorphic and may arise from more than one source. More work is required in this important area, using techniques that are more appropriate than histology.

One of the striking differences between the ovaries of mammals and nonmammals is the size of the ovary relative to that of the body, yet this is entirely accounted for by the presence of yolk in the oocytes of the latter. It is much less pronounced if the comparison is made between the mammal ovary and the previtellogenic ovary of other vertebrates. During vitellogenesis, the follicular investments become squamous as yolk continues to accumulate, whereas in mammals the granulosa becomes multilayered although the number of granulosa cells relative to the size of the oocyte may not be very different in the two cases. Also, in nonmammals, the stromal elements appear to be sparse, but this again is relative.

It seems unlikely that the corpora lutea of nonmammalian species are ever as well developed as those of mammals, though the granulosa cells of the postovulatory follicles of some viviparous elasmobranchs and reptiles, widely separated groups, produce glandular bodies which luteinize and may be steroidogenic. These appear to be "true" corpora lutea, and their presence in such disparate groups illustrates one of the most striking attributes of the vertebrate ovary: its plasticity.

The fluid-filled antrum is a mammalian specialization which plays an important part in the control of follicular activities; in vitellogenic follicles, it may be that the spaces between contiguous granulosa cells have a similar role. There is no doubt that in nonmammals, intraovarian mechanisms are present that control ovulation, follicular growth and selection, follicular hierarchies, and atresia, though they are insufficiently well understood to allow detailed comparisons with their mammalian counterparts.

The series of volumes to which this chapter is a contribution is particularly concerned with the contributions comparative studies have made to human endocrinology and biomedicine. Unfortunately, such contributions have been small in respect of ovarian morphology and are likely to remain so. This is due almost entirely to the differences imposed on the ovaries of nonmammals by the presence of yolk. As a result, in spite of the close basic similarities, there appear to be few, if any, problems of ovarian morphology in mammals that can be illuminated from what is known of nonmammals.

REFERENCES

Andersen, D. H. (1926). Lymphatics and blood-vessels of the ovary of the sow. *Contrib. Embryol. Carnegie Inst.* **17,** 107–123.

Anderson, E. (1967). The formation of the primary envelope during oocyte differentiation in teleosts. *J. Cell. Biol.* **35,** 193–212.

Anderson, E. (1968). Cortical alveoli formation and vitellogenesis during oocyte differentiation in the pipefish, *Syngnathus fuscus,* and the killifish, *Fundulus heteroclitus. J. Morphol.* **125,** 23–60.

Anderson, E. (1974). Comparative aspects of the ultrastructure of the female gamete. *Int. Rev. Cytol., Suppl.* **4,** 1–70.

Anderson, J. W., and Yatvin, M. B. (1970). Metabolic and ultrastructural changes in the frog ovarian follicle in response to pituitary stimulation. *J. Cell Biol.* **46,** 491–504.

Andreuccetti, P., Taddei, C., and Filosa, S. (1978). Intercellular bridges between follicle cells and oocyte during the differentiation of follicular epithelium in *Lacerta sicula*. *J. Cell. Sci.* **33**, 341–350.

Azevedo, C. (1974). Évolution des enveloppes ovocytaires au cours de l'ovogénèse, chez un téléostéen vivipare, *Xiphophorus helleri*. *J. Microsc.* (*Paris*) **21**, 43–54.

Badir, N. (1968). Structure and function of corpus luteum during gestation in the viviparous lizard *Chalcides ocellatus*. *Anat. Anz.* **122**, 1–10.

Barr, W. A. (1963). Endocrine control of the sexual cycle in the plaice *Pleuronectes platessa*. 1. Cyclical changes in the normal ovary. *Gen. Comp. Endocrinol.* **3**, 197–204.

Barr, W. A. (1969). Patterns of ovarian activity. *In* "Perspectives in Endocrinology" (E. J. W. Barrington and C. Barker Jørgensen, eds.), pp. 164–238. Academic Press, New York.

Beach, A. W. (1959). Seasonal changes in the cytology of the ovary and of the pituitary gland of the goldfish. *Can. J. Zool.* **37**, 615–626.

Beams, H. W., and Kessel, R. G. (1973). Oocyte structure and early vitellogenesis in the trout *Salmo gairdneri*. *Am. J. Anat.* **136**, 105–122.

Bellairs, R. (1964). Biological aspects of the yolk of the hen's egg. *Adv. Morphog.* **4**, 217–272.

Bellairs, R. (1965). The relationship between oocyte and follicle in the hen's ovary as shown by electron microscopy. *J. Embryol. Exp. Morphol.* **13**, 215–233.

Belterman, R. (1965). Elektronenmikroskopische Befunde bei beginnender Follikelatresie im Ovar der Maus. *Arch. Gynekol.* **200**, 601–609.

Bennett, T., and Malmfors, T. (1970). The adrenergic nervous system of the domestic fowl. *Z. Zellforsch. Mikrosk. Anat.* **106**, 22–50.

Betz, T. W. (1963). The ovarian histology of the diamond-backed water snake, *Natrix rhombifera* during the reproductive cycle. *J. Morphol.* **113**, 245–260.

Bowers, A. B., and Holliday, F. G. T. (1961). Histological changes in the gonad associated with the reproductive cycle of the herring *Clupea harengus* L. *Mar. Res. Ser. Scott. Home Dep.* **5**, 1–16.

Boyd, M. M. M. (1940). The structure of the ovary and the formation of the corpus luteum in *Hoplodactylus maculatus* Gray. *Q. J. Microsc. Sci.* [N.S.] **82**, 337–376.

Bradley, O. C., and Grahame, T. (1961). "The Structure of the Fowl," 4th ed. Oliver & Boyd, Edinburgh.

Braekevelt, G. R., and McMillan, D. B. (1967). Cyclic changes in the ovary of the brook stickleback *Eucalia inconstans* (Kirtland). *J. Morphol.* **123**, 373–396.

Bragdon, D. E. (1952). Corpus luteum formation and follicular atresia in the common garter snake, *Thamnophis sirtalis*. *J. Morphol.* **91**, 413–445.

Brambell, F. W. R. (1925). The oogenesis of the fowl (*Gallus bankiva*). *Philos. Trans. R. Soc. London, Ser. B* **214**, 113–151.

Brambell, F. W. R. (1956). Ovarian changes. *In* "Marshall's Physiology of Reproduction" (A. S. Parkes, ed.), 3rd ed., Vol. 1, Part I, pp. 397–542. Longmans Green, London and New York.

Branca, A. (1925). L'ovocyte atrésique et son involution. *Arch. Biol.* **35**, 325–440.

Bretschneider, L. H., and Duyvené de Wit, J. J. (1947). "Sexual Endocrinology of Non-Mammalian Vertebrates." Elsevier, Amsterdam.

Brock, J. (1878). Beiträge zur Anatomie und Histologie der Geschlechtsorgane der Knochenfische. *Morphol. Jahrb.* **4**, 505–572.

Browning, H. C. (1973). The evolutionary history of the corpus luteum. *Biol. Reprod.* **8**, 128–157.

Burden, H. W. (1972). Ultrastructural observations on ovarian perifollicular smooth muscle in the cat, guinea pig and rabbit. *Am. J. Anat.* **133**, 125–142.

Busson-Mabillot, S. (1967a). Structure ovarienne de la Lamproie adulte (*Lampetra planeri* Bloch). I. Zone pellucide, Morphogénèse et constitution chimique. *J. Microsc.* (*Paris*) **6**, 577–598.

Busson-Mabillot, S. (1967b). Structure ovarienne de la lamproie adulte (*Lampetra planeri* Bloch). II. Les enveloppes de l'ovocyte: Cellules folliculaires et stroma ovarien. *J. Microsc. (Paris)* **6**, 807–838.

Busson-Mabillot, S. (1973). Évolution des enveloppes de l'ovocyte et de l'oeuf chez un poisson téléostéen. *J. Microsc. (Paris)* **18**, 23–44.

Byskov, A. G. (1978). Follicular atresia. In "The Vertebrate Ovary: Comparative Biology and Evolution" (R. E. Jones, ed.), pp. 533–562. Plenum, New York.

Byskov, A. G., Skakkebaek, N. E., Stafanger, G., and Peters, H. (1977). Influence of ovarian surface epithelium and rete ovarii on follicle formation. *J. Anat.* **123**, 77–86.

Castro, J. I. (1983). "The Sharks of North American Waters." Texas A & M Univ. Press, College Station.

Chan, S. T. H. (1970). Natural sex reversal in vertebrates. *Philos. Trans. R. Soc. London, Ser. B* **259**, 59–71.

Chieffi, G., and Botte, V. (1963). Osservazioni istochimiche sul attività della steroide—3β-olodeidrogenasi nell'interrenale e nelle gonadi di girini adulti de *Rana esculenta*. *Riv. Istol. Norm. Patol.* **9**, 172–174.

Chubb, C. F., and Potter I. C. (1984). The reproductive biology and estuarine movements of the gizzard shad, *Nematalosa vlaminghi* (Munro). *J. Fish Biol.* **25**, 527–543.

Cunningham, J. T., and Smart, W. A. M. (1934). The structure and origin of the corpora lutea in some of the lower vertebrata. *Proc. R. Soc. London, Ser. B* **116**, 258–281.

Cyrus, R. V., Mahmoud, I. Y., and Klicka, J. (1978). Fine structure of the corpus luteum of the snapping turtle, *Chelydra serpentina*. *Copeia*, No. 4, pp. 622–627.

Dahl, E. (1970). Studies on the fine structure of ovarian interstitial tissue. 3. The innervation of the thecal gland of the domestic fowl. *Z. Zellforsch. Mikrosk. Anat.* **109**, 212–226.

Das, R. S. (1931). On the cytoplasmic inclusions in the oogenesis of birds. *Russ. Arkh. Anat. Gistol. Embriol.* **10**, 309–321.

Dodd, J. M. (1960). Gonadal and gonadotrophic hormones in lower vertebrates. In "Marshall's Physiology of Reproduction" (A. S. Parkes, ed.), 3rd ed., Vol. I, Part 2, pp. 417–582. Longmans Green, London and New York.

Dodd, J. M. (1977). The structure of the ovary of non-mammalian vertebrates. In "The Ovary" (S. Zuckerman and B. J. Weir, eds.), 2nd ed., Vol. 1, pp. 219–263. Academic Press, New York.

Dodd, J. M. (1983). Reproduction in cartilaginous fishes (Chondrichthyes). In "Fish Physiology" (W. S. Hoar, D. J. Randall, and E. M. Donaldson, eds.), Vol. 9, Part A, pp. 31–95. Academic Press, New York.

Dodd, J. M., and Dodd, M. H. I. (1986). Evolutionary aspects of reproduction in cyclostomes, and cartilaginous fishes. In "Evolutionary Biology of Primitive Fishes" (R. E. Foreman, A. Gorbman, J. M. Dodd, and R. Olsson, eds.), pp. 295–319. Plenum, New York.

Dodd, J. M., and Sumpter, J. P. (1984). Fishes. In "Marshall's Physiology of Reproduction" (G. E. Lamming, ed.), 4th ed., Vol. 1, pp. 1–126. Churchill-Livingstone, Edinburgh and London.

Dodd, J. M., Evennett, P. J., and Goddard, C. K. (1960). Reproductive endocrinology in cyclostomes and elasmobranchs. *Symp. Zool. Soc. London* **1**, 77–103.

Dodd, M. H. I., and Dodd, J. M. (1980). Ultrastructure of the ovarian follicle of the dogfish *Scyliorhinus canicula*. *Gen. Comp. Endocrinol.* **40**, 330–331.

Duke, K. L. (1978). Nonfollicular ovarian components. In "The Vertebrate Ovary: Comparative Biology and Evolution" (R. E. Jones, ed.), pp. 563–582. Plenum, New York.

Dumont, J. N. (1972). Oogenesis in *Xenopus laevis* (Daudin). I. Stages of oocyte development in laboratory maintained animals. *J. Morphol.* **136**, 153–180.

Ellinwood, W. E., Nett, T. M., and Niswender, G. D. (1978). Ovarian vasculature: Structure and function. In "The Vertebrate Ovary: Comparative Biology and Evolution" (R. E. Jones, ed.), pp. 583–614. Plenum, New York.

Evennett, P. J. (1963). The endocrine control of reproduction in the river lamprey *Lampetra fluviatilis* L. Ph. D. Thesis, University of St. Andrews.

Eyeson, K. N. (1971). Pituitary control of ovarian activity in the lizard *Agama agama. J. Zool.* **165**, 367–372.

Filosa, S., Taddei, C., and Andreuccetti, P. (1979). The differentiation and proliferation of follicle cells during oocyte growth in *Lacerta sicula. J. Embryol. Exp. Morphol.* **54**, 5–15.

Flügel, H. (1967). Licht und elektronenmikroskopische Untersuchungen an Oozyten und Eiern einiger Knochenfische. *Z. Zellforsch. Mikrosk. Anat.* **83**, 82–116.

Flynn, T. T., and Hill, J. P. (1939). The development of the Monotremata. IV. Growth of the ovarian ovum, maturation, fertilization, and early cleavage. *Trans. Zool. Soc. London* **24**, 445–622.

Fox, H. (1977). The urinogenital system of reptiles. *In* "Biology of the Reptilia" (C. Gans and T. S. Parsons, eds.), Vol. 6, pp. 1–157. Academic Press, New York.

Franchi, L. L., Mandl, A. M., and Zuckerman, S. (1962). The development of the ovary and the process of oogenesis. *In* "The Ovary" (S. Zuckerman, ed.), Vol. 1, pp. 1–88. Academic Press, New York.

Fukayama, S., and Takahashi, H. (1982). Sex differentiation and development of the gonad in the Japanese river lamprey, *Lampetra japonica. Bull. Fac. Fish., Hokkaido Univ.* **33**, 206–216.

Fukayama, S., and Takahashi, H. (1983). Sex differentiation and development of the gonad in the sand lamprey. *Lampetra reissneri. Bull. Fac. Fish., Hokkaido Univ.* **34**, 279–290.

Garde, M. L. (1930). The ovary of *Ornithorhynchus*, with special reference to follicular atresia. *J. Anat.* **64**, 422–453.

Gardiner, D. M. (1978). Cyclic changes in fine structure of the epithelium lining the ovary of the viviparous teleost *Cymatogaster aggregata* (Perciformes: Embiotocidae). *J. Morphol.* **156**, 367–380.

Gerrard, A. M., Jones, R. E., and Roth, J. J. (1973). Thecal vascularity in ovarian follicles of different size and rank in the lizard *Anolis carolinensis. J. Morphol.* **141**, 227–234.

Gilbert, A. B. (1965). Innervation of the ovarian follicle of the domestic hen. *Q. J. Exp. Physiol. Cogn. Med. Sci.* **50**, 437–445.

Gilbert, A. B. (1967). The formation of the egg in the domestic chicken. *Adv. Reprod. Physiol.* **1**, 111–180.

Gilbert, A. B. (1969). Innervation of the ovary of the domestic hen. *Q. J. Exp. Physiol. Cogn. Med. Sci.* **54**, 404–411.

Gilbert, A. B. (1971). The Ovary. *In* "Physiology and Biochemistry of the Domestic Fowl" (D. J. Bell and B. M. Freeman, eds.), Vol. 3, pp. 1163–1208. Academic Press, New York.

Gilbert, A. B. (1979). Female genital organs. *In* "Form and Function in Birds" (A. S. King and J. McLelland, eds.), Vol. 1, Chapter 5, pp. 237–360. Academic Press, New York.

Gilbert, A. B., and Wells, J. W. (1984). Structure and function of the ovary. *In* "Reproductive Biology of Poultry" (F. J. Cunningham, P. E. Lake, and D. Hewitt, eds.), pp. 15–27. Longmans Group, Harlow.

Gilbert, A. B., Hardie, M. A., Perry, M. M., Dick, H. R., and Wells, J. W. (1980). Cellular changes in the granulosa layer of the maturing ovarian follicle of the domestic fowl. *Br. Poult. Sci.* **21**, 257–263.

Gokhale, S. V. (1957). Seasonal histological changes in the gonads of the whiting *Gadus merlangus* L. and the Norway pout *G. esmarkii* Nilsson. *Indian J. Fish* **4**, 92–112.

Gondos, G. (1978). Oogonia and oocytes in mammals. *In* "The Vertebrate Ovary: Comparative Biology and Evolution" (R. E. Jones, ed.), pp. 83–120. Plenum, New York.

Gorbman, A. (1983). Reproduction in cyclostome fishes and its regulation. *In* "Fish Physiology" (W. S. Hoar, D. J. Randall, and E. M. Donaldson, eds.), Vol. 9, Part A, pp. 1–29. Academic Press, New York.

Gorbman, A., and Dickhoff, W. W. (1978). Endocrine control of reproduction in hagfish. *In*

"Comparative Endocrinology" (P. J. Gaillard and H. H. Boer, eds.), pp. 29–57. Elsevier/ North-Holland Biomedical Press, Amsterdam.

Gupta, N. N., and Yamamoto, K. (1972). Electron microscope study on the fine structural changes in the oocytes of the goldfish, *Carassius auratus*, during yolk formation stage. *Bull. Fac. Fish., Hokkaido Univ.* **22**, 187–206.

Guraya, S. S. (1965a). A comparative histochemical study of fish (*Channa marulius*) and amphibian (*Bufo stomaticus*) oogenesis. *Z. Zellforsch. Mikrosk. Anat.* **65**, 662–700.

Guraya, S. S. (1965b). A histochemical study of follicular atresia in the snake ovary. *J. Morphol.* **117**, 151–170.

Guraya, S. S. (1969). Histochemical observations on the interstitial gland cells of amphibian ovary (frog, toad). *Gen. Comp. Endocrinol.* **13**, 173–178.

Guraya, S. S. (1976). Recent advances in the morphology, histochemistry and biochemistry of steroid-synthesising cellular sites in the non-mammalian vertebrate ovary. *Int. Rev. Cytol.* **44**, 365–409.

Guraya, S. S. (1978). Maturation of the follicular wall of non-mammalian vertebrates. *In* "The Vertebrate Ovary: Comparative Biology and Evolution" (R. E. Jones, ed.), pp. 261–329. Plenum, New York.

Guraya, S. S., and Varma, S. K. (1976). Morphology of ovarian changes during the reproductive cycle of the house lizard, *Hemidactylus flaviviridis*. *Acta Morphol. Neerl. Scand.* **14**, 165–192.

Hardisty, M. W. (1965). Sex differentiation and gonadogenesis in lampreys. I. The ammocoete gonads of the brook lamprey, *Lampetra planeri*. *J. Zool.* **146**, 305–345.

Hardisty, M. W. (1971). Gonadogenesis, sex differentiation and gametogenesis. *In* "The Biology of Lampreys" (M. W. Hardisty and I. C. Potter, eds.), Vol. 1, pp. 295–360. Academic Press, New York.

Harrington, R. W., Jr. (1974). Sex determination and differentiation in fishes. *In* "Control of Sex in Fishes" (C. B. Schreck, ed.), pp. 3–12. Extension Division, Virginia Polytechnic Institute and State University, Blacksburg.

Harrison, R. J. (1948). The development and fate of the corpus luteum in the vertebrate series. *Biol. Rev. Cambridge Philos. Soc.* **23**, 269–331.

Harrison, R. J. (1962). The structure of the ovary in mammals. *In* "The Ovary" (S. Zuckerman, ed.), Vol. 1, Chapter 2c, pp. 143–187. Academic Press, New York.

Harrison, R. J., and Weir, B. J. (1977). Structure of the mammalian ovary. *In* "The Ovary" (S. Zuckerman and B. J. Weir, eds.), Vol. 1, Chapter 4, pp. 113–217. Academic Press, New York.

Henderson, N. E. (1963). Extent of atresia in mature ovaries of the Eastern brook trout, *Salvelinus fontinalis*. *J. Fish. Res. Board Can.* **20**, 899–908.

Hertig, A. T., and Adams, E. C. (1967). Studies on the human oocyte and its follicle. I. Ultrastructural and histochemical observations on the primordial follicle stage. *J. Cell Biol.* **34**, 647–676.

Hett, J. (1923). Das Corpus luteum des Molches (*Triton vulgaris*). *Z. Anat. Entwicklungsgesch.* **68**, 243–272.

Hill, J. P. (1910). The early development of the Marsupialia, with special reference to the native cat (*Dasyurus viverrinus*). *Q. J. Microsc. Sci.* **56**, 1–36.

Hill, J. P., and Gatenby, J. B. (1926). The corpus luteum of the Monotremata. *Proc. Zool. Soc. London* **47**, 715–763.

Hisaw, F. L., and Albert, A. (1947). Observations on the reproduction of the spiny dogfish (*Squalus acanthias*). *Biol. Bull.* (*Woods Hole, Mass*) **92**, 187–199.

Hisaw, F. L., and Hisaw, F. L., Jr. (1959). Corpora lutea of elasmobranch fishes. *Anat. Rec.* **135**, 269–277.

Hoar, W. S. (1957). The gonads and reproduction. *In* "The Physiology of Fishes" (M. E. Brown, ed.), Vol. 1, pp. 287–321. Academic Press, New York.

Hoar, W. S. (1965). Hormones and reproduction in fishes. *Annu. Rev. Physiol.* **27**, 51–70.

Hoar, W. S. (1969). Reproduction. *In* "Fish Physiology" (W. S. Hoar and D. J. Randall, eds.), Vol. 3, pp. 1–72. Academic Press, New York.

Hoar, W. S., and Nagahama, Y. (1978). The cellular sources of sex steroids in teleost gonads. *Ann. Biol. Anim., Biochim., Biophys.* **18,** 893–898.

Hope, J. (1965). The fine structure of the developing follicle of the rhesus ovary. *J. Ultrastruct. Res.* **12,** 592–610.

Hope, J., Humphries, A. A., Jr., and Bourne, G. H. (1963). Ultrastructural studies on developing oocytes of the salamander *Triturus viridescens*. 1. The relationship between follicle cells and developing oocytes. *J. Ultrastruct. Res.* **9,** 302–324.

Hope, J., Humphries, A. A., Jr., and Bourne, G. H. (1964a). Ultrastructural studies on developing oocytes of the salamander *Triturus viridescens*. II. The formation of yolk. *J. Ultrastruct. Res.* **10,** 547–556.

Hope, J., Humphries, A. A., Jr., and Bourne, G. H. (1964b). Ultrastructural studies on developing oocytes of the salamander *Triturus viridescens*. III. Early cytoplasmic changes and the formation of pigment. *J. Ultrastruct. Res.* **10,** 557–566.

Hubert, J. (1971a). Étude histologique et ultrastructurale de la granulosa à certains stades de développement du follicule ovarien chez un lézard, *Lacerta vivipara* Jacquin. *Z. Zellforsch. Mikrosk. Anat.* **116,** 46–59.

Hubert, J. (1971b). Aspects ultrastructuraux des relations entre les couches folliculaires et l'ovocyte depuis la formation du follicule jusqu'au début de la vitellogenèse chez le lézard *Lacerta vivipara* Jacquin. *Z. Zellforsch. Mikrosk. Anat.* **116,** 240–249.

Hubert, J. (1973). Les cellules piriformes du follicule ovarien de certains reptiles. *Arch. Anat. Histol. Embryol.* **56,** 5–18.

Hubert, J. (1975). Étude descriptive, cytochimique et autoradiographique en microscopie electronique sur les sphères granulaires nucléolaires et les micronucléoles des cellules piriformes chez le lézard, *Lacerta vivipara* Jacquin. *C. R. Hebd. Seances Acad. Sci.* **280,** 2555–2557.

Hubert, J. (1976). Étude ultrastructurale des cellules piriformes du follicule ovarien chez 5 sauriens. *Arch. Anat. Histol. Embryol.* **65,** 47–58.

Hutt, F. B. (1949). "Genetics of the Fowl," McGraw-Hill, New York.

Ingram, D. L. (1962). Atresia. *In* "The Ovary" (S. Zuckerman, ed.), Vol. 1, pp. 247–273. Academic Press, New York.

Joly, J., and Picheral, B. (1972). Ultrastructure, histochimie et physiologie du follicule pre-ovulatoire et du corps jaune de l'urodèle ovo-vivipare *Salamandra salamandra* (L). *Gen. Comp. Endocrinol.* **18,** 235–259.

Jones, R. E., ed. (1978). "The Vertebrate Ovary: Comparative Biology and Evolution." Plenum, New York.

Jones, R. E., Tokarz, R. R., and La Greek, F. T. (1975). Endocrine control of clutch size in reptiles. V. FSH-induced follicular formation and growth in immature ovaries of *Anolis carolinensis*. *Gen. Comp. Endocrinol.* **26,** 354–367.

Jones, R. E., Summers, C. H., Austin, H. B., Smith, H. M., and Gleeson, T. T. (1983). Ovarian, oviductal, and adrenal vascular connections in female lizards (genus *Anolis*). *Anat. Rec.* **206,** 247–255.

Kessel, R. G., and Panje, W. R. (1968). Organization and activity in the pre- and post-ovulatory follicle of *Necturus maculosus*. *J. Cell Biol.* **39,** 1–34.

Kessel, R. G., Beams, H. W., and Tung, H. N. (1984). Relationships between annulate lamellae and filament bundles in oocytes of the zebrafish, *Brachydanio rerio*. *Cell Tissue Res.* **236,** 725–727.

Kobayashi, H., Ichikawa, T., Suzuki, H., and Sekimoto, M. (1972). Seasonal migration of the hagfish *Eptatretus burgeri*. *Jpn. J. Ichthyol.* **19,** 191–194.

Krarup, T., Pedersen, T., and Faber, M. (1969). Regulation of oocyte growth in the mouse ovary. *Nature (London)* **224,** 187–188.

Kuchnow, K. P., and Scott, J. R. (1977). Ultrastructure of the chorion and its micropyle apparatus in the mature *Fundulus heteroclitus* (Wallbaum) ovum. *J. Fish Biol.* **10**, 197–201.

Lambert, J. G. D. (1970a). The ovary of the guppy *Poecilia reticulata*. The granulosa cells as sites of steroid biosynthesis. *Gen. Comp. Endocrinol.* **15**, 464–476.

Lambert, J. G. D. (1970b). The ovary of the guppy *Poecilia reticulata*. The atretic follicle, a *corpus atreticum* or a *corpus luteum praeovulationis*. *Z. Zellforsch. Mikrosk, Anat.* **107**, 54–67.

Lance, V., and Callard, I. P. (1969). A histochemical study of ovarian function in the ovoviviparous elasmobranch *Squalus acanthias*. *Gen. Comp. Endocrinol.* **13**, 255–267.

Lance, V., and Callard, I. P. (1978). Hormonal control of ovarian steroidogenesis in nonmammalian vertebrates. In "The Vertebrate Ovary: Comparative Biology and Evolution" (R. E. Jones, ed.), pp. 361–407. Plenum, New York.

Lanzing, W. J. R. (1959). Studies on the River Lamprey, *Lampetra fluviatilis* during its anadromous migration. Thesis, University of Utrecht.

Larsen, L. O. (1978). Hormonal control of sexual maturation in lampreys. In "Comparative Endocrinology" (P. J. Gaillard and H. H. Boer, eds.), pp. 105–108. Elsevier/North-Holland Biomedical Press, Amsterdam.

Laughran, L. J., Larsen, J. H., Jr., and Schroeder, P. C. (1981). Ultrastructure of developing ovarian follicles and ovulation in the lizard *Anolis carolinensis* (Reptilia). *Zoomorphology* **98**, 191–208.

Licht, P. (1984). Reptiles. In "Marshall's Physiology of Reproduction" (G. E. Lamming, ed.), 4th ed., Vol. 1, pp. 206–282. Churchill-Livingstone, Edinburgh and London.

Lintern-Moore, S., Moore, G. P. M., Tyndale Biscoe, C. H., and Poole, W. E. (1976). The growth of the oocyte and follicle in the ovaries of monotremes and marsupials. *Anat. Rec.* **185**, 325–332.

Lofts, B. (1974). Reproduction. In "Physiology of the Amphibia" (B. Lofts, ed.), Vol. 2, pp. 107–218. Academic Press, New York.

Lofts, B., and Bern, H. A. (1972). The functional morphology of steroidogenic tissues. In "Steroids in Nonmammalian Vertebrates" (D. R. Idler, ed.), pp. 37–125. Academic Press, New York.

Lofts, B., and Murton, R. K. (1973). Reproduction in birds. In "Avian Biology" (D. S. Farner and J. R. King, eds.), Vol. 3, pp. 1–107. Academic Press, New York.

Loyez, M. (1906). Recherches sur le développement ovarien des oeufs méroblastiques à vitellus nutritif abondant. *Arch. Anat. Microsc. Morphol. Exp.* **8**, 69–397.

Lyngnes, R. (1936). Rückbildung der ovulierten und nicht ovulierten Follikel im Ovarium der *Myxine glutinosa* L. *Skr. Nor. Vidensk.-Akad. [Kl.] 1: Mat.-Naturvidensk. Kl.* **4**, 1–116.

McNatty, K. P. (1978). Follicular fluid. In "The Vertebrate Ovary: Comparative Biology and Evolution" (R. E. Jones, ed.), pp. 215–259. Plenum, New York.

Matthews, L. H. (1950). Reproduction in the basking shark *Cetorhinus maximus* (Gunner). *Philos. Trans. R. Soc. London, Ser. B* **234**, 247–316.

Mayhew, W. W. (1963). Reproduction in the granite spiny lizard *Sceloporus orcutti*. *Copeia*, pp. 144–153.

Mayhew, W. W. (1966). Reproduction in the psammophilous lizard *Uma scoparia*. *Copeia* **1**, 114–122.

Mendoza, G. (1940). The reproductive cycle of the viviparous teleost, *Neotoca bilineata*, a member of the family Goodeidae. II. The cyclic changes in the ovarian soma during gestation. *Biol. Bull. (Woods Hole, Mass)* **84**, 87–97.

Miller, M. R. (1948). The seasonal histological changes occurring in the ovary, corpus luteum, and testis of the viviparous lizard, *Xantusia vigilis*. *Univ. Calif., Berkeley, Publ. Zool.* **47**, 197–224.

Mossman, H. W., and Duke, K. L. (1973). "Comparative morphology of the mammalian ovary." Univ. of Wisconsin Press, Madison.

Nagahama, Y. (1983). The functional morphology of teleost gonads. *In* "Fish Physiology" (W. S. Hoar, D. J. Randall, and E. M. Donaldson, eds.), Vol. 9, pp. 223–275. Academic Press, New York.

Nagahama, Y., Clarke, W. C., and Hoar, W. S. (1978). Ultrastructure of the putative steroid-producing cells in the gonads of coho (*Oncorhynchus kisutch*) and pink salmon (*Oncorhynchus gorbuscha*). *Can. J. Zool.* **56**, 2508–2519.

Neaves, W. B. (1971). Intercellular bridges between follicle cells and oocyte in the lizard *Anolis carolinensis. Anat. Rec.* **170**, 285–302.

Nicholls, T. J., and Maple, G. (1972). Ultrastructural observations on possible sites of steroid biosynthesis in the ovarian follicular epithelium of two species of cichlid fish, *Cichlasoma nigrofasciatum* and *Haplochromis multicolor. Z. Zellforsch. Mikrosk. Anat.* **128**, 317–335.

Ohta, H., and Takano, K. (1982). Ultrastructure of micropylar cells in the pre ovulatory follicles of Pacific herring. *Clupea pallasi* (Valenciennes). *Bull. Fac. Fish., Hokkaido Univ.* **33**, 57–64.

Ohta, H., and Teranishi, T. (1982). Ultrastructure and histochemistry of granulosa and micropylar cells in the ovary of the loach *Misgurnus anguillicaudatus* (Cantor). *Bull. Fac. Fish., Hokkaido Univ.* **33**, 1–8.

O'Shea, J. D. (1970). Ultrastructural study of smooth muscle-like cells in the theca externa of ovarian follicles in the rat. *Anat. Rec.* **167**, 127–139.

O'Shea, J. D. (1971). Smooth muscle-like cells in the theca externa of ovarian follicles in the sheep. *J. Reprod. Fertil.* **24**, 283–285.

Owman, C., and Sjöberg, N.-O. (1966). Adrenergic nerves in the female genital tract of the rabbit: With remarks on cholinesterase-containing structures. *Z. Zellforsch. Mikrosk. Anat.* **74**, 182–197.

Patzner, R. A. (1978). Cyclical changes in the ovary of the hagfish *Eptatretus burgeri* (Cyclostomata). *Acta Zool. (Stockholm)* **59**, 57–61.

Pearl, R., and Schoppe, W. F. (1921). Studies on the physiology of reproduction in the domestic fowl. XVIII. Further observations on the anatomical basis of fecundity. *J. Exp. Zool.* **34**, 101–118.

Pedersen, T. (1970). Determination of follicle growth rate in the ovary of the immature mouse. *J. Reprod. Fertil.* **21**, 81–93.

Pendergrass, P., and Schroeder, P. (1976). The ultrastructure of the thecal cell of the teleost *Oryzias latipes,* during ovulation *in vitro. J. Reprod. Fertil.* **47**, 229–233.

Perry, M. M., and Gilbert, A. B. (1979). Yolk transport in the ovarian follicle of the hen. (*Gallus domesticus*): Lipoprotein-like particles at the periphery of the oocyte in the rapid growth phase. *J. Cell Sci.* **39**, 257–272.

Perry, M. M., Gilbert, A. B., and Evans, A. J. (1978). Electron microscope observations on the ovarian follicle of the domestic fowl during the rapid growth phase. *J. Anat.* **125**, 481–497.

Polder, J. J. W. (1961). Cyclical changes in testis and ovary related to maturity stages in the North sea herring *Clupea harengus* L. *Arch. Neerl. Zool.* **14**, 45–60.

Polder, J. J. W. (1964). Occurrence and significance of atretic follicles (pre-ovulatory corpora lutea) in ovaries of the bitterling, *Rhodeus amarus* (Bloch). *Proc. K. Ned. Akad. Wet., Ser. C* **67**, 218–222.

Press, N. (1964). An unusual organelle in avian ovaries. *J. Ultrastruct. Res.* **10**, 528–546.

Rai, B. P. (1966). Corpora atretica and the so-called corpora lutea in the ovary of *Tor* (*Barbus*) *tor. Anat. Anz.* **119**, 459–465.

Rajalakshmi, M. (1966). Atresia of oocytes and ruptured follicles in *Gobius giuris. Gen. Comp. Endocrinol.* **6**, 378–385.

Ranzi, S. (1934). Le basi fisio-morfologiche dello sviluppo embrionale dei Selaci. *Pubbl. Stn. Zool. Napoli* **13**, 331–437.

Redshaw, M. R. (1972). The hormonal control of the amphibian ovary. *Am. Zool.* **12**, 289–306.

Redshaw, M. R., and Nicholls, T. J. (1971). Oestrogen biosynthesis by ovarian tissue of the South African clawed toad, *Xenopus laevis* Daudin. *Gen. Comp. Endocrinol.* **16**, 85–96.

Remacle, C. (1970). Contribution à l'étude de la sexualité chez certains labridae et sparidae (téléostéens perciformes). *Bull. Inst. R. Sci. Nat. Belg.* **46**, 1–13.

Saidapur, S. K. (1978). Follicular atresia in the ovaries of nonmammalian vertebrates. *Int. Rev. Cytol.* **54**, 225–244.

Saidapur, S. K., and Nadkarni, V. B. (1974). Steroid-synthesising cellular sites in amphibian ovary, a histochemical study. *Gen. Comp. Endocrinol.* **22**, 459–462.

Saidapur, S. K., and Nadkarni, V. B. (1976). Steroid synthesising cellular sites in the ovary of catfish *Mystus cavassius:* A histochemical study. *Gen. Comp. Endocrinol.* **30**, 457–461.

Scanes, C. G., Mozolic, H., Kavanagh, E., Merrill, G., and Rabii, J. (1982). Distribution of blood flow in the ovary of domestic fowl (*Gallus domesticus*) and changes after postaglandin F-2 treatment. *J. Reprod. Fertil.* **64**, 227–231.

Schulz, R., and Blüm, V. (1983). Elimination of the nucleus in preovulatory oocytes of the rainbow trout, *Salmo gairdneri* Richardson (Teleostei). *Cell Tissue Res.* **232**, 685–689.

Solomons, B., and Gatenby, J. W. B. (1924). Notes on the formation structure and physiology of the corpus luteum of man, the pig and the duck-billed platypus. *J. Obstet. Gynaecol. Br. Emp.* **31**, 580–594.

Stanley, H. P. (1963). Urogenital morphology of the chimaeroid fish *Hydrolagus colliei* (Lay and Bennett). *J. Morphol.* **112**, 99–128.

Stieve, H. (1918). Über experimentell durch veränderte äussere Bedingungen hervorgerufene Rückbildungsvorgänge am Eierstock des Haushuhnes (*Gallus domesticus*). *Arch. Entwicklungsmech. Org.* **44**, 531–588.

Stieve, H. (1919). Die entwicklung des Eierstockseies der Dohle (*Colaeus monedula*). *Arch. Mikrosk. Anat.* **92**, 137–288.

Strauss, F. (1938). Die Befruchtung und der Vorgang der Ovulation bei *Ericulus* aus der Familie der Centetiden. *Biomorphosis* **1**, 281–312.

Sund, O. (1943). Et Brugdebarsel. *Naturen* **67**, 285–286.

Szöllösi, D., and Jalabert, B. (1974). La thèque du follicule ovarien de la Truite. *J. Microsc. (Paris)* **20**, 92a.

Takano, K. (1964). On the egg formation and the follicular changes in *Lebistes reticulatus. Bull. Fac. Fish., Hokkaido Univ.* **15**, 147–155.

Tan, E. S. P. (1976). Some aspects of the reproductive physiology of teleosts with special reference to the sea bass, *Dicentrarchus labrax* (L.). Ph.D. Thesis, University of Wales.

Teshima, K., and Mizue, K. (1972). Studies on sharks. I. Reproduction in the female sumitsuki shark *Carcharhinus dussumieri. Mar. Biol.* **14**, 222–231.

Te Winkel, L. E. (1950). Notes on ovulation, ova, and early development in the smooth dogfish *Mustelus canis. Biol. Bull.* (*Woods Hole, Mass.*) **99**, 474–486.

Thornton, V. F., and Evennett, P. J. (1973). Changes in the fine structure of the ovarian follicle of the toad (*Bufo bufo*) prior to induced ovulation. *Gen. Comp. Endocrinol.* **20**, 413–423.

Tokarz, R. R. (1977). An autoradiographic study of the effects of FSH and estradiol 17-β on early ovarian follicular maturation in adult *Anolis carolinensis. Gen. Comp. Endocrinol.* **31**, 17–28.

Treasurer, J. W., and Holliday, F. G. T. (1981). Some aspects of the reproductive biology of perch (*Perca fluviatilis* L.). A histological description of the reproductive cycle. *J. Fish Biol.* **18**, 359–376.

Turner, C. L. (1933). Viviparity superimposed upon ovo-viviparity in the Goodeidae, a family of cyprinodont teleost fishes of the Mexican plateau. *J. Morphol.* **55**, 207–251.

Turner, C. L. (1937). Reproductive cycles and superfetation in poeciliid fishes. *Biol. Bull.* (*Woods Hole, Mass.*) **72**, 145–164.

Turner, C. L. (1940). Superfetation in viviparous cyprinodont fishes. *Copeia*, pp. 88–91.

Turner, C. L. (1947). Viviparity in teleost fishes. *Sci. Mon.* **65**, 508–518.

Turner, R. T., Dickhoff, W. W., and Gorbman, A. (1982). Estrogen binding to hepatic nuclei of Pacific hagfish *Eptatretus stouti*. *Gen. Comp. Endocrinol.* **45**, 26–29.

van den Hurk, R., and Peute, J. (1979). Cyclic changes in the ovary of the rainbow trout, *Salmo gairdneri*, with special reference to sites of steroidogenesis. *Cell Tissue Res.* **199**, 289–306.

Vazquez-Nin, G. H., and Sotelo, J. R. (1967). Electron microscope study of the atretic oocytes of the rat. *Z. Zellforsch. Mikrosk. Anat.* **80**, 518–533.

Vladykov, V. D. (1956). Fecundity of wild speckled trout (*Salvelinus fontinalis*) in Quebec lakes. *J. Fish Res. Board. Can.* **13**, 799–841.

Vu Tan Tue (1972). Variations cycliques des gonades et de quelques glandes endocrines chez *Chimaera monstrosa* Linne (Pisces, Holocephali). *Ann. Sci. Nat., Zool. Biol. Anim.* [12] **14**, 49–94.

Walvig, F. (1963). The gonads and the formation of the sexual cells. In "The Biology of *Myxine*" (A. Brodal and R. Fänge, eds.), pp. 530–580. Oslo Univ. Press, Oslo.

Wartenberg, H. (1962). Elektronenmikroskopische und histochemische studien über die Oogenese der Amphibieneizelle. *Z. Zellforsch. Mikrosk. Anat.* **58**, 427–486.

Weekes, H. C. (1934). The corpus luteum in certain oviparous and viviparous reptiles. *Proc. Linn. Soc. N.S.W.* **59**, 380–391.

Weir, B. J. (1971). Some observations on reproduction in the female agouti, *Dasyprocta aguti*. *J. Reprod. Fertil.* **24**, 203–211.

Weir, B. J., and Rowlands, I. W. (1977). Ovulation and atresia. In "The Ovary" (S. Zuckerman and B. J. Weir, eds.), Vol. 1, Chapter 6, pp. 265–301. Academic Press, New York.

Wiebe, J. P. (1968). The reproductive cycle of the viviparous sea perch *Cymatogaster aggregata* Gibbons. *Can. J. Zool.* **46**, 1221–1234.

Wischnitzer, A. (1963). The ultrastructure of the layers enveloping yolk-forming oocytes from *Triturus viridescens*. *Z. Zellforsch. Mikrosk. Anat.* **60**, 452–462.

Wischnitzer, S. (1964a). Ultrastructural changes in the cytoplasm of developing amphibian oocytes. *J. Ultrastruct. Res.* **10**, 14–26.

Wischnitzer, S. (1964b). An electron microscope study of the formation of the zona pellucida in oocytes from *Triturus viridescens*. *Z. Zellforsch. Mikrosk. Anat.* **64**, 196–209.

Wischnitzer, S. (1966). The ultrastructure of the cytoplasm of the developing amphibian egg. *Adv. Morphog.* **5**, 131–179.

Wourms, J. P. (1977). Reproduction and development in chondrichthyan fishes. *Am. Zool.* **17**, 379–410.

Wourms, J. P. (1981). Viviparity: The maternal-fetal relationship in fishes. *Am. Zool.* **21**, 473–515.

Wyburn, G. M., Aitken, R. N. C., and Johnston, H. S. (1965). The ultrastructure of the zona radiata of the ovarian follicle of the domestic fowl. *J. Anat.* **99**, 469–484.

Wyburn, G. M., Johnston, H. S., and Aitken, R. N. C. (1966). Fate of the granulosa cells in the hen's follicle. *Z. Zellforsch. Mikrosk. Anat.* **72**, 53–65.

Yamamoto, K., and Onozato, H. (1965). Electron microscope study on the growing oocyte of the goldfish during the first growth phase. *Mem. Fac. Fish., Hokkaido Univ.* **13**, 79–106.

Yamamoto, K., and Onozato, H. (1968). Steroid producing cells in the ovary of the zebrafish, *Brachydanio rerio*. *Annot. Zool. Jpn.* **41**, 119–128.

Yamamoto, K., and Oota, I. (1967). Fine structure of the yolk globules in the oocyte of the zebra fish, *Brachydanio rerio*. *Annot. Zool. Jpn.* **40**, 20–27.

Yaron, Z. (1971). Observations on the granulosa cells of *Acanthobrama terrae-sanctae* and *Tilapia nilotica* (Teleostei). *Gen. Comp. Endocrinol.* **17**, 247–252.

York, M. A., and McMillan, D. B. (1980). Structural aspects of ovulation in the lamprey *Petromyzon marinus*. *Biol. Reprod.* **22**, 897–912.

Zuckerman, S., and Baker, T. G. (1977). The development of the ovary and the process of oogenesis. In "The Ovary" (S. Zuckerman and B. J. Weir, eds.), Vol. 1, Chapter 2, pp. 41–67. Academic Press, New York.

Zuckerman, S., and Weir, B. J. (1977). "The Ovary," 3 vol. Academic Press, New York.

13

Testis

YOSHITAKA NAGAHAMA
Laboratory of Reproductive Biology
National Institute for Basic Biology
Okazaki 444, Japan

I. INTRODUCTION

The two basic tasks of the vertebrate testis are to produce (1) the male gamete, the spermatozoon, and (2) steroid hormones, principally androgens, which are necessary for both the production of spermatozoa and the maintenance of the differentiated state of the various excretory ducts and accessory glands. This chapter reviews the general structure of the testis and morphological aspects of male germ cell development. This is followed by a histochemical and ultrastructural review of the appearances of two major testicular somatic cells, i.e., Sertoli cells and interstitial or Leydig cells. Both of these types of cells often possess characteristic features of steroid-producing cells, including an extensive agranular endoplasmic reticulum, plentiful mitochondria with tubular cristae, a prominent Golgi apparatus, and a number of lipid droplets (Christensen and Gillim, 1969; Lofts and Bern, 1972; Guraya, 1976). In each section, the mammalian system will be used as a basis for understanding the relevance of the nonmammalian systems. The following detailed descriptions will emphasize what is currently known about these topics in the other classes of vertebrates. Because of the complexity of modern biomedical questions, no vertebrate system should be considered alone. Based on this synthesis of the extensive literature concerning vertebrate testicular morphology, investigators should be able to creatively design their studies by matching the most appropriate experimental system to their particular problem.

II. GENERAL STRUCTURE

A. Mammals, Birds, and Reptiles

The basic structural organization of the testes is similar in all amniotes; they are compact, more or less ovoid organs which are closely surrounded

399

by a capsule rich in collagen fibers, which is known as the tunica albuginea (cf. Roosen-Runge, 1977; Lance and Callard, 1980; Pilsworth and Setchell, 1981; van Tienhoven, 1983). Outside is a looser membrane, vascular in nature, called the tunica vaginalis. The testes of most mammals are contained in a scrotum, maintained at a temperature lower than the rest of the body. The testicular parenchyma consists of seminiferous tubules which are embedded in a connective tissue matrix containing interspersed Leydig cells (interstitial cells), blood vessels, lymph vessels, and nerves, none of which penetrates the tubules. The relationship between the walls of the tubules, Leydig cells, and blood and lymph vessels varies between species (Fawcett *et al.*, 1973). The blood supply to the testes of a range of mammals has been reviewed by Setchell (1970) and Free (1977). Seminiferous tubules are basically cylindrical structures composed of a stratified epithelium surrounded by a thick basal lamina and alternate layers of collagen and flattened cells called myoid cells. The epithelium which lines the seminiferous tubules consists of two types of cells, the somatic Sertoli cells and the various germinal cells which surround a lumen (Fig. 1). Spermatogonia are located at the outer surface of the seminiferous tubules next to the basal lamina, spermatocytes are located more medially, and spermatids are found nearest the lumen. As they mature, spermatozoa are released into the lumen of the tubules. No cysts are present within the tubules. In birds, the seminiferous tubules are extensively branched (Lofts and Murton, 1973).

B. Amphibians

The testes of most anuran amphibians are ovoid or elongated paired organs (Lofts, 1984). In some urodele species, however, the organs are distinctly lobed due to the occurrence of several regions displaying different stages of germ cell development (Fig. 2A). The lobes are joined together by narrow bridges of tissue. The number of lobes constituting each testis varies from species to species. The testicular structure of amphibians is basically similar to that of the amniote vertebrates in consisting of numerous seminiferous tubules or lobules, but it differs from that of amniotes in that the development of germ cells occurs in a coordinated manner within cysts (Fig. 2B).

C. Teleosts

In most teleosts, the testes are elongated paired organs attached to the dorsal body wall; in some species (e.g., poeciliids) they are combined into a single sac. Testicular structure in teleosts varies from species to species, although two basic types, lobular and tubular, can be identified according to the differentiation of the germinal tissue (Hoar, 1969; Billard *et al.*, 1982; Nagahama, 1983) (Fig. 3A,B).

The testis of the lobular type, which is typical of most teleosts, is com-

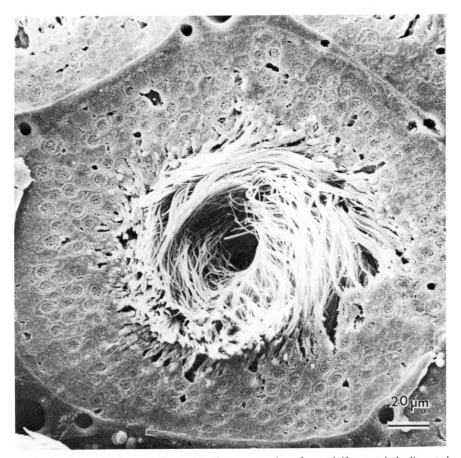

Fig. 1.   Scanning electron micrograph of a cross section of a seminiferous tubule dissected from a testis of an adult rat. [Courtesy of Dr. T. Nagano.]

posed of numerous separate lobules jointed by a thin layer of fibrous connective tissue; the arrangement of the lobules varies considerably (Roosen-Runge, 1977). Within the lobules, primary spermatogonia undergo numerous mitotic divisions, producing cysts containing several spermatogonial cells. During maturation, all of the germ cells within each cyst are at approximately the same stage of development (Fig. 3C). As spermatogenesis and then spermiogenesis proceed, the cysts expand and eventually rupture, liberating sperm into the lobular lumen which is continuous with the sperm duct. In salmonids, the cells in most cysts composing the same testes are synchronized and contain germ cells at the same spermatogenic stage. This is especially advantageous in studies where biochemical data are required, since large amounts of similar cytoplasm or nucleoplasm can be obtained.

The alternative testicular structure, the tubular type, is restricted to the

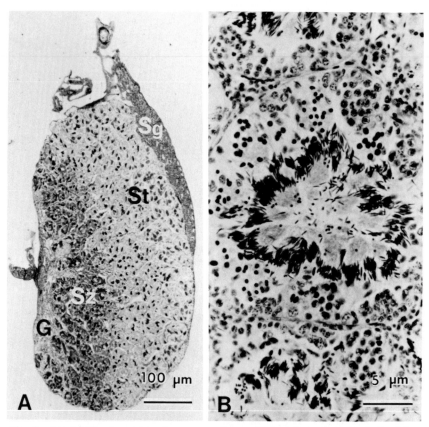

Fig. 2. (A) Light micrograph of a paramedian sagittal section of the testicular lobe of
Cynops pyrrhogaster pyrrhogaster in October. Note several regions displaying different stages
of germ cell development. G, glandular tissue; Sg, spermatogonium; St, spermatid; Sz, spermatozoon. [Tanaka and Iwasawa, 1979.] (B) Light micrograph of a portion of the testis of *Rana
catesbeina* in October, showing cysts containing germ cells at various stages of development.
[Courtesy of Dr. H. Yonezawa.]

atheriniform species, such as the guppy, *Poecilia reticulata* (Billiard *et al.,*
1982). In this situation, the tubules are regularly oriented between the external tunica propria (blind end) and a central cavity into which spermatozoa
are released. Primary spermatogonia are located only at the blind end of
each tubule. As spermatogenesis and spermiogenesis proceed, the germinal
cyst moves centrally within the testis toward the vas efferens (efferent duct);
there is no structure corresponding to the lobular lumen (Roosen-Runge,
1977; Pilsworth and Setchell, 1981; Billard *et al.,* 1982). More recently, Grier
et al. (1980) examined testes of four orders of teleosts (Salmoniformes,
Perciformes, Cypriniformes, and Atheriniformes) using conventional light
microscopy and scanning and transmission electron microscopy. According

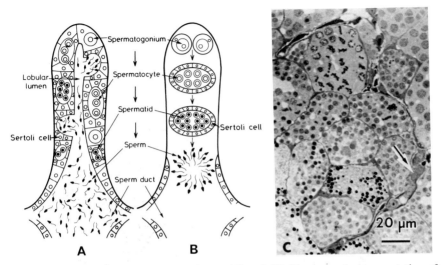

Fig. 3. Testicular structure of the teleost. (A) and (B) Diagrammatic representation of lobular type (A) and tubular type (B). [From Nagahama, 1983.] (C) Light micrograph (1 μm epon-embedded section) of a portion of the testis of *Oryzias latipes,* showing cysts containing germ cells at various stages of development. Note a cluster of Leydig cells (arrow) present in the interstitium. [Courtesy of Dr. A. Kanamori.]

to the distribution of spermatogonia, they divided testicular structure in these orders into two basic tubular designs—unrestricted spermatogonial testes and restricted spermatogonial testis. The former type is common to most teleosts, and the latter type (in which spermatogonia are totally restricted to the distal terminus of the tubule immediately beneath the tunica albuginea) is characteristic of the Atheriniformes (Grier, 1981; cf. Billard *et al.*, 1982).

D. Elasmobranchs

Testes of elasmobranchs are paired and suspended from the dorsal body wall by mesorchia. In the spotted dogfish, *Scyliorhinus canicula,* the testes are elongated, subcylindrical, and extend almost the full length of the body cavity (Dodd, 1983). The structure of testes in elasmobranchs differs from that of other vertebrates in having closed ampullae rather than the tubule or lobule (Dodd, 1983; Dodd and Sumpter, 1984). The ampullae are arranged in bands which radiate out from an ampullogenic zone (where new ampullae are produced) and which is located along the ventrolateral border of the testis. Ampullae nearest to this zone contain spermatogonia, whereas those occupying the dorsal regions of the testis furthest away from the ampullae contain spermatozoa (Stanley, 1966). All spermatocysts within a single ampulla are at approximately the same stage of spermatogenesis. Each fully mature ampulla of *S. canicula* contains about 32,000 spermatozoa (Stanley, 1966).

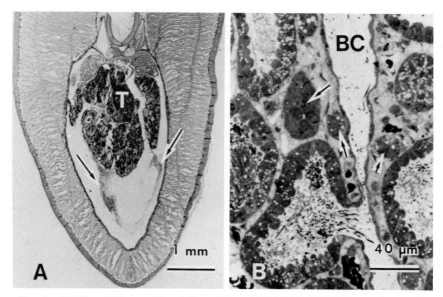

Fig. 4. (A) Transverse section of the middle of the body of sexually mature *Lampetra japonica*. Sperm are seen not only in the testis (T) but also in the body cavity. (B) Light micrograph (1 μm epon-embedded section) of a portion of the testis of sexually mature *L. japonica*. Both Sertoli cells and interstitial cells (arrows) contain numerous lipid droplets. Sperm are seen in the body cavity (BC). [Courtesy of Dr. S. Fukayama.]

E. Cyclostomes

Lampreys become sexually mature and carry out gametogenesis once and die soon after spawning (Hardisty, 1971, 1979). In contrast, it is commonly assumed that most hagfishes are repetitively cyclic breeders without seasonally defined reproductive cycles (Gorbman, 1983). The cyclostomes have a single, medially placed gonad. The male gametes of the cyclostomes develop in cysts. The cysts gradually transform into spermatozoan masses and eventually rupture, shedding spermatozoa directly into the body cavity (Gorbman, 1983; Dodd and Sumpter, 1984) (Fig. 4A,B).

III. SPERMATOGENESIS

A. Spermatogenesis, Spermiogenesis, and Spermiation

During spermatogenesis, i.e., the transition from mitotically active primordial germ cells to mature spermatozoa, many unique aspects of cell differentiation occur; these include spermatogonial stem cell renewal, the two mitotic reduction divisions, and spermiogenesis. Spermatogenesis is of specialized cytological interest and is beyond the scope of this chapter, although further information may be found in a number of recent reviews (e.g., Lofts, 1968; Roosen-Runge, 1977; Dym and Cavicchia, 1978; Setchell, 1978; Guraya, 1980; Pilsworth and Setchell, 1981; van Tienhoven, 1983). The following section reviews representative examples of spermatogenesis that are potentially useful for our basic understanding of germ cell development in vertebrates.

As stated earlier, spermatogonia of amniotes are found next to the basal membranes of the seminiferous tubules. Some of these spermatogonia enter the next phase of spermatogenesis and are called primary spermatocytes. Germ cells slowly move centripetally along the sides of the Sertoli cells as they differentiate and approach the lumen of the tubules.

Groups of spermatogonia, spermatocytes, or spermatids at the same stage of differentiation are connected by intercellular bridges (0.5–1 μm in diameter) (Fawcett, 1961) resulting from incomplete cytokinesis during all but the earliest spermatogonial divisions. Incomplete meiotic divisions expand these cellular clones even more. Widespread occurrence of intercellular bridges throughout the animal kingdom suggests that this phenomenon is of fundamental importance in spermatogenesis. It is likely that their presence facilitates the precise synchrony of differentiation of germ cells.

Although spermatids have a haploid set of chromosomes, they are still not capable of functioning as male gametes. They must undergo a series of morphological and biochemical transformations, leading to the production of the highly differentiated germ cell, the spermatozoon. This process is called spermatid maturation or spermiogenesis and does not involve cell division. Whereas spermatogenesis refers to the entire process of forming spermatozoa from spermatogonia, spermiogenesis refers only to the differentiation of spermatozoa from spermatids. Details of the extraordinarily complex process of spermiogenesis have been previously reviewed (e.g., Phillips, 1974; Fawcett, 1975b; Dym and Cavicchia, 1978; Holstein and Roosen-Runge, 1981).

The term "spermiation" describes the terminal phenomenon of the spermatogenic cycle, in which embedded bundles of spermatozoa are released from enveloping Sertoli cells and are swept into the efferent ducts (cf. Lofts, 1984). Certain morphological aspects of the process of spermiation have been investigated in mammals (Sapsford and Rae, 1969; Fawcett and Phillips, 1969; Vitale-Calpe and Burgos, 1970a,b; Fouquet, 1974; Fawcett, 1975b). It has been shown that spermatid detachment involves at least two separate events. The first of these is the partial extrusion of the spermatid head from the Sertoli cell. The second is the separation of the thin stalk connecting the neck region of the spermatid to the residual cytoplasm. It has been suggested that the tubulobulbar complexes serve as anchoring devices that retain the spermatids at the surface of the seminiferous epithelium, whereas their dissolution contributes in part to the process of spermiation (Russell and Clermont, 1976).

Gonadotropin-induced spermiation is particularly interesting and has been investigated in detail in an anuran species, the toad, *Bufo arenarum,* using electron microscopy (Burgos and Vitale-Calpe, 1967). Three major sequential changes are recognized in Sertoli cells during gonadotropin-induced spermiation: (1) swelling of the endoplasmic reticulum, (2) swelling of the apical cytoplasm, and (3) apical detachment. These morphological changes are accompanied by an increase in water and sodium content of the testis.

Spermiation in other classes of vertebrates has not yet been studied in ultrastructural detail. Nonetheless some remarkable circumstances are obtained. In some teleosts with tubular-type testes, spermatids or sperm nuclei become embedded within recesses in the Sertoli cell cytoplasm near or at the termination of spermiogenesis (Billard, 1970a; van den Hurk *et al.,* 1974b; Grier, 1975). A similar association of germ cells with Sertoli cells is found in the Goodeidae, but in this case the developing flagellum of the spermatid becomes associated with the Sertoli cell layer (e.g., Grier *et al.,* 1978; cf. Grier, 1981). No such associations occur between germ cells and Sertoli cells in species with lobular-type testes. A fascinating relationship between sperm and somatic cells occurs in the ovarian cavity of viviparous teleost species, which is almost certainly associated with sperm storage and internal fertilization. In this case a localized expansion of the epithelium of the ovarian cavity overlying each developing oocyte forms a "delle" and a similarly formed, specialized single "seminal receptacle" exists in the anterodorsal ovarian cavity (Jalabert and Billard, 1969; K. Takano, unpublished). Sperm are often found embedded head first in the apical cytoplasm of the epithelium lining the ovarian cavity.

Spermatozoa of vertebrates exhibit remarkable differences in size, form, and internal structure. Underlying such diversity there exists a fundamental structural pattern common to most vertebrates. A typical spermatozoon consists of two parts: (1) a head consisting of the acrosome and nucleus and (2) the tail, which can be further subdivided into the neck, the middle piece, the principal piece, and the end piece. Teleost spermatozoa lack an acrosome which occurs in all other vertebrate groups; this may be related to the presence of a micropyle in teleost eggs which insures sperm penetration (Grier, 1981; Nagahama, 1983).

Sperm morphology reflects the mode of fertilization. The so-called primitive type of sperm belongs to aquatic species with external fertilization; these sperm typically have a spheroidal nucleus with a short middle piece (Grier, 1981). Sperm in species with internal fertilization are typically slender and elongate in form. An example demonstrating that diversity in sperm morphology is related to the mode of reproduction has been noted in two closely related teleost species. In the medaka, *Oryzias latipes,* with external fertilization, the sperm exhibits a primitive morphology, with a rounded nucleus and short midpiece. In the guppy, *Poecilia reticulata,* with internal fertilization, an elongation of both nucleus and midpiece is observed (Grier,

1981). More detailed information on the comparative morphology of vertebrate spermatozoa may be found in a number of reviews (e.g., Fawcett,
1970; Nicander, 1970; Billard, 1970b).

It should be noted that mammalian spermatids contain a well-developed
system of smooth endoplasmic reticulum. It would be of interest to know
whether this smooth endoplasmic reticulum possesses steroid-metabolizing
enzymes. It has been suggested that rat germ cells at the primary spermatocyte stage are able to produce androgen not only from external progesterone
but also from endogenous steroid precursors (Salhanick and Terner, 1979).
Ozon and Collenot (1965) have also found 3β-hydroxysteroid dehydrogenase
(3β-HSD) and 17β-HSD in spermatozoa of an elasmobranch, *Scyliorhinus
canicula*. Furthermore, Ueda *et al.* (1984) recently reported that the sperm
of rainbow trout, *Salmo gairdneri*, possess 20β-HSD, the enzyme controlling the conversion of 17α-hydroxyprogesterone to 17α,20β-dihydroxy-4-
pregnen-3-one. Future studies should define the subcellular localization of
20β-HSD in *S. gairdneri* sperm and should determine the first appearance of
this enzyme during germ cell development. These cellular systems from
teleosts and elasmobranchs should serve as useful models in studying the
contribution of steroids from germ cells to their own differentiation.

B. Spermatogenesis *in Vitro*

Sperm development involves a complex sequence of genetic, biochemical,
and morphological differentiation events. Despite the efforts of many investigators, much remains to be learned about the regulation of this process. A
major reason for this lies in the complex structure of the testis. This makes it
difficult to separate inherent control mechanisms of germinal cells from
effects of hormones or influences of Sertoli cells. One approach to this
problem is to develop a suitable *in vitro* cell culture system. However, most
attempts to maintain spermatogenesis in cultures of dissociated mammalian
spermatogenic cells have had limited success. This may be related to the
occurrence of intimacy of the relationship between male germ cells and
Sertoli cells in mammals. However, the development of techniques for the
isolation and culture of Sertoli cells has facilitated the assessment of several
hormone-dependent processes *in vitro* (Steinberger, 1975). In a more recent
study by Tres and Kierszenbaum (1983), spermatogenic cells from 20- to 35-
day-old rats were grown *in vitro* in the presence of Sertoli cells maintained in
serum-free hormone/growth-factor-supplemented medium and alternating
high/low concentrations of follicle-stimulating hormone in the medium. Under these conditions, proliferation of spermatogonia and differentiation of
meiotic prophase spermatocytes into secondary spermatocytes occurred *in
vitro*.

In contrast to the situation in mammals, amphibian systems appear to be
more promising for the study of spermatogenesis *in vitro*. Meiotic prophase,

meiosis, and spermatid development occur in cultures of dissociated spermatogenic cells from the frog, *Xenopus laevis,* which lacks Sertoli cells (Risley and Eckhardt, 1979; Kalt, 1979). Similarly, an *in vitro* development of spermatids from the late meiotic prophase stages of the newt, *Cynops pyrrhogaster,* has been demonstrated (Abe, 1981). *In vitro* differentiation of amphibian male germ cells, however, seems to be blocked at two points in the spermatogenetic pathway: the development of spermatogonia to meiotic prophase and the development of elongate spermatids into spermatozoa. Therefore, an association with somatic cells may be necessary for these two processes of spermatogenesis to occur. In this connection, it is pertinent to recall the establishment of a close relationship between Sertoli cells and late spermatids during the time of spermatid elongation in amphibians. Furthermore, the failure of spermatogonia to enter meiotic prophase may result from the lack of a hormonal stimulus. It should be noted that in certain teleosts, testicular Leydig cells proliferate immediately prior to the initiation of meiosis (see Section V,E). Cell culture systems developed for amphibian testicular cells are certainly useful for the study of the cellular and molecular processes fundamental to meiosis and spermatid development (cf. Risley, 1983).

IV. SERTOLI CELLS

The sertoli cells are the only nongerminal elements within the seminiferous epithelium. Sertoli cells have also been referred to as nurse cells, sustentacular cells, and supporting cells. Recent morphological, physiological, and biochemical studies of the testes of various vertebrates indicate that the Sertoli cells have a wide spectrum of junctions (Fawcett, 1975a; Hudson and Burger, 1979; Ewing *et al.,* 1980; Ritzen *et al.,* 1981). Sertoli cells clearly provide mechanical support for the developing germ cells at all stages of differentiation; their function includes participation in sperm release. The phagocytotic activity of the Sertoli cell has also been demonstrated; Sertoli cells are responsible for removing residual bodies and degenerated germ cells from the seminiferous epithelium. The Sertoli cell secretes (1) seminiferous tubule fluid, (2) androgen-binding protein (ABP), and (3) inhibin. It has also been suggested that the Sertoli cells are one source of steroid hormones. Finally, electron microscopical studies have clearly demonstrated that adjacent Sertoli cells form the blood–testis barrier which divides the germinal epithelium into basal and adluminal compartments (Setchell and Waites, 1975). The following sections review the morphology of Sertoli cells and emphasize evidence for their function in steroidogenesis.

A. Mammals

Sertoli cells in mammals are distributed radially within the seminiferous tubules and share the surface of the basal boundary tissue with the spermatogonia. The germ cells remain associated with Sertoli cells during the entire spermatogenic process. The Sertoli cells are columnar, with broad bases and long, narrow processes that extend into the lumen of the tubule. These somatic cells cease to proliferate before puberty and live for a long time.

Mammalian Sertoli cells have both generalized and specialized ultrastructural features that reflects their varied functions (cf. Burgos *et al.*, 1970; Setchell and Waites, 1975; Steinberger and Steinberger, 1977; Kerr and de Kretser, 1981; Schulze, 1984). Striking cytological features of the mature Sertoli cell are its large irregular nucleus and prominent central nucleolus flanked by satellite heterochromatin bodies. The basal and intermediate regions of Sertoli cells contain various cytoplasmic organelles and inclusions like numerous ovoid and slender mitochondria, lysosomes, and autophagic multivesicular bodies. The presence of these inclusions is consistent with the view that mammalian Sertoli cells are phagocytotic in nature (Carr *et al.*, 1968). Concentric membranes of smooth endoplasmic reticulum are also observed within the basal cytoplasm of the Sertoli cell and are often associated with or surrounding large lipid inclusions.

The presence of an elaborate smooth endoplasmic reticulum, lipid droplets, prominent Golgi bodies, and abundant mitochondria in the Sertoli cells of certain mammalian species is consistent with the view that the Sertoli cells are steroidogenic. Pure preparations of rat Sertoli cells have the capacity to metabolize steroids *in vitro* (Welsh and Wiebe, 1975, 1976; Wiebe and Tilbe, 1979). Although Sertoli cells lack the capacity to form estradiol-17β from either cholesterol or pregnenolone, they can convert testosterone to estradiol (Dorrington and Armstrong, 1975). A two-cell type model involving Leydig cells and Sertoli cells has been proposed for testicular synthesis of estradiol-17β (Dorrington *et al.*, 1978). In this model, testosterone formed by the Leydig cells under the influence of luteinizing hormone (LH) is transported to Sertoli cells, where it is converted to estradiol-17β under the influence of follicle-stimulating hormone (FSH). In this connection, it is of interest that a similar two-cell type model involving cooperation between theca interna cells and the granulosa cells has been proposed for follicular production of estradiol-17β in the ovaries of certain mammalian species (Dorrington and Armstrong, 1979; Channing *et al.*, 1980). These findings support the concept that Sertoli cells of the testes and granulosa cells of the ovary are of equivalent embryological origin.

ABP has been localized at the light microscopical level in the cytoplasm of Sertoli cells of the rat seminiferous tubule (Pelliniemi *et al.*, 1981). Anti-Mullerian hormone (AMH) was localized, with the aid of ultrastructural immunocytochemical techniques, in the cisternae of the rough endoplasmic reticulum of bovine Sertoli cells (Tran and Josso, 1982).

B. Birds

Avian Sertoli cells have a morphology similar to that of the Sertoli cells of mammals described above. Using a fluorescent antibody technique, Woods and Domn (1966) localized the presence of the enzyme 3β-HSD in Sertoli cells of the domestic fowl, *Gallus domesticus*. Cooksey and Rothwell (1973) have shown that avian Sertoli cells contain abundant tubular smooth endoplasmic reticulum, prominent Golgi complexes, and numerous mitochondria, which are features characteristic of steroid-producing cells; however, the mitochondria cristae are lamellar rather than tubular. Similar ultrastructural features have also been described in Sertoli cells of normal and photostimulated Japanese quail, *Coturnix coturnix* (Lofts, 1972). Specific binding of radioiodinated rat FSH has been demonstrated in the Sertoli cells of the newly hatched chick (Ishii *et al.*, 1978).

C. Reptiles

Seasonal changes in the histology of reptilian Sertoli cells have been reviewed by Fox (1977). Sertoli cells accumulate lipids at the end of the spermatogenic cycle and then disappear, coincident with the renewal of spermatogenesis at the start of the next cycle (cf. Licht, 1984). In most species of reptiles, there is one time of year when spermatogenic activity is high, but adjacent interstitial cells are atrophic (Lofts, 1980). At this stage, strongly positive reactions for 3β- and 17β-HSD occur within the tubules, and the Sertoli cells appear to be steroidogenic (Lofts and Tsui, 1977; Lofts, 1980). Some of the ultrastructural characteristics of steroid-producing cells have also been observed in the Sertoli cells of *Naja* (Lofts, 1972) and *Lacerta* (Dufaure, 1971). In addition, in *Chrysemys picta* the only morphological changes observed after ovine FSH treatment were confined to Sertoli cells and included release of germ cells and discharge of Sertoli cells granules and vacuoles from the apical cytoplasm (Callard *et al.*, 1976). Taken together, these results are consistent with the hypothesis that Sertoli cells in certain reptilian species represent an important source of endogenous androgen within the testicular tubules.

D. Amphibians

The presence of 3β-HSD has been noted in the Sertoli cells of several anuran species (van Oordt and Brands, 1970; Saidapur and Nadkarni, 1975). During the period of maturation of the germinal cysts, Sertoli cells of *Rana temporaria* possess all the general ultrastructural features associated with steroidogenic tissues (Brökelmann, 1964), as do those of *Rana nigromaculata* (H. Iwasawa, unpublished) (Fig. 5). Thus, it seems likely that anuran Sertoli cells produce steroid hormones. The picture is not as clear in urodeles. In the axolotl, *Ambystoma mexicanum*, Sertoli cell 3β-HSD activity is closely associated with the spermatogonial divisions (Lazard, 1976). In contrast, the Sertoli cells of *Cynops p. pyrrhogaster* show no 3β-HSD activity at any stage of testicular development (Imai and Tanaka, 1978).

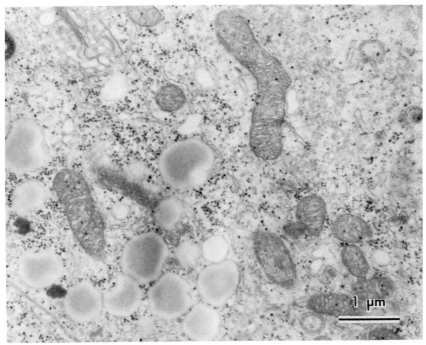

Fig. 5. Electron micrograph of a portion of the Sertoli cell of *Rana nigromaculata* in October. Cytoplasmic organelles include an abundant smooth endoplasmic reticulum, mitochondria with tubular cristae, and numerous lipid droplets [Courtesy of Dr. H. Iwasawa.]

E. Teleosts

In certain species of teleosts during the spawning season, numerous lipid droplets are contained in the cytoplasm of the Sertoli cells forming the walls of testicular cysts (Fig. 6). Unfortunately, such lipid droplets were originally considered evidence for the homology of these cells with mammalian Leydig cells (Marshall and Lofts, 1956). This interpretation, however, is no longer tenable in light of recent ultrastructural investigations (Grier and Linton, 1977; Grier et al., 1980; Nagahama, 1983). These authors conclude that Sertoli cells have been misnamed "lobule boundary cells" in the testes of several teleosts. Boundary cells in the teleost testis are not lipophilic Leydig cell homologs described by Marshall and Lofts (1956). The true teleost boundary cells reside immediately outside of the tubule basement membrane, have been described in several teleost species, and may be myoid in nature, as they are in mammalian testes. In the teleosts, it is still uncertain whether Sertoli cells can produce steroids. Ultrastructural evidence generally does not indicate steroid synthetic capacity for these cells (Grier et al., 1980; Grier, 1981). However, certain species do seem to have some of the appropriate ultrastructural features (Nicholls and Graham, 1972a; van den

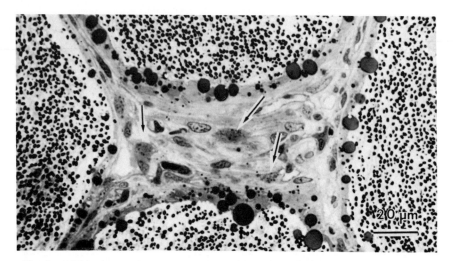

Fig. 6. Light micrograph (1 μm epon-embedded section) of a portion of the testis of a mature *Oncorhynchus gorbuscha*. Lobular lumen are filled with sperm. Sertoli cells have numerous large lipid droplets. Leydig cells (arrows) are seen in the interlobular space. [From Hoar and Nagahama, 1978.]

Hurk *et al.*, 1974b, 1978a,b; Nagahama *et al.*, 1978). Also, histochemical studies have revealed the presence of 3β-HSD activity in the Sertoli cells of *Cymatogaster aggregata* (Wiebe, 1969), *Fundulus heteroclitus* (Bara, 1969), *Salmo gairdneri* (van den Hurk *et al.*, 1978a,b), *Tilapia mossambica* (Yaron, 1966, and *Salmo salar* (O'Halloran and Idler, 1970). In the latter two species, Sertoli cells were misnamed "lobule boundary cells."

Sertoli cells can phagocytize residual bodies and degenerating germ cells (Fig. 7) and are involved in the transport of metabolites (e.g., Billard *et al.*, 1972; Gresik *et al.*, 1973b; Grier, 1975). In some teleosts, Sertoli cells are transformed into efferent duct cells following spermiation (Grier *et al.*, 1980).

F. Elasmobranchs

Sertoli cells of elasmobranchs yield a positive histochemical reaction for 3β-HSD activity (Collenot and Ozon, 1964; Simpson and Wardle, 1967). Smooth endoplasmic reticulum has been described in the Sertoli cells of two species of selachians (Holstein, 1969; Collenot and Dames, 1980). Recently, Pudney and Callard (1984a) show remarkable proliferation of the smooth endoplasmic reticulum during various stages of spermatid differentiation in the Sertoli cells of *Squalus acanthias*. Other organelles characteristic of steroidogenic tissues, such as lipid droplets and mitochondria with tubulovesicular cristae, have also been observed to increase in number concomitant with membrane proliferation in the smooth endoplasmic reticulum. Leydig

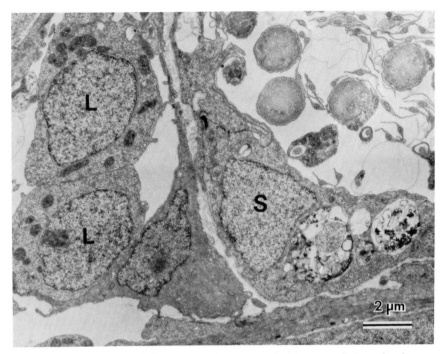

Fig. 7. Electron micrograph of a portion of the testis of adult *Oryzias latipes*, showing germ cells undergoing spermiogenesis, Sertoli cells (S), and Leydig cells (L). Note various degrees of phagocytotic activity in Sertoli cells. [Courtesy of Dr. A. Kanamori.]

cells of this species are small, sparse, and contain a poorly developed agranular reticulum (see Section V,F). In *Squalus*, Sertoli cells are apparently responsible for androgens present in circulation fluid of this species (Pudney and Callard, 1984a). This is apparently the best documented morphological evidence for androgen production by Sertoli cells throughout the vertebrates.

G. Cyclostomes

The Sertoli cells of the hagfish, *Eptatretus stouti*, only occasionally contain mitochondria with tubular cristae and smooth endoplasmic reticulum. This seems to indicate that they have a structure-supporting function for the germ cells rather than a secretory role (Tsuneki and Gorbman, 1977). Characteristic multivesiculate bodies have been reported in the cells of the lobule wall (Sertoli cells) of the river lamprey, *Lampetra fluviatilis*, during the final stages of sexual ripening (Barnes and Hardisty, 1972). Perhaps these bodies have a autophagic function. In the period immediately preceding spermiation, cells of the lobule wall (Sertoli cells) in *L. fluviatilis* (Barnes and Hardisty, 1972) and *L. japonica* (S. Fukayama, unpublished) develop extensively lipid droplets (Fig. 4B) and some ultrastructural characteristics of steroidogenic tissues (Barnes and Hardisty, 1972).

H. Blood-Testis Barrier

The existence of a permeability barrier surrounding the seminiferous tubules of the mammalian testis was initially suggested by the physiological studies of Setchell (1967). The injection of electron-dense extracellular tracer into the vascular system of the testes, coupled with electron microscopy after fixation, has demonstrated abrupt restriction in penetration of these markers by tight junctions between adjacent Sertoli cells (Fawcett *et al.*, 1970; Dym and Fawcett, 1970; Setchell and Waites, 1975; Neaves, 1977; Waites and Gladwell, 1982). These junctions effectively subdivide the seminiferous epithelium into basal and adluminal compartments. Spermatogonia and preleptotene and leptotene spermatocytes reside in the basal compartment of the seminiferous epithelium, and more differentiated germ cells are associated with the adluminal compartment (Dym and Cavicchia, 1977; Cavicchia and Dym, 1978; Russell, 1978; Connell, 1980). The junctional complex between adjacent Sertoli cells consists of fused membranes, bundles of fine 50-Å-diameter filaments, and a series of profiles of the endoplasmic reticulum (cf. Setchell and Waites, 1975; Fawcett, 1975b; Setchell, 1978; Furuya *et al.*, 1980). In freeze–fracture preparations these inter-Sertoli cell structures have been characterized as parallel occluding or tight junctions (Gilula *et al.*, 1976; Nagano and Suzuki, 1976a,b; McGinley *et al.*, 1977; Meyer *et al.*, 1977; cf. Nagano and Suzuki, 1983) (Fig. 8). The blood–testis barrier is established in postnatal rats between 16 and 19 days of age (Vitale *et al.*, 1973) and in humans at either 4 (Hadziselimovic, 1977) or 5 years of age (Camatini *et al.*, 1982).

More recent investigations, using techniques similar to those for mammalian studies, have demonstrated the presence of the blood–testis barrier in a number of nonmammalian species such as birds (Osman *et al.*, 1980; Bergman *et al.*, 1984), reptiles (Baccetti *et al.*, 1983; Bergman *et al.*, 1984), amphibians (Franchi *et al.*, 1982; Bergman *et al.*, 1983, 1984; Cavicchia and Moviglia, 1983), the teleosts (Abraham *et al.*, 1980; Marcaillou and Szöllösi, 1980). Bergman *et al.* (1984), who studied seven species of nonmammalian vertebrates, demonstrated that tight junctions are the structural basis of the blood-testis barrier in these species. In addition to tight junctions, desmosomes can be found in the testes of teleosts and anuran amphibians. The septate-like junctions are also observed in the testes of two species of birds, *Taenopygia guttata* and *Lonchura striata*. The developmental stage at which germ cells enter the tight compartment appears to differ among the vertebrates. In teleosts and amphibians which have cystic testes, only the haploid germ cells appear to be within the adluminal compartment, while the germ cells of amniotes traverse these junctional complexes immediately after the

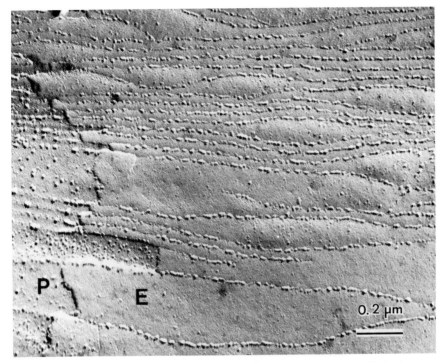

Fig. 8. Freeze–fracture replica of occluding junctions between Sertoli cells in the mouse testis. Note numerous glandular particles occurring on the ridges and in the grooves of the fractured membrane surfaces. Two fracture faces, i.e., the P and E phases, are indicated. [From Nagano and Suzuki, 1983.]

beginning of meiosis. The stage of life at which Sertoli cell junctional complexes develop in different organisms depends in part upon the unique structure of the testis, which differs according to the phylogenic relationship of each species.

V. INTERSTITIAL CELLS (LEYDIG CELLS)

The interstitial cells of Leydig are present in the testes of all major vertebrate groups. Although an endocrine function has been suspected for interstitial cells for a long time, direct evidence that Leydig cells are the main source of testicular androgens has emerged comparatively recently as a result of observations using electron microscopy, histochemistry, and biochemistry (Guraya, 1976, 1980). The application of these techniques and more recently developed methods for isolation of Leydig cells from the remaining testicular components have revealed that the major function of Leydig cells is to secrete androgens in response to pituitary gonadotropins, specifically LH, in higher vertebrates.

A. Mammals

Mammalian testicular Leydig cells are generally polygonal in shape and contain round to oval nuclei. The seasonal changes in morphology of Leydig cells in annually bleeding mammals have been reviewed by Christensen (1975). The relative number and size of mammalian Leydig cells also varies with species. In humans, Leydig cells are about 15 μm in diameter and occupy about 12% of the total testicular volume; in guinea pigs, they comprise about 0.9% of testicular volume, whereas in boars almost 37% of the testis is occupied by these cells (cf. Christensen, 1975; Mori, 1984).

The ultrastructure of mammalian Leydig cells is consistent with their function in steroid synthesis (Burgos et al., 1970; Christensen, 1975; Connell and Connell, 1977; Guraya, 1980; Ewing and Zirkin, 1983; Schulze, 1984; Hall, 1984; Mori, 1984). A number of previous reports have attempted to relate changes in testosterone levels to changes in Leydig cell morphology under both natural and experimental conditions. Zirkin et al. (1980) correlated densities of various cellular organelles in the cytoplasm of Leydig cells of five mammalian species with testosterone secretion by in vitro perfused contralateral testes. A significant linear, positive correlation was found between testosterone secretion and the volume density of smooth endoplasmic reticulum. In contrast, no correlations were found between testosterone and the volume density of other Leydig cell organelles.

In a variety of mammalian species, the degree of development of the Leydig cells follows a biphasic pattern. The first developmental phase occurs in the fetus, and the second takes place at puberty. Such biphasic development of Leydig cells is a common phenomenon among mammalian species, but is not often as distinct as in the human (Christensen, 1975). There is considerable variation in the degree of Leydig cell involution that occurs between the fetal and pubertal phases. It is generally thought that in the human testis, Leydig cells are not recognizable during the prepubertal period. Leydig cells reappear at the age of 11–14 years. Although partially differentiated Leydig cells have been demonstrated in the testes of the rabbit (Gondos et al., 1976) and the mouse (Aoki, 1970) during the period between fetal and adult life, the source of adult human Leydig cells is unknown. Recently, Prince (1984) observed a population of small cells with cytoplasm exhibiting steroid-producing features, classified as immature human Leydig cells. He found these in the prepubertal testicular interstitium and suggested that immature Leydig cells are the progenitors of the adult Leydig cell population. It is unknown whether these cells are remnants of the fetal Leydig cells or have differentiated neonatally from stem fibroblastic cells.

The availability of partially purified Leydig cells, obtained in a number of laboratories using density gradient centrifugation, has facilitated the analysis of mechanisms of gonadotropin action in vitro. Using this procedure, Payne et al. (1980a,b) have suggested that in rat testes there are two populations of

Leydig cells with similar LH receptor numbers but different *in vitro* testosterone secretory responses to human chorionic gonadotropin (hCG); Leydig cell responsiveness to markedly greater in population II than I. These results suggest that the Leydig cell population of the rat testis is functionally heterogeneous. More recently, however, different results were obtained with rats of the same strain and age as those used by Payne *et al.* (1980a,b). Aquilano and Dufau (1984), using centrifugal elutriation techniques, purified an active Leydig cell population from testicular interstitial tissue. The active population was composed of cells of different densities and a defined range of sedimentation velocities, which showed similar morphology and biological activity. This suggests that there is only one population of Leydig cells with comparable LH receptor numbers, steroidogenic activity, and susceptibility to desensitization by gonadotropins.

Hodson (1970) has reviewed the innervation of the mammalian testis and has shown it to be present in several mammalian species. It has recently been reported that bilateral denervation of the testes of mature male rats blocks an acute stress-induced rise of plasma testosterone, although this operation has no effect on basal levels of plasma testosterone (Frankel and Ryan, 1981). Thus, testicular innervation appears to be important for the functioning of Leydig cells in the rat.

B. Birds

In common with mammals, the interstitial tissue of birds contains Leydig cells which show all the histochemical (Arvy, 1962; Boucek *et al.*, 1966; Woods and Domm, 1966; Scheib and Haffen, 1969; Garnier *et al.*, 1973; Tingari, 1973) and ultrastructural (Narbaitz and Adler, 1966; Connell, 1972; Nicholls and Graham, 1972b; Rothwell, 1973; Humphreys, 1975; Lam and Farner, 1976) features of steroid-secreting cells. The development of the ultrastructural steroidogenic characteristics of testicular Leydig cells has been described in LH-stimulated chick (Connell, 1972), Japanese quail, *Coturnix coturnix japonica* (Brown *et al.*, 1975), and photoperiodically stimulated *C. coturnix japonica* (Nicholls and Graham, 1972b) and white-crowned sparrow, *Zonotrichia leucophrys gambelli* (Lam and Farner, 1976).

Cells which are cytologically intermediate between fibroblasts and fully differentiated Leydig cells have been described in the interstitial tissue of the 2-day-old posthatching chick testis (Connell, 1972). Because mitotic figures are rarely seen in the interstitium and since a cytological continuum exists between the fibroblast-like cell and the interstitial cell, Connell (1972) suggested that these transitional cells are the most likely precursors of the Leydig cell in the 2-day-old chicks. Similarly, fibroblast-like precursors of Leydig cells have also been reported in the domestic fowl, *Gallus domesticus* (Rothwell, 1973), and in *C. coturnix japonica* (Nicholls and Graham,

1972b). Leydig cells of the swan, *Cygnus olor*, have an extensive supply of bare axons containing dense core vesicles (Baumgarten and Holstein, 1968).

C. Reptiles

The interstitial cells in the testes of turtles, snakes, and lizards have been shown to possess 3β-HSD activity (Arvy, 1962; Callard, 1967; Botte and Delrio, 1967; Mesure, 1968; Licht *et al.*, 1969; Erpino, 1971) as well as the ultrastructural features of steroid-producing cells (Della Corte *et al.*, 1969; Dufaure, 1970; Pearson *et al.*, 1976).

Testes of teiid lizards (*Cnemidophorus* and *Ameiva*) are exceptional in having a compact layer of Leydig cells between the tunica albuginea and the seminiferous tubules (DeWolfe and Telford, 1966; Goldberg and Lowe, 1966; Lowe and Goldberg, 1966). During the annual breeding season, the testes of the lizard *Cnemidophorus gularis* are yellow-orange, ovoid organs measuring almost 1 cm in diameter. The pigment is confined to the testicular tunic, which contains a zone of Leydig cells and vascular channels more than 50 μm thick. Leydig cells constitute approximately 60% of the zone, with the remaining space occupied by capillaries, sinusoids, and lymphatic vessels. In this species, the interstitium is poorly developed and contains only a few widely scattered interstitial Leydig cells (Neaves, 1976). These "circumtesticular" Leydig cells have histochemical and ultrastructural features characteristic of steroid-secreting cells (Currie and Taylor, 1970; Tsui, 1976; Neaves, 1976). Licht and Midgley (1977) demonstrated specific binding of radioiodinated human FSH in the "circumtesticular" Leydig cell capsule of the lizard *C. tigris* (see also Licht *et al.*, 1977). This characteristic distribution of reptilian Leydig cells has proven useful for the investigation of Leydig cell function (Neaves, 1976). For example, Tsui (1976) has found that the Leydig cells are the major source of androgen in the lizard *Cnemidophorus,* with ovine FSH and LH being equipotent in stimulating androgen production *in vitro*.

The innervation of the interstitial tissue in reptiles has been studied in detail by Unsicker (1973). Unmyelinated axons and axon terminals make contact with Leydig cells. The relative amount and distribution of these fibers vary with species and the reproductive stage.

D. Amphibians

The testes of anurans have a distribution of intertubular Leydig cells similar to that of mammals. Histochemical (Pesonen and Rapola, 1962; Botte and Lupo, 1965; Biswas, 1969; Saidapur and Nadkarmi, 1973, 1975) and ultrastructural (Doerr-Schott, 1964; Aoki *et al.*, 1969) studies point to the interstitial cells as possible sites of steroid synthesis. Seasonal changes in the ultrastructure of anuran Leydig cells have also been reported (Brökelmann, 1964; Schulze, 1972; Unsicker, 1975). A characteristic feature of ranid Leydig cells is the occurrence of granular vesicles (Brökelmann, 1964; Unsicker, 1975; Kera and Iwasawa, 1981) (Fig. 9), which probably correspond to the fuchsinophilic granules in frog Leydig cells described by Lofts and Boswell (1960). Morphologically these fuchsinophilic granules undergo marked seasonal variations. Although the granular vesicles have often been considered lysosomal, their exact function in relation to steroidogenesis is unknown. A radioautographic study by Adachi *et al.* (1979) revealed that the binding of radioiodinated rat FSH was predominantly restricted to the testicular interstitium of the frog, *Xenopus laevis*.

The location of interstitial cells in urodeles has been the subject of much controversy. Typical interstitial cells seem to be absent in urodele testes. A positive 3β-HSD reaction was demonstrated in the glandular tissue of several urodele species (Della Corte *et al.*, 1962; Picheral, 1968; Joly, 1971; Tso and Lofts, 1977a; Imai and Tanaka, 1978; Tanaka and Iwasawa, 1979). These glandular cells (pericystic cells) possess all of the ultrastructural features generally associated with steroidogenic tissues (Picheral, 1966, 1968, 1970; Tso and Lofts, 1977b; Imai and Tanaka, 1978; Callard *et al.*, 1980; Pudney *et al.*, 1983). The glandular cells of certain urodele species also

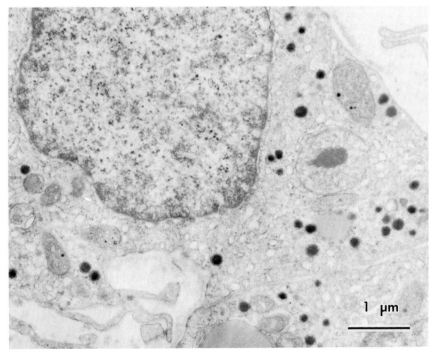

Fig. 9. Electron micrograph of portions of Leydig cells of *Rana nigromaculata* in October. Note large mitochondia with cristae arranged in a crystalline pattern and numerous granular vesicles. [Courtesy of Dr. H. Iwasawa.]

exhibit an unusual development of their mitochondria (Imai and Tanaka, 1978; Pudney et al., 1983). In addition to the ordinary mitochondrial cristae, there are clusters of regularly arranged parallel tubules, often in a crystalline pattern. To account for the absence of typical Leydig cells in some urodele species, Marshall and Lofts (1956) named these glandular cells "lobule boundary cells" and considered them to be the source of the sex steroids in these animals. However, Pudney and Callard (1984c) who studied the organization of the interstitial tissue of *Necturus maculosus,* recommended that the term "lobule boundary cells" implied homology with mammalian Leydig cells and was therefore improper. According to them, Leydig cells do not, at any stage of differentiation, form part of the lamina propria.

The glandular tissue of *N. maculosus* is highly vascular, and groups of Leydig cells appeared closely associated with blood vessels (Pudney and Callard, 1984c). This observation is consistent with an endocrine function of this tissue. Numerous myelinated axons have also been shown to occur in close proximity to *Necturus* Leydig cells. Since their preliminary investigations revealed that immunoreactive substance P and neurotensin increased in amount as the glandular tissue developed, Pudney and Callard (1984c) suggest that *Necturus* Leydig cells are modulated locally by neurons, possibly by the production of neuropeptides.

E. Teleosts

In the teleost testis, interstitial cells are usually distributed singly or in small groups in the connective tissue matrix between lobules of germinal elements (Billard et al., 1982; Grier, 1981; Nagahama et al., 1982; Nagahama, 1983) (Figs. 3, 6, and 7). In *Poecilia latipinna,* there are interstitial cells around the efferent duct and at the periphery of the testis, but not between testicular lobules (van den Hurk, 1973, 1974; van den Hurk et al., 1974a).

Stanley et al. (1965) first noted a unique distribution of testicular interstitial cells in *Gobius paganellus.* The testis of this species is composed of two parts, the seminiferous and the glandular portions. Later, their observation was confirmed with the aid of electron microscopy in two other gobiid species, *Gobius jozo* (Colombo and Burighel, 1974) and *Glossogobius olivaceus* (Asahina et al., 1983, 1985) (Fig. 10A–C). Strong evidence of steroidogenesis has been obtained recently in *G. olivaceus* by microscopically dissecting glandular tissue from the seminiferous tissue. By incubating the glandular tissue with several radioactive steroids, it has been shown that the tissue produces various labeled steroid metabolites (Asahina et al., 1985).

In some species, such as the pike, *Esox lucius,* and the char, *Salvelinus willughbii,* interstitial cells have not been identified (Marshall and Lofts, 1956). In these species, lobule boundary cells have been described as homologous with mammalian Leydig cells. According to these authors, such cells arise, not in the interstices, but within the walls of the lobule. Similarly,

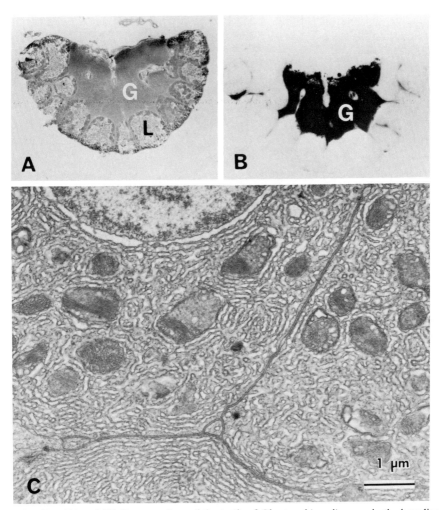

Fig. 10.   (A) and (B) Cross sections of the testis of *Glossogobius olivaceus* in the breeding season. [From Asahina *et al.*, 1983.] (A) Frozen section. Note that the glandular tissue (G) can be clearly distinguished from the swelled testicular lobules (L). (B) Frozen section. Note strong Δ^5-3β-HSD activity in the glandular cells (G). (C) Electron micrograph of portions of Leydig cells of *G. olivaceus* in the breeding season. Note many mitochondria with tubular cristae and well-developed smooth endoplasmic reticulum. [Courtesy of Dr. K. Asahina.]

O'Halloran and Idler (1970) proposed homology between the lobule boundary cells of the testis of Atlantic salmon, *Salmo salar,* and mammalian Leydig cells. The term lobule boundary cell originally proposed by Marshall and Lofts (1956) was introduced for those species which do not have typical interstitial cells. However, recent electron microscopical observations clearly indicate that there are some teleost species whose testes appear to have both interstitial cells and lobule boundary cells. Thus, the term lobule boundary cell with its implied homology to mammalian Leydig cells is misleading, and these cells actually may correspond to Sertoli cells (see Section IV,E).

Various enzymes involved in steroid hormone synthesis have been demonstrated by histochemistry in the interstitial cells of the testes of a number of teleosts (*Blennius* spp. (Chieffi and Botte, 1964), *Gobius paganellus* (Stanley *et al.*, 1965), *Tilapia mossambica* (Yaron, 1966; Hyder, 1970), *Cymatogaster aggregata* (Wiebe, 1969), *Carassius auratus* (Yamazaki and Donaldson, 1969), *Poecilia latipinna* (Takahashi and Iwasaki, 1973a), *Oryzias latipes* (Takahashi and Iwasaki, 1973b), *Salmo gairdneri* (van den Hurk *et al.*, 1978a,b), *Xiphophorus maculatus* (Schreibman *et al.*, 1982), *Glossogobius olivaceus* (Asahina *et al.*, 1983, 1985)]. Electron microscopical observations have shown that testicular interstitial cells of teleosts possess ultrastructural features commonly found in steroid-producing cells [*P. latipinna* (Follénius and Porte, 1960), *Gasterosteus aculaetus* (Follénius, 1968), *S. gairdneri* (Oota and Yamamoto, 1966; van den Hurk *et al.*, 1978a,b), *Cichlasoma nigrofasciatum* (Nicholls and Graham, 1972a), *O. latipes* (Gresik *et al.*, 1973a), *Gobius jozo* (Colombo and Burighel, 1974), *Mollienisia latipinna* (van den Hurk *et al.*, 1974b), *Oncorhynchus kisutch* and *O. gorbuscha* (Nagahama *et al.*, 1978), *Auguilla japonica* (Sugimoto and Takahashi, 1979), *G. olivaceus* (Asahina *et al.*, 1983, 1985)]. The interstitial cells of the immature silver eel, *Anguilla japonica*, were markedly stimulated by hCG, which resulted in a pronounced increase in the size and number of mitochondria and an increased organization of agranular endoplasmic reticulum (Sugimoto and Takahashi, 1979). These morphological changes in the interstitial cells may be associated with enhanced steroid production observed in testicular tissues of the European eel, *Anguilla anguilla*, after administration of hCG (Eckstein *et al.*, 1982). Some nerve fibers have been shown to provide specific innervation to the Leydig cells of several species of teleosts (e.g., Follénius, 1964a; Gresik, 1973; van den Hurk *et al.*, 1974b). The functional significance of these fibers is unknown.

Nicholls and Graham (1972a), who studied the ultrastructure of the testis of *Cichlasoma nigrofasciatus,* found evidence for the origin of interstitial Leydig cells from fibroblast-like connective tissue elements. In certain teleosts, it has been reported that the first appearance of interstitial cells with typical steroidogenic features occurs later than primary gonadal sex differentiation (Satoh, 1974; van den Hurk *et al.*, 1982). Thus, steroid hormones produced by the testes do not appear to be involved in the induction of gonadal differentiation in these species. In a tilapia, *Sarotherodon niloticus,* however, testicular Leydig cells first appear at the same time as morphological testicular differentiation occurs at 23–26 days after hatching (M. Nakamura and Y. Nagahama, unpublished). In the male medaka, *Oryzias latipes,* there is a striking increase in the number of testicular interstitial cells around the time of the initiation of meiosis (Kanamori *et al.*, 1985). This observation strongly suggests that testicular androgens play a role in initiating meiosis in this species.

In the protogynous, rice field eel, *Monopterus albus,* extensive development of the interstitial Leydig cells precedes the formation of testicular lobules during natural sex reversal (Chan and Phillips, 1967; Chan and Yeung, 1983). This proliferation of the Leydig cells is extensive in the interstitium of the germinal cords, and the cells are slightly positive in the histoenzymological reaction for 3β-HSD. Similar proliferation of Leydig cells is observed in parallel with the progress of sex reversal from the functional female phase to the functional male phase in two species of wrasses, *Halichoeres tenuispinis* and *Thalassoma duperrey* (M. Nakamura and Y. Nagahama, unpublished).

F. Elasmobranchs

Whether or not interstitial Leydig cells are present in the elasmobranch testis has been the subject of much discussion (Dodd, 1983). There is controversy as to the identity of the cells between species and even between individuals of the same species. A positive histochemical reaction for 3β-HSD has been demonstrated in interstitial cells lying in nests between neighboring, testicular ampullae in *Torpedo marmorata* (Chieffi *et al.,* 1961) and *Scyliorhinus stellaris* (Della Corte *et al.,* 1961). Holstein (1969) failed to detect Leydig cells in *S. acanthias* using electron microscopy. In contrast, Pudney and Callard (1984b), studying the same species, described interstitial cells which form an exceedingly loose assemblage lying within large interstitial spaces. These cells possess a smooth endoplasmic reticulum, mitochondria with tubular cristae, and lipid droplets. These interstitial cells, therefore, can be considered at least morphologically analogous to Leydig cells occurring in the testes of higher vertebrates. The discrepancies between these two studies probably result from the method of fixation; in the former study the testes were fixed by immersion, whereas in the latter, vascular perfusion was used. Compared to those in other higher vertebrates, however, interstitial cells of *S. acanthias* are not particularly well differentiated and do not occupy a large volume of the interstitium through all stages of the spermatogenic cycle. As a result, Pudney and Callard (1984b) call these cells Leydig-like.

G. Cyclostomes

Although progesterone, testosterone, and estradiol-17β have been reported in the male hagfish, *Eptatretus stouti* (Matty *et al.,* 1976), very few morphological studies are available to suggest the presence of interstitial cells in the hagfish testis. Tsuneki and Gorbman (1977) observed nonfibroblastic cells among the spermatogenic follicles of the testis of *E. stouti.*

These cells exhibit ultrastructural characteristics of steroid-producing cells and may be considered homologous to the Leydig cells of other vertebrates.

In lampreys the available information on testicular interstitial cells is extensive (Fig. 4B). Histochemically, 3β-HSD has been demonstrated in the interstitium of the testis of the river lamprey, *Lampetra fluviatilis* (Hardisty and Barnes, 1968), and the sea lamprey, *Petromyzon marinus* (Weisbart *et al.*, 1978). Ultrastructurally, the interstitial cells of *L. fluviatilis* (Barnes and Hardistry, 1972) and *L. japonica* (S. Fukayama, unpublished) show all of the features characteristic of steroidogenic tissues. Similar ultrastructural features were observed in the interstitial cells of *Lampetra planeri* (Follénius, 1964b) and *P. marinus,* except that mitochondria in this species do not possess tubular cristae (Weisbart *et al.,* 1978).

VI. CONCLUDING REMARKS

In this chapter, we consider diversity of testicular morphology among vertebrates, especially the features which suggest steroidogenic or other endocrine functions. Two major somatic cell types, Sertoli cells and Leydig cells, have been described throughout the vertebrate group. All major vertebrate groups possess testicular Leydig cells which display all characteristic morphological features of steroid-producing cells, indicating that these cells are the main source of testicular androgens. Although some histochemical and ultrastructural evidence is available, it is still uncertain in some animals that Sertoli cells are steroidogenic.

Although many structural features are similar throughout different vertebrate classes, significant variations exist, particularly between mammals and other vertebrates. Investigations which take advantage of the unique morphological characteristics of nonmammalian vertebrates may be especially useful in developing a more basic understanding of germ cell maturation and somatic cell function in mammals.

Examples of special morphological features which should receive emphasis are found in a variety of nonmammalian vertebrates. The development and use of a cell culture system for amphibian germ cells is already yielding important results. This system exploits the difference between amniotes and anamniotes in the relationship of germ cells and Sertoli cells. Spermatogenesis is extremely difficult to maintain in cultures of isolated mammalian spermatogenic cells, probably because of the intimate association which exists between germ cells and Sertoli cells. In amphibians, isolated germ cells can be successfully maintained in the absence of Sertoli cells. The study of the role of specific hormones, in addition to the analysis of Sertoli and Leydig cell influences on germ cell development is possible by means of coculture experiments. This system is of course potentially applicable to other anamniotes.

In many anamniotes having cystic testes, germ cell development occurs synchronously within each cyst. In addition, some groups of animals exhibits a breeding cycle during which germ cells in different stages of development are topographically segregated within the testis. In salmonids, the cells in most cysts composing the same testes are synchronized and contain germ cells at the same spermatogenic stage. Under the influence of pituitary gonadotropins, growth of testes in these animals can be substantially accelerated. Therefore, it is possible to obtain germ cells at any stage during the spermatogenic process. This is especially advantageous in studies where biochemical data are required (i.e., Louie and Dixon, 1972).

The best documented morphological evidence for steroidogenesis by Sertoli cells was obtained from elasmobranchs. However, the role of Sertoli cells at various stages of germ cell development in mammals and birds is still confusing, since each lobule contains germ cells at various stages, thus rendering it difficult to correlate Sertoli cell morphology with particular spermatogenic stages. On the other hand, the testes of some anamniotes differ from those of amniotes in having encysted germ and Sertoli cells. Pudney and Callard (1984b,c) point out that in salachians and urodeles, Sertoli cells are associated with a single generation of germinal cells, which gives rise to the cystic testis. This situation differs radically from the more complex mammalian testis, in which each Sertoli cell is simultaneously associated with numerous generations of germ cells. Hence, these species provide an excellent natural animal model for investigating the morphological and physiological characteristics of Sertoli cells in relation to spermatogenesis.

For studies of Leydig cell development, steroidogenesis, and gonadotropin control, it is necessary to obtain large populations of highly pure Leydig cells. The use of two nonmammalian groups offers attractive alternatives. The demanding separation techniques and enzyme treatments currently employed for the isolation of Leydig cells in mammals can be avoided by using fish and reptiles. In the testes of both gobiid teleosts and reptiles, such as *Cnemidophorus,* Leydig cells occur as a naturally segregated population which could potentially be separated with a technique as simple as microdissection.

The attention of students of Sertoli cell and Leydig cell morphology should be drawn to a potentially valuable group of animals for their studies— teleosts such as the wrasses, which undergo natural sex reversal from functional female to functional male. During the reverse process, it is not known whether ovarian somatic cells differentiate into testicular somatic cells directly or not.

Classically, comparative morphology was used as a means of tracing evolutionary relationships. The preceding discussion offers a more significant role for comparative studies of the vertebrates based on the cellular reproductive morphology of their testes. Such studies bring to the attention of the biomedial researcher, alternative, often structurally simpler systems, based on nonmammalian vertebrates, which may provide considerable advantages over the use of the highly complex mammalian spermatogenic system.

ACKNOWLEDGMENTS

Thanks are extended to Drs. G. Young and C. W. Walker for reading the manuscript and to Drs. T. Nagano and H. Iwasawa for their valuable discussion and for providing photographs. I also wish to thank Dr. H. Takahashi for his valuable discussion, and Drs. K. Asahina, S. Fukayama, A. Kanamori, S. Tanaka, and H. Yonezawa for providing photographs.

REFERENCES

Abe, S. (1981). Meiosis of primary spermatocytes and early spermiogenesis in the resultant spermatids in newt, *Cynops pyrrhogaster* in vitro. *Differentiation* **20**, 65–70.

Abraham, M., Rahamin, E., Tibika, H., Golenser, E., and Kieselstein, M. (1980). The blood-testis barrier in *Aphanius dispar* (Teleostei). *Cell Tissue Res.* **211**, 207–214.

Adachi, T., Pandey, A. K., and Ishii, S. (1979). Follicle-stimulating-hormone receptors in the testis of the frog, *Xenopus laevis. Gen. Comp. Endocrinol.* **37**, 177–185.

Aoki, A. (1970). Hormonal control of Leydig cell differentiation. *Protoplasma* **71**, 209–225.

Aoki, A., Vitale-Calpe, R., and Pisano, A. (1969). The testicular interstitial tissue of the amphibian *Physalaemus fuscumaculatus. Z. Zellforsch. Mikrosk. Anat.* **98**, 9–16.

Aquilano, D. R., and Dufau, M. L. (1984). Functional and morphological studies on isolated Leydig cells: Purification by centrifugal elutriation and metrizamide fractionation. *Endocrinology (Baltimore)* **114**, 499–510.

Arvy, L. (1962). Présence d'une activite steroido-3β-ol-deshydrogénasique chez quelques Sauropsides. *C. R. Hebd. Seances Acad. Sci.* **255**, 1803–1804.

Asahina, K., Uematsu, K., and Aida, K. (1983). Structure of the testis of the goby *Glossogobius olivaceus. Bull. Jpn. Soc. Sci. Fish.* **49**, 1493–1498.

Asahina, K., Suzuki, K., Aida, K., Hibiya, T., and Tamaoki, B. (1985). Relationship between the structures and steroidogenic functions of the testes of the urohaze-goby (*Glossogobius olivaceus*). *Gen. Comp. Endocrinol.* **57**, 281–292.

Baccetti, B., Bigliardi, E., Talluri, M. V., and Burrini, A. G. (1983). The Sertoli cell in lizards. *J. Ultrastruct. Res.* **85**, 11–23.

Bara, G. (1969). Histochemical demonstration of 3β-, 3α-, 11β-, and 17β-hydroxysteroid dehydrogenases in the testis of *Fundulus heteroclitus. Gen. Comp. Endocrinol.* **13**, 189–200.

Barnes, K., and Hardisty, M. W. (1972). Ultrastructural and histochemical studies on the testis of the river lamprey, *Lampetra fluviatilis* (L.). *J. Endocrinol.* **53**, 59–69.

Baumgarten, H. G., and Holstein, A.-F. (1968). Adrenerge Innervation im Hoden und Nebenhoden vom Schwan (*Cygnus olor*). *Z. Zellforsch. Mikrosk. Anat.* **91**, 402–410.

Bergman, M., Greven, H., and Schindelmeiser, J. (1983). Observations on the blood-testis barrier in a frog and a salamander. *Cell Tissue Res.* **232**, 189–200.

Bergman, M., Schindelmeiser, J., and Greven, H. (1984). The blood-testis barrier in vertebrates having different testicular organization. *Cell Tissue Res.* **238**, 145–150.

Billard, R. (1970a). La spermatogénèse de *Poecilia reticulata*. IV. La spermiogénèse. Etude ultrastructurale. *Ann. Biol. Anim., Biochim., Biophys.* **10**, 493–510.

Billard, R. (1970b). Ultrastructure comparée de spermatozoïdes de quelques poissons téléostéens. In "Comparative Spermatology" (B. Baccetti, ed.), pp. 71–79. Academic Press, New York.

Billard, R., Jalabert, B., and Breton, B. (1972). Les cellules de Sertoli des poissons téléostéens. I. Etude ultrastructurale. *Ann. Biol. Anim., Biochim., Biophys.* **12,** 19–32.

Billard, R., Fostier, A., Weil, C., and Breton, B. (1982). Endocrine control of spermatogenesis in teleost fish. *Can. J. Fish. Aquat. Sci.* **39,** 65–79.

Biswas, N. M. (1969). Δ^5-3β-hydroxysteroid dehydrogenase in toad testis: synergistic action of ascorbic acid and luteinizing hormone. *Endocrinology (Baltimore)* **85,** 981–983.

Botte, V., and Delrio, G. (1967). Effect of estradiol-17β on the distribution of 3β-hydroxysteroid dehydrogenase in the testes of *Rana esculenta* and *Lacerta sicula*. *Gen. Comp. Endocrinol.* **9,** 110–115.

Botte, V., and Lupo, C. (1965). The Δ^5-3β-hydroxysteroid dehydrogenase of the amphibian testicular tissue. *Gen. Comp. Endocrinol.* **5,** 665–666.

Boucek, R. J., Gyori, E., and Alvarez, R. (1966). Steroid dehydrogenase reactions in developing chick adrenal and gonadal tissue. *Gen. Comp. Endocrinol.* **7,** 292–303.

Brökelmann, J. (1964). Über die Stütz- und Zwischenzellen des Froschhodens wahrend des spermatogenetischen Zyklus. *Z. Zellforsch. Mikrosk. Anat.* **64,** 429–461.

Brown, N. L., Boylé, J.-D., Scanes, C. G., and Follett, B. K. (1975). Chicken gonadotrophins: their effects on the testes of immature and hypophysectomized Japanese quail. *Cell Tissue Res.* **156,** 499–520.

Burgos, M. H., and Vitale-Calpe, R. (1967). The mechanism of spermiation in the toad. *Am. J. Anat.* **120,** 227–252.

Burgos, M. H., Vitale-Calpe, R., and Aoki, A. (1970). Fine structure of the testis and its functional significance. In "Testis" (A. D. Johnson, W. R. Gomes and N. L. Vandemark, eds.), Vol. 1, pp. 551–649. Academic Press, New York.

Callard, I. P. (1967). Testicular steroid synthesis in the snake, *Natrix sipedon pictiventris*. *J. Endocrinol.* **37,** 105–106.

Callard, G. V., Canick, J. A., and Pudney, J. (1980). Estrogen synthesis in Leydig cells: Structural-functional correlations in Necturus testis. *Biol. Reprod.* **23,** 461–479.

Callard, I. P., Callard, G. V., Lance, V., and Eccles, S. (1976). Seasonal changes in testicular structure and function and the effects of gonadotropins in the freshwater turtle, *Chrysemys picta*. *Gen. Comp. Endocrinol.* **30,** 347–356.

Camatini, M., Franchi, E., de Curtis, I., Anelli, G., and Masera, G. (1982). Chemotherapy does not affect development of inter-Sertoli junctions in childhood leukaemia. *Anat. Rec.* **203,** 353–363.

Carr, I., Clegg, E. J., and Meek, G. A. (1968). Sertoli cells as phagocytes: an electron microscopic study. *J. Anat.* **102,** 501–509.

Cavicchia, J. C., and Dym, M. (1978). Ultrastructural characteristics of monkey spermatogonia and preleptotene spermatocytes. *Biol. Reprod.* **18,** 219–228.

Cavicchia, J. C., and Moviglia, G. A. (1983). The blood-testis barrier in the toad (*Bufo arenarus* Hensel): A freeze fracture and lanthanum tracer study. *Anat. Rec.* **205,** 387–396.

Chan, S. T. H., and Phillips, J. G. (1967). The structure of the gonad during natural sex reversal in *Monopterus albus* (Pisces: Teleostei). *J. Zool.* **151,** 129–141.

Chan, S. T. H., and Yeung, W. S. B. (1983). Sex control and sex reversal in fish under natural conditions. In "Fish Physiology" (W. S. Hoar, D. J. Randall, and E. M. Donaldson, eds.), Vol. 9, Part B, pp. 171–222. Academic Press, New York.

Channing, C. P., Schaerf, F. W., Anderson, L. D., and Tsafriri, A. (1980). Ovarian follicular and luteal physiology. *Int. Rev. Physiol.* **22,** 117–201.

Chieffi, G., and Botte, H. (1964). Osservazioni sul significato funzionale della geiandola annessa del testicolo dei Blennidii. *Boll. Zool.* **31,** 471–477.

Chieffi, G., Della Corte, F., and Botte, V. (1961). Osservazioni sul tessuto interstiziale del testicolo dei Selaci. *Boll. Zool.* **28,** 211–217.

Christensen, A. K. (1975). Leydig cells. In "Handbook of Physiology" (D. W. Hamilton and R. O. Greep, eds.), Sect. 7, Vol. V, pp. 57–94. Physiol. Soc., Washington, D.C.

Christensen, A. K., and Gillim, S. W. (1969). The correlation of fine structure and function in steroid-secreting cells with emphasis on those of the gonads. In "The Gonads" (K. W. McKerns, ed.), pp. 415–488. North-Holland Publ., Amsterdam.

Collenot, G., and Dames, D. (1980). Etude ultrastructurale de la cellule de Sertoli au cours de le spermiogénèse chez Scyliorhinus canicula L. Cah. Biol. Mar. **21**, 209–219.

Collenot, G., and Ozon, R. (1964). Mises en évidence biochimique et histochimique d'une Δ⁵-3β-hydroxystéroide deshydrogénase dans le testicule de Scyliorhinus canicula L. Bull. Soc. Zool. Fr. **89**, 577–587.

Colombo, L., and Burighel, P. (1974). Fine structure of the testicular gland of the black goby, Gobius jozo L. Cell Tissue Res. **154**, 39–49.

Connell, C. J. (1972). The effect of luteinizing hormone on the ultrastructure of the Leydig cell of the chick. Z. Zellforsch. Mikrosk. Anat. **128**, 139–151.

Connell, C. J. (1980). Blood-testis barrier formation and the initiation of meiosis in the dog. In "Testicular Development, Structure, and Function" (A. Steinberger and E. Steinberger, eds.), pp. 71–78. Raven Press, New York.

Connell, C. J., and Connell, G. M. (1977). The interstitial tissue of the testis. In "The Testis" (A. D. Johnson and W. R. Gomes, eds.), Vol. 4, pp. 333–369. Academic Press, New York.

Cooksey, E. J., and Rothwell, B. (1973). The ultrastructure of the Sertoli cell and its differentiation in the domestic fowl (Gallus domesticus). J. Anat. **114**, 329–345.

Currie, C., and Taylor, H. L. (1970). A histochemical study of the circumtesticular Leydig cells of a teiid lizard, Cnemidophorus tigris. J. Morphol. **132**, 101–108.

Della Corte, F., Botte, V., and Chieffi, G. (1961). Ricerca istochimica dell' attivita della steroide 3β-olo-deidrogenasi nel testicolo de Torpedo marmorata Risso e di Scyliorhinus stellaris (L). Atti Soc. Peloritana Sci. Fis., Mat. Nat. **7**, 393–397.

Della Corte, F., Galgano, M., and Cosenza, L. (1962). Su alcune reazioni isotochimiche pergli steroide nel testiculo di Triturus cristatus carnifex (Laur). Arch. Zool. Ital. **47**, 353–363.

Della Corte, F., Galgano, M., and Varano, L. (1969). Osservazioni ultrastructurali sulle cellule di Leydig di Lacerta s. sicula Raf. in esemplari di gernnaio e di maggio. Z. Zellforsch. Mikrosk. Anat. **98**, 561–575.

DeWolfe, B. B., and Telford, S. R. (1966). Lipid positive cells in the testis of the lizard, Cnemidophorus tigris. Copeia, pp. 590–592.

Dodd, J. M. (1983). Reproduction in cartilaginous fishes (Chondrichthyes). In "Fish Physiology" (W. S. Hoar, D. J. Randall, and E. M. Donaldson, eds.), Vol. 9, Part A, pp. 31–95. Academic Press, New York.

Dodd, J. M., and Sumpter, J. P. (1984). Fishes. In "Marshall's Physiology of Reproduction" (G. E. Lamming, ed.), Vol. 1, pp. 1–126. Churchill-Livingstone, Edinburgh and London.

Doerr-Schott, J. (1964). Étude au microscope électronique des cellules interstitielles de la Grenouille rousse Rana temporaria. C. R. Hebd. Seances Acad. Sci. **258**, 2896–2898.

Dorrington, J. H., and Armstrong, D. T. (1975). Follicle stimulating hormone stimulates estradiol-17β synthesis in cultured Sertoli cells. Proc. Natl. Acad. Sci. U.S.A. **72**, 2677–2681.

Dorrington, J. H., and Armstrong, D. T. (1979). Effect of FSH on gonadal functions. Recent Prog. Horm. Res. **35**, 301–342.

Dorrington, J. H., Fritz, I. B., and Armstrong, D. T. (1978). Control of testicular estrogen synthesis. Biol. Reprod. **18**, 55–64.

Dufaure, J. P. (1970). L'ultrastructure de testicule de lézard vivipare (Reptile, Lacertilien). I. Les cellules interstitielles. Z. Zellforsch. Mikrosk. Anat. **109**, 33–45.

Dufaure, J. P. (1971). L'ultrastructure du testicule de lézard vivipare (Reptile, Lacertilien). II. Les cellules de Sertoli. Etude du glycogène. Z. Zellforsch. Mikrosk. Anat. **115**, 565–578.

Dym, M., and Cavicchia, J. C. (1977). Further observations on the blood-testis barrier in monkeys. Biol. Reprod. **17**, 390–403.

Dym, M., and Cavicchia, J. C. (1978). Functional morphology of the testis. *Biol. Reprod.* **18,** 1–15.

Dym, M., and Fawcett, D. W. (1970). The blood-testis barrier in the rat and the physiological compartmentation of the seminiferous epithelium. *Biol. Reprod.* **3,** 308–326.

Eckstein, B., Cohen, S., and Hilge, V. (1982). Steroid production in testicular tissue of the Europian eel. *Endocrinology (Baltimore)* **110,** 916–919.

Erpino, M. J. (1971). Effect of substrate on histochemistry of 3β-hydroxysteroid dehydrogenase in lizard testis. *Gen. Comp. Endocrinol.* **17,** 563–566.

Ewing, L. L., Davis, J. C., and Zirkin, B. R. (1980). Regulation of testicular function: A spatial and temporal view. *Int. Rev. Physiol.* **22,** 41–115.

Ewing, L. L., and Zirkin, B. (1983). Leydig cell structure and steroidogenic function. *Recent Prog. Horm. Res.* **39,** 599–635.

Fawcett, D. W. (1961). Intercellular bridges. *Exp. Cell Res., Suppl.* **8,** 174–187.

Fawcett, D. W. (1970). A comparative view of sperm ultrastructure. *Biol. Reprod.* **2,** 90–127.

Fawcett, D. W. (1975a). Ultrastructure and function of the Sertoli cell. *In* "Handbook of Physiology" (D. W. Hamilton and R. O. Greep, eds.), Sect. 7, Vol. V, pp. 21–55. Am. Physiol. Soc., Washington, D.C.

Fawcett, D. W. (1975b). Gametogenesis in the male: Prospects for its control. *In* "The Developmental Biology of Reproduction" (C. L. Markert and J. Papaconstantinou, eds.), pp. 25–53. Academic Press, New York.

Fawcett, D. W., and Phillips, D. M. (1969). Observations on the release of spermatozoa and on changes in the head during pasage through the epididymis. *J. Reprod. Fertil., Suppl.* **6,** 405–418.

Fawcett, D. W., Leak, L. V., and Heidger, P. M. (1970). Electron microscopic observations on the structural components of the blood-testis barrier. *J. Reprod. Fertil., Suppl.* **10,** 105–122.

Fawcett, D. W., Neaves, W. B., and Flores, M. N. (1973). Comparative observations on intertubular lymphatics and the organization of the interstitial tissue of the mammalian testis. *Biol. Reprod.* **9,** 500–532.

Follénius, E. (1964a). Innervation des cellules interstitielles chez un poisson teléostéen *Lebistes reticulatus* R. Etude au microscope électronique. *C. R. Hebd. Seances Acad. Sci.* **259,** 228–230.

Follénius, E. (1964b). Structure fine des cellules interstitielles du cyclostome, *Lampetra planeri*. *C. R. Hebd. Seances Acad. Sci.* **259,** 450–452.

Follénius, E. (1968). Cytologie et cytophysiologie des cellules interstitielles de l'Epinoche: *Gasterosteus aculeatus* L. Etude au microscope électronique. *Gen. Comp. Endocrinol.* **11,** 198–219.

Follénius, E., and Porte, A. (1960). Cytologie fine des cellules interstitielle du testicule du poisson *Lebistes reticulatus* R. *Experientia* **16,** 190–192.

Fouquet, J. P. (1974). La spermiation et la formation des corps residuels chez le hamster: Role des cellules de Sertoli. *J. Microsc. (Paris)* **19,** 161–168.

Fox, H. (1977). The urogenital system of reptiles. *In* "Biology of the Reptilia" (C. Gans and T. S. Parsons, eds.), Vol. 6, pp. 1–157. Academic Press, New York.

Franchi, E., Camatini, M., and de Curtis, I. (1982). Morphological evidence of a permeability barrier in urodele testis. *J. Ultrastruct. Res.* **80,** 253–263.

Frankel, A. I., and Ryan, E. L. (1981). Testicular innervation is necessary for the response of plasma testosterone levels to acute stress. *Biol. Reprod.* **24,** 491–495.

Free, M. J. (1977). Blood supply to the testis and its role in local exchange and transport of hormones. *In* "The Testis" (A. D. Johnson and W. R. Gomes, eds.), Vol. 4, pp. 39–90. Academic Press, New York.

Furuya, S., Kumamoto, Y., Mori, M., and Sugiyama, S. (1980). The blood-testis barrier. *In*

"Normal and Cryptorchid Testis" (E. S. E. Hafez, ed.), pp. 73–93. Martinus Nijhoff, The Hague.

Garnier, D. H., Tixier-Vital, A., Gourdji, D., and Picart, R. (1973). Ultrastructure des cellules de Leydig et des cellules de Sertoli au cours du cycle testiculaire du Canard Pékin. Z. Zellforsch. Mikrosk. Anat. **144**, 369–394.

Gilula, N. B., Fawcett, D. W., and Aoki, A. (1976). The Sertoli cell occluding junctions and gap junctions in mature and developing mammalian testis. Dev. Biol. **59**, 142–168.

Goldberg, S. R., and Lowe, C. H. (1966). The reproductive cycle of the western whiptail lizard (Cnemidophorus tigris) in southern Arizona. J. Morhpol. **118**, 543–548.

Gondos, B., Renston, R., and Goldstein, D. (1976). Postnatal differentiation of Leydig cells in the rabbit testis. Am. J. Anat. **145**, 167–182.

Gorbman, A. (1983). Reproduction in cyclostome fishes and its regulation. In "Fish Physiology" (W. S. Hoar, D. J. Randall, and E. M. Donaldson, eds.), Vol. 9, Part A, pp. 1–29. Academic Press, New York.

Gresik, E. W. (1973). Fine structural evidence for the presence of nerve terminals in the testis of the teleost, Oryzias latipes. Gen. Comp. Endocrinol. **21**, 210–213.

Gresik, E. W., Quirk, J. G., and Hamilton, J. B. (1973a). A fine structural and histochemical study of the Leydig cell in the testis of the teleost, Oryzias latipes (Cyprinidontiformes). Gen. Comp. Endocrinol. **20**, 86–98.

Gresik, E. W., Quirk, J. G., and Hamilton, J. B. (1973b). Fine structure of the Sertoli cell of the testis of the teleost Oryzias letipes. Gen. Comp. Endocrinol. **21**, 341–352.

Grier, H. J. (1975). Aspects of germinal cyst and sperm development in Poecilia latipinna (Teleostei: Poeciliidae). J. Morphol. **146**, 229–250.

Grier, H. J. (1981). Cellular organization of the testis and spermatogenesis in fishes. Am. Zool. **21**, 345–357.

Grier, H. J., and Linton, J. R. (1977). Ultrastructural identification of the Sertoli cell in the testis of the northern pike, Esox lucius. Am. J. Anat. **149**, 283–288.

Grier, H. J., Fitzsimons, J. M., and Linton, J. R. (1978). Structure and ultrastructure of the testis and sperm formation in goodeid teleosts. J. Morphol. **156**, 419–438.

Grier, H. J., Linton, J. R., Leatherland, J. F., and de Vlaming, V. L. (1980). Structural evidence for two different testicular types in teleost fishes. Am. J. Anat. **159**, 331–345.

Guraya, S. S. (1976). Recent advances in the morphology, histochemistry, and biochemistry of steroid-synthesizing cellular sites in the testes of nonmammalian vertebrates. Int. Rev. Cytol. **47**, 99–136.

Guraya, S. S. (1980). Recent progress in the morphology, histochemistry, biochemistry, and physiology of developing and maturing mammalian testis. Int. Rev. Cytol. **62**, 187–309.

Hadziselimovic, F. (1977). Ultrastructure of normal and cryptorchid testes. Adv. Anat., Embryol. Cell Biol. **53**, 1–69.

Hall, P. F. (1984). Cellular organization for steroidogenesis. Int. Rev. Cytol. **86**, 53–95.

Hardisty, M. W. (1971). Gonadogenesis, sex differentiation and gametogenesis. In "The Biology of Lampreys" (M. W. Hardisty and I. C. Potter, eds.), Vol. 1, pp. 295–359. Academic Press, London.

Hardisty, M. W. (1979). "Biology of the Cyclostomes." Chapman & Hall, London.

Hardisty, M. W., and Barnes, K. (1968). Steroid Δ^5-3β-dehydrogenase activity in the cyclostome gonad. Nature (London) **218**, 880.

Hoar, W. S. (1969). Reproduction. In "Fish Physiology" (W. S. Hoar and D. J. Randall, eds.), Vol. 3, pp. 1–72. Academic Press, New York.

Hoar, W. S., and Nagahama, Y. (1978). The cellular sources of sex steroids in teleost gonads. Ann. Biol. Anim., Biochim., Biophys. **18**, 893–898.

Hodson, N. (1970). The nerves of the testis, epididymis, and scrotum. In "The Testis" (A. D. Johnson, N. L. Vandemark, and W. R. Gomes, eds.), Vol. 1, pp. 47–99. Academic Press, New York.

Holstein, A. F. (1969). Zur Frage der lokalen Steuerung der spermatogenese beim Dornhai (*Squalus acanthias* L.). *Z. Zellforsch. Mikrosk. Anat.* **93**, 265–281.

Holstein, A. F., and Roosen-Runge, E. C. (1981). "Atlas of Human Spermatogenesis." Grosse Verlag, Berlin.

Hudson, B., and Burger, H. G. (1979). Physiology and function of the testes. *In* "Human Reproductive Physiology" (R. P. Sheaman, ed.), pp. 73–96. Blackwell, Oxford.

Humphreys, P. N. (1975). Ultrastructure of the budgerigar testis during a photoperiodically induced cycle. *Cell. Tissue Res.* **159**, 541–550.

Hyder, M. (1970). Histological studies on the testes of pond specimens of *Tilapia nigra* (Gunther) (Pisces: Cichlidae) and their implications of the pituitary-testis relationship. *Gen. Comp. Endocrinol.* **14**, 198–211.

Imai, K., and Tanaka, S. (1978). Histochemical and electron microscopic observations on the steroid hormone-secreting cells in the testis of the Japanese red-bellied newt, *Cynops pyrrhogaster pyrrhogaster*. *Dev., Growth Differ.* **20**, 151–167.

Ishii, S., Tsutsui, K., and Adachi, T. (1978). Effects of gonadotropins on elements of testes of birds. *In* "Comparative Endocrinology" (P. J. Gaillard and H. H. Boar, eds.), pp. 73–76. Elsevier/North-Holland Biomedical Press, Amsterdam.

Jalabert, B., and Billard, R. (1969). Etude ultrastructurale du site de conservation des spermatozoïdes dans l'ovaire de *Poecilia reticulata* (Poisson Teleosteen). *Ann. Biol. Anim., Biochim., Biophys.* **9**, 273–280.

Joly, J. (1971). Les cycles sexuels de Salamandra (L.). I. Cycle sexuel des males. *Ann. Sci. Nat., Zool. Biol. Anim.* [12] **13**, 451–504.

Kalt, M. R. (1979). In vitro synthesis of RNA by Xenopus spermatogenic cells. I. Evidence for polyadenylated and non-polyadenylated RNA syntheis in different cell population. *J. Exp. Zool.* **208**, 77–96.

Kanamori, A., Nagahama, Y., and Egami, N. (1985). Development of the tissue architecture in the gonads of the medaka *Oryzias latipes*. *Zool. Sci.* **2**, 695–706.

Kera, Y., and Iwasawa, H. (1981). Functional histology of the testis in the process of sexual maturation in the frog, *Rana nigromaculata*. *Zool. Mag.* **90**, 6–14 (in Japanese with English abstract).

Kerr, J. B., and de Kretser, D. M. (1981). The cytology of the human testis. *In* "The Testis" (H. Burger and D. de Kretser, eds.), pp. 141–169. Raven Press, New York.

Lam, F., and Farner, D. S. (1976). The ultrastructure of the cells of Leydig in the white-crowned sparrow (*Zonotrichia leucophrys gambelli*) in relation to plasma levels of luteinizing hormone and testosterone. *Cell Tissue Res.* **169**, 93–109.

Lance, V., and Callard, I. P. (1980). Phylogenetic trends in hormonal control of gonadal steroidogenesis. *In* "Evolution of Vertebrate Endocrine Systems" (P. K. T. Pang and A. Epple, eds.), pp. 167–231. Texas Tech Univ. Press, Lubbock.

Lazard, L. (1976). Spermatogenesis and 3β-HSDH activity in the testis of the axolotl. *Nature (London)* **264**, 796–797.

Licht, P. (1984). Reptiles. *In* "Marshall's Physiology of Reproduction" (G. E. Lamming, ed.), Vol. 1, pp. 206–282. Churchill-Livingstone, Edinburgh and London.

Licht, P., and Midgley, A. R., Jr. (1977). Autoradiographic localization of binding sites for human follicle-stimulating hormone in reptilian testes and ovaries. *Biol. Reprod.* **16**, 117–121.

Licht, P., Hoyer, H. E., and van Oordt, P. G. W. J. (1969). Influence of photoperiod and temperature on testicular recrudescence and body growth in the lizards, *Lacerta sicula* and *Lacerta muralis*. *J. Zool.* **157**, 469–501.

Licht, P., Papkoff, H., Farner, S. W., Muller, C. H., Tsui, H. W., and Crews, D. (1977). Evolution in gonadotrophin structure and function. *Recent Prog. Horm. Res.* **33**, 169–248.

Lofts, B. (1968). Pattern of testicular activity. *In* "Perspectives in Endocrinology" (E. J. W. Barrington and C. B. Jorgensen, eds.), pp. 239–304. Academic Press, New York.

Lofts, B. (1972). The Sertoli cell. *Gen. Comp. Endocrinol., Suppl.* **3**, 636–648.

Lofts, B. (1980). Regulation of reproduction in male vertebrates. *In* "Hormones, Adaptations and Evolution" (S. Ishii, T. Hirano, and M. Wada, eds.), pp. 175–184. Jpn. Sci. Soc. Press, Tokyo.

Lofts, B. (1984). Amphibians. *In* "Marshall's Physiology of Reproduction" (G. E. Lamming, ed.), Vol. 1, pp. 127–205. Churchill-Livingstone, Edinburgh and London.

Lofts, B., and Bern, H. A. (1972). The functional morphology of steroidogenic tissues. *In* "Steroids in Nonmammalian Vertebrates" (D. R. Idler, ed.), pp. 37–125. Academic Press, New York.

Lofts, B., and Boswell, C. (1960). Cyclical changes in the distribution of the testis lipids in the common frog *Rana temporaria*. *Nature (London)* **187**, 708–709.

Lofts, B., and Murton, R. K. (1973). Reproduction in birds. *In* "Avian Biology" (D. S. Farner and J. R. King, eds.), Vol. 3, pp. 1–107. Academic Press, New York.

Lofts, B., and Tsui, H. W. (1977). Histological and histochemical changes in the gonads and epididymides of the male softshelled turtle, *Trionyx sinensis*. *J. Zool.* **181**, 57–68.

Louie, A. J., and Dixon, G. H. (1972). Trout testis. I. Characterization by deoxyribonucleic acid and protein analysis of cells separated by velocity sedimentation. *J. Biol. Chem.* **247**, 5490–5497.

Lowe, C. H., and Goldberg, S. R. (1966). Variation in the circumtesticular Leydig cell tunic of teiid lizards. (*Cnemidophorus* and *Ameiva*). *J. Morphol.* **119**, 277–282.

McGinley, D., Posalaky, Z., and Porvaznik, M. (1977). Intercellular junctional complexes of the rat seminiferous tubules: A freeze-fracture study. *Anat. Rec.* **189**, 211–232.

Marcaillou, C., and Szöllösi, A. (1980). The "blood-testis" barrier in a nematode and a fish: A generalizable concept. *J. Ultrastruct. Res.* **70**, 128–136.

Marshall, A. J., and Lofts, B. (1956). The Leydig-cell homologue in certain teleost fishes. *Nature (London)* **177**, 704–705.

Matty, A. J., Tsuneki, K., Dickhoff, W. W., and Gorbman, A. (1976). Thyroid and gonadal function in hypophysectomized hagfish, *Eptatretus stouti*. *Gen. Comp. Endocrinol.* **30**, 500–516.

Mesúre, M. (1968). Mise en évidence d'une variation annule de l'activité Δ^5-3β-hydroxysté-roiide deshydrogénesique dans le Lezard vivipare (*Lacerta vivipara* J.). *C. R. Seances Soc. Biol. Ses Fil.* **162**, 422–425.

Meyer, R., Posalaky, Z., and McGinley, S. (1977). Intercellular junction development in maturing rat seminiferous tubules. *J. Ultrastruct. Res.* **61**, 271–283.

Mori, H. (1984). Ultrastructure and stereological analysis of Leydig cells. *In* "Ultrastructure of Endocrine Cells and Tissues" (P. M. Motta, ed.), pp. 225–237. Martinus Nijhoff, The Hague.

Nagahama, Y. (1983). The functional morphology of teleost gonads. *In* "Fish Physiology" (W. S. Hoar, D. J. Randall, and E. M. Donaldson, eds.), Vol. 9, Part A, pp. 223–275. Academic Press, New York.

Nagahama, Y., Clarke, W. C., and Hoar, W. S. (1978). Ultrastructure of putative steroid-producing cells in the gonads of coho (*Oncorhynchus kisutch*) and pink salmon (*Oncorhynchus gorbuscha*). *Can. J. Zool.* **56**, 2508–2519.

Nagahama, Y., Kagawa, H., and Young, G. (1982). Cellular sources of sex steroids in teleost glands. *Can. J. Fish. Aquat. Sci.* **39**, 56–64.

Nagano, T., and Suzuki, F. (1976a). Freeze-fracture observations on the intercellular junctions of Sertoli cells and the Leydig cells in the human testis. *Cell Tissue Res.* **166**, 37–48.

Nagano, T., and Suzuki, F. (1976b). Postnatal development of the junctional complexes of mouse Sertoli cells as revealed by freeze fracture. *Anat. Rec.* **185**, 403–418.

Nagano, T., and Suzuki, F. (1983). Cell junctions in the seminiferous tuble and the excurrent duct of the testis: Freeze-fracture studies. *Int. Rev. Cytol.* **81**, 163–190.

Narbaitz, R., and Adler, R. (1966). Submicroscopic observations on the differentiation of the chick gonads. *J. Embryol. Exp. Morphol.* **16**, 41–47.

Neaves, W. B. (1976). Structural characterization and rapid manual isolation of a reptilian testicular tunic rich in Leydig cells. *Anat. Rec.* **186**, 553–564.

Neaves, W. B. (1977). The blood-testis barrier. In "The Testis" (A. D. Johnson and W. R. Gomes, eds.), Vol. 4, pp. 126–162. Academic Press, New York.

Nicander, L. (1970). Comparative studies on the fine structure of vertebrate spermatozoa. In "Comparative Spermatology" (B. Baccetti, ed.), pp. 47–56. Academic Press, New York.

Nicholls, T. J., and Graham, G. P. (1972a). The ultrastructure of lobule boundary cells and Leydig cell homologs in the testis of a cichlid fish, *Cichlasoma nigrofasciatus. Gen. Comp. Endocrinol.* **19**, 133–146.

Nicholls, T. J., and Graham, G. P. (1972b). Observations on the ultrastructure and differentiation of Leydig cells in the testis of the Japanese quail (*Coturnix coturnix japonica*). *Biol. Reprod.* **6**, 179–192.

O'Halloran, M. J., and Idler, D. R. (1970). Identification and distribution of the Leydig cell homolog in the testis of sexually mature Atlantic salmon (*Salmo salar*). *Gen. Comp. Endocrinol.* **15**, 361–364.

Oota, I., and Yamamoto, K. (1966). Interstitial cells in the immature testes of the rainbow trout. *Annot. Zool. Jpn.* **39**, 142–148.

Osman, D. I., Ekwall, H., and Ploen, L. (1980). Specialized cell contacts and the blood-testis barrier in the seminiferous tubules of the domestic fowl (*Gallas domesticus*). *Int. J. Androl.* **31**, 553–562.

Ozon, R., and Collenot, G. (1965). Transformation, in vitro, de la dehydroepiandrosterone par les spermatozoïdes de *Scyliorhinus* canicular L. *C. R. Hebd. Seances Acad. Sci.* **261**, 3204–3206.

Payne, A. H., Downing, J. R., and Wong, K. (1980a). Luteinizing hormone receptors and testosterone synthesis in two distinct populations of Leydig cells. *Endocrinology (Baltimore)* **106**, 1424–1429.

Payne, A. H., Wong, K., and Vega, M. M. (1980b). Differential effects of single and repeated administrations of gonadotropins on luteinizing hormone receptors and testosterone synthesis in two populations of Leydig cells. *J. Biol. Chem.* **255**, 7118–7122.

Pearson, A. K., Tsui, H. W., and Licht, P. (1976). Effects of temperature on spermatogenesis on the production and action of androgens and on the ultrastructure of gonadotropic cells in the lizard *Anolis carolinensis. J. Exp. Zool.* **195**, 291–303.

Pelliniemi, L. J., Dym, M., Gunsalus, G. L., Musto, N. A., Bardin, C. W., and Fawcett, D. W. (1981). Immunocytochemical localization of androgen binding protein in the male rat reproductive tract. *Endocrinology (Baltimore)* **108**, 925–931.

Pesonen, S., and Rapola, J. (1962). Observations on the metabolism of adrenal and gonadal steroids in *Xenopus laevis* and *Bufo bufo. Gen. Comp. Endocrinol.* **2**, 425–432.

Phillips, D. M. (1974). "Spermiogenesis." Academic Press, New York.

Picheral, B. (1966). Sur l'ultrastructure des mitochondries des cellules du tissu glandulaire de testicule de *Pleurodeles waltlii* Michah. (Amphibien Urodele). *C. R. Hebd. Seances Acad. Sci., Ser. D* **262**, 1769–1772.

Picheral, B. (1968). Les tissues élaborateurs d'hormones stéroides chez les amphibiens urodèles. I. Ultrastructure des cellules du tissu glandulaire du testicule de *Pleurodeles waltlii* Michah. *J. Microsc. (Paris)* **7**, 115–134.

Picheral, B. (1970). Les tissus élaborateurs d'hormones stéroïdes chez les amphibiens urodèles. IV. Étude en microscopie électronique et photonique du tissu glandulaire du testicule et de la glande interrénale après hypophysectomie, chez *Pleurodeles waltlii* Michah. *Z. Zellforsch. Mikrosk. Anat.* **107**, 68–86.

Pilsworth, L. M., and Setchell, B. P. (1981). Spermatogenic and endocrine functions of the testes of invertebrate and vertebrate animals. *In* "The Testis" (H. Burger and D. de Kretser, eds.), pp. 9–38. Raven Press, New York.

Prince, F. P. (1984). Ultrastructure of immature Leydig cells in the human prepubertal testis. *Anat. Rec.* **209**, 165–176.

Pudney, J., and Callard, G. V. (1984a). Development of agranular reticulum in Sertoli cells of the testis of the dogfish *Squalus acanthias* during spermatogenesis. *Anat. Rec.* **209**, 311–321.

Pudney, J., and Callard, G. V. (1984b). Identification of Leydig-like cells in the testis of the dogfish *Squalus acanthias*. *Anat. Rec.* **209**, 323–330.

Pudney, J., and Callard, G. V. (1984c). Organization of interstitial tissue in the testis of the salamander *Necturus maculosus* (Caudata: Proteidae). *J. Morphol.* **181**, 87–95.

Pudney, J., Canick, J. A., Mak, P. M., and Callard, G. V. (1983). The differentiation of Leydig cells, steroidogenesis and the spermatogenetic wave in the testis of *Necturus maculosus*. *Gen. Comp. Endocrinol.* **50**, 43–66.

Risley, M. S. (1983). Spermatogenic cell differentiation in vitro. *Gamete Res.* **4**, 331–346.

Risley, M. S., and Eckhardt, R. A. (1979). Evidence for the continuation of meiosis and spermiogenesis in *in vitro* cultures of spermatogenic cells from *Xenopus laevis*. *J. Exp. Zool.* **207**, 513–520.

Ritzen, E. M., Hansson, V., and French, F. S. (1981). The Sertoli cell. *In* "The Testis" (H. Burger and D. de Kretser, eds.), pp. 171–194. Raven Press, New York.

Roosen-Runge, E. C. (1977). "The Process of Spermatogenesis in Animals." Cambridge Univ. Press, London and New York.

Rothwell, B. (1973). The ultrastructure of Leydig cells in the testis of the domestic fowl. *J. Anat.* **116**, 245–253.

Russell, L. D. (1978). The blood-testis barrier and its formation relative to spermatocyte maturation in the adult rat: A lanthanum tracer study. *Anat. Rec.* **190**, 99–112.

Russell, L. D., and Clermont, Y. (1976). Anchoring device between Sertoli cells and late spermatids in rat seminiferous tubules. *Anat. Rec.* **185**, 259–278.

Saidapur, S. K., and Nadkarni, V. B. (1973). Histochemical localization of Δ^5-3β-hydroxysteroid dehydrogenase and glucose-6-phosphate dehydrogenase in the tesis of indian skipper frog *Rana cyanophlyctis* (Schneider). *Gen. Comp. Endocrinol.* **21**, 225–230.

Saidapur, S. K., and Nadkarni, V. B. (1975). Histochemical localization of Δ^5-3β-hydroxysteroid dehydrogenase, glucose-6-phosphate dehydrogenase, and NADH- and NADPH-diaphorase activities in the testis of *Rana hexadactyla* and *Cacopus systoma*. *Histochem. J.* **7**, 557–561.

Salhanick, A. I., and Terner, C. (1979). Androgen synthesis in absence of Leydig and Sertoli cells in a germ cell fraction from rat seminiferous tubules. *Biol. Reprod.* **21**, 293–300.

Sapsford, C. S., and Rae, C. A. (1969). Ultrastructural studies on Sertoli cells and spermatids in the bandicoot and ram during the movement of mature spermatids into the lumen of the seminiferous tubule. *Aust. J. Zool.* **17**, 415–445.

Satoh, N. (1974). An ultrastructural study of sex differentiation in the teleost *Oryzias latipes*. *J. Embryol. Exp. Morphol.* **32**, 195–215.

Scheib, D., and Haffen, K. (1969). Apparition et localisation des hydroxystéroide déshydrogénase (Δ^5, 3β et 17β) dans les gonades de l'embryon et du poussin de la caille (*Coturnix coturnix japonica*). Etude histoenzymologique et comparaisons avec le poulet (*Gallus gallus domesticus*). *Gen. Comp. Endocrinol.* **12**, 586–597.

Schreibman, M. P., Berkowitz, E. J., and van den Hurk, R. (1982). Histology and histochemistry of the testis and ovary of the platyfish, *Xiphophorus maculatus,* from birth to sexual maturity. *Cell Tissue Res.* **224**, 81–87.

Schulze, C. (1972). Saisonbedingte Veranderungen in der Morphologie der Leydigzellen von *Rana esculenta*. *Z. Zellforsch. Mikrosk. Anat.* **142**, 367–386.

Schulze, C. (1984). Sertoli cells and Leydig cells in man. *Adv. Anat., Embryol. Cell Biol.* **88,** 1–101.

Setchell, B. P. (1967). The blood-testicular fluid barrier in sheep. *J. Physiol. (London)* **189,** 63p–65p.

Setchell, B. P. (1970). Testicular blood supply, lymphatic drainage and secretion of fluid. *In* "The Testis" (A. D. Johnson, W. R. Gomes, and N. L. VanDemark, eds.), Vol. 1, pp. 101–239. Academic Press, New York.

Setchell, B. P. (1978). "The Mammalian Testis." Elek, London.

Setchell, B. P., and Waites, G. M. H. (1975). The blood-testis barrier. *In* "Handbook of Physiology" (D. W. Hamilton and R. O. Greep, eds.), Sect. 7. Vol. V, pp. 143–172. Am. Physiol. Soc., Washington, D.C.

Simpson, T. H., and Wardle, C. S. (1967). A seasonal cycle in the testis of the spurdog, *Squalus acanthias* and the sites of 3β-hydroxysteroid dehydrogenase activity. *J. Mar. Biol. Assoc. U.K.* **47,** 699–708.

Stanley, H. P. (1966). The structure and development of the seminiferous follicle in *Scyliorhinus caniculus* and *Torpedo marmorata* (Elasmobranchii). *Z. Zellforsch. Mikrosk. Anat.* **75,** 453–468.

Stanley, H. P., Chieffi, G., and Batte, V. (1965). Histological and histochemical observations on the tesis of *Gobius paganellus. Z. Zellforsch. Mikrosk. Anat.* **65,** 350–362.

Steinberger, A. (1975). In vitro techniques for the study of spermatogenesis. *In* "Methods in Enzymology" (J. G. Hardman and B. W. O'Malley, eds.), Vol. 39, Part D, pp. 283–296. Academic Press, New York.

Steinberber, A., and Steinberber, E. (1977). The Sertoli cells. *In* "The Testis" (A. D. Johnson and W. R. Gomes, eds.), Vol. 4, pp. 371–399. Academic Press, New York.

Sugimoto, Y., and Takahashi, H. (1979). Ultrastructural changes of testicular interstitial cells of silver Japanese eels, *Anguilla japonica,* treated with human chorionic gonadotropin. *Bull. Fac. Fish., Hokkaido Univ.* **30,** 23–33.

Takahashi, H., and Iwasaki, Y. (1973a). The occurrence of histochemical activity of 3β-hydroxysteroid dehydrogenase in the developing testes of *Poecilia reticulata. Dev., Growth Differ.* **15,** 241–253.

Takahashi, H., and Iwasaki, Y. (1973b). Histochemical demonstration of Δ⁵-3β-hydroxysteroid dehydrogenase activity in the testis of the medaka, *Oryzias latipes. Endocrinol. Jpn.* **20,** 529–534.

Tanaka, S., and Iwasawa, H. (1979). Annual changes in testicular structure and sexual character of the Japanese red-bellied newt *Cynop pyrrhogaster pyrrhogaster. Zool. Mag.* **88,** 295–305 (in Japanese with English abstract).

Tingari, M. D. (1973). Histochemical localization of 3β- and 17β-hydroxysteroid dehydrogenases in the male reproductive tract of the domestic fowl, *Gallus domesticus. Histochem. J.* **5,** 57–65.

Tran, D., and Josso, N. (1982). Localization of anti-Mullerian hormone in the rough endoplasmic reticulum of the developing bovine Sertoli cell using immunocytochemistry with a monoclonal antibody. *Endocrinology (Baltimore)* **111,** 1562–1567.

Tres, L. L., and Kierszenbaum, A. L. (1983). Viability of rat spermatogenic cells *in vitro* is facilitated by their coculture with Sertoli cells in serum-free hormone-supplemented medium. *Proc. Natl. Acad. Sci. U.S.A.* **80,** 3377–3381.

Tso, E. C. F., and Lofts, B. (1977a). Seasonal changes in the newt, *Trituroides hongkongensis,* testis. I. A histological and histochemical study. *Acta Zool. (Stockholm)* **58,** 1–8.

Tso, E. C. F., and Lofts, B. (1977b). Seasonal changes in the newt, *Trituroides hongkongensis,* testis. II. An ultrastructural study on the lobule boundary cell. *Acta Zool. (Stockholm)* **58,** 9–15.

Tsui, H. W. (1976). Stimulation of androgen production by the lizard testis: Site of action of ovine FSH and LH. *Gen. Comp. Endocrinol.* **28,** 386–394.

Tsuneki, K., and Gorbman, A. (1977). Ultrastructure of the testicular interstitial tissue of the hagfish *Eptatretus stouti*. *Acta Zool.* (*Stockholm*) **58**, 17–25.

Ueda, H., Kambegawa, A., and Nagahama, Y. (1984). In vitro 11-ketotestosterone and 17α,20β-dihydroxy-4-pregnen-3-one production by testicular fragments and isolated sperm of rainbow trout, *Salmo gairdneri*. *J. Exp. Zool.* **231**, 435–439.

Unsicker, K. (1973). Innervation of the testicular interstitial tissue in reptiles. *Z. Zellforsch. Mikrosk. Anat.* **146**, 123–138.

Unsicker, K. (1975). Fine structure of the male genital tract and kidney in the anura *Xenopus laevis* Daudin, *Rana temporaria* L. and *Bufo bufo* L. under normal and experimental conditions. I. Testicular interstitial tissue and seminal efferent ducts. *Cell Tissue Res.* **158**, 215–240.

van den Hurk, R. (1973). The localization of steroidogenesis in the testis of oviparous and viviparous teleosts. *Proc. K. Ned. Akad. Wet., Ser. C* **76**, 270–279.

van den Hurk, R. (1974). Steroidogenesis in the testis and gonadotropin activity in the pituitary during postnatal development of the black molly (*Mollienisia latipinna*). *Proc. K. Ned. Akad. Wet., Ser. C* **77**, 193–200.

van den Hurk, R., Meek, J., and Peute, J. (1974a). Ultrastructural study of the testis of the black molly (*Mollienisia latipinna*). I. The intratesticular efferent duct system. *Proc. K. Ned. Akad. Wet., Ser. C* **77**, 460–469.

van den Hurk, R., Meek, J., and Peute, J. (1974b). Ultrastructural study of the testis of the black molly (*Mollienisia latipinna*). II. Sertoli cells and Leydig cells. *Proc. K. Ned. Akad. Wet., Ser. C* **77**, 470–476.

van den Hurk, R., Peute, J., and Vermeij, J. A. J. (1978a). Morphological and enzyme cytochemical aspects of the testis and vas deferens of the rainbow trout, *Salmo gairdneri*. *Cell Tissue Res.* **186**, 309–325.

van den Hurk, R., Vermeij, J. A. J., Stegenga, J., Peute, J., and van Oordt, P. G. W. J. (1978b). Cyclic changes in the testis and vas deferens of the rainbow trout (*Salmo gairdneri*) with special reference to sites of steroidogenesis. *Ann. Biol. Anim., Biochim., Biophys.* **18**, 899–904.

van den Hurk, R., Lambert, J. G. D., and Peute, J. (1982). Steroidogenesis in the gonads of rainbow trout fry (*Salmo gairdneri*) before and after the onset of gonadal sex differentiation. *Reprod. Nutr. Dev.* **22**, 413–425.

van Oordt, P. G. W. J., and Brands, F. (1970). The Sertoli cells in the testis of the common frog, *Rana temporaria*. *J. Endocrinol.* **48**, 1.

van Tienhoven, A. (1983). "Reproductive Physiology of Vertebrates." Cambridge Univ. Press, London and New York.

Vitale, R., Fawcett, D. W., and Dym, M. (1973). The normal development of the blood-testis barrier and the effects of clomiphene and estrogen treatment. *Anat. Rec.* **176**, 333–344.

Vitale-Calpe, R., and Burgos, M. H. (1970a). The mechanism of spermiation in the hamster. I. Ultrastructure of spontaneous spermiation. *J. Ultrastruct. Res.* **31**, 381–393.

Vitale-Calpe, R., and Burgos, M. H. (1970b). The mechanism of spermiation in the hamster. II. The ultrastructural effects of coitus and LH administration. *J. Ultrastruct. Res.* **31**, 394–406.

Waites, G. M. H., and Gladwell, R. I. (1982). Physiological significance of fluid secretion in the testis and blood-testis barrier. *Physiol. Rev.* **62**, 624–671.

Weisbart, M., Youson, J. H., and Wiebe, J. P. (1978). Biochemical, histochemical, and ultrastructural analyses of presumed steroid-producing tissues in the sexually mature sea lamprey, *Petromyzon marinus* L. *Gen. Comp. Endocrinol.* **34**, 26–37.

Welsh, M. J., and Wiebe, J. P. (1975). Rat Sertoli cells; a rapid method for obtaining viable cells. *Endocrinology* (*Baltimore*) **96**, 618–624.

Welsh, M. J., and Wiebe, J. P. (1976). Sertoli cells from immature rats: In vitro stimulation of steroid metabolism by FSH. *Biochem. Biophys. Res. Commun.* **69**, 936–941.

Wiebe, J. P. (1969). Steroid dehydrogenases and steroids in gonads of the seaperch *Cymatogaster aggregata* Gibbons. *Gen. Comp. Endocrinol.* **12,** 256–266.

Wiebe, J. P., and Tilbe, K. S. (1979). De novo synthesis of steroids (from acetate) by isolated rat Sertoli cells. *Biochem. Biophys. Res. Commun.* **89,** 1107–1113.

Woods, J. E., and Domm, L. V. (1966). A histochemical identification of the androgen-producing cells of the gonads of the domestic fowl and the albino rat. *Gen. Comp. Endocrinol.* **7,** 559–570.

Yamazaki, F., and Donaldson, E. M. (1969). Involvement of gonadotropin and steroid hormones in the spermiation of the goldfish (*Carassius auratus*). *Gen. Comp. Endocrinol.* **12,** 491–497.

Yaron, Z. (1966). Demonstration of 3β-hydroxysteroid dehydrogenase in the testis of *Tilapia mossambica. J. Endocrinol.* **34,** 127–128.

Zirkin, B. R., Ewing, L. L., Kromann, N., and Cochran, R. C. (1980). Testosterone secretion by rat, rabbit, guinea pig, dog, and hamster testes perfused *in vitro:* Correlation with Leydig cell ultrastructure. *Endocrinology (Baltimore)* **107,** 1867–1874.

14

Stannius Corpuscles

SJOERD WENDELAAR BONGA
Faculteit der Wiskunde en Natuurwetenschappen
Katholieke Universiteit
6525 ED Nijmegen, The Netherlands

PETER K. T. PANG
Department of Physiology
School of Medicine
University of Alberta
Edmonton, Alberta, Canada T6G 2H7

I. INTRODUCTION

Whereas most endocrine glands and tissues are found in all major vertebrate groups, the corpuscles of Stannius (CS) are unique among the vertebrate endocrine glands in that they only occur in holostean and teleostean fishes. After their discovery by Stannius in 1839 (De Smet, 1962), they were long considered as homologs of the adrenal glands of the terrestrial vertebrates, but following the identification of the interrenal tissue in the head kidney of fishes as the true adrenal homolog (Youson *et al.*, 1976), the CS were recognized as endocrine glands of a specific nature.

Of the two main functions attributed to the CS, control of calcium metabolism and a cardiovascular action, the first seems to be best founded, and the CS are now generally considered as endocrine glands producing one or more hypocalcemic proteins or glycoproteins that have been denoted as hypocalcin (Pang *et al.*, 1974), teleocalcin (Ma and Copp, 1978), and parathyrin (Milet *et al.*, 1980).

Similar to the terrestrial vertebrates, fish are able to regulate their blood calcium levels very efficiently. This enables them to live in both seawater, with a calcium concentration of around 10 mM, and in fresh water, with calcium levels that may be lower than 0.01 mM. The aquatic environment, which represents the main calcium source for fish (Berg, 1970; Flik *et al.*,

439

1985), as well as the presence of gills, in which the main calcium uptake mechanisms are located, make calcium regulation in fish essentially different from that of terrestrial vertebrates. It is therefore not surprising that the hormonal control of calcium metabolism in fish and, possibly, other aquatic lower vertebrates, has some unique characteristics. Although in terrestrial vertebrates the calcium content of blood plasma and other body fluids is mainly controlled by the hypercalcemic actions of the parathyroid glands and by the hypocalcemic effects of calcitonin, there is now growing evidence that in fish and some aquatic amphibians the pituitary gland is responsible for hypercalcemic control, and the CS, more than calcitonin, account for hypocalcemic regulation (Pang *et al.,* 1980; Pang and Yee, 1980).

This chapter deals with various structural and some cytophysiological aspects of the CS. Occurrence of these glands, their embryological origin, morphology, vascularization, and innervation, as well as the ultrastructure of the gland cells, are described. The CS are compared with some other endocrine glands and tissues, and since the possibility was raised recently that the CS of fish are homologous to the parathyroids of terrestrial vertebrates, this aspect will receive special attention.

II. DISTRIBUTION AND EMBRYOLOGY OF THE CS

A. Distribution of the CS

CS have only been found in the Holostei and Teleostei. In the holosteans, the number of CS is high, varying from several hundred in the bowfin (*Amia calva*) to between 4 and 8 in the garpike (*Lepisosteus platyrhynchus*) (Garrett, 1942; De Smet, 1962; Youson *et al.,* 1976). In the Teleostei, the numbers are usually lower, although up to 14 have been reported for salmonids (Krishnamurthy, 1976). In most species of teleosts only one pair is present, but for many species, deviant numbers have been reported (Fig. 1).

After a comparative study of the CS of 47 species, Bauchot (1953) concluded that a phylogenetic significance can be attributed to the number and location of the CS. In more primitive fishes the numbers tend to be high, and the glands are often located in the middle of the mesonephric kidneys. In more advanced species the number of CS is usually reduced to one pair of relatively large bodies, located in the posterior part of the mesonephric kidneys. However, many exceptions to these rules have been reported. In a rather primitive species like *Notopterus notopterus* (Order Clupeiformes) only one CS is present, located in a posterior position (Belsare, 1973a), whereas in the more advanced representatives of the Cypriniformes large numbers may occur, often in more anterior positions. Up to 10 may be present in *Clarias batrachus,* up to 8 in *Heteropneustes fossilis,* and between 2 and 4 in *Mystus vittatus* (Belsare, 1973a; Krishnamurthy, 1976; Subhedar and Rao, 1976: Ahmad and Swarup, 1979). Moreover, within one species,

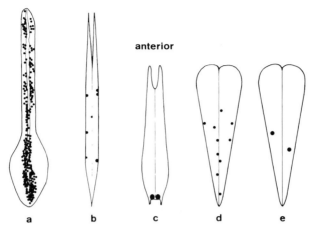

Fig. 1.   Location of the CS in fish kidneys. (a) The bowfin *Amia calva*. [After Youson *et al.*, 1976.] (b) Trout (*Salmo gairdneri*) (personal observation). (c) Stickleback (*Gasterosteus acu-leatus*) (personal observation). (d) and (e) Two specimens of catfish (*Clarias batrachus*) (P. Strivastav and K. Swarup, personal communication).

individual differences may occur that are related to age or sex. In *Esox lucius* and *Colisa lalia* the CS may be reduced in number from 5 to 2 during growth of the fish (Krishamurthy, 1976), whereas in the Chilean clingfish *Sicyases sanguineus,* the males have 2 corpuscles while females usually have 3 (Galli-Gallardo *et al.*, 1977).

B. Embryological Origin of the CS

The embryology of the CS has been described in two holostean and several teleostean species. The CS were long considered to be the homolog of the adrenal cortex of the higher vertebrates because of their location and their superficial resemblance to the elasmobranch interrenal. Huot (1898) was the first to show that the embryological origin of the CS was completely different from that of the interrenals. He found that in lophobranch teleosts the CS arise as buds from the pronephric ducts. In the holostean *Amia calva* the CS appear as outgrowths of the pronephric ducts, although some of the 40–50 corpuscles in this species may arise from the mesonephric (Garrett, 1942) or even the opisthonephric ducts (De Smet, 1962). In adults of this species the CS are still closely associated to the distal segments of the renal tubules (Fig. 2). In the holostean *Lepisosteus platyrhynchus* the CS originate from the pronephric duct only (Garrett, 1942). For teleosts it was reported that the CS are derived from the pronephric and mesonephric ducts in several salmonids and in more advanced species like *Platypoecilus maculatus* (Garrett, 1942; Ford, 1959). In the catfish (*Channa punctatus*) (Belsare, 1973b) as well as the anabantid *Colisa lalia* (Krishnamurthy, 1967) the CS

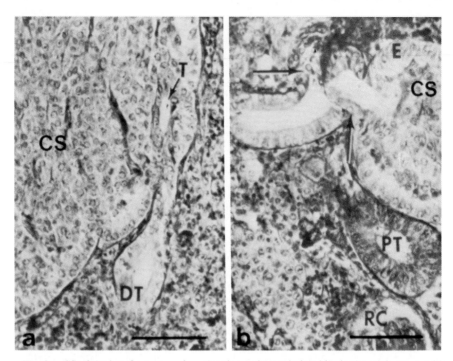

Fig. 2. CS of the bowfin *Amia calva,* showing a close relationship, in the adult stage, with the renal tubules. (a) Part of a corpuscle (CS) with a tubule (T) which appears closely associated to a distal tubule (DT) of a nephron. (b) A distal tubule of the opisthonephric kidney appears to be connected to two lobules of a CS (arrows); the glandular epithelium (E) of the corpuscle seems continuous with the epithelium of the distal tubule. PT, proximal tubule; RC, renal corpuscle; bars represent 50 μm; Mallory-Heidenhain stain. [From Youson *et al.,* 1976.]

are formed from evaginations of the mesonephric ducts. This embryological origin makes the CS unique among the endocrine glands of the vertebrates.

III. GROSS ANATOMY, VASCULARIZATION, AND INNERVATION

The CS are, in general, oval structures, although in salmonids their shape may be highly irregular. The cells of the gland are usually arranged in strands, lobes, or lobules that are separated by septa of connective tissue continuous with the outer fibrous capsule. The septa contain all the vascular and nervous elements. After a survey of the organization of the CS in 29 species, Krishnamurthy and Bern (1969) divided the CS into four categories on the basis of the arrangement of the gland cells (varying from major lobes to small aggregates of gland cells) and the extensiveness of the connective tissue septa. Later observations have shown that in some species the arrangement of the gland cells may differ between the CS of individual fish

(Ahmad and Swarup, 1979) and, more importantly, that in many fish the arrangement of the gland cells and the extensiveness of the septa depend on the state of activity of the glands (Lopez, 1969; Heyl, 1970; Ahmad and Swarup, 1979).

Although the vascularization of the CS is usually very rich (Johnson, 1972; Belsare, 1973a; Wendelaar Bonga *et al.,* 1977; Bhattacharya and Butler, 1978) it may become even more elaborate when the secretory activity of the CS is increased (Lopez, 1969; Johnson, 1972; Ahmad and Swarup, 1979). This leads to an extension of the capillary network in the connective tissue septa and a subdivision of the glandular lobes, which results in a better access of the gland cells to the blood circulation (Ahmad and Swarup, 1979). The blood capillaries in the septa are usually of the fenestrated type (Tomasulo *et al.,* 1970; Wendelaar Bonga *et al.,* 1977; Bhattacharya and Butler, 1978; Meats *et al.,* 1978).

Data on the vascular supply to the CS are limited. Johnson (1972) concluded that in the mullet (*Mugil cephalus*) the blood supply of the CS is associated with the renal arteries and the cardinal veins. In the toadfish (*Opsanus tau*) the CS are vascularized by branches from the genital arteries and are drained by small veins running to the cardinal veins (Bhattacharya and Butler, 1978). In the CS of sticklebacks (*Gasterosteus aculeatus*) the vascular elements form a portal system, similar to the blood supply of the kidneys (Fig. 3). The CS receive their blood from the dorsal caudal veins and from segmental veins from the dorsal musculature. They branch into an extensive network that is drained by veins opening into the renal blood circulation. These studies indicate that the products released by the CS first pass through the kidneys before they enter the general blood circulation.

The CS also have a rich nervous supply, as was already noted by Huot (1898). Young (1931) described a sympathetic type of innervation in *Uranoscopus scaber.* In a light microscope study of eight teleost species, Krishnamurthy and Bern (1971) demonstrated a rich autonomic innervation of the CS, but they were unable to show whether it was sympathetic or parasympathetic. In all species examined, nerves and ganglia were found in close proximity to the CS. Ganglialike accumulations of neurons, usually in close

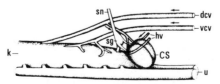

Fig. 3. Diagrammatic lateral view of the caudal region of the kidneys (k) of the stickleback *Gasterosteus aculeatus.* CS, Stannius corpuscles; sn, sympathetic nerve; sg, sympathetic ganglion sending two nerve branches to the CS and a major nerve to the kidneys; u, ureters; dcv, dorsocaudal vein; vcv, ventrocaudal vein; hv, veins from the hypaxial musculature; arrows, direction of blood flow. The venous connections between CS and kidneys cannot be observed. [From Wendelaar Bonga *et al.,* 1977.]

444 Sjoerd Wendelaar Bonga and Peter K. T. Pang

proximity to the CS but occasionally within the CS capsule, have further been described by Heyl (1970), Belsare (1973a), Schreibman and Pang (1975), Wendelaar Bonga et al. (1977), and Unsicker et al. (1977), and their occurrence seems to be a general phenomenon. The ganglia are usually associated with the blood vessels that enter the CS. Such associations have been called vasculoganglionic units by Heyl (1970), who found one close to every CS in Atlantic salmon. One unit is found in sticklebacks, located between the two closely associated CS. This unit could be equated to a sympathetic ganglion (Wendelaar Bonga et al., 1977). From this ganglion, nerve fibers penetrate the bodies along the blood vessels and branch into small fibers that run into the septa, in close association with the capillary network (Fig. 3). In many species the septa also contain single neurons (Krishnamurthy and Bern, 1971; Belsare, 1973a; Unsicker et al., 1977).

The nerve fibers are always confined to the septa. They do not seem to penetrate between the glandular cells, and synaptic nerve endings on gland cells have never been reported. The lack of evidence for direct nervous innervation of the gland cells, together with the close association invariably found between the nervous and vascular supply of the CS, have led most authors to conclude that the nervous supply of the CS is primarily connected with the control of the blood flow through the bodies. Any effect on the gland cells may be indirect (Heyl, 1970; Krishnamurthy and Bern, 1971; Wende-

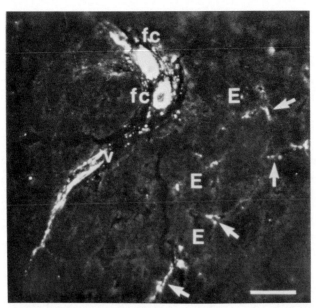

Fig. 4. Demonstration of monoamines in the CS of *Salmo irideus,* according to the method of Falck and Hillarp. (Green) fluorescent nerve fibers accompany blood vessels (v) and are also found (arrows) close to the surface of endocrine cells (E) but not within the endocrine lobes. fc, Intensely (yellow) fluorescent neurones; bar represents 100 μm; for reconstruction see Fig. 5. [From Unsicker et al., 1977.]

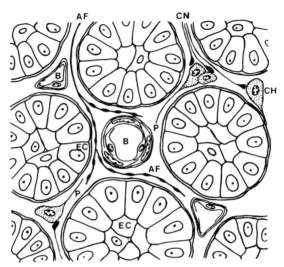

Fig. 5. Diagram of the nervous innervation of the CS in *Salmo irideus*. Adrenergic nerve fibers (AF), which could also store adrenalin and serotonin in addition to noradrenalin, run in the interlobular connective tissue of the CS and approach both blood vessels (B) and endocrine cells (EC). Moreover, the CS contain noradrenalin-storing "chromaffin" cells (CH), which are innervated by cholinergic nerves (CN) and give rise to processes (P). [From Unsicker *et al.*, 1977).]

laar Bonga *et al.*, 1977; Bhattacharya and Butler, 1978). In the most detailed study of the nervous supply of the CS available so far, Unsicker *et al.* (1977) came to the same conclusion for rainbow trout (*Salmo irideus*). With fluorimetric techniques, they showed that the CS contain considerable amounts of catecholamines and 5-hydroxytryptamine (Fig. 4). They also showed that the nerve fibers running in the septa are adrenergic, although the fibers also might contain some 5-hydroxytryptamine. The neurons dispersed in the connective tissue appeared to produce noradrenaline. These cells were synaptically innervated by cholinergic nerves (Fig. 5). The authors did not completely exclude direct innervation of the gland cells, but considered it most likely that any effect of the nervous system on production and release of the CS hormones is mediated by vascular mechanisms.

The rich autonomic innervation of the CS resembles the innervation of the adrenal medulla but contrasts with the nervous supply of calcium-regulating glands like the ultimobranchial bodies and parathyroid glands, which are sparsely innervated.

IV. STRUCTURE AND CYTOPHYSIOLOGY OF THE GLAND CELLS

A. Light and Electron Microscopy of the Gland Cells

There are many reports, based on both light and electron microscopy, that demonstrate a histochemical and structural diversity in the gland cells of the

CS. The main question raised by these reports is the following: Do these structurally different cell types reflect different phases in the cell cycle of one functional cell type, or do they represent functionally different cells, each producing a specific type of hormone?

Light microscope studies have not contributed significantly to the solution of this problem. Moreover, no staining procedure is known to be specific for the secretory material of the CS in general. With the three staining procedures that have been applied to the CS of a large number of teleosts [periodic acid Schiff (PAS), paraldehyde fuchsin (PAF), and Bowie's stain] marked species-specific differences in staining affinity have been observed. All combinations of PAS and PAF positive or negative staining have been reported for the secretory substance of the CS (Krishnamurthy and Bern, 1969; Oguri and Sokabe, 1974). Within the CS of one species, different cell types have been described on the basis of the staining affinity of their cytoplasm (Nadkarni and Gorbman, 1966; Krishnamurthy, 1976). Due to the limitations of light microscopical techniques, it was not possible to show whether these differences reflected the secretion by the CS of more than one hormonal substance. In the last 10 years only a few light microscopic studies have been published on this topic, and these were concerned with morphometric aspects of the glands rather than the nature of the gland cells and their secretory products (Olivereau and Olivereau, 1978; Swarup and Ahmad, 1978; Ahmad and Swarup, 1979; Subhedar and Rao, 1979). Conversely, several electron microscopic studies that appeared in the last decade have greatly extended our knowledge of the structure of the CS and our understanding of the cytophysiology of their gland cells. Based on a survey of the literature on the ultrastructure of the CS we hope to demonstrate that there is now good evidence that the CS may contain two functionally different endocrine cell types.

Krishnamurthy and Bern (1969) were the first to show that the PAS-positive and PAS-negative cells in the CS of trout (*Salmo gairdneri*) represented two ultrastructurally distinct cell types, with the characteristics of what we later called the type-1 cells and type-2 cells in the CS of sticklebacks (*Gasterosteus aculeatus*) and European eels (*Anguilla anguilla*) (Wendelaar Bonga and Greven, 1975; Wendelaar Bonga et al., 1977). In sticklebacks, both cell types appear very prominent. The type-1 cells are more abundant and are generally ovoid in shape, with large secretory granules, extensive granular endoplasmic reticulum arranged usually in stacks, and large Golgi areas. These cells alternate with type-2 cells, with slender, irregular cell bodies that may contain cytoplasmic processes extending between the type-1 cells. Type-2 cells are characterized by small secretory granules, the presence of strands of granular endoplasmic reticulum sparsely distributed over the cytoplasm, and several small Golgi areas. In other species similar cell types have been described. Although there may be minor differences between the species in cell shape and size of the Golgi areas or the extent of the granular endoplasmic reticulum of the cell types, their resemblance to the cells of the CS in sticklebacks qualifies them to be considered as type-1 cells and type-2 cells.

Type-1 cells predominate in the CS of all holostean and teleostean species examined so far and were the only cell type described in older literature, e.g., descriptions of European eels (Ristow and Piepho, 1963), goldfish (*Carassius auratus*) (Oguri, 1966; Ogawa, 1967), Japanese eel (*Anguilla japonica*) (Fujita and Honma, 1967), guppy (*Lebistes reticulatus*) (Tomasulo *et al.*, 1970), Atlantic salmon (*Salmo salar*) (Carpenter and Heyl, 1974), and killifish (*Fundulus heteroclitus*) (Cohen *et al.*, 1975). Type-2 cells were found not only in rainbow trout (Krishnamurthy and Bern, 1969; Meats *et al.*, 1978), European eels, and sticklebacks (Wendelaar Bonga and Greven, 1975), but also in killifish and goldfish (Wendelaar Bonga *et al.*, 1980) (Fig. 6), coho salmon (*Oncorhynchus kisutch*); (Aida *et al.*, 1980a), and the tilapia

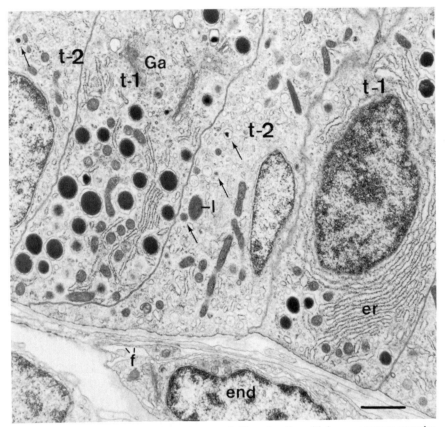

Fig. 6. CS of goldfish, *Carassius auratus*. t-1, Type-1 cells, with large secretory granules, extensive granular endoplasmic reticulum (er), and large golgi areas (Ga). t-2, Type-2 cells, which alternate with the type-1 cells (t-1); they contain small secretory granules (arrows); granular endoplasmic reticulum is scarce and consists of short strands; l, lysosome; end, endothelial layer of a compressed blood capillary, partly with fenestrations (f); bar represents 1 μm.

[*Oreochromis* (formerly *Saratherodon*) *mossambicus*] (Urasa and Wendelaar Bonga, 1985). Type-2 cells have further been described as light cells in the CS of chum salmon (*Oncorhynchus keta*) and as atypical cells in the guppy *Lebistes reticulatus* (Tomasulo *et al.*, 1970). Interestingly, the CS of the holostean *Lepisosteus platyrhynchus* (Fig. 7) also contain typical type-2 cells in addition to type-1 cells (Bhattacharya *et al.*, 1982).

In the holostean bowfin *Amia calva* (Youson and Butler, 1978), teleosts like the cod and plaice, *Gadus morhua* and *Pleuronectes platessa,* respectively (Wendelaar Bonga and Greven, 1975), and the toadfish, *Opsanus tau* (Bhattacharya and Butler, 1978), only type-1 cells have been found. It is interesting to note that all the preceding are marine fish except for the bowfin, whereas the species containing both cell types are typically freshwater fish or euryhaline fish spending part of their life cycle in fresh water. That type-1 cells are the only cell type so far described in marine teleosts is of interest, since this cell type becomes activated upon transfer of euryhaline fish from fresh water to seawater (Cohen *et al.*, 1975; Wendelaar Bonga *et al.*, 1976; Meats *et al.*, 1978). A converse response is found for type-2 cells, which are actively secreting in fresh water but are reduced in activity follow-

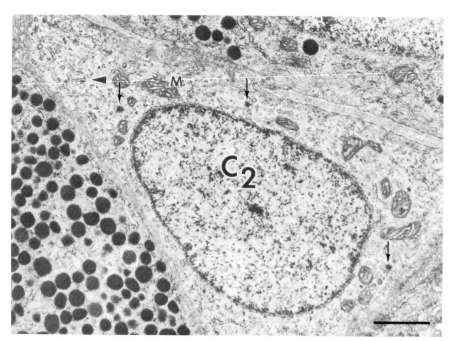

Fig. 7.   CS of garpike, *Lepisosteus platyrhynchus*. C_2, Type-2-like cell, with small secretory granules (arrows) and granular endoplasmic reticulum organized in short isolated cisternae (arrowhead); adjacent type-1-like cells contain numerous large granules; M, mitochondrion; bar represent 1 μm. [From Bhattacharya *et al.*, 1982.]

ing transfer of the fish to seawater in the stickleback *Gasterosteu aculeatus* (Wendelaar Bonga *et al.*, 1976), the trout *Salmo gairdneri* (Meats *et al.*, 1978), and the killifish *F. heteroclitus* (Wendelaar Bonga *et al.*, 1980). However, in coho salmon (*Oncorynchus kisutch*) type-2 cells seem to be more active in seawater than in fresh water (Aida *et al.*, 1980a), whereas in goldfish the type-2 cells do not noticeably respond to transfer from fresh water to (one-third) seawater (Wendelaar Bonga *et al.*, 1980). The high secretory activity of type-1 cells in sea water is likely due to its high calcium concentration. The activation of these cells shown after transfer of fish from fresh water to seawater does occur after an increase in the calcium content of fresh water (Wendelaar Bonga *et al.*, 1976, 1980; Meats *et al.*, 1978) but is prevented when fish are transferred from fresh water to low-calcium seawater (Cohen *et al.*, 1975; Wendelaar Bonga *et al.*, 1980).

Type-2 cells are not affected by changes in the calcium content of the ambient water (Wendelaar Bonga *et al.*, 1976, 1980; Meats *et al.*, 1978). It has been reported for sticklebacks that these cells respond to the sodium and potassium concentrations of the water (Wendelaar Bonga *et al.*, 1976), but this observation has not been confirmed for other species (Meats *et al.*, 1978). Thus, although the type-2 cells appear to be active in media of low ionic and osmotic strength, the specific environmental factor that controls the activity of these cells is unknown.

We consider the type-1 and type-2 cells of the CS as two functionally different types of cell, each presumably producing a different kind of hormone. This conclusion is based on the structural distinctness of both cell types and their specific responses under experimental conditions. On the basis of the same arguments, Wendelaar Bonga *et al.*, (1976, 1980), Meats *et al.* (1978), and Bhattacharya *et al.* (1982) came to the same conclusion. In a recent study on the European eel, Lopez *et al.* (1984a) raised doubts about this conclusion. Staining the CS with an antiserum against mammalian parathyroid hormone PTH showed cross-reactivity of the anti-serum with the secretory substances of all the gland cells (Figs. 4 and 8), whereas overloading the blood with calcium chloride led to the release of secretory material from both cell types. However, as will be discussed in Section V,A, cross-reactivity of proteins or polypeptides with an antiserum may require the presence of a homologous sequence of only three amino acids and does not allow the conclusion that the substances concerned are identical. Moreover, stimulation of the release of secretory material by high calcium concentrations is a phenomenon that the cells of the CS share with the cells of many exocrine and endocrine glands (Case, 1984).

Not all of the structural diversity observed in the gland cells of the CS can be ascribed to the presence of two cell types. In their detailed study of the CS of the toadfish, Bhattacharya and Butler (1978) distinguished between two "types" of agranular cells, in addition to the predominant cell type, which had all the characteristic of type-1 cells. One "type" of agranular cell

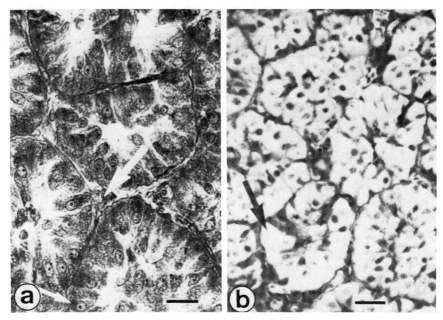

Fig. 8.   Details of CS of the eel *Anguilla anguilla,* showing the follicular arrangement of the gland cells. (a) Cleveland-Wolfe staining; large arrow, connective tissue with blood capillaries; small arrow, nucleus of gland cell. (b) Staining with anti bPTH-(1-84) serum (indirect immunof-luorescence technique); fluorescence is visible in the cytoplasm of all gland cells (arrow) but not in the interfollicular spaces; bars represent 10 μm. [From Lopez *et al.,* 1984a.]

showed an abundance of granular endoplasmic reticulum and extensive Golgi areas, the other a paucity of cell organelles. These types were inter-preted as metabolically highly active and exhausted phases, respectively, of the predominent cell type. In addition, Bhattacharyya and Butler noticed the presence of a few dark cells, recognizable by their secretory granules and other cell organelles as type-1 cells. The authors suggested that the dark cells represent fixation artifacts, perhaps resulting from anoxia during circu-latory interruption. Similar dark cells have been described in the CS of the guppy (Tomasulo *et al.,* 1970), Atlantic salmon (Carpenter and Heyl, 1974), bowfin (Youson and Butler, 1978), trout (Meats *et al.,* 1978), and garpike (Bhattacharya *et al.,* 1982). The structure of these cells characterizes them as degenerating by a process called *apoptosis* by Wyllie *et al.* (1980). This process is a mode of cellular degeneration that is structurally distinct from necrosis, the mode of cell death following anoxia. Apoptosis is a process of progressive condensation of the nuclear and cytoplasmic matrix, widening of the space enclosed by the nuclear envelope and the membrane-limiting secretory granules, distension of the cisternae of the granular endoplasmic reticulum, reduction of the Golgi areas, and, often, swelling of the mitochon-dria. These features are noticeable in the dark cells of the trout and in comparable cells described in the aforementioned reports (Fig. 9). Apoptosis

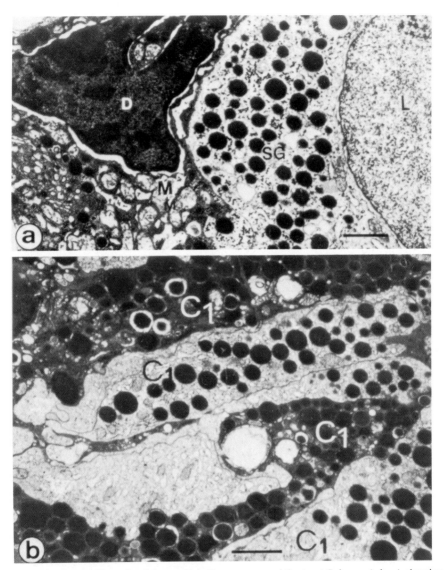

Fig. 9. Type-1-like cells of the toadfish *Opsanus tau* and the trout *Salmo gairdneri,* showing degenerative phenomena that are typical for apoptotic cell death. (a) Normal cell, with euchromatic nucleus (L) and many secretory granules (SG), and an apoptotic cell (left), with a dense, heterochromatic nucleus (D) and condensed cytoplasm with some swollen mitochondria (M). [From Bhattacharya and Butler, 1978.] (b) Normal (light) and degenerative (dense) type-1-like cells (C_1) in trout after transfer from fresh water to 25% seawater for 3 days; the degenerative cells show the cytoplasmic densification typical for apoptosis; bars represent 1 μm. [From Meats *et al.,* 1978.]

is commonly observed in gland cells as well as in many other cell types, and cell death by apoptosis is considered to balance mitosis under normal conditions (Wyllie *et al.*, 1980). It may become more prominent when glandular activity is decreased or stimulated. In the latter condition, apoptosis becomes more frequent due to accelerated turnover of the gland cells. This phenomenon has in fact been observed in trout (Meats *et al.*, 1978). Activation of the type-1 cells in the CS of trout following transfer from fresh water to seawater was accompanied by degenerative reduction of part of the cells, which showed characteristics of apoptotic cell death (Fig. 9). Degeneration of CS cells has often been observed during light microscopic studies. It can be very prominent in the CS of salmonids, in which groups of degenerating cells may accumulate in the center of the glandular lobes, forming a kind of pseudolumen (Olivereau and Fontaine, 1965; Lopez, 1969; Carpenter and Heyl, 1974).

B. Cytophysiology of the CS Gland Cells

Type-1 and type-2 cells show minor differences from species to species in cell shape, location and shape of the nucleus, extent of the granular endoplasmic reticulum and Golgi areas, and size of the secretory granules. However, there is a general consensus among the authors of ultrastructural studies of both holosteans and teleosteans that the CS gland cells invariably show all the structural characteristics of cells producing polypeptides, proteins, or glyco-proteins (Ristow and Piepho, 1963; Tomasulo *et al.*, 1970; Cohen *et al.*, 1975; Youson and Butler, 1978; Wendelaar Bonga *et al.*, 1977, 1980; Bhattacharya *et al.*, 1982). The nature of the secretory products, as suggested by ultrastructural evidence, has been confirmed by biochemical data on the biologically active factors produced by the CS. Although in the older literature the CS were considered as potentially steroid-secreting glands, histochemically detectable key enzymes for steroid biosynthesis appeared to be lacking (Bara, 1968). Biochemical analysis demonstrated that the steroid synthesizing and transforming capacity of the CS is negligible (Colombo *et al.*, 1971; Chester Jones *et al.*, 1974). For a detailed summary of the pertinent literature the reader is referred to Krishnamurthy (1976). The hypocalcemic principle isolated from salmon CS turned out to be a 3000-Da glycoprotein (Ma and Copp, 1978). The reninlike and hypocalcemic substances identified in several other teleosts are proteins or glycoproteins with molecular weights over 10,000 (Pang *et al.*, 1981a, Fenwick, 1982). The main substances produced and released by trout CS during *in vitro* incubation are glycoproteins of 10,000 and 28,000 Da, one or both showing hypocalcemia in eels (Lafeber *et al.*, 1984).

The mode of synthesis of the secretory granules is typical for protein secreting cells: material produced in the granular endoplasmic reticulum is transported via vesicles to the Golgi saccules, where it becomes visible as electron-dense material that accumulates at the distal ends and is budded off as membrane-bound granules (Fig. 10) (Tomasulo *et al.*, 1970; Youson and Butler, 1978; Wendelaar Bonga *et al.*, 1977; Bhattacharya and Butler, 1978).

As in many other glandular cell types, the mode of release of the contents of the granules has long been disputed. Holocrine as well as merocrine (exocytosis) modes of release have been reported. However, authors of almost all ultrastructural studies agree on exocytosis as the main or only mode of release of the granular contents, as has been concluded for protein-secreting endocrine glands in general (Wendelaar Bonga *et al.*, 1977, 1980; Meats *et al.*, 1978; Bhattacharya and Butler, 1978; Aida *et al.*, 1980a,b; Bhattacharya *et al.*, 1982). We recently demonstrated exocytosis in the CS cells of the cichlid *Oreochromis mossambicus* (Fig. 11), using the tannic acid method developed by Buma *et al.* (1984).

The occurrence of holocrine secretion has been concluded for European eels and Atlantic salmon on the basis of light microscopic observations of degenerating cells in the gland lobes (Olivereau and Fontaine, 1965; Lopez, 1969). It has also more recently been suggested by Meats *et al.* (1978) as an

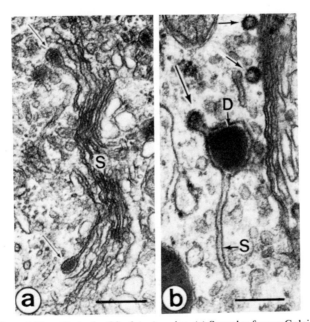

Fig. 10. Type-1-like cells in the CS of *Amia calva*. (a) Saccules from a Golgi apparatus (S), showing accumulations of electron-dense material at the ends of those saccules (arrows), located at the distal face. (b) Portion of a Golgi apparatus with free, coated vesicles (small arrows) and a dilated part of a saccule (S) with electron-dense material (D). The membrane of a coated vesicle is continuous with that of the saccule (large arrow). Both (a) and (b) indicate formation of secretory granules by budding from Golgi saccules; bars represent 0.2 μm. [From Youson and Butler, 1976.]

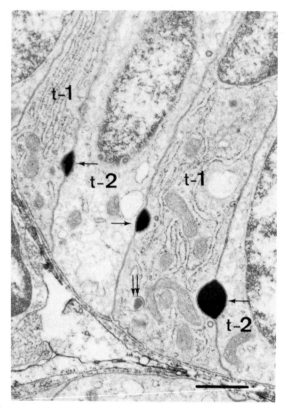

Fig. 11. CS of the cichlid *Oreochromis mossambicus*, fixed after immersion in tannic acid according to Buma *et al.* (1984). t-1, Type-1 cells; t-2, type-2 cells. Tannic acid stains and fixes secretory material immediately after its release from gland cells. The accumulation of highly electron-dense (tannic acid stained) material between the cells is suggestive of a release of granular material by exocytosis. The immersion solution contained a high calcium concentration (3 m*M*), to stimulate release of the type-1 cells, and secretory granules in the cytoplasm are therefore scarce (double arrow); bar represents 1 μm.

additional mode of release under stressful conditions for trout CS. In their electron microscopic study, Meats *et al.* observed cellular debris, including membrane-bound secretory granules, within blood capillaries. This was considered as a type of cellular degeneration that might contribute to a release of hormonal substances into the blood circulation. In a study of the bowfin (Youson and Butler, 1976), the presence of intact secretory granules in the blood vessels was attributed to mechanical damage of the glands, which seems a more plausible explanation. There is no evidence that degeneration of cells, whether occurring by apoptosis or not, may have any significance for the total amount of secretory products released by the CS. True holocrine secretion is unlikely to occur in the CS, since well-documented ultrastructural reports on holocrine secretion in the vertebrates do not pertain to endocrine glands and are restricted for the vertebrates to the sebaceous glands.

V. RELATIONSHIP BETWEEN CS AND OTHER ENDOCRINE TISSUES

When Stannius discovered the corpuscles that now bear his name, he believed that they represented the homolog of the tetrapod adrenal glands. This interpretation was no longer tenable after the identification of the interrenal cells as the adrenal homolog in fish. Any functional similarity between the CS and adrenal glands could be excluded after the demonstration that the hormonal products of the CS were proteins or glycoproteins instead of steroids or bioamines. On the basis of the pressor activity observed after injection of CS homogenates in rats, Chester Jones et al. (1966) suggested that the CS might belong to a renin-angiotensin system. The similarity between CS homogenates and the principle of the juxtaglomerular cells, renin, was stressed by the finding that CS homogenates stimulate the production of true angiotensins when incubated with blood plasma (Sokabe et al., 1970; Pang et al., 1981b). However, since the CS homogenates produce angiotensin-like substances that are hypocalcemic in killifish bioassays (Pang et al., 1981a), the CS probably are functionally distinct from juxtaglomerular cells.

Recently the CS have been equated with the parathyroid (PT) glands of the terrestrial vertebrates (Milet et al., 1980; Lopez et al., 1984a). True PT glands have been considered absent in fish (Pang, 1973; Roth and Schiller, 1976). The presumption of homology between CS and PT glands is based on immunological, structural, and functional arguments. We will consider these arguments and will further compare the embryological origin of both glands and their distribution among the vertebrates.

A. Immunological Arguments

Milet et al. (1980, 1982) identified a substance in the CS of European eels that cross-reacted with an antibody raised against the N-terminal part of bovine parathyroid hormone bPTH-(1–84) (Fig. 8). This substance was present in plasma of normal eels but disappeared after stanniectomy. After infusion of $CaCl_2$ in eels, the cross-reacting substance decreased dramatically in the CS but increased in the blood. It may therefore represent a hypocalcemic factor released by the CS in response to the heavy calcium load. Orimo et al. (1982) demonstrated PTH-like immunoreactivity in the blood of the Japanese eel with an antibody cross-reacting with the C-terminal fragment of bPTH-(1–84). The plasma level of this substance also increased after calcium infusion but remained unchanged following stanniectomy. Its source was tentatively localized in the head kidney. On the other hand, Parsons *et al* (1978) demonstrated immunoreactivity, with antisera raised against mammalian bPTH-(1–84) and hPTH-(1–34), in extracts of eel (*Anguilla anguilla*) and cod (*Gadus morhua*) pituitary glands that induced hypercalcemia in trout (*Salmo gairdneri*).

The discrepancy between the aforementioned results illustrates that different antibodies raised against mammalian PTH may cross-react with different substances in the fish body. The hypocalcemic product of the CS is probably one out of many substances in the fish body that have some immunological resemblance to PTH. Since antigenic determinants may be composed of a sequence of only three amino acids, many peptides and proteins have homologies in their amino acid sequence and therefore exhibit immunological cross-reactivity. Such homologies have frequently been reported for secretory proteins of endocrine and neuroendocrine tissues and certainly do not point to identify or similarity of the molecules concerned (De Mey, 1983). Thus, phylogenetic homology of the CS and the PT glands cannot be claimed on the basis of immunological resemblance alone.

B. Structural Arguments

The gland cells of the parathyroids of amphibians, reptiles, birds, and mammals show some similarity in their ultrastructure. They often contain light and dark chief cells and low numbers of oxyphilic cells (Roth and Schiller, 1976). Light and dark chief cells are structurally very similar and represent active and inactive phases of one single PTH-producing cell type. Both contain small secretory granules with a maximum diameter that hardly ever exceeds 0.4 μm, in contrast to those of CS type-1 cells, which may attain 2 μm. The oxyphilic cells may contain many mitochondria but are poor in other organelles and often have a pycnotic nucleus (Roth, 1978). Transitional stages between chief cells and oxyphils have been described (Capen *et al.*, 1968; Munger and Roth, 1963), but the function of oxyphils is unclear. In turtles, oxyphils have been suggested to represent degenerative chief cells (Clark and Khairallah, 1972).

A comparison of the CS with the PT glands in general does not reveal more than the superficial ultrastructural resemblance that is shared by many protein-secreting gland cells. On the basis of the structure and the distribution of the granular endoplasmic reticulum and the size of the secretory granules, the chief cells may be compared to some extent with type-2 cells, but not to the type-1 cells, the predominant cells and probably the only cell type in the CS of saltwater fish. Moreover, a very rich autonomic innervation like that found in the CS does not occur in the PT glands, although the latter receive some autonomic fibers (Shoumura *et al.*, 1983). Typical oxyphils as is found in parathyroids (Munger and Roth, 1963; Capen *et al.*, 1968) have so far not been reported for the CS. The similarity suggested by Lopez *et al.* (1984b) between clear and dark chief cells of mammalian parathyroids and type-1 and type-2 cells, respectively, in the CS of European eels does not apply to the CS of teleosts in general. The only cells reminiscent of oxyphils in the CS are apoptotic cells, and these are not specific for either the CS or the PT gland.

However, it is doubtful whether ultrastructural evidence can prove or disprove any phylogenetic homology between CS and PT glands. There are, for example, marked structural differences between the ultrastructure and organization of the calcitonin-producing cells among the vertebrates, whereas the homology of these cells is no longer disputed.

C. Distribution and Embryology

Giacomini (quoted by De Smet, 1962) stated, around the beginning of the century, that CS are found in those animals that are lacking PT glands. This generalization is not justified, however. CS are probably restricted to the Holostei and Teleostei and have not been found in the Chondrostei, Dipnoi, or Crossopterygii, which do not seem to have PT glands either. Another confusing factor is the absence of PT glands in some amphibians. Whereas PT glands have been found in all anurans examined, they do not occur in some typical aquatic urodeles, such as *Necturus,* and other neotenous urodeles and do not arise before metamorphosis in other urodeles (Roth and Schiller, 1976). This distribution pattern indicates that the PT glands are typical for terrestrial vertebrates. The absence of both CS and PT glands in several groups of fish and in some urodeles does not support the view that both glands are homologous.

The embryological data do not point to a close relationship between both types of glands. According to Le Lievre and Le Douarin (1975) the PT glands originate from the pharyngeal epithelium, from which the gland cells are derived, and from the neural crest, which contributes to the mesenchymal components of the glands. Pearse and Takor (1976) concluded that for the frog, the gland cells may originate from neural crest cells, but they further agreed on the participation, by mammals, birds, and amphibians, of both pharyngeal and ectodermal elements in the formation of the PT glands. Since the CS gland cells originate from the epithelium of the nephric tubules (Section II,B), the embryological origin of both types of glands is completely different.

D. Functional Arguments

In the last 25 years many PTH preparations, which have hypercalcemic and bone-calcium releasing effects in all tetrapods except for some aquatic amphibians, have been tested in fish. No consistent actions on either plasma calcium or bone have been found (Pang, 1973). In a recent study, however, a rapid hypocalcemic response was obtained in European eels following administration of bPTH-(1–84). This effect was comparable to that of the trout

CS homogenates (Lafeber *et al.*, 1984). CS preparations have been tested in amphibians and mammals. In the clawed toad (*Xenopus laevis*), CS extracts are hypocalcemic and hyperphosphatemic, similar to injections of hPTH-(1–34) (Milet *et al.*, 1984). Similarly, both homogenates of both trout CS and bPTH-(1–84) had a calcium-releasing effect on mouse embryonic bone *in vitro* (Lafeber *et al.*, 1984). Although these results are interesting and, we hope, will stimulate more research in this area, so far they are preliminary and incomplete and must be interpreted with caution. They seem to suggest only that there may be receptors in the vertebrates that have some affinity in common with both CS principles and PTH.

VI. CONCLUSIONS AND BIOMEDICAL IMPLICATIONS

After reviewing the pertinent literature on the CS, we are well aware of the fact that while the teleost fishes form a highly diversified and very large group of vertebrates, our knowledge of the CS is based on investigations of a limited number of species, not even including representatives of all of the major families. Nevertheless, the similarities encountered in the literature, especially that of the last 15 years, including the studies on the two holostean species, are such that some general conclusions can be drawn.

1. The CS are unique among the vertebrate endocrine organs on the basis of their embryological origin. They have a rich autonomic innervation, which is uncommon for the other calcium-regulating endocrine glands of the vertebrates.

2. Structurally, the CS gland cells have their own identity. On the basis of ultrastructural distinctiveness and differential responses to environmental changes, we conclude that there may occur two functionally different cell types in the CS: type-1 and type-2 cells. Both are found in the CS of freshwater fish and euryhaline fish spending part of their life cycle in fresh water. The only exception reported so far is the holostean bowfin, in which only type-1 cells could be demonstrated. Whereas type-1 cells predominate in freshwater fish, they are the only cell type found so far in the CS of saltwater fish. Hypocalcemic activity has been demonstrated in the CS of both freshwater and saltwater fish, and therefore type-1 cells are the likely source of the hypocalcemic principle of the CS. The function of the second cell type is not known.

3. The CS seem to represent a specific endocrine adaptation for the control of calcium metabolism in an aquatic environment.

The morphological characteristics of the CS are such that they can be rightfully considered as an endocrine gland secreting a peptide hormone or hormones. So far they have been considered as unique in some fishes only. Nevertheless, these corpuscles are playing an important role in calcium regulation. The next question is whether such an endocrine system is present in higher vertebrates and can be identified in mammals. The evolution of endocrine systems has been most conservative. In fact, perhaps with the exception of the parathyroid gland, there are thus very few examples of endocrine systems missing or present only in one major vertebrate group. There are situations where major changes in anatomical distribution occurred. For example, calcitonin is produced by the ultimobranchial body of all vertebrates except mammals in which the ultimobranchial cells become incorporated into the thyroid as parafollicular or C cells. Is it possible that the CS are present in other vertebrates in a different anatomical position? A systematic search can perhaps identify the tissue. Such findings would have significant biomedical implications. It may identify a new as yet unnoticed endocrine system in mammals, including humans. It is also possible that the CS are homologous to some existing mammalian endocrine system. Such findings may reveal some functions of the mammalian hormone, different from their known functions. The data showing the possible similarities between PTH and the CS principle are enticing. Although from the preceding discussion, the differences in embryonic origin and morphological characteristics make it rather unlikely that the CS represent the fish PT gland, these two endocrine tissues may produce different hormones of the same family. This is a distinct possibility which should not be forgotten.

No matter whether the CS are present in mammals or not, the CS principle is sufficiently different from the known mammalian hormones. There is some evidence that the CS principle is active in mammals. With extracts of eel CS, pressor activity (Chester Jones et al., 1966) and hypocalcemic activity (Leung and Fenwick, 1978) have been demonstrated in rats. Bone resorption was stimulated with eel CS extracts in rats (Milet et al., 1980) and with trout CS extracts in mice (Lafeber et al., 1984). It is important to bear in mind that although the CS principle is a potent hypocalcemic agent in fish, it does not necessarily follow that it should only affect calcium regulation in mammals. Many peptide hormones have multiple actions. It is dangerous to be blindfolded by data from one vertebrate group when studying the action in another vertebrate group. Irrespective of what the action may be, demonstration of some actions by the CS principle in mammals may have important physiological, pharamacological, or even pharmaceutical implications.

It has been pointed out that the endocrine control of calcium regulation in fish is very different from that of tetrapods and that such a difference may be related to the constant exchanges with the aquatic environment in fish. The CS may be a special adaptation to a water habitat by some fishes which are primarily aquatic vertebrates. If that is the case, these glands will not be present in the terrestrial or secondarily aquatic tetrapods. Its absence in adult mammals will then be understandable. Fetal mammals live essentially in an aquatic environment of amniotic fluid. Some of their endocrine systems take on the feature of the more primitive lower vertebrates. An example of this is the presence of a considerable amount of arginine vasotocin rather than the mammalian arginine vasopressin in fetal mammals (Perks and Vizsolyi, 1973). Will CS-like tissues be found in fetal mammals? Will the tissues be found to play important roles in the "aquatic" adaptation of the fetus? These are interesting questions to be answered.

In addition, that the CS contains a pressor substance similar to renin is a distinct possibility. The control of mammalian blood pressure is an important area of biomedical research, for obvious reasons. Any information that helps elucidate blood pressure control is invaluable for our understanding of normal blood pressure control and hypertension. The CS principle may be similar to or different from true renin. Whether there is a mammalian counterpart will have a significant biomedical meaning.

To further elucidate the biomedical significance of the CS system, one must first anatomically search for such a system in higher vertebrates including mammals. One should pay attention to unidentified as well as known endocrine tissues. A second important aspect is the chemical identification of the active principle(s) which would make the anatomical search much easier. The biomedical implications can then also be more easily realized.

ACKNOWLEDGMENTS

This work is supported in part by a research grant from the National Science Foundation, PCM 105 425.

REFERENCES

Ahmad, N., and Swarup, K. (1979). Corpuscles of Stannius of *Mystus vittatus* in relation to calcium and sodium rich environments. *Arch. Biol.* **90**, 1–22.

Aida, K., Nishioka, R. S., and Bern, H. A. (1980a). Changes in the Corpuscles of Stannius of Coho salmon (*Oncorhynchus kisutch*) during smoltification and seawater adaptation. *Gen. Comp. Endocrinol.* **41**, 296–304.

Aida, K., Nishioka, R. S., and Bern, H. A. (1980b). Degranulation of the Stannius corpuscles of Coho salmon (*Oncorhynchus kisutch*) in response to ionic changes *in vitro*. *Gen. Comp. Endocrinol.* **41**, 305–313.

Bara, G. (1968). Histochemical study of 3β and 3α, 11β and 17β hydroxysteroid dehydrogenases in the adrenocortical tissue and the corpuscles of Stannius of *Fundulus heteroclitus*. *Gen. Comp. Endocrinol.* **10**, 126–137.

Bauchot, R. (1953). Anatomie comparée des corpuscles de Stannius chez les Teleosteens. *Arch. Zool. Exp. Gen.* **89**, 147–169.

Belsare, D. K. (1973a). Comparative anatomy and histology of the corpuscles of Stannius in teleosts. *Z. Mikrosk. Anat. Forsch.* **87**, 445–456.

Belsare, D. K. (1973b). Histogenesis of the interrenal corpuscles, the chromaffin tissue and the corpuscles of Stannius in *Channa punctatus* Bloch. (Actinopterygii, Percomorphi). *Cellule* **6**, 343–349.

Berg, A. (1970). Studies on the metabolism of calcium and strontium in freshwater fish. II. Relative contribution of direct and intestinal absorption in growth conditions. *Mem. Ist. Ital. Idrobiol.* **26**, 241–255.

Bhattacharya, T. K., and Butler, D. B. (1978). Fine structure of the corpuscles of Stannius in the toadfish. *J. Morphol.* **155**, 271–286.

Bhattacharya, T. K., Butler, D. G., and Youson, J. H. (1982). Ultrastructure of the corpuscles of Stannius in the garpike (*Lepisosteus platyrhynchus*). *Gen. Comp. Endocrinol.* **46**, 29–41.

Buma, P., Roubos, E. W., and Buijs, R. M. (1984). Ultrastructural demonstration of exocytosis of neural, neuroendocrine and endocrine secretions with an in vitro tannic acid (TARI-) method. *Histochemistry* **80**, 247–256.

Capen, C. C., Cole, C. R., and Hibbs, J. W. (1968). Influence of vitamin D on calcium metabolism and the parathyroid glands of cattle. *Fed. Proc., Fed. Am. Soc. Exp. Biol.* **27**, 142–152.

Carpenter, S. J., and Heyl, H. L. (1974). Fine structure of the corpuscles of Stannius of Atlantic salmon during the freshwater spawning journey. *Gen. Comp. Endocrinol.* **23**, 212–223.

Case, R. M. (1984). The role of Ca^{2+} stores in secretion. *Cell Calcium* **5**, 89–110.

Chester Jones, I., Henderson, I. W., Chan, D. K. O., Rankin, J. C. Mosley, W., Brown, J. J., Lever, A. F., Robertson, J. I. S., and Tree, M. (1966). Pressor activity in extracts of the corpuscles of Stannius from the European eel, *Anguilla anguilla* L. *J. Endocrinol.* **34**, 393–408.

Chester Jones, I., Ball, J. N., Henderson, I. W., Sandor, T., and Baker, B. I. (1974). Endocrincology of fishes. *In* "Chemical Zoology" (M. Florkin and B. T. Scheer, eds.), Vol. 8, pp. 523–593. Academic Press, New York.

Clark, N. B., and Khairallah, L. H. (1972). Ultrastructure of the parathyroid gland of freshwater turtles. *J. Morphol.* **138**, 131–140.

Cohen, R. S., Pang, P. K. T., and Clark, N. B. (1975). Ultrastructure of the Stannius corpuscles of the killifish, *Fundulus heteroclitus,* and its relation to calcium regulation. *Gen. Comp. Endocrinol.* **27**, 413–423.

Colombo, L., Bern, H. A., and Pieprzyk, J. (1971). Steroid transformations by the corpuscles of Stannius and the body kidney of *Salmo gairdnerii* (Teleostei). *Gen. Comp. Endocrinol.* **16**, 74–84.

De Mey, J. (1983). Raising and testing antibodies for immunocytochemistry. *In* "Immunocytochemistry. Practical Applications in Pathology and Biology" (J. M. Polak and S. van Noorden, eds.). pp. 43–81. Wright & Sons Ltd., Bristol.

De Smet, D. (1962). Considerations on the Stannius corpuscles and interrenal tissue of bony tissues, especially based on researches into *Amia. Acta Zool. (Stockholm)* **43**, 201–219.

Fenwick, J. C. (1982). Some evidence concerning the nature of the hypocalcemic factor in the Stannius corpuscles. *In* "Comparative Endocrinology of Calcium Regulation" (C. Oguro and P. K. T. Pang, eds.), pp. 167–172. Jpn. Sci. Soc. Press, Tokyo.

Flik, G., Fenwick, J. C., Kolar, Z., Mayer-Gostan, N., and Wendelaar Bonga, S. E. (1985). Whole body calcium flux rates in the cichlid teleost fish *Oreochromis mossambicus* adapted to fresh water. *Am. J. Physiol.* **249**, R432–R437.

Ford, P. (1959). Some observations on the corpuscles of Stannius. *In* "Comparative Endocrinology" (A. Gorbman, ed.), pp. 728–734. Wiley, New York.

Fujita, H., and Honma, Y. (1967). On the fine structure of corpuscles of Stannius of the eel, *Anguilla japonica. Z. Zellforsch.* **77**, Mikrosk. Anat. 175–187.

Galli-Gallardo, S. M., Marusic, E. T., and Pang, P. K. T. (1977). Studies on the Stannius corpuscles of the Chilean clingfish, *Sicyases sanguineus. Gen. Comp. Endocrinol.* **32**, 316–320.

Garrett, F. D. (1942). The development and phylogeny of the corpuscles of Stannius in ganoid and teleostean fishes. *J. Morphol.* **70**, 41–67.

Heyl, H. L. (1970). Changes in the corpuscles of Stannius during the spawning journey of Atlantic salmon (*Salmo salar*). *Gen. Comp. Endocrinol.* **14**, 43–52.

Huot, E. (1898). Préliminaire sur l'origine des capsules surrénales des poissons lophobranches. *C. R. Hebd. Seances Acad. Sci., Ser. D* **126**, 49–50.

Johnson, D. W. (1972). Variations in the interrenal and corpuscles of Stannius of *Mugil cephalus* from the Colorado river and its estuary. *Gen. Comp. Endocrinol.* **19**, 7-25.

Krishnamurthy, V. G. (1967). Development of the corpuscles of Stannius in the anabantid teleost, *Colisa lalia. J. Morphol.* **123**, 109–120.

Krishnamurthy, V. G. (1976). Cytophysiology of corpuscles of Stannius. *Int. Rev. Cytol.* **46**, 177–246.

Krishnamurthy, V. G., and Bern, H. A. (1969). Correlative histological study of the corpuscles of Stannius and the juxtaglomerular cells of teleost fishes. *Gen. Comp. Endocrinol.* **13**, 313–335.

Krishnamurthy, V. G., and Bern, H. A. (1971). Innervation of the corpuscles of Stannius. *Gen. Comp. Endocrinol.* **16**, 162–165.

Lafeber, F., Hermann-Erlee, M. P. M., Van der Meer, J. M., Flik, G., Verbost, P., and Wendelaar Bonga, S. E. (1984). *Calcif. Tissue Int.* **36**, Suppl. 2, 16.

Le Lievre, C. S., and Le Douarin, N. M. (1975). Mesenchymal derivatives of the neural crest: Analysis of chimaeric quail and chick embryos. *Dev. Biol.* **47**, 215–221.

Leung, E., and Fenwick, J. C. (1978). Hypocalcemic action of eel Stannius corpuscle in rats. *Can. J. Zool.* **56**, 2333-2335.

Lopez, E. (1969). Étude histophysiologique des corpuscles de Stannius de *Salmo salar* L., au cours des diverses étapes de son cycle vital. *Gen. Comp. Endocrinol.* **12**, 339–349.

Lopez, E., Tisserand-Jochem, E.-M., Eyquem, A., Milet, C., Hillyard, C., Lallier, F., Vidal, B., and MacIntyre, I. (1984a). Immunocytochemical detection in eel corpuscles of Stannius of a mammalian parathyroid-like hormone. *Gen. Comp. Endocrinol.* **53**, 28–36.

Lopez, E., Tisserand-Jochem, E.-M., Vidal, B., Milet, C., Lallier, F., and MacIntyre, I. (1984b). Are corpuscles of Stannius the parathyroid glands in fish? Immunocytochemical and ultrastructural arguments. *In* "Endocrine Control of Bone and Calcium Metabolism" (D. V. Cohn, J. T. Potts, Jr., and T. Fujita, eds.), Vol. 8B, pp. 181–184. Am. Elsevier, New York.

Ma, S. W. Y., and Copp, D. H. (1978). Purification, properties and action of a glycopeptide from the corpuscles of Stannius which affect calcium metabolism in the teleost. *In* "Comparative Endocrinology" (P. H. Gaillard and H. H. Boer, eds.), pp. 283–286. Elsevier/North-Holland Biomedical Press, Amsterdam.

Meats, M., Ingleton, P. M., Chester Jones, I., Garland, H. O., and Kenyon, C. J. (1978). Fine structure of the corpuscles of Stannius of the trout, *Salmo gairdneri.* Structural changes in response to increased environmental salinity and calcium ions. *Gen. Comp. Endocrinol.* **36**, 451–461.

Milet, C., Hillyard, C. J., Martelly, E., Girgis, G., MacIntyre, I., and Lopez, E. (1980). Similitudes structurales entre l'hormone hypocalcémiante des corpuscules de Stannius (PCS) de l'anguille (*Anguilla anguilla* L.) et l'hormone parathyroidienne mammalienne. *C. R. Hebd. Seances Acad. Sci. Ser. D* **291**, 977–980.

Milet, C., Hillyard, C. J., Martelly, E., Chartier, M. M., Girgis, S., McIntyre, I., and Lopez, E. (1982). A parathyroid-like hormone from eel corpuscles of Stannius which exhibits hypocalcemic action. *In* "Comparative Endocrinology of Calcium Regulation" (C. Oguro and P. K. T. Pang, eds.), pp. 181–185). Jpn. Sci. Soc. Press, Tokyo.

Milet, C., Buscaglia, M., Chartier, M. M., Martelly, E., and Lopez, E. (1984). Comparative effect of Stannius corpuscles extracts and 1-34 hPTH in the Anura, *Xenopus laevis. Gen. Comp. Endocrinol.* **53**, 497.

Munger, B. L., and Roth, S. I. (1963). The cytology of the normal parathyroid glands of man and Virginia deer. A light and electron microscopic study with morphologic evidence of secretory activity. *J. Cell Biol.* **16**, 379–400.

Nadkarni, V. B., and Gorbman, A. (1966). Structure of the corpuscles of Stannius in normal and radiothyroidectomized chinook fingerlings and spawning Pacific salmon. *Acta Zool. (Stockholm)* **47,** 61–66.

Ogawa, M. (1967). Fine structure of the corpuscles of Stannius and the inter-renal tissue in goldfish, *Carassius auratus. Z. Zellforsch Mikrosk. Anat.* **81,** 174–189.

Oguri, M. (1966). Electron-microscopic observations on the corpuscles of Stannius in goldfish. *Bull. Jpn. Soc. Sci. Fish.* **32,** 903–908.

Oguri, M., and Sokabe, H. (1974). Comparative histology of the corpuscles of Stannius and the juxtaglomerular cells in the kidneys of teleosts. *Bull. Jpn. Soc. Sci. Fish.* **40,** 545–549.

Olivereau, M., and Fontaine, M. (1965). Effet de l'hypophysectomie sur les corpuscles de Stannius de l'anguille. *C. R. Hebd. Seances Adad. Sci., Ser.* D **261,** 2003–2008.

Olivereau, M., and Olivereau, J. (1978). Prolactin, hypercalcemia and corpuscles of Stannius in seawater eel. *Cell Tissue Res.* **186,** 81–96.

Orimo, H., Shiraki, M., Hasegawa, S., and Hirano, T. (1982). Parathyroid hormone-like immunoreactivity in the eel plasma. *In* "Comparative Endocrinology of Calcium Regulation" (C. Oguro and P. K. T. Pang, eds.), pp. 51–55. Jpn. Sci. Soc. Press, Tokyo.

Pang, P. K. T. (1973). Endocrine control of calcium metabolism in teleosts. *Am. Zool.* **13,** 775–792.

Pang, P. K. T., and Yee, J. A. (1980). Endocrine control of hypocalcemic regulation. *In* "Hormones, Adaptation and Evolution" (S. Ishii, T. Hirano, and M. Wada, eds.), pp. 103–111. Jpn. Sci. Soc. Press, Tokyo.

Pang, P. K. T., Pang, R. K., and Sawyer, W. H. (1974). Environmental calcium and the sensitivity of killifish (*Fundulus heteroclitus*) in bioassays for the hypocalcemic response to Stannius corpuscles from killifish and cod (*Gadus morrhua*). *Endocrinology (Baltimore)* **94,** 548–555.

Pang, P. K. T., Pang, R. K., Liu, V. K. Y., and Sokabe, H. (1981a). Effect of fish angiotensins and angiotensin-like substances on killifish calcium regulation. *Gen. Comp. Endocrinol.* **43,** 292–298.

Pang, P. K. T., Yang, M. C. M., and Sokabe, H. (1981b). Effects of a converting enzyme inhibitor and angiotensin antagonists on the rat pressor effect of angiotensin-like substance formed by goosefish Stannius corpuscles and homologous plasma. *Gen. Comp. Endocrinol.* **45,** 402–405.

Pang, P. K. T., Kenny, A. D., and Oguro, C. (1980). The evolution of the endocrine control of calcium metabolism. *In* "Evolution of Vertebrate Endocrine Systems" (P. K. T. Pang and A. Epple, eds.), pp. 323–356. Texas Tech Univ. Press, Lubbock.

Parsons, J. A., Gray, D., Rafferty, B., and Zanelly, J. M. (1978). Evidence for a hypercalcemic factor in the fish pituitary, immunologically related to mammlian parathyroid hormone. *In* "Endocrinology of Calcium Metabolism" (D. H. Copp and R. V. Talmage, eds.), pp. 111–114. Excerpta Medica, Amsterdam.

Pearse, A. G. E., and Takor, T. T. (1976). Neuroendocrine embryology and the APUD concept. *Clin. Endocrinol.* **5,** Suppl., 229S–244S.

Perks, A. M., and Vizsolyi, E. (1973). Studies of the neurohypophysis in foetal mammals. *In* "Foetal and Neonatal Physiology" (K. S. Comline, K. W. Cross, G. S. Dawes, and P. W. Nathanielsz, eds.), pp. 430–438. Cambridge Univ. Press, London and New York.

Ristow, H., and Piepho, H. (1963). Uber die Bildung der Sekretgranula in den Stanniuschen Korperchen des Flussales. *Naturwiss. Dtsch.* **50,** 382–383.

Roth, S. I. (1978). Anatomy of the parathyroid gland. *In* "Endocrinology" (L. J. De Groot, ed.), Vol. 2, pp. 587–592. Grune & Stratton, New York.

Roth, S. I., and Schiller, A. L. (1976). Comparative anatomy of the parathyroid glands. *In* "Handbook of Physiology" (G. D. Aurbach, ed.), Sect. 7, Vol. VII, pp. 281–312. Am. Physiol. Soc., Washington, D. C.

Schreibman, M. P., and Pang, P. K. T. (1975). The histophysiology of transplanted corpuscles of Stannius in the killifish *Fundulus heteroclitus. Gen. Comp. Endocrinol.* **26,** 186–191.

Shoumura, S., Iwasaki, Y., Ishizaki, N., Emura, S. Hayashi, K., Yamahira, T., Shoumura, K., and Isono, H. (1983). Origin of autonomic fibers innervating the parathyroid gland in the rabbit. *Acta Anat.* **115,** 289–295.

Sokabe, H., Nishimura, H., Ogawa, M., and Oguri, M. (1970). Determination of renin in the corpuscles of Stannius of the teleost. *Gen. Comp. Endocrinol.* **14,** 510–516.

Subhedar, N., and Rao, P. D. P. (1976). On the cytoarchitecture of the corpuscles of Stannius of the catfish, *Heteropneustes fossilis* (Bloch). *Z. Mikrosk.- Anat. Forsch.* **90,** 737–748.

Subhedar, N., and Rao, P. D. P. (1979). Seasonal changes in the corpuscles of Stannius and the gonads of the catfish, *Heteropneustes fossilis* (Bloch). *Z. Mikrosk. -Anat. Forsch.* **93,** 74–90.

Swarup, K., and Ahmad, N. (1978). Studies of corpuscles of Stannius of *Notopterus notopterus* in relation to calcium and sodium rich environments. *Natl. Acad. Sci. Lett.* **1,** 239–240.

Tomasulo, J. A., Belt, W. D., and Hayes, E. R. (1970). The fine structure of the Stannius body of the guppy and its response to deionized water. *Am. J. Anat.* **129,** 307–328.

Unsicker, K., Plonius, T., Lindmar, R., Loffelholz, K., and Wolf, U. (1977). Catecholamines and 5-hydroxytryptamine in corpuscles of Stannius of the salmonid, *Salmo irideus* L. A study correlating electron microscopical, histochemical and chemical findings. *Gen. Comp. Endocrinol.* **31,** 121–132.

Urasa, F., and Wendelaar Bonga, S. E. (1985). Stannius corpuscles and plasma calcium levels during the reproductive cycle in the cichlid teleost fish *Oreochromis mossambicus. Cell Tissue Res.* **241,** 219–227.

Wendelaar Bonga, S. E., and Greven, J. A. A. (1975). A second cell type in Stannius bodies of two euryhaline teleost species. *Cell Tissue Res.* **159,** 287–290.

Wendelaar Bonga, S. E., Greven, J. A. A., and Veenhuis, M. (1976). The relationship between ionic composition of the environment and secretory activity of the endocrine cell types of Stannius corpuscles in the teleost *Gasterosteus aculeatus. Cell Tissue Res.* **175,** 297–312.

Wendelaar Bonga, S. E., Greven, J. A. A., and Veenhuis, M. (1977). Vascularization, innervation, and ultrastructure of the endocrine cell types of Stannius corpuscles in the teleost *Gasterosteus aculeatus. J. Morphol.* **153,** 225–244.

Wendelaar Bonga, S. E., Van der Meij, J. C. A., and Pang, P. K. T. (1980). Evidence for two secretory cell types in the Stannius bodies of the teleosts *Fundulus heteroclitus* and *Carassius auratus. Cell Tissue Res.* **212,** 295–306.

Wyllie, A. H., Kerr, J. F. R., and Currie, A. R. (1980). Cell death: The significance of apoptosis. *Int. Rev. Cytol.* **68,** 251–306.

Young, J. Z. (1931). On the autonomic nervous system of the teleostean fish *Uranoscopus scaber. Q. J. Microsc. Sci.* **74,** 491–535.

Youson, J. H., and Butler, D. G. (1976). Fine structure of the adrenocortical homolog and the corpuscles of Stannius of *Amia calva* L. *Acta Zool.* (*Stockholm*) **57,** 217–230.

Youson, J. H., Butler, D. G., and Chan, A. T. C. (1976). Identification and distribution of the adrenocortical homolog, chromaffin tissue and corpuscles of Stannius in *Amia calva* L. *Gen. Comp. Endocrinol.* **29,** 198–211.

15

Evolutionary Morphology of Endocrine Glands

AUBREY GORBMAN
Department of Zoology
University of Washington
Seattle, Washington 98195

In the context of what we may call modern endocrinology, what can we conceive as the role of a book devoted to comparative morphology of the endocrine glands? Indeed, which are the endocrine glands? What is the significance of such discussions in the human or medical sphere?

Attempting to answer the first question, we can say at the outset that there is no single reference source at this time in which we can find an organized account of the comparative morphology of all of the "classical" endocrine organs. Without a frame of phylogenic reference, mammalian endocrinologists have no way to assess the adapted character of the systems with which they work. At this time endocrinologists face a need to redefine their field or to decide at least whether it is a definable field (Tata, 1984). Adherence to a classical concept of a few glands as the sources of the hormones with which they work is no longer tenable. The brain, liver, gut, kidney, thymus, skin, lung, and spleen are as much sources of humoral regulatory factors as are any of the classical endocrine structures. The classical endocrine structures (e.g., thyroid, gonads), once thought to manufacture only one or two characteristic hormones, are now found to make still other substances, generally of a peptide nature. Findings have been accumulating faster than explanations for them. Can any of them be explained by comparative endocrinology? Unfortunately, comparative endocrinology, instead of providing answers, is a source of still more questions. With a few interesting exceptions, vertebrate hormones are found throughout the vertebrates and even in many invertebrates that lack the endocrine organ usually associated (in the vertebrates) with that hormone.

465

VERTEBRATE ENDOCRINOLOGY:
FUNDAMENTALS AND BIOMEDICAL IMPLICATIONS
Volume 1

I. THE SCOPE OF THIS VOLUME

The multiplicity of tissue sources for hormones, the chemical and there-fore generic relatedness between hormones and nonhormonal biological molecules, the distribution of hormone-like molecules by nonvascular routes, all have obscured the distinctions between endocrine and nonendo-crine phenomena and systems. Nevertheless, we can lay out a definable area for this volume, since it will be limited to the consideration of the anatomical structures of a group of organs known in vertebrates to manufacture and to secrete into the blood a series of well-characterized hormonal substances. Thus, the authors discuss here the vertebrate pituitary gland and its principal neurosecretory afferents (neuronal and vascular) from the brain (Chapters 2 and 3). Chapters 4 and 5 deal with endocrines of a more purely neural origin, the pineal and urophysis. Endodermal endocrine glands of the neck region are the thyroid, parathyroids, and ultimobranchials (Chapters 6, 7, and 8). Endodermal endocrines in the abdominal region include the pancreas and gut. Morphological studies of gut endocrine cells have been remarkably few and confined largely to mammals. This is unfortunate, since there has been an interesting evolution and redistribution of endocrine cells between the stomach, intestine, pancreas, and even the bile ducts, from the earliest to the most highly evolved vertebrates (see Chapters 9 and 10). Biochemically related endocrine tissues derived from retroperitoneal dorsal trunk meso-derm include the adrenal cortex, gonads, and corpuscles of Stannius (Chap-ters 11, 12, 13, and 14).

This list ignores a variety of other organs, known sources of hormones and hormone-like substances. For well-established hormones like angiotensin and vitamin D, the skin, liver, and kidney—nontraditional endocrine or-gans—are involved, but morphological studies of these organs from their endocrine aspect have barely begun, and comparative morphological consid-erations are out of the question, for lack of information from nonmammalian species. The list also fails to include the thymus and other leucocytogenic organs that produce an array of humoral peptidic factors that are involved in the immune response, in hemopoiesis and in blood cell extravascular migra-tion. Finally, we cannot include the tissues that produce the so-called growth factors that may act locally in an autocrine or paracrine fashion, as well as through the vascular system. Here again, morphology, and especially comparative morphology, constitutes an undeveloped field.

II. EVOLUTION OF ENDOCRINE ORGANS

Comparative endocrinologists are familiar with the adage (attributed vari-ously to either P. B. Medawar or F. L. Hisaw), "It is not the hormones that evolve, but the uses to which they are put." While this appears to be quite true, we can now add to it, according to the foregoing discussion, "and the organs and tissues in which they are synthesized." Accordingly, it is worth bearing in mind in a volume devoted to comparative morphology of endocrine glands that these structures may not be the only possible sources of the hormones classically associated with them. Multiplicity of sources of a hormone would appear to have some influence on the pattern of evolution of the organs that secrete it.

Another generalization that can be made about the morphological evolution of endocrine glands is that it is relatively trivial, with the obvious exception of the brain–pituitary structural relationship. The function of an endocrine gland is most commonly served by secretion of a hormone under the regulatory influence of a blood-borne or nervous signal. Thus, as long as the glandular tissue is well vascularized it performs equally well regardless of the shape of its structural unit. For example, the thyroid gland of all vertebrates is follicular. The follicles perform the function of thyroid hormone synthesis in the same way regardless of the shape of the gland as a whole (lobulated, divided, dispersed, etc.). Thus, it is difficult to assign any particular adaptive evolutionary significance to the fact that thyroid follicles in bony fishes are dispersed, that in amphibians and birds they are gathered into two or more widely separate rounded lobes, and that in reptiles and mammals they are most often in two elongated lobes next to the trachea, with or without a connecting isthmus. In some vertebrates (certain fishes and turtles) the gland is a single mass. Similarly, ultimobranchial tissue as follicular or compact units, free or embedded in the thyroid, secretes calcitonin regardless of its gross morphology.

III. ENDOCRINE ORGANS OF LIMITED PHYLETIC OCCURRENCE

Although it is easy to make the point that morphological evolution of endocrine glands is a relatively trivial phenomenon, there are some aspects that clearly are not trivial, and they deserve special consideration. First, is the situation in which particular endocrines are present in some vertebrates but lacking in others (Fig. 1).

A. Caudal Neurosecretory System

Neurosecretory cells (Dahlgren cells) in the caudal spinal cord with (teleosts) or without (elasmobranchs) a differentiated neurohemal structure, the urophysis, are an obvious example. Not all cartilaginous fishes (e.g., chimaeroids) have Dahlgren cells (Bern and Takasugi, 1962), and not all teleosts have a urophysis (see Chapter 5). From the evolutionary point of view this creates a puzzle. Cephalochordate invertebrates and cyclostomes

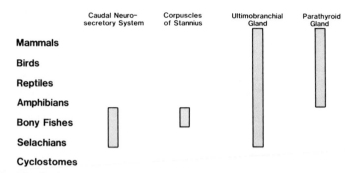

Fig. 1. Distribution among the vertebrate groups of four endocrine organs of limited phyletic occurrence.

(hagfish, lamprey), which are lower in the evolutionary scheme than the fishes, have no caudal neurosecretory system, nor do Amphibia, which are higher. How are we to explain the appearance of Dahlgren cells in the cartilaginous and bony fishes and the prompt evolutionary extinction of these cells? The hormones or hormonelike peptides of the teleostean urophysis, urophysins I and II, are somatostatin-like and CRF-like. Somatostatins are broadly distributed peptides in the vertebrate organism; CRF is found in the brain in all vertebrates. Of what adaptive value would it be to evolve sources of these peptides at the lower extremity of the spinal cord of fishes? Our current inability to suggest an answer to this question merely reveals lack of understanding of the adaptive functional value of the urophysial hormones.

B. Corpuscles of Stannius

The corpuscles of Stannius (see Chapter 14) would seem to be the most restricted in distribution of all vertebrate endocrine organs, being found only in bony fishes. They are variable in number and are always in anatomical association with the kidney. They are derived during development from buds from pronephric and/or mesonephric tubules. Their presumptive secretory product, extractable from the tissue, is a peptidic hypocalcemic factor (hypocalcin), and there is disagreement as to whether the corpuscles also produce an angiotensin-like substance as well. One of the evolutionary puzzles presented by the corpuscles of Stannius is why fishes produce calcitonin as well as hypocalcin. Calcitonin in fish (e.g., salmonids) is relatively ineffectual with respect to plasma calcium regulation, though it is a potent hypocalcemic agent in mammals! As in the case of the urophysis, the apparent duplication of sources of regulatory factors for a single function in fishes is difficult to explain. To find the adaptive value to the fish for having specifically evolved corpuscles of Stannius, Wenderlaar Bonga and Pang (Chapter

14) suggest that it may be necessary to search for additional functions for both calcitonin and hypocalcin, and probably for the corpuscular angiotensin-like substance. Here again, Medawar's aphorism "it is not the hormones that evolve, but the uses to which they are put" offers one of the few alternatives for investigation of the evolution of the special piscine endocrine organs.

C. Parathyroid Glands

The parathyroid glands are lateral pharyngeal derivatives in terrestrial vertebrates (reviewed in Chapter 7). Thus, even more clearly than the urophysis or ultimobranchial glands, the parathyroid glands have evolved within the vertebrate group. Neither the glands nor their hormone, the hypercalcemic factor parathormone, occur in fishes or cyclostomes. Thus, the evolution of hypercalcemic regulation of plasma calcium levels through parathormone and morphologically differentiated parathyroid glands apparently was a relatively rapid phylogenic occurrence. Hypercalcemic regulation in fishes, if it occurs at all, is considered by workers in this field to be a pituitary function (see discussion in Chapters 5, 7, and 14). If this is so, then there may have been an additional requirement for the management of calcium metabolism that was part of the assumption of terrestrial life. This requirement, if it could not be met by hypophyseal regulation, would have created the adaptive need for secretion of a hypercalcemic factor by cells directly sensitive to plasma calcium levels—as parathyroid cells are. Unfortunately, there are not anatomical clues available to indicate how multiple derivatives of the pharyngeal pouch epithelium evolved to assume the function of hypercalcemic regulation. The question is a most interesting one, and it deserves further direct investigation. Perhaps by exploration with antibodies to parathormone, a structural parathyroid antecedent might be found in the pharyngeal epithelium or on the gills of larval amphibians or dipnoan fishes.

D. Ultimobranchial Glands

The ultimobranchial glands (reviewed in Chapter 8), like the parathyroids, are derived during development from the pharyngeal pouch epithelium. Like the parathyroids, ultimobranchial glands are absent in cyclostomes. However, unlike the parathyroids, they do occur in cartilaginous and bony fishes and in all higher vertebrates. Thus, the ultimobranchials evolve within the vertebrate group like the caudal neurosecretory system, the corpuscles of Stannius, and the parathyroids. Generally, ultimobranchials are small, inconspicuous follicular structures which migrate away from their pharyngeal sites of origin. In fishes they commonly may be found on the pericardium at the anterior end of the heart. Characteristically, in mammalian embryos, however, ultimobranchial tissues fuse with median thyroid tissue and form a second type of intrathyroidal follicle, distinct from thyroid follicles, and/or give rise to interfollicular scattered or grouped "C-cells." There is no known functional value to the intimate thyroid–ultimobranchial anatomical fusion. Yet, virtually all mammals that have been studied have this composite endocrine organ. We are obliged to assume that there must have been an adaptive value to this close association, though we do not understand what it may have been.

For that matter, the parathyroid glands of mammals also fuse with the thyroids, though they generally remain superficial. Here again, there is no known functional value to the glandular association, though there must have been some value to cause its evolution.

The hormone of the ultimobranchials is calcitonin, a glycoprotein hypocalcemic factor. As pointed out earlier, this provides the bony fish with two hypocalcemic hormones, the other being hypocalcin from the corpuscles of Stannius. The need for such duplication of function is an evolutionary puzzle. We may assume on the basis of meager, currently available evidence that calcitonin may be the more ancient hormone, since it is found in cartilaginous as well as bony fishes, and since an immunoreactive calcitonin-like substance has been reported by Thorndyke and Probert (1979) in the pharynx of a protochordate invertebrate, *Styela*. In fishes with both hypocalcemic hormones, calcitonin is relatively ineffectual in lowering plasma calcium levels. What is needed at this point, for better understanding, is a study of the binding of the two hormones by presumptive target tissues in an effort to decide their relative functions.

IV. EVOLUTION OF THE ADRENAL CORTEX

A second example of endocrine glandular anatomical association is that between the corticosteroidogenic tissues (adrenal cortex, interrenal) and some of the catechecholamine-secreting tissue (medullary adrenal, chromaffin tissue, etc.) in the lower trunk region (reviewed in Chapter 11). These separate elements have widely different embryologic origins, the steroid-secreting cells from somite-derived dorsal mesoderm near the kidney, and the adrenergic cells from neural crests in that region. Throughout the vertebrate series the two endocrine tissues display a progressively evolving and increasing intimacy of association. In cyclostomes and elasmobranchs they are completely separated. In fact, in cyclostomes the steroidogenic tissues have not even been identified definitively, and plasma corticosteroid levels are extremely low. In bony fishes the chromaffin and steroidogenic tissues occur in the same dorsal abdominal position in the pronephric area. In amphibians there are multiple islets of steroidogenic and adrenergic tissues on the surface of the mesonephros. In reptiles, birds, and mammals there are separate adrenal glands containing both kinds of tissue. In reptiles and birds the two endocrine tissues form interweaving cords of cells. In mammals the steroidogenic tissue characteristically forms an enveloping cortex that encloses a medulla of catecholaminergic tissue. Thus, the different vertebrate groups present a picture of a progressive and continuous evolution of the relationship between the two different endocrine tissues. Various proposals have been advanced, suggesting how the cortical and medullary endocrine functions influence each other, and some of these ideas are supported by the patterns of centripetal and centrifugal flow of blood in the mammalian adrenal gland. However, the conclusion by Chester Jones and Phillips (Chapter 11) is that none of these proposals is yet completely persuasive. Nevertheless, we must say once more that a progressive and continuous direction of evolution, such as that seen in the corticosteroidogenic–catecholaminergic tissue association, must have definite functional value or it would not have occurred.

Another feature of the morphological evolution of the adrenal gland is in the zonation of the cortex. In mammals there is a readily recognizable morphological differentiation of at least three zones in the radiating cords of cortical cells. These are specialized physiologically as well, secreting characteristic and different steroids in the several zones. Furthermore, the cortical zones are under separate regulatory control of their secretory activity (i.e., whether or not they are responsive to ACTH, angiotensin, or plasma ionic concentrations).

To what extent does adrenal cortical zonation represent a progressive evolution? Chester Jones and Phillips (Chapter 11) indicate that evolution of cortical zonation in vertebrates may be minimal. They make the point that even in Amphibia the corticosteroid-secreting cells are in cords, but that these cords are not systematically oriented as they are in mammals, making the zonation difficult to recognize. Thus, corticoadrenal evolution in vertebrates may have been mostly a process of more orderly arrangement of already zonally differentiated cell cords. This interesting thesis certainly deserves further study by currently available experimental approaches.

V. EVOLUTION OF THE HYPOTHALAMO–HYPOPHYSIAL SYSTEM

The evolution of the endocrine anatomical associations seen in the thyroid–parathyroid–ultimobranchial and adrenal cortex–medulla systems are both remarkable and difficult to understand. It may merely reveal a bias of this author when he states that the evolutionary association of the brain and pituitary gland is even more remarkable and not at all obscure as to its physiological and adaptive value. Peter in Chapter 3 reviews the anatomical organization of the neurosecretory neurons in the brain that project to the neurohypophysis (the median eminence and pars nervosa). Schreibman in Chapter 2 reviews the structures of the pituitary gland itself.

The neurohypophyseal part of the pituitary gland develops and remains as a downgrowth of the diencephalic brain. Structurally it is the neurohemal terminus of the two types of neurosecretory neurons. Those neurons that end upon blood vessels in the pars nervosa secrete the octapeptide hormones that enter the general circulation at this point. Those neurons that terminate upon blood vessels in the median eminence of the neurohypophysis secrete an array of peptides that regulate the release of the six or more hormones of the pars distalis. This is possible because blood from the median eminence is channeled through portal vessels into the pars distalis.

The pars distalis and pars intermedia (together, the adenohypophysis) develop from adjacent epithelial tissue that comes in contact with the formative neurohypophysis in the embryo. It is of interest that this epithelium can be quite different in the embryos of various vertebrate groups. In most higher vertebrates the adenohypophysis develops from oral epithelium; in frogs, the epithelium near the olfactory structures; in lampreys and hagfishes, the adenohypophysis forms from a part of the nasopharyngeal duct. In hagfishes the nasopharyngeal duct is endodermal rather than ectodermal. From the evolutionary point of view it is significant that the developing neurohypophysis appears to retain an inductive potential to form adenohypophysis from whatever head epithelium it contacts.

The obvious evolutionary adaptive value of the brain–pituitary relationship is that it serves as a way to transduce signals from the external (and internal) environment, which are detected by peripheral and central nervous elements, into endocrine events that are fundamental for regulating the metabolism, reproduction, and survival of the entire organism. This being so, it is regrettable that the basic anatomical organization of the brain–pituitary relationship is already well established in the most primitive vertebrates available for study, the cyclostomes. This leaves very little to discuss concerning structural evolution of the hypothalamo–pituitary system, aside from minor and secondary features. These are dealt with in some detail in Chapters 2 and 3.

Fossil evidence provides no clues concerning the evolution of the brain–pituitary relationship. Many reviewers have proposed as anatomical homologues of the neurohypophysis–adenohypophysis association certain ganglionic structures that are near adjacent epithelial structures in a variety of invertebrates. It is quite possible that such simple structures may have been the structural antecedents of the pituitary gland. However, there is so great a difference in complexity between even the most comparable of these proposed homologues (e.g., the neural ganglion–subneural gland complex of ascidians) and the pituitary that they are not useful models and offer little understanding of evolution of the pituitary.

One important hypophysial anatomical feature does appear to evolve within the vertebrates, and it is useful to discuss it here. In the cyclostomes there is no vascular portal system (Tsuneki and Gorbman, 1975a,b) to convey regulatory peptides from the brain to the adenohypophysis. There may be a limited degree of direct neurosecretory innervation of adenohypophysial cells in the posterior part of the gland of *Myxine,* according to Schultz and Adam (1975). In all other vertebrates, however, there is a fully developed system of portal vessels that performs this function. Can we assume that the lack of a hypophysial portal system in cyclostomes is a primitive rather than degenerate feature? There are at least two reasons for believing that this is the case. First, it is unlikely that the ancestors of both the hagfishes and the lampreys would have had a hypophysial portal system which degenerated later in both cyclostome groups. Second, there are three primitive fish groups in which there is a *partial* hypophysial portal system (Gorbman, 1980a). In each of these (the holocephalians, the selachians, and in *Latimeria*) there is an extended region of the pars distalis which does not receive venous blood from the brain, while the major part of the pars distalis is irrigated by blood from a well-developed median eminence and numerous hypophysial portal vessels. All three of these primitive fish groups arose (about 400,000,000 years ago) shortly after the presumed divergence of the cyclostomes from the vertebrate line (see Gorbman, 1980b). The partial hypophysial portal system may represent a midphase in the evolution of the median-eminence/hypophysial-portal system. This implies, then, that the lack of such a system in cyclostomes is indeed the primitive condition.

If the lack of a vascular link between the brain and pituitary is primitive then we must look further for an explanation of the evolution of the original brain–pituitary apposition. This apposition could only have evolved if it allowed a reasonably efficient movement of neurosecretory regulating substances between the two structures. Nozaki *et al.* (1975) have proposed that such movement could have taken place originally by diffusion across the connective tissue layer between the brain and adenohypophysis. Recently, Tsukahara *et al.* (1986) tested this hypothesis by injecting substances of different molecular size into the third ventricle of hagfish and timing their rates of diffusion across the brain–pituitary connective tissue septum. They found that both a protein (horseradish peroxidase) and ferric ion readily diffuse from the ventricular lumen, across the wall of the neurohypophysis (probably through the tanycytes in the case of the peroxidase), across the connective tissue septum to the adenohypophysial follicles within 5–10 min. Furthermore, the protein continued to diffuse *between* the follicles, following the bars of connective tissue that separate the follicles (Fig. 2). Thus, the connective tissue between brain and pituitary is not only not a barrier to diffusion, it is actually a conduit that fairly quickly distributes materials between the hagfish brain and the adenohypophysis.

With this experimental information available it is now possible to understand the functional value of other features of the structure of the neurohypophysis–adenohypophysis system of both lampreys and hagfishes. In these species the neurohypophysis is a thin, extended structure where it faces the

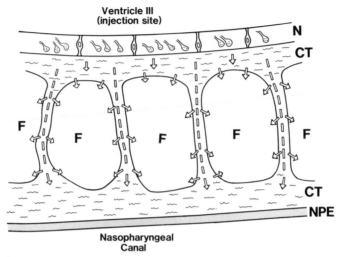

Fig. 2. Summary of the results of injection experiments into the hagfish neurohypophysial space (ventricle III) of several chemical markers (Tsukahara *et al.,* 1986). A protein, horseradish peroxidase (HRP) and ferric ion preferentially diffused through the connective tissue, following the septa between follicles of adenohypophyseal tissue. Broad dashed lines and arrows indicate these diffusional routes. HRP remained in the connective tissue, but ferric ion accumulated in the adenohypophyseal cells. Abbreviations: N, neurohypophysis, ventral wall; CT, connective tissue; F, adenohypophyseal follicle. Neurosecretory endings and tanycytes are indicated diagramatically in the neurohypophysis.

adenohypophysis (pars distalis). The adenohypophysis of hagfish (which lacks a pars intermedia) and the pars distalis of lampreys in turn are thin, disc-shaped structures that are broken up into follicles. Between the follicles are tongues of loose connective tissue which are continuous with the lamina that separates the neurohypophysis from the adenohypophysis. The perihypophyseal connective tissue has a much less dense character than the more dense collagenous layers elsewhere (see Fig. 3) and may be specialized to serve as a facilitated route for diffusion. It is understandable that if diffusion of regulatory substances from the brain were too slow, only the most dorsal adenohypophysial cells would be exposed to the highest concentrations of brain peptides. The interfollicular septa bring these peptides to the more ventral cells of the pars distalis almost as rapidly as to the dorsal region.

No other vertebrate pituitary complex has the parallel thin laminar morphology that is seen in the cyclostomes. The evolution of a vascular conduction system between the brain (median eminence) and pars distalis made possible the more globular compact pituitary structures found in higher vertebrates. In this sense the elasmobranchs form an intermediate group, since their pars distalis is elongated and coextensive with a thin neurohypophysis–median eminence. However, the presence of hypophyseal portal vessels in elasmobranchs makes possible the evolution of a much thicker and more

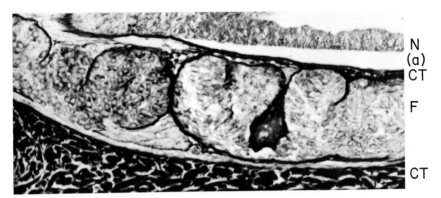

Fig. 3. Section through the pituitary region of an ammocoete of the lamprey. *Petromyzon marinus,* showing structural similarities to hagfish and adaptation for diffusional control from the neurohypophysis. Neurohypophysis (N) is thin and horizontally coextensive with the follicular adenohypophyseal follicles (F). There is an artefactual break in the tissue (a) between the neurohypophysis and the connective tissue below it. There is only a single layer of glandular follicles embedded in a loose spongy connective tissue (CT) layer. Note that the periadenohypophyseal CT is structurally different from the densely collagenous CT ventral to it. This can be interpreted, according to the Tsukahara *et al.* (1986), as an adaptation for diffusional distribution of neurohypophyseal regulatory factors to the adenohypophysis. [Photograph provided by D. B. McMillan, University of Western Ontario.]

compact pars distalis. The neurohypophysis–pars distalis relationship in teleosts also is reminiscent of the cyclostome pituitary morphlogy. Teleosts retain a pars distalis which is highly lobulated by many tongues of connective tissue, and they lack a true median eminence. However, in teleosts the connective tissue septa carry both blood vessels and extended neurosecretory neurones. In a number of teleost species (see Chapter 2) the neurosecretory neurones end in the connective tissue septa, and the regulatory peptides must diffuse, as they do in cyclostomes, to the nearby cells of the pars distalis. In other species of teleosts the neurosecretory neurons extend beyond the connective tissue septa to the cells of the pars distalis. Here again, the presence of blood vessels and extended neurosecretory axones in the interlobular connective tissue septa of teleosts makes possible a much more compact pars distalis than in cyclostomes.

In sum, these speculations concerning evolution of the vertebrate brain–pituitary relationship are based on the morphology of this system in cyclostomes and on recent injection experiments that demonstrate the rapidity of movement of diffusible materials between the brain and the folliculate adenohypophysis of the hagfish. In this conceptualization the cyclostomes have a "diffusional median eminence" that requires an organization of tissues in which the neurohypophysis and adenohypophysis are broad and thin and apposed to each other coextensively. In further evolution of this primitive arrangement, vascular conduction of neurosecretory materials across this small distance and/or extension of neurosecretory axones over it (as in the pars intermedia, and in the pars distalis of teleosts) made possible the more compact and larger pars distalis of higher vertebrates.

VI. CONCLUSIONS

In this chapter, placed at the conclusion of this volume on the comparative morphology of the classical vertebrate endocrine organs, there has been a minimum of reference to the published literature. The detailed factual data and direct literature citations can be found within the separate 14 chapters of this book. On the other hand, an attempt has been made here to place in an evolutionary perspective some of the information in the separate chapters. The length of the perspective has varied considerably with respect to particular evolutionary questions.

One kind of question that can be raised concerns those vertebrate endocrine organs that do not occur in all vertebrates. These would seem to have evolved within the vertebrate series, but the adaptive basis for their limited phyletic occurrence remains relatively obscure and a subject that is highly speculative. This group of organs includes the corpuscles of Stannius, found only in bony fishes; the caudal neurosecretory system, found only in bony and cartilaginous fishes; the parathyroid glands, limited largely to terrestrial vertebrates; and the ultimobranchial glands, lacking only in cyclostomes.

A second type of evolutionary question concerns the endocrine anatomical associations that can be found in vertebrates: the thyroid–parathyroid–ultimobranchial complex, the adrenal cortex and medulla, and the brain–pituitary relationship. For the first two, little beyond the barest speculation can be offered. This is a frustrating situation, because the anatomical associations have been progressive and continuous through phylogeny. According to the principles known to guide evolution, the changes seen must have been in response to some kind of adaptive "pressure."

Somewhat more can be said about the evolution of the brain–pituitary relationship, though the initial vertebrate brain–pituitary associations are lost in the missing prevertebrate organisms which left no fossil record. However, arguments can be presented that the structure of the neurohypophysis–adenohypophysis complex of cyclostomes is a primitive one in which brain peptides gain access to the pars distalis by diffusion through favored connective tissue paths. The later addition of a median eminence and portal blood vessel system, as seen in holocephalians, selachians, and *Latimeria,* partially vascularized the brain–pars distalis link. The full portal vascularization of the pars distalis followed. The definitive median-eminence/pars-distalis relationship, since it assures full access of all pars distalis cells to regulatory brain neurosecretions, made possible the large variety of structural configurations of the pars distalis seen in higher vertebrates.

REFERENCES

Bern, H. A., and Takasugi, N. (1962). The caudal neurosecretory system of fishes. *Gen. Comp. Endocrinol.* **2**, 96–110.

Gorbman, A. (1980a). Evolution of brain-pituitary relationship: Evidence from the Agnatha. *Can. J. Fish. Aquat. Sci.* **37**, 1680–1686.

Gorbman, A. (1980b). Endocrine regulation in agnatha: primitive or degenerate? *In* "Hormones, Adaptation and Evolution" (S. Ishii, T. Hirano, and M. Wada, eds.), pp. 81–92. Jpn. Sci. Soc. Press, Tokyo.

Nozaki, M., Fernholm, B., and Kobayashi, H. (1975). Ependymal absorption of peroxidase into the third ventricle of the hagfish *Eptatretus burgeri* (Girard). *Acta Zool. (Stockholm)* **56**, 265–269.

Schultz, H. J., and Adam, H. (1975). Elektronenoptische Hinweise auf eine nervöse Verbinding zwischen Neuro- und Adenohypophyse bie *Myxine glutinosa* L. (Cyclostomata). *Norw. J. Zool.* **23**, 297–306.

Tata, J. R. (1984). What is so unique about hormone action? *Mol. Cell. Endocrinol.* **36**, 17–27.

Thorndyke, M., and Probert, L. (1979). Calcitonin-like cells in the pharynx of an ascidian, *Styela clava. Cell Tissue Res.*, **203**, 301–309.

Tsukahara, T., Gorbman, A., and Kobayashi, H. (1986). Median eminence equivalence of the neurohypophysis of the hagfish *Eptatretus burgeri. Gen. Comp. Endocrinol.* **61**, 348–354.

Tsuneki, K., and Gorbman, A. (1975a). Ultrastructure of the anterior neurohypophysis and pars distalis of the lamprey, *Lampetra tridentata. Gen. Comp. Endocrinol.* **25**, 487–508.

Tsuneki, K., and Gorbman, A. (1975b). Ultrastructure of pars nervosa and pars intermedia of the lamprey, *Lampetra tridentata. Cell Tissue Res.* **157**, 165–184.

Index

A

A cell, *see* Glucagon, A cell

Acetylcholine
  caudal neurosecretory system content, 154, 167
  pituitary gland content, 87

N-Acetyltransferase, in melatonin synthesis, 4, 116

Acidophil, 20–21, 26, 27, 28, 34, 36

Acipenseriformes, pituitary gland, 27–28

Actinopterygia
  caudal neurosecretory system, 149
  pancreatic islet organ, 294–298
  pituitary gland, 26–33

Addison's disease, 347, 348

Adenohypophysis
  comparative morphology, 13, 20–21, 472
    Agnatha, 13–14, 15, 17, 22–23, 24
    Amphibia, 36
    Aves, 18
    Chondrichthyes, 25
    Mammalia, 40, 42–43
    Osteichthyes, 28, 34
    Teleostei, 16, 29
  cytology, 18
  development, 12, 13
  neurohypophysis association, 15–18, 472–476
  neurosecretory product transport, 15, 18

Adrenal cortex
  evolution, 470–471
  zonal theory, 345, 346
  zonation, 320, 321, 323, 471
    X zone, 328–329
    zona fasciculata, 321, 323–324, 325, 329, 343, 345–346
    zona glomerulosa, 321, 323, 324, 329, 343, 345–346
    zona intermedia, 345–346
    zona reticularis, 321, 323, 324–326, 329, 343, 345–346

Adrenal gland, 319–350, *see also* Interrenal gland
  aldosterone secretion, 7–8
  cell migration theory, 343–346
  comparative morphology, 7
    Aves, 330–332
    biomedical implications, 346–348
    Mammalia, 319–330
    Reptilia, 332–335
  corticotropin-releasing factor secretion, 8
  cortisol secretion, 7–8
  epinephrine secretion, 8, 321, 323, 326
  fetal, 329
  norepinephrine secretion, 8, 321, 323, 326
  somatostatin secretion, 8
  vasculature, 320, 322, 323, 333, 334, 336, 337
  vasoactive intestinal peptide secretion, 8

Adrenal medulla, 320–321, 343

Adrenocorticotropin, pituitary, 20, 23, 25, 38, 40, 42

Adrenogenital syndrome, 347, 348

Agnatha, *see also* Myxinoidea; Petromyzonidae
  adenohypophysis, 13–14, 15, 17, 22–23, 24
  caudal neurosecretory system, 148–149
  endostyle, 192–193
  interrenal gland, 341–342, 346
  ovary, 353–357
  pancreatic islet organ, 289–292
  pituitary gland, 21–24, 472–474, 475–476
    development, 13–14
  stomach absence, 272–273
  testis, 404
    Leydig cells, 423–424
    Sertoli cells, 413–414
  thyroid gland, 188, 192–195, 196, 198

Aldosterone
  adrenal, 7–8, 332, 346
  interrenal, 339–340